Computer Algebra Recipes

An Introductory Guide to the Mathematical Models of Science

Richard H. Enns
George C. McGuire

Computer Algebra Recipes

An Introductory Guide to the Mathematical Models of Science

Richard H. Enns
Simon Fraser University
Department of Physics
Burnaby, B.C. V5A 1S6
Canada

George C. McGuire
University College of Fraser Valley
Department of Physics
Abbotsford, BC V2S 7M9
Canada

Cover design by Mary Burgess.

Library of Congress Control Number: 2005937149

ISBN-10 0-387-25767-5
ISBN-13 978-387-25767-9
e-ISBN 0-387-31262-5

Printed on acid-free paper.

Printed in the United States of America. (EB)

9 8 7 6 5 4 3 2 1

springeronline.com

Preface

A computer algebra system (CAS) not only has the ability to "crunch numbers" and plot results, like traditional computing languages such as Fortran and C, but it can also perform the symbolic manipulations and analytic derivations required in most undergraduate and graduate science and engineering courses. To introduce students in these disciplines to mathematical modeling and computation using a CAS, the authors have previously published *Computer Algebra Recipes: A Gourmet's Guide to the Mathematical Models of Science*, based on the Maple CAS. Judging by course evaluations and reader feedback, the response to this book and the CAS approach has been quite favorable. After observing students' enthusiasm, their higher quality of work, their ability to solve more realistic problems, and *best of all*, their ability to answer "what if?" questions, we believe that the importance of using a CAS in learning and exploring mathematically based science subjects cannot be overstated.

With the release of new, more powerful, versions of the Maple CAS since the first edition was published and the accumulation of many insightful comments and helpful suggestions from readers of the text, it seemed timely to produce a second edition. However, incorporating the necessary changes and suggestions would make an already lengthy book even longer, so the topics of the first edition have been reorganized and expanded into two new standalone volumes based on the expected mathematical level of the reader.

In this first volume, we assume the reader's familiarity with linear algebra, vectors, and elementary calculus, and knowledge of (but not necessarily expertise at) linear ordinary differential equations. The second volume (*Computer Algebra Recipes: An Advanced Guide to the Mathematical Models of Science*) deals with more advanced differential equation models, both ordinary and partial, and nonlinear as well as linear.

Each volume, which may be used either as a course text or for self-study, features an eclectic collection of Maple computer algebra worksheets or "recipes" drawn from a wide variety of disciplines, including biology, economics, medicine, engineering, game theory, physics, mathematics, and chemistry. These recipes are systematically organized to illustrate graphical, analytical, and numerical techniques applied to scientific modeling. No prior knowledge of Maple is assumed in either volume, the early recipes of each book introducing the reader to the basic Maple syntax, and the subsequent recipes introducing further Maple commands and structure on a need-to-know basis.

The recipes are fully annotated in the text and typically presented as "stories" or in a historical context. Each recipe takes the reader from the analytic formulation or statement of an interesting mathematical model to its analytic or numerical solution, and to a graphical visualization of the answer where relevant. The graphical representations vary from static 2-dimensional pictures, to contour and vector field plots, to 3-dimensional graphs that can be rotated, and to animations of analytic and numerical solutions.

Every recipe is followed by a set of problems that readers can use to check their understanding or develop the topic further. For your convenience in solving these problems and to facilitate further exploration of a given topic, the unannotated recipes for each volume are included on an accompanying CD.

Contents

Preface v

INTRODUCTION **1**

A. Computer Algebra Systems . . . 1
B. Computer Algebra Recipes . . . 3
C. Introductory Recipe: Bridge Design 101 . . . 5
D. Maple Help . . . 9
E. How to Use This Text . . . 10

I THE APPETIZERS **11**

1 The Pictures of Science **13**
1.1 Data and Function Plots . . . 13
1.1.1 Correcting for Inflation . . . 15
1.1.2 The Plummeting Badminton Bird . . . 22
1.1.3 Minimizing the Travel Time . . . 31
1.2 Log-Log (Power Law) Plots . . . 36
1.2.1 Chimpanzee Brain Size . . . 36
1.2.2 Scaling Arguments and Gulliver's Travels . . . 40
1.3 Contour and Gradient Plots . . . 46
1.3.1 The Secret Message . . . 46
1.3.2 Designing a Ski Hill . . . 49
1.4 Animated Plots . . . 56
1.4.1 Waves Are Dynamic . . . 56
1.4.2 The Sands of Time . . . 59
1.4.3 These Arrows Are Useful . . . 61

2 Deriving Model Equations **65**
2.1 Linear Correlation . . . 66
2.1.1 The Corn Palace . . . 67
2.2 Least Squares Derivations . . . 69
2.2.1 Will You Be Better Off Than Your Parents? . . . 71
2.2.2 What Was the Heart Rate of a Brachiosaur? . . . 76

2.2.3 Senate Renewal . . . 84
2.2.4 Bikini Sales and the Logistic Curve . . . 87
2.2.5 Following the Dow Jones Index . . . 91
2.2.6 Variation of "g" with Latitude . . . 98
2.2.7 Finding Romeo a Juliet . . . 103
2.3 Multiple Regression Equations . . . 106
2.3.1 Real Estate Appraisals . . . 107
2.3.2 And the Winner Is? . . . 113

II THE ENTREES 119

3 Algebraic Models. Part I 121
3.1 Scalar Models . . . 121
3.1.1 Bombs Versus Schools . . . 122
3.1.2 Kirchhoff Rules the Electrical World . . . 129
3.1.3 The Window Washer's Secret . . . 136
3.1.4 The Science Student's Summer Job Interview . . . 142
3.1.5 Envelope of Safety . . . 148
3.1.6 Rainbow County . . . 152
3.2 Integral Examples . . . 156
3.2.1 The Great Pyramid of Cheops . . . 156
3.2.2 Noah's Ark . . . 162

4 Algebraic Models. Part II 173
4.1 Vector Models . . . 173
4.1.1 Vectoria's Mathematical Heritage . . . 174
4.1.2 Vectoria and Fowles's Fly . . . 179
4.1.3 Ain't She Sweet . . . 183
4.1.4 Born Curl-Free . . . 188
4.1.5 Of Coordinates and Circulation Too . . . 194
4.1.6 All Is Flux . . . 199
4.2 Matrix Models . . . 202
4.2.1 Secret Message Revisited . . . 202
4.2.2 A Fishy Tale . . . 205
4.2.3 Population Waves . . . 208

5 Linear ODE Models 213
5.1 Phase-Plane Portraits . . . 214
5.1.1 Tenure Policy at Erehwon University . . . 216
5.1.2 Vectoria Investigates the RLC Circuit . . . 221
5.2 First-Order ODE Models . . . 229
5.2.1 There Goes Louie's Alibi . . . 229
5.2.2 The Water Skier . . . 238
5.2.3 Ready to Charge . . . 242

5.3 Second-Order ODE Models . . . 245
5.3.1 Shrinking the Safety Envelope . . . 245
5.3.2 Frank N. Stein Is Not Heartless . . . 251
5.3.3 Halley's Comet . . . 255
5.3.4 Wheel of misFortune . . . 260
5.3.5 The Weedeater . . . 266
5.3.6 Can an Unstable Spring Find Stability? . . . 269

6 Difference Equation Models 271
6.1 Linear Models . . . 272
6.1.1 Those Dratted Gnats . . . 272
6.1.2 Gone Fishing . . . 276
6.1.3 Fibonacci's Adam and Eve Rabbit . . . 279
6.1.4 How Red Is Your Blood? . . . 283
6.1.5 Fermi–Pasta–Ulam Is Not a Spaghetti Western . . . 285
6.2 Nonlinear Models . . . 292
6.2.1 Competition for Available Resources . . . 292
6.2.2 The Logistic Map and Cobweb Diagrams . . . 299
6.2.3 The Bouncing Ball Art Gallery . . . 306
6.2.4 Onset of Chaos: A Model for the Outbreak of War . . . 310

III THE DESSERTS 317

7 Monte Carlo Methods 319
7.1 Random Walks . . . 321
7.1.1 The Soccer Fan's Drunken Walk . . . 324
7.1.2 Blowin' in the Wind . . . 329
7.1.3 Flight of Penelope Jitter Bug . . . 333
7.1.4 That Meandering Perfume Molecule . . . 335
7.2 Monte Carlo Integration . . . 338
7.2.1 Numerical Integration Methods . . . 339
7.2.2 Wait and Buy Later! . . . 344
7.2.3 Wait and Buy Later! The Sequel . . . 348
7.2.4 Estimating π . . . 353
7.2.5 Chariot of Fire and Destruction . . . 355
7.3 Probability Distributions . . . 361
7.3.1 Of Nuts and Bolts and Hospital Beds Too . . . 361
7.3.2 The Ice Wines of Rainbow County . . . 367
7.4 Monte Carlo Statistical Distributions . . . 372
7.4.1 Estimating e . . . 372
7.4.2 Vapor Deposition . . . 376

8 Fractal Patterns **381**
8.1 Difference Equations 384
8.1.1 Wallpaper for the Mind 384
8.1.2 Sierpinski's Fractal Gasket 386
8.1.3 Barnsley's Fern 391
8.1.4 Douady's Rabbit and Other Fauna and Flora 396
8.1.5 The Rings of Saturn 400
8.2 Cellular Automata 408
8.2.1 A Navaho Rug Design 408
8.2.2 The One out of Eight Rule 411
8.3 Strange Attractors 414
8.3.1 Lorenz's Butterfly 414
Epilogue 416

Bibliography **417**

Index **421**

INTRODUCTION

A. Computer Algebra Systems

The purpose of computing is insight, not numbers.
R.W. Hamming, *Numerical Methods for Scientists and Engineers* (1973)

Although modern scientific models are usually not difficult to understand qualitatively, the task of deriving the relevant model equations and finding, visualizing, and interpreting the associated solutions may be too demanding or too tedious to realistically carry out without the aid of a computer. As a consequence, over the last several years a new branch of science, referred to as computational science, has evolved to deal with this issue. Traditionally, the approach of most computational science texts [PFTV89], [DeV94], [LP97] has been to introduce engineering and science students to the art of creating efficient computer programs in languages such as Fortran and C to carry out a multitude of numerical tasks ranging from finding the solutions to ordinary and partial differential equations (ODEs and PDEs) to performing Monte Carlo simulations and so on. Although some scientists and engineers may still wish to learn one or more of these programming languages for certain specialized research objectives, an even more powerful general computer algebra approach is being developed. This new approach is changing the way that complex mathematical modeling problems are tackled by science and engineering students as well as by practitioners in these fields. It not only allows the user to handle such problems numerically, but also permits him or her to carry out analytic differentiation, integration, and other symbolic manipulations, and easily create a wide variety of two- and three-dimensional static, as well as animated, plots. Computer software systems, such as Maple, that can carry out all of these diverse functions in a unified and cohesive fashion, are referred to as symbolic computation systems or computer algebra systems (CASs).

As personal computers become smaller, cheaper, faster, and possess greater memory, it is already clear that CASs, which are also rapidly increasing in sophistication and ease of use, are making many of the traditional topics and approaches covered in existing computational science texts less relevant, since many of the same tasks can be executed more easily with a CAS. With a CAS not only can complex model equations be analytically derived, they can be solved either analytically or numerically and then plotted in two or three dimensions, all steps executed with short, simple, transparent commands. The aim of this text is to introduce you, the reader, to what we shall call the "new" computational science based on the utilization of a CAS. More specifically, this text is designed to demonstrate how a CAS can serve as a valuable adjunct tool in easily deriving, solving, plotting, and exploring interesting, modern scientific models chosen from a wide variety of disciplines ranging from the physical and biological sciences to the social sciences and engineering.

Associated with each of the investigated mathematical models of a physical phenomenon is an accompanying computer algebra worksheet or *recipe* based on the **Maple 10** software system. Useful reference books to this CAS are the *Maple 10 User Manual* [Map05] and the *Maple 10 Introductory* and *Advanced Programming Guides* [MGH$^+$05]. "Classic" Maple is employed in this text and the classic worksheet interface[1] used to produce all Maple output (including figures) displayed here.

A common fear that prevails among some instructors who have not used a CAS is that in solving assigned problems, their students will rely on the computer and do calculations with, say, Maple, that could be easily done with pen and paper or even in their heads. But often this trepidation arises because the problems are oversimplified models of reality that have been simplified precisely so that they can be solved by hand or in one's head.[2] For example, in studying the flight of a golf ball,[3] the student is usually told to ignore the viscous drag of the air and the lift on the ball due to its dimpled surface and the back spin imparted to it by the grooved golf club. Heaven forbid if, on top of all of this, the effects of a cross wind, or even a swirling wind, have to be included. Neglecting all of these effects, a nice parabolic trajectory results, the path being readily derived with pen and paper and shown in every elementary physics text. Unfortunately, as any ardent golfer will attest, this is not usually the trajectory that a golf ball actually follows, even when the ball is not "shanked," "sliced," or otherwise badly hit. But, says the timorous instructor, the more realistic model that includes drag and lift is too difficult for my students to solve. We think otherwise. The more realistic model is easily set up and the underlying dynamical equations can be readily understood by any second-year undergraduate physics or engineering student. With a CAS, the solution to the resulting ODE for the more complex golf ball model is easy to obtain and the golf ball's motion is readily animated.

Is it really important that a second-year student know the details of the numerical algorithm that Maple uses in solving the relevant ODE? Our experience has shown the answer to be no. These mathematical details can be learned at a later stage in the student's academic development. It is more important initially that he or she be able to explore how the motion of the golf ball is affected by greater or lesser spin, by varying air density, and so on. In fact,

[1] The alternative "standard" worksheet interface has some additional features, such as a math palette and a math dictionary, so requires more computer memory.

[2] The authors of this text plead guilty to having engaged in the same practice!

[3] This example, which involves nonlinear ODEs, is dealt with in the *Advanced Guide*.

developing the model and exploring the behavior that it predicts is more important at any stage of the student's career than the mechanics of solving the relevant equations. With a CAS the student can take an interactive approach to problems such as the one involving the golf ball and ask "What happens if...?" questions and, most importantly, use the complete computer algebra package to answer these questions. Within seconds or minutes of changing the model parameters or altering the model assumptions, new solutions can be produced by executing the modified recipe. With this CAS approach to scientific computation, we have found that students are able to investigate complex mathematical and physical models early in their educational careers and gain a much deeper understanding of the models and the effects of changing assumptions and/or parameters. Most importantly, we have seen our students turn into budding scientific researchers by this interactive approach; rather than quitting thinking, their thought processes are accelerated and their imagination stimulated.

B. Computer Algebra Recipes

Science is a collection of successful recipes.
Paul Valéry, French poet and essayist (1871–1945)

The recipes in our computer algebra *menu* have been organized into **Appetizers**, **Entrees**, and **Desserts**. The **Appetizers** consist of mathematically tasty light fare, featuring recipes dealing with aspects of *graphical analysis*. Chapter 1 shows how to create different types of data and function plots, while Chapter 2 illustrates how model equations can be obtained from data by least squares fitting techniques. The **Entrees** contain the "meat and potatoes" of a CAS, involving topics in *symbolic computation*. Chapters 3 and 4 feature algebraic (scalar, vector, and matrix) models, while Chapters 5 and 6 deal with linear ODEs and linear and nonlinear difference equation models, respectively. The **Desserts** feature two "scrumptious" *numerical* topics, viz., Monte Carlo methods and fractal patterns.

The recipes have been designed to fulfill not only a useful and serious pedagogical purpose but also to titillate and stimulate the reader's intellect and imagination. Associated with each recipe is an intrinsically important scientific model or method and often an interesting or amusing story[4] featuring a fellow engineering or science student who will guide YOU, the reader, through the steps of the recipe. These storybook students are fictitious composites of the many delightful individuals that the authors have had the privilege of teaching over the years. Admittedly, a few students were not so delightful and we have declined to include them in our stories. They will have to write their own books! Some of the stories have been deliberately given enigmatic titles. For example, one of the recipes in the first chapter is called **The Secret Message**, with the contents of the message being revealed only when the computer code or recipe

[4]If a story elicits a groan, rather than a chuckle, feel free to make up your own story line!

is run. Still other recipes, particularly those that are animated, will truly reveal their power and/or beauty only when viewed on the computer screen.

Every topic or story in the text contains the Maple code or recipe to explore that particular topic. To make life easier for you, all recipes have been placed on the CD-ROM enclosed within the back cover of this text. The recipes are ordered according to the chapter number, the section number, and the subsection (story) number. For example, the recipe **01-2-2**, entitled **Scaling Arguments and Gulliver's Travels**, is associated with Chapter 1, Section 2, Subsection 2. Although the recipes can be directly accessed on the CD by clicking on the appropriate worksheet number, it is strongly recommended that you access them through the menu index file, **00menu**. All recipes can be conveniently accessed from this menu using the provided hyperlinks. Complete instructions on how to do this may be found in the menu file.

The computer code on the CD is unannotated, so you will have to read the text in order to understand what the code is trying to accomplish. The code has been imported into the text and here is accompanied by detailed explanations of the underlying modeling concepts and computational methods.

The recommended procedure for using this text is first to read a given topic/story for overall understanding and enjoyment. If you are having any difficulty in understanding a piece of the text code, then you should execute the corresponding Maple worksheet and try variations on the code. Keep in mind that the same objective can often be achieved by a different combination of Maple commands from those used by the authors. After reading the topic, you should execute the worksheet (if you have not already done so) to make sure the code works as expected. At this point feel free to explore the topic. Try rotating any three-dimensional graphs or running any animations in the file. See what happens when changes in the model or Maple code are made and then try to interpret any new results. This book is intended to be open-ended and merely serve as a guide to what is possible in mathematical modeling using a CAS, the possibilities being limited only by your own background and desires.

Each topic or story is self-contained and generally done completely, from the derivation to the solution to the plot, and accompanied by a thorough discussion of the steps and results. Since arriving at the answer is more important in our opinion than the method used, one will encounter some recipes where an analytic derivation of the model equations occurs, followed by a numerical solution because an analytic solution doesn't exist. Although brief introductions, which generally include some definitions of terminology and short explanations of underlying concepts, are given for each main topic area, this text is not intended to teach you everything that you want to know, for example, about vector operators, or working with matrices. Neither is it intended to teach you about the myriad subject areas of science or engineering. Instead, it is meant to serve as a guide to how these topics and areas can be handled using a CAS. However, this book is not just any ordinary guide. It is a *gourmet's guide*! It presupposes that the reader has learned or is about to learn about various scientific models and/or methods, and we are providing the computer algebra tools to enable you to solve complex scientific problems more easily, to attain greater understanding, and to explore the frontiers of science that interest you.

At the end of most recipe subsections there are related problems where you, the reader, can check your mastery of the scientific computation and computer algebra techniques presented in the recipes. The problems also allow you to explore new frontiers and challenge you to invent and solve "What happens if...?" problems. The purpose of this text is not only to teach computer-assisted computational techniques useful to engineering and science students, but to whet the student's curiosity and put some fun back into the pursuit of a science education. For maximum satisfaction and learning, it demands an interactive approach by the reader. Although the stories were designed to be interesting or amusing to read, the Maple recipes must be run, the models explored, and the problems solved. Some things never change in the learning process!

C. Introductory Recipe: Bridge Design 101

...there is nothing in embankments and railways and iron bridges... to oblige them to be ugly. Ugliness is the measure of imperfection.
H. G. Wells, British writer, *A Modern Utopia* (1905)

To give you some idea of what a typical computer algebra recipe looks like and to introduce some basic Maple syntax, consider the problem that follows. The recipe that solves this problem is not on the CD, so after reading this section you should open up "classic" Maple 10 on your computer, type in the recipe, and execute it. You might then wish, for example, to change the bridge design or the cost equations and see how quickly new results can be obtained.

Russell, an engineer, is to design a steel bridge that crosses a river 300 meters wide in such a way that the total cost is a minimum. Assume that the bridge has a supporting pier at each end in addition to the intermediate piers and that the length x of each span between adjacent piers is the same. The cost per span goes up with length but fewer supporting piers would then be needed. The cost (in dollars) per span is given by the formula

$$C_s = 50\,x^2 + 5000\,x - 100000,$$

while the cost (in dollars) per pier is

$$C_p = 200000 + 1000\,x.$$

(a) Derive a formula for the total cost of the bridge as a function of x.

(b) What is the cost of the bridge if it has 6 spans?

(c) Plot the total cost formula over a suitable range of x.

(d) Determine the value of x that minimizes the total cost.

(e) How many spans and piers would be needed?

(f) What is the minimum total cost?

To solve this problem, Russell first clears Maple's internal memory of any previously assigned values (other worksheets may be open with numerical values given to some of the same symbols being used in the present recipe). This is done by typing in the `restart` command after the opening prompt (>) symbol, ending the command with a colon (`:`), and pressing Enter (which generates a new prompt symbol) on the computer keyboard.

```
> restart:
```

All Maple command lines must be ended with either a colon, which suppresses any output, or a semicolon (`;`), which allows the output to be viewed.

The cost per span formula is now entered.

```
> Cs:=50*x^2+5000*x-100000; #cost per span
```

$$Cs := 50\,x^2 + 5000\,x - 100000$$

Use has been made of the assignment (`:=`) operator, placing `Cs` (`C` for cost, `s` for span) on the left-hand side (lhs) of the operator and its analytic form on the right-hand side (rhs). Assigned quantities can be mathematically manipulated. The symbols `+`, `-`, `*`, and `^` are used for addition, subtraction, multiplication, and exponentiation, respectively. The symbol `/` is used for division. Russell has added a comment to the command line, which is useful for later reference or for other people to read. Comments are prefixed by the sharp character `#`.

Similarly, Russell inputs the cost per pier formula, again adding a comment.

```
> Cp:=200000+1000*x; #cost per pier
```

$$Cp := 200000 + 1000\,x$$

Because they are short, Russell now enters two commands on the same line. Although not necessary, he separates the commands with an intervening space for reading clarity. He first enters the width $W = 300$ of the river, and then the expression $n = W/x$ for calculating the number n of spans. Note how the assigned value of W is automatically substituted into n.

```
> W:=300: n:=W/x; #n=number of spans
```

$$n := \frac{300}{x}$$

Since it is stated in the problem wording that the bridge has a supporting pier at each end, the number of piers must be be one more than the number of spans. Therefore, the total cost of the bridge must be given by $Ct = n\,Cs + (n+1)\,Cp$.

```
> Ct:=n*Cs+(n+1)*Cp; #total cost
```

$$Ct := \frac{300\,(50\,x^2 + 5000\,x - 100000)}{x} + \left(\frac{300}{x} + 1\right)(200000 + 1000\,x)$$

Again, the expressions for n, Cs, and Cp have been automatically substituted into the total cost expression. To reduce Ct to a simpler form, Russell applies the `simplify` command to it.

```
> total_cost:=simplify(Ct);
```

$$total\_cost := \frac{16000\,(x^2 + 125\,x + 1875)}{x}$$

For variety, Russell has decided to use words for labeling the total cost, using an underscore to separate the two words on the lhs. Words are more tedious to type than symbols, but have the advantage of being easier for someone else to read. Maple has certainly simplified the total cost expression. However, be warned. Maple doesn't always simplify a complicated expression in the form that you may desire. As will be utilized in later recipes, the `simplify` command comes with various possible options that can be included in the command as an additional argument, such as `simplify(expression,trig)` for simplifying a trigonometric expression.

As requested, Russell calculates the total cost of the bridge assuming that it has 6 spans. This is done by applying the evaluation (`eval`) command to the total cost, taking the span length to be $x = W/6$, i.e., $x = 300/6 = 50$ meters.

```
> cost_example:=eval(total_cost,x=W/6); #cost for 6 spans
```

$$cost\_example := 3400000$$

In this case the total cost would be \$3,400,000, i.e., 3.4 million dollars.

Russell next plots[5] the total cost, expressed in millions of dollars, over the range $x = 10$ (which would correspond to 30 spans) to $x = 150$ (2 spans). The resulting computer picture is reproduced in Figure 1.

```
> plot(total_cost/10^6,x=10..150,tickmarks=[3,3],
  labels=["x","total_cost"]);
```

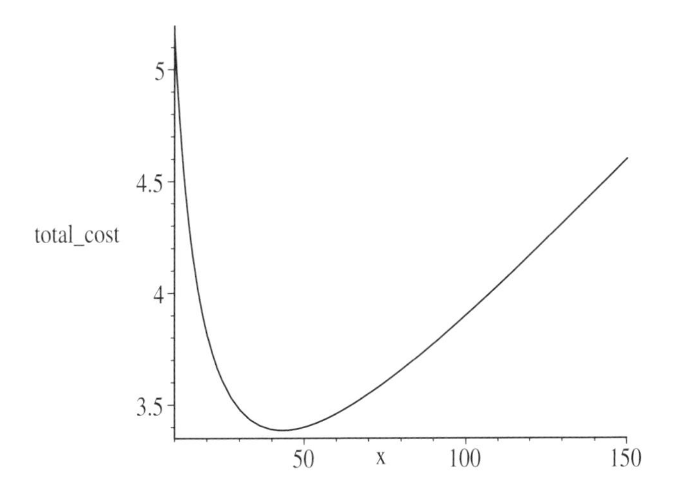

Figure 1: Total cost (millions of dollars) of bridge vs. span length x.

The `plot` command comes with various optional arguments. Russell has controlled the minimum number (3 here) of tickmarks along the horizontal and

[5] To fit into the page width, the plot command is artificially split into two lines here. Other long Maple commands will be similarly split in the text.

vertical axes, and has added labels to both axes. Note that the axis label names have been enclosed in double quotes. Each enclosed item is a Maple "string." A string is a sequence of characters that has no value other than itself. It cannot be assigned to, and will always evaluate to itself. Omit the double quotes and see what happens. Then omit the axis labels and the `tickmarks` option entirely, and again see what picture Maple produces.

If you wish to learn more about the `plot` command and its optional arguments, click the left mouse button on the `plot` command and then on **Help** at the top of the computer screen. Clicking on the entry **Help on plot** opens up a help page with information about this command structure. The various plotting options available can be found by clicking on the underlined hyperlink **plot/details** that appears on the help page and then on **plot[options]**.

Returning to the problem at hand, Russell observes that the curve has a minimum in the neighborhood of $x=40$. Clicking on the computer plot with the left mouse button, then placing the cursor on the minimum of the curve, and clicking again opens up a small window at the top left of the computer screen, which gives the horizontal and vertical coordinates of the cursor location. Try it and see what approximate values you get for span length and total cost.

To find the minimum in the curve quantitatively, Russell uses the `diff` command to analytically differentiate the total cost with respect to x, and assigns the result the name *derx* (derivative with respect to x).

```
> derx:=diff(total_cost,x);
```

$$derx := \frac{16000\,(2\,x+125)}{x} - \frac{16000\,(x^2+125\,x+1875)}{x^2}$$

The minimum value of x can be found by setting *derx* equal to 0, and solving for x. This is done numerically by using the floating-point solve command, `fsolve`, which is based on Newton's method. Guided by the information already obtained, the search range is taken to be between $x = 30$ and 60. It should be noted that if Russell had not explicitly set *derx* equal to zero in the `fsolve` command, Maple would have by default assumed that this is what was intended.

```
> X:=fsolve(derx=0,x=30..60);
```

$$X := 43.30127019$$

The answer is given to 10 digits, Maple's default accuracy for the `fsolve` command. The minimum in the curve occurs at about $x=X=43.3$. The number N of spans is obtained by evaluating n at $x=X$,

```
> N:=eval(n,x=X);
```

$$N := 6.928203231$$

yielding $N \approx 6.9$. But the number of spans (and piers) must be an integer. The `round` command now rounds N to the nearest integer,

```
> Ns:=round(N); Np:=Ns+1;
```

$$Ns := 7 \qquad Np := 8$$

yielding $Ns=7$ spans. The number Np of piers is one more than the number of spans, i.e., $Np=8$.

Russell then uses the floating-point evaluation command, `evalf`, to numerically evaluate the span length $Xmin = W/Ns$ that minimizes the total cost. If `evalf` is omitted from the following command line, $Xmin$ will appear as a ratio of whole numbers.

```
> Xmin:=evalf(W/Ns);
```

$$Xmin := 42.85714286$$

The minimum total cost in millions of dollars follows on evaluating the total cost, divided by 10^6, at $x = Xmin$.

```
> min_cost:=eval(total_cost/10^6,x=Xmin); #millions of dollars
```

$$min\_cost := 3.385714286$$

The minimum total cost is about 3.386 million dollars. This is slightly less than the 3.4 million dollars that Russell obtained earlier for 6 spans.

D. Maple Help

Public money is like holy water; everyone helps himself to it.
Italian proverb

We have already seen in the introductory recipe one approach to accessing Maple's **Help**. The reader may use this method to learn more about the various Maple commands that appear in the text recipes.

If you wish to learn what other help approaches are available, click on **Help** at the top of the computer screen and then on the entry **Using Help**. A help page opens with a number of hyperlinks that you should explore. Two of the more important hyperlinks are entitled **Perform a Topic Search** and **Perform a Full Text Search**. Here we shall give two simple examples of these searches, leaving the full descriptions of the search types for you to read. It should be noted that neither type of search is case-sensitive.

Our first example illustrates a topic search, which locates help based on a keyword that you specify. For example, suppose that you wanted the correct form of the command for taking a square root. Click on **Help**, then on **Topic Search**, making sure that the **Auto-search** box is selected. Depending on the programming language, the square root command could be `sqr`, `sqrt`,. . .,. . . On typing `sq` in the **Topic** box, Maple will display all the commands starting with `sq`. Double click on `sqrt` or, alternatively, single click on `sqrt` and then on OK. A description of the square root command will appear on the computer screen.

The second example illustrates a full text search. Suppose, for example, that you wish to find the command for analytically or numerically solving an ODE. In the Help window, click on **Full Text Search**. Then type **ode** in the **Word(s)** box and click on **Search**. When you double click on **dsolve**, a description of the `dsolve` command for solving ODEs will appear along with several examples. If you want to know how to find a numerical solution, click on the hyperlink **dsolve,numeric**.

The approach employed in the introductory recipe can also be used to find information about unfamiliar mathematical functions that appear in the Maple output. If, for example, the output contained the word "EllipticF," you can find out what this function is by clicking on the word to highlight it, then on **Help**, and finally on **Help on EllipticF**. You will find that EllipticF refers to the incomplete elliptic integral of the first kind, which is defined in the Help page. The same Help window can also be opened by typing in a question mark followed by the word and a semicolon, e.g., `?EllipticF;`

Maple's Help is not perfect, and on occasion you might feel frustrated, but generally it is helpful. It should be consulted whenever you do not fully understand the Maple syntax, or the options available, in one of the text recipes or are seeking just the right command to accomplish a certain mathematical task in your own recipe.

E. How to Use This Text

Begin at the beginning ... and go on till you come to the end.
Lewis Carroll, *Alice's Adventures in Wonderland* (1865)

The recommended procedure for most readers, particularly for someone who is new to CASs in general and to Maple in particular, is to follow the advice given in the quotation from Lewis Carroll. Start with the **Appetizers**, then go on to the **Entrees**, and finish off with the **Desserts**. In the early recipes of the **Appetizers** you will be introduced to more of the basic features of the Maple system and see further examples of Maple's Help.

Of course, if you are already a Maple expert, feel free to pick and choose those topics and recipes that interest you or are relevant to your own scientific tastes or goals.

No matter what approach to using this text is taken, we hope that you will enjoy the wide range of interdisciplinary topics and stories, which range from the stock market to zoological scaling and from the world of sports to chaos and the outbreak of war. Before beginning your journey through this text, let us paraphrase a well-known saying from the world of sports with these words of advice:

You can't learn the great game of scientific modeling
by being a spectator. You must play the game!

We trust that as you sample and explore the various recipes on which our menu is based, you will enjoy the "intellectual feast" that we have prepared and presented in this introductory guide to the mathematical models of science.

Bon Appetit!
Richard and George,
Your CAS chefs

Part I

THE APPETIZERS

Some books are to be tasted, others to be swallowed,
and some few to be chewed and digested.
Francis Bacon, English monk (1561–1626)

I did toy with the idea of doing a cook-book.
The recipes were to be the routine ones:
how to make dry toast, instant coffee,....
But as an added attraction...
my idea was to put a fried egg on the cover.
I think a lot of people who hate literature but
love fried eggs would buy it if the price was right.
Groucho Marx, American comic actor (1895–1977)

It's red hot, mate. I hate to think of this sort of book
getting into the wrong hands. As soon as
I've finished this, I shall recommend they ban it.
Tony Hancock, British comedian (1924–1968)

Chapter 1

The Pictures of Science

The great tragedy of science ...
the slaying of a beautiful theory by an ugly fact.
Thomas Huxley, English biologist (1825–1895)

In experimental science, the detailed analysis of accurate data is of paramount importance in checking existing theories and formulating new ones. Since the laws of science are more useful and more easily understood when they are expressed as mathematical relationships rather than as a collection of numbers, methods are needed for deriving these relationships. One of the simplest approaches to determining whether a mathematical relationship exists between one variable and another is to create a graph of the data and establish whether the data can be fitted by a model equation. This is the central subject matter of the two *graphical analysis* chapters that form the **Appetizers**. A model equation is, of course, valid only in so far as it is supported by new data, the history of science being dotted with instances in which model equations have had to be revised, sometimes substantially, because of new facts coming to light.

In this first chapter we look at how to create a variety of computer-generated graphs, while in the second chapter the least squares method of deriving model equations, using the Maple command structure, is the central topic. Our coverage of graphing techniques at this stage is only an introduction to the topic, since many other important types of graphs will be encountered in ensuing chapters. For your convenience as well as for future reference, a complete list of all the Maple command structures used in this text, including graphical commands, can be found in the index under the heading "Maple Command."

1.1 Data and Function Plots

The old adage that a picture is worth a thousand words is certainly applicable to the world of graphs. Graphs are the pictures of science. It is rare to read a scientific paper or text that contains no graphs. Graphs are important because they help fulfill a number of useful functions:

(1) Graphs present the data in a unified, concise, and visual manner and can stimulate insight and aid in remembrance of the data structure.

(2) Graphs can be created that show the data averaged by a smooth continuous line. It is the scientist's hope that the line forms an easily recognized shape that can be described by a simple model equation. Some commonly encountered shapes and associated model equations are schematically illustrated in Figure 1.1. The exact shapes depend on the signs and values of the parameters a, b, c, and n.

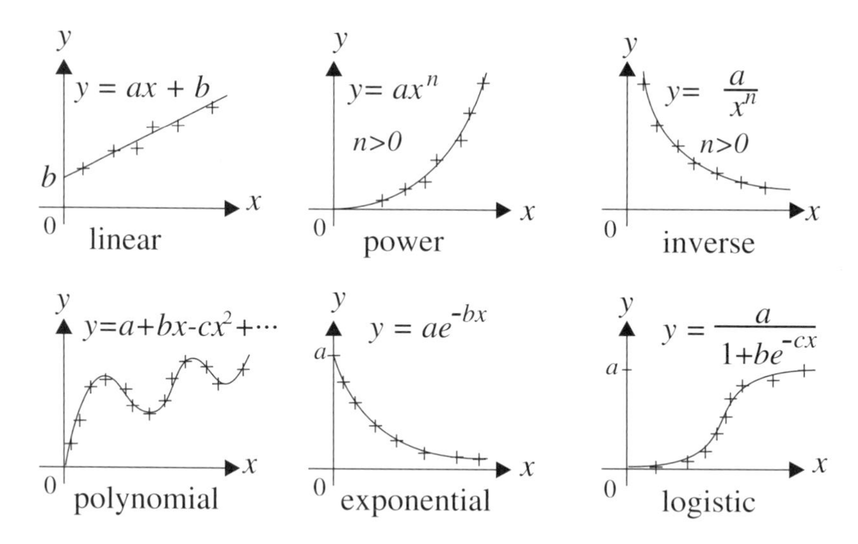

Figure 1.1: Commonly encountered shapes and model equations.

(3) Graphs permit an individual data point to be compared with the curve of the proposed model equation. A data point that has a large deviation from the curve might indicate a measurement error, a blunder in entering the data, or the need for a more realistic mathematical expression.

(4) Graphs help one to remember visually the mathematical expression derived from the graph.

PROBLEMS:

Problem 1-1: Physical laws

Suggest some examples of physical laws or data sets that correspond to each of the shapes shown in Figure 1.1.

Problem 1-2: Ideal gas law

For n moles of gas, the ideal gas law is given by $PV = nRT$, where P is the pressure, V the volume, R the gas constant, and T the absolute temperature. Sketch the possible gas law graphs relating any two of the thermodynamic variables P, V, and T, holding the third variable fixed. Relate the graphs to the shapes in Figure 1.1.

Problem 1-3: Kepler's third law
Kepler's third law states that the square of the period T for elliptical planetary motion about the sun is proportional to the cube of the ellipse's semimajor axis a. Sketch possible third law graphs, relating them to the shapes in Figure 1.1.

Problem 1-4: Potential Curves
For each of the following relationships, sketch the potential $V(r)$ as a function of the radial distance r and relate the shape to one or more of the graphs in Figure 1.1 (k, a, and b are positive constants):

(a) Coulomb potential: $V(r) = k/r$,

(b) Lennard–Jones potential: $V(r) = a/r^{12} - b/r^{6}$.

1.1.1 Correcting for Inflation

If all economists were laid end to end, they would not reach a conclusion.
George Bernard Shaw, Anglo-Irish playwright, critic (1856–1950)

When presented with observational or experimental data, it is recommended that a science or engineering student first plot the data to get a qualitative feeling for any possible trend or, perhaps, cyclic behavior. The pictorial representation of data is also the normal starting point for further quantitative analysis. For example, it may be possible to describe the data by a model equation that can in turn be used to make predictions of future trends. Data come in all sorts of shapes and sizes and from a wide variety of sources, such as scientific journals and newspapers. For example, the business sections of newspapers and magazines often give the prices of various commodities as a function of time. In some cases, the prices are corrected for changes in purchasing power and in other cases not. Purchasing power refers to the value of money as measured by the services it can buy [DG95] and is inversely related to the consumer price index. It is usually referenced to the purchasing power in a particular year. For example, one dollar in the mid-1990s had only 67% the purchasing power of a dollar in 1982 due to inflation. As an illustration of how to correct prices for changes in purchasing power, consider the following example.

Colleen is a science graduate who recently received an MBA degree. She has just been hired as a sales manager at the Glitz department store located in a suburban shopping mall in the city of Metropolis. To get to work from her high-rise apartment in the city center, she is considering buying a new car but is concerned with the cost and scarcity of rental spaces as well as the cost of running the car. In particular, she speculates on whether the current price of gasoline will undergo a substantial inflationary increase. She intends on doing an Internet search to get gasoline price statistics covering the last few years, but while waiting for her friend Vectoria to drive her to work she plays around with the data presented in Table 1.1, gleaned from an old economics text.

Table 1.1: Gasoline prices and purchasing power of the dollar.

Year	**Price** (dollars)	**Purchasing power**	**Year**	**Price** (dollars)	**Purchasing power**
1983	1.24	1.00	1988	0.95	0.85
1984	1.21	0.96	1989	1.02	0.81
1985	1.20	0.93	1990	1.16	0.77
1986	0.93	0.91	1991	1.14	0.73
1987	0.95	0.88	1992	1.13	0.71

This table, extracted from the *Statistical Abstract of the U.S.* [SAU94], shows the average price in dollars of a gallon of regular unleaded gasoline in the United States for the years 1983 to 1992. The purchasing power of the dollar, as measured by consumer prices, relative to that in 1983 is also given.

As an exercise, she decides to first plot the uncorrected data of Table 1.1 and then calculate and plot the gasoline prices adjusted for the decrease in purchasing power. Influenced by her engineering and mathematics friends, she has learned to use the Maple CAS. For your benefit, we will eavesdrop on what Colleen is doing and explain the structure and purpose of each command line that she uses in her worksheet.

Colleen knows that it is a good idea to start each worksheet with the `restart` command, which clears Maple's internal memory and removes all assigned values from the variables. Since she intends to create a statistical plot, she also enters the command `with(Statistics)`, which accesses the statistics library package. Library packages are extremely important, since they contain approximately 90% of Maple's mathematical knowledge. The preface `with` always indicates that a Maple library package is being "loaded" into the worksheet. Because the command is ended here with a semicolon, a partial[1] list of the contents of this library package is revealed in the output shown below.

```
> restart: with(Statistics);
```

[*AbsoluteDeviation*, *AreaChart*, *BarChart*, *Bootstrap*, *BoxPlot*, *BubblePlot*, ..., *ScatterPlot*, ..., *WeightedMovingAverage*, *Winsorize*, *WinsorizedMean*]

Colleen will use the `ScatterPlot` command to plot the relevant data points, using the entries of Table 1.1.

[1]Because the list of contents of this library package is very long, not all output entries are shown here in the text.

For convenience in entering and plotting the data, Colleen labels the year 1983 as 0, 1984 as 1, and so on. The years are then entered as a Maple "list" and assigned the name *year*. In a list the items are separated by commas and enclosed with square brackets. Maple preserves the order of the items in a list, a property that is obviously important for plotting data. Colleen also adds a comment to the end of the command line, indicating that the data entries refer to the year since 1983.

```
> year:=[0,1,2,3,4,5,6,7,8,9]; #year since 1983
```

$$year := [0, 1, 2, 3, 4, 5, 6, 7, 8, 9]$$

Similarly, Colleen creates separate lists for the uncorrected gasoline prices and the purchasing power, attaches appropriate names, and suppresses the redundant output with line-ending colons.

```
> price:=[1.24,1.21,1.20,0.93,0.95,0.95,1.02,1.16,1.14,1.13]:
> purchasing_power:=[1.00,0.96,0.93,0.91,0.88,0.85,0.81,
                     0.77,0.73,0.71]:
```

Again, to fit into the page width the purchasing power data have been artificially split over two text lines. In some cases, it will be necessary for us to split the Maple command over even more text lines, so remember that the command is not complete until you see a colon or a semicolon.

The `ScatterPlot` command is used in `graph1` to create a plot of price vs. year. The first argument, `year`, will be plotted horizontally and the second argument, `price`, vertically.

```
> graph1:=ScatterPlot(year,price,symbol=box,symbolsize=16,
        view=[0..9.1,0..1.25],tickmarks=[3,2],
        labels=["year","price"]):
```

Colleen has included various plot options. For example, she has decided to use size-16 boxes[2] to represent the data points and has also included a `view` command option for the horizontal and vertical axes. The numbers `0..9.1` in the view option span the entire horizontal plotting range, while `0..1.25` specifies the vertical range. If this option is omitted, Maple will choose its own vertical range, which excludes the origin. Omit this `view` option and see what vertical range Maple gives. This option is very useful and will often be invoked throughout this text.

Continuing with the other plot options in the above command line, the minimum number of `tickmarks` on the horizontal and vertical axes was specified and labels added to these axes. Since she wants ultimately to superimpose the "raw" gasoline price data on the same graph as the inflationary adjusted data, Colleen has suppressed the output in the last command line by using a colon and attaching the name `graph1` to the plot.

To actually see the graph, she types the name again followed by a semicolon,

[2]The default symbol size is 10. Circles, crosses, and diamonds are other symbols that can be used. Employing different symbols is obviously useful when more than one set of data is to be plotted in the same figure.

```
> graph1;
```

thus producing the picture shown in Figure 1.2.

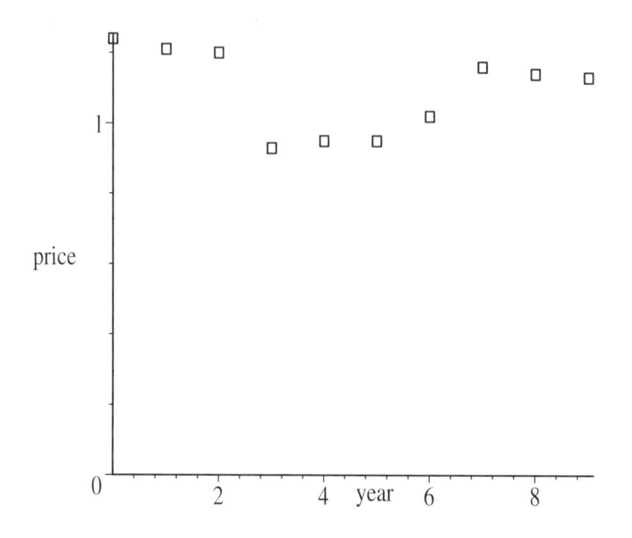

Figure 1.2: Uncorrected gasoline prices in dollars per gallon.

The reader can, and should, experiment with other stylistic options. This can be easily done by clicking the left mouse button on the computer plot, which opens up some style options on the uppermost tool bar on the computer screen. For example, you might click on the Style box and then change the Symbol from a Box to a Cross and the Symbol Size to, say, 20. Similarly, the axes style may also be changed, e.g., from Normal to Boxed. Clicking on the Projection box opens up two options, Constrained and Unconstrained. The default picture produced here was unconstrained. What happens to the picture if you choose the Constrained projection?

After one has viewed the above possibilities, more permanent changes can be produced, if desired, by changing the style options in the code. To find out what options are available, you can click[3] on the `ScatterPlot` command, on **Help**, and then on **Help on ScatterPlot**. This opens a help page with a hyperlink to **plot[options]**. Using the information provided there, you could try adding a title of your own choice to the plot.

To correct the gasoline prices for the decrease in purchasing power, the entries in the price column of Table 1.1 should be multiplied by the corresponding entries in the purchasing power column. In the following command line, Colleen uses the command `price[i]` to extract the ith entry from the Maple `price` list. A similar command structure is used to select the ith entry for the purchasing power, and then the multiplication is performed. A sequence command, `seq`,

[3]Alternatively, you could access this hyperlink by typing **ScatterPlot** in **Topic Search**.

is added to the command structure so as to perform the above operation on all 10 entries, i=1..10. Finally, the result is enclosed in square brackets to make a Maple list, and Colleen attaches a new name, using the acronym *ap* to stand for (inflation) "adjusted price."

```
> ap:=[seq(price[i]*purchasing_power[i],i=1..10)];
```

$ap := [1.2400, 1.1616, 1.1160, .8463, .8360, .8075, .8262, .8932, .8322, .8023]$

A plot (graph2) of the gasoline prices adjusted for decreasing purchasing power is now created, the data points being represented by size-16 crosses.

```
> graph2:=ScatterPlot(year,ap,symbol=cross,symbolsize=16,
        view=[0..9.1,0..1.6],tickmarks=[3,2],
        labels=["year","price"]):
```

Examining the adjusted gasoline prices in *ap*, Colleen notices that the data appear to be approximated by a "piecewise linear" curve consisting of three straight-line segments. She takes the first line segment to be from the point $(0, 1.24)$ to the point $(2, 1.11)$, the second line from $(2, 1.11)$ to $(3, 0.83)$, and the third line to be horizontal between $(3, 0.83)$ and $(10, 0.83)$. To plot the three line segments, she uses the following plot command to produce graph3.

```
> graph3:=plot({[[0,1.24],[2,1.11]],[[2,1.11],[3,0.83]],
        [[3,0.83],[10,0.83]]},color=magenta,thickness=2):
```

To form the first line segment, Colleen has placed the two coordinates of the line's endpoints into Maple lists, viz., [0,1.24] and [2,1.11]. These two lists are then separated by a comma and enclosed in square brackets, thus forming a "list of lists." Two more lists of lists are formed for the endpoints of the other two straight-line segments. The three lists of lists are then separated by commas and enclosed within curly brackets. Such curly brackets indicate a Maple "set." Maple sets have the same properties as sets in mathematics. Unlike a list, a set does not preserve order or repetition. In some later recipes, the order in which the curves or pictures are plotted will matter. Then, square brackets should be used instead of curly brackets. Finally, the three line segments were given a magenta color and a thickness of 2 (the default, 0, produces a fairly thin line).

The three graphs can be superimposed in the same figure using the display command. First the plots library package must be accessed, because it contains this command. Replace the colon with a semicolon to see the very lengthy list of specialized plot commands contained in the plots package.

```
> with(plots):
```

Warning, the name changecoords has been redefined

Note that a warning message is produced here, which informs us that the name changecoords has been redefined in the current release of Maple. If desired, warnings can be removed by inserting the command interface(warnlevel=0) prior to loading the library package. From now on, all such warnings will generally be artificially removed in the text.

Using the display command, the three graphs are now superimposed in the same figure, which is reproduced in Figure 1.3.

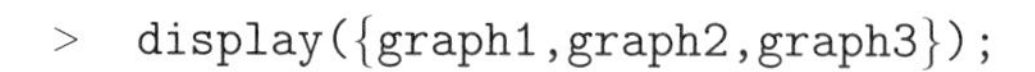

```
> display({graph1,graph2,graph3});
```

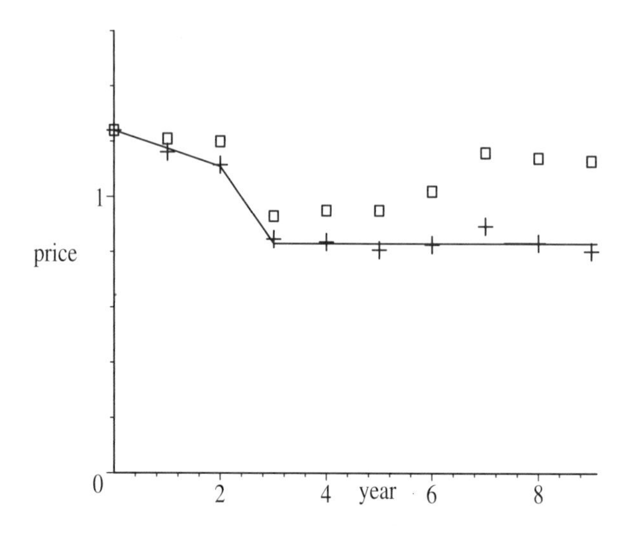

Figure 1.3: Inflation corrected (+) and uncorrected (□) gasoline prices.

Colleen's piecewise linear curve does a reasonable job of fitting the adjusted data. In 1986 (year 3), there was a sizeable drop in the adjusted prices, the data thereafter oscillating around the 83-cent level up to 1992 (year 10).

Colleen wonders what the trend was after 1992. What do you think? Much higher? Remember that one is talking about inflation-adjusted gasoline prices here, not the unadjusted prices. Also remember that for a scientist or engineer, it's not good enough to state an unsupported opinion without some backup evidence. This might mean a library search or a search on the Internet for relevant statistics, followed by a calculation similar to that carried out by Colleen.

We will leave Colleen now as her young friend Vectoria has arrived to give her a ride to work. As to her earlier trepidation on whether to buy a new car or not, she has decided, in a fit of "irrational exuberance,"[4] that the sporty red convertible that she saw the other day would suit her just fine!

PROBLEMS:

Problem 1-5: Purchasing power

Plot the relative purchasing power for the years 1983 to 1992 given in Table 1.1, choosing your own stylistic options.

Problem 1-6: More recent gasoline data

Go to the Internet and obtain more recent data on gasoline prices and on the purchasing power of the dollar and carry out a calculation similar to that in the text. Discuss your results.

[4]With due apologies to the U.S. Federal Reserve chairman Alan Greenspan, who used this phrase to describe the mood underlying the U.S. stock market surge of the late 1990s.

Problem 1-7: Colleen's restaurant job

While going to college, Colleen worked part-time at a local restaurant, the Hungry Heifer Steak House, to pay for her tuition. As part of her job, she kept tabs on the number of customers that the restaurant served each hour as well as the total revenue (dollars spent) taken in during that hour. Table 1.2 shows, for example, that in the hour centered on 7:00 a.m., 30 customers had been served, with a total of 120 dollars having been taken in. Your task is to plot Colleen's restaurant data in various ways.

Table 1.2: Colleen's restaurant data.

Time	Number	$ Spent	Time	Number	$ Spent
7 a.m.	30	120	3 p.m.	40	205
8 a.m.	45	230	4 p.m.	50	310
9 a.m.	35	170	5 p.m.	75	680
10 a.m.	45	200	6 p.m.	100	1150
11 a.m.	62	320	7 p.m.	95	1215
12 p.m.	80	650	8 p.m.	60	500
1 p.m.	78	546	9 p.m.	30	200
2 p.m.	55	280	10 p.m.	20	100

(a) Using Colleen's data, plot the number of customers against the time, setting 7:00 a.m. as zero. Use circles and add appropriate labels and tickmarks. Choose a view that includes the origin and the whole range of customer data.

(b) Plot the dollars taken in against the time. Use boxes and add appropriate labels and tickmarks. Choose an appropriate view.

(c) Create a Maple list showing the average amount spent per customer as a function of time.

(d) Plot the above Maple list, using crosses and appropriate labels and tickmarks. Choose a view that includes the origin.

(e) Display the first and third graphs in the same picture, adding an appropriate title that distinguishes the two sets of data.

Problem 1-8: The great Spanish flu epidemic of 1918

In the fall of 1918, just as World War I was ending, a flu epidemic (the Spanish flu) began in the U.S. navy, spread to the U.S army, then to the American civilian population, and finally to the rest of the world, resulting ultimately in some 20 million deaths by 1920. Table 1.3 shows U.S. death statistics [LFHC95] due to the flu for the fall of 1918. The numbers refer to the cumulative (total) deaths up to the end of the week indicated for the navy, army, and civilian (Civ.) populations. The civilian deaths are for 45 major cities.

Table 1.3: Cumulative deaths for the 1918 great flu.

Week	Navy	Army	Civ.	Week	Navy	Army	Civ.
Aug. 31	2			Oct. 19	2670	15,319	37,853
Sept. 7	13	40		Oct. 26	2820	17,943	58,659
Sept. 14	56	76	68	Nov. 2	2919	19,126	73,477
Sept. 21	292	174	517	Nov. 9	2990	20,034	81,919
Sept. 28	1172	1146	1970	Nov. 16	3047	20,553	86,957
Oct. 5	1823	3590	6528	Nov. 23	3104	20,865	90,449
Oct. 12	2338	9760	17,914	Nov. 30	3137	21,184	93,641

(a) Plot three separate graphs for the naval, army, and civilian death numbers. Take the same number of tickmarks in each case and a view that includes the origin. Use different symbols for the data points of the different graphs. To keep the number of list entries the same for later comparison, set the army and civilian entries that are blank in the table equal to zero.

(b) Create a list for the total number of deaths for the navy, army, and civilian populations combined. Plot the new list data, with a view that includes the origin and a different symbol for the data points from those chosen for the first three graphs.

(c) Use the `display` command to show all four graphs in the same plot. Add the title "number of deaths versus week" to your plot.

(d) The curves in each case are examples of what type of model equation?

1.1.2 The Plummeting Badminton Bird

Physics tries to discover the pattern of events which controls the phenomena we observe.
Sir James Jeans, British physicist (1877–1946)

In a delightful reprint collection entitled *The Physics of Sports*, edited by Angelo Armenti, Jr. [PLA92], the sports-minded reader can learn about the aerodynamics of a knuckleball,[5] the physics of drag racing, the stability of a bicycle, and the physics of karate, to name just a few topics that are covered.

Vectoria,[6] a physics major at the Metropolis Institute of Technology (MIT), enjoys playing badminton and is delighted to find an article in Armenti's book

[5]A knuckleball is a type of baseball pitch. The baseball is held with the first knuckles or the fingertips and thrown in such a way as to virtually eliminate the spin. Because of the stitching on the baseball, this leads to an erratic trajectory of the ball.

[6]The origin of this unusual first name will be the subject of a later story.

on the aerodynamics of a badminton bird. Peastrel et al. report on their investigation of the effect of air resistance on a badminton bird falling vertically from rest. Taking y to be the distance fallen (in meters) and t to be the elapsed time (in seconds), their experimental data are given in Table 1.4. Vectoria decides to plot these data and investigate the behavior of the falling bird.

Table 1.4: Distance that the badminton bird falls in t seconds.

y (meters)	t (seconds)	y (meters)	t (seconds)
0.61	0.347	2.74	0.823
1.00	0.470	3.00	0.870
1.22	0.519	4.00	1.031
1.52	0.582	5.00	1.193
1.83	0.650	6.00	1.354
2.00	0.674	7.00	1.501
2.13	0.717	8.50	1.726
2.44	0.766	9.50	1.873

She accesses the same Maple library packages as used by her good friend Colleen in the previous example, but suppresses all of the output by using colons.

```
> restart: with(plots): with(Statistics):
```

Maple protects certain symbols and names, `time` being one of them. The command `time()` returns the total CPU time in seconds used since the start of a Maple session.[7] Since Vectoria intends to use this name for a list of numbers, the word `time` must be first unprotected from its Maple meaning.

```
> unprotect(time):
```

To learn more about the `unprotect` command, consult the relevant help page, opened by clicking on `unprotect`, **Help**, and **Help on unprotect**.

The data are entered as separate Maple lists for the time and the distance.

```
> time:=[0.347,0.470,0.519,0.582,0.650,0.674,0.717,0.766,
        0.823,0.870,1.031,1.193,1.354,1.501,1.726,1.873]:
> distance:=[0.61,1.00,1.22,1.52,1.83,2.00,2.13,2.44,
        2.74,3.00,4.00,5.00,6.00,7.00,8.50,9.50]:
```

A graph of the data points (represented by size-16 circles) is created

```
> pts:=ScatterPlot(time,distance,symbol=circle,symbolsize=16,
     tickmarks=[2,4]):
```

using the `ScatterPlot` command, and then displayed in Figure 1.4, the circles being colored blue on the computer screen.

```
> display(pts,color=blue);
```

[7] This use of the `time` command will be exploited in Chapters 7 and 8, dealing with Monte Carlo numerical methods and fractal patterns, respectively.

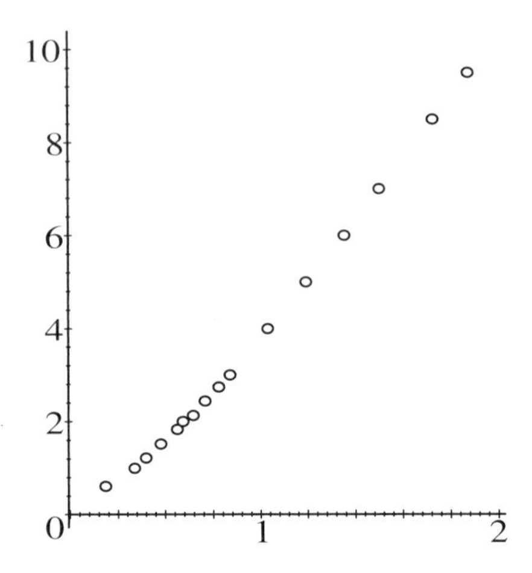

Figure 1.4: Plot of distance versus time data points.

Qualitatively, Vectoria notes that the shape of the graph looks like a power law for small times and a straight line for larger times. Thus, she anticipates a mathematical form that has these limiting behaviors. Actually, a model equation can be analytically derived directly from Newton's second law. The interested reader should consult the **Advanced Guide**, where Vectoria carries out the nontrivial derivation. Here she will only outline the underlying physics.

Assuming that the resistive force on the falling badminton bird is described by *Newton's law of resistance*, $F_{\rm res} = -k\, m\, |v|\, v$, where k is a positive constant, m the mass of the bird, and v its velocity, the distance that the bird falls from rest is given by

$$y = \frac{v_T^2}{g} \ln\left(\cosh\left(\frac{g\,t}{v_T}\right)\right). \tag{1.1}$$

Here g is the acceleration due to gravity, v_T is the terminal velocity, and cosh is the hyperbolic cosine function, i.e., $\cosh(g\,t/v_T) = (e^{g\,t/v_T} + e^{-g\,t/v_T})/2$. When air resistance is present, a falling object does not continually accelerate but instead reaches a terminal velocity, v_T. At this speed, there is a balance between the gravitational force downward and the resistive force of the air upward, viz., $m\,g = k\,m\,v_T^2$, so that $v_T = \sqrt{g/k}$. A sky diver, with arms and legs spread-eagled, will hit a terminal speed of about 200 km/h (125 miles/hour).

A name can be subscripted using the Maple syntax, `name[subscript]`. Thus to create v_T in the Maple output, one enters `v[T]`. Making use of this result, formula (1.1) is now entered and displayed.

```
> y:=(v[T]^2/g)*ln(cosh(g*t/v[T]));
```

$$y := \frac{{v_T}^2 \ln\left(\cosh\left(\frac{g\,t}{v_T}\right)\right)}{g}$$

To confirm that y reduces to a straight-line equation for large t, Vectoria first assumes[8] that $g > 0$, $v_T > 0$. Next, she Taylor expands y about $t = \infty$, to second order.

```
> assume(g>0,v[T]>0): y_asymptotic:=taylor(y,t=infinity,2);
```

$$y\_asymptotic := t\, v_T + \mathrm{O}(1)$$

The "order of" term, O(1), is removed with the following `convert` command.

```
> y_asymptotic:=convert(y_asymptotic,polynom);
```

$$y\_asymptotic := t\, v_T$$

The asymptotic form of y is the straight-line equation $y = v_T\, t$. The slope of the graph at large t gives the terminal velocity. Using the last pair of data points in Table 1.4, Vectoria finds that

$$v_T = \frac{(9.50 - 8.50)}{(1.873 - 1.726)} = 6.80 \text{ m/s}.$$

To plot the analytic expression for y, the parameter values must be given. The badminton investigators took the gravitational acceleration to be $g = 9.81$ m/s^2, while, as shown above, their data yields $v_T = 6.80$ m/s.

```
> g:=9.81: v[T]:=6.80:
```

Vectoria now displays the distance formula with the parameter values inserted.

```
> y:=y;
```

$$y := 4.713557594 \ln(\cosh(1.442647059\, t))$$

The default setting of Maple is to give 10-digit accuracy. The formula in the last line should not be quoted to more significant figures than in the original data. How many significant figures are relevant here? The number of digits, e.g., 4, can be controlled by inserting the command `Digits:=4;` at the beginning of the Maple program. If you do specify the number of digits, be sure to carry enough digits in your calculation to avoid round-off error. In the recipes that follow, the default setting on digits will generally be used. Thus, in discussing the results, one should remember to round off the answers to the number of digits that are significant.

A graph, `Gr`, of y is now created over the time interval $t = 0$ to 2 seconds.

```
> Gr:=plot(y,t=0..2):
```

Using the `display` command, both the data points and distance formula are now superimposed in the same graph. For variety, Vectoria chooses size-12 Times italic symbols for the axis label fonts. In the plot options, the number of tickmarks is controlled and a title entered as a Maple string.

```
> display({pts,Gr},tickmarks=[3,2],labels=["t","y"],
  labelfont=[TIMES,ITALIC,12],title="distance vs time");
```

---

[8] The `assume` command will apply the assumption throughout the worksheet. Assumed quantities will generally appear in the output with "trailing tildes," e.g., g~. These trailing tildes can be removed from all worksheets by clicking on File, then Preferences, I/O Display, No Annotation, and Apply Globally. This has been done here.

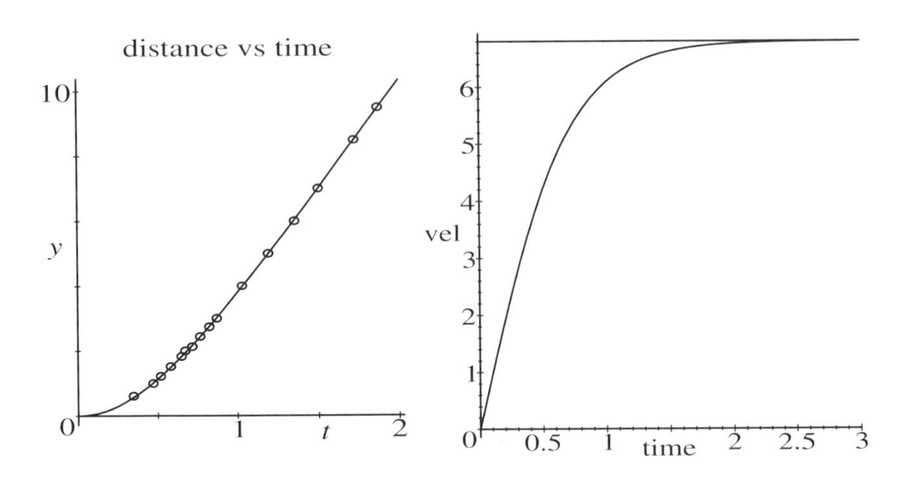

Figure 1.5: Left: data points and formula. Right: Approach of velocity to v_T.

The resulting picture is shown on the lhs of Figure 1.5. The theoretically derived formula fits the experimental data extremely well.

To see how the velocity of the falling bird approaches the terminal velocity, Vectoria analytically differentiates the solution y with respect to t,

```
> vel:=diff(y,t);
```

$$vel := 6.800000000 \frac{\sinh(1.442647059\,t)}{\cosh(1.442647059\,t)}$$

and forms a plot (coloring the curve black) of *vel* for t from 0 to 3 seconds.

```
> V:=plot(vel,t=0..3,color=black):
```

She also plots a horizontal line corresponding to the terminal velocity (6.80 m/s),

```
> terminal_vel:=plot([[0,6.80],[3,6.80]],color=blue):
```

and displays the two curves on the rhs of Figure 1.5.

```
> display({terminal_vel,V},labels=["time","vel"]);
```

From this figure, Vectoria can see that the terminal velocity of 6.8 m/s was reached at about 2 s. How much time, she wonders, did the bird take to reach a speed of, say, 5.0 m/s? From the plot, the answer must clearly lie between 1/2 and 1 s. However, she can produce a more accurate result by first clicking the mouse on the computer plot. This action results in the lowest context bar at the top of the screen being replaced with a new bar displaying a small window on the far left. Then placing the tip of the cursor arrow on the velocity curve near 5 m/s and clicking once again produces the horizontal and vertical coordinates (e.g., 0.64, 4.97) of the arrow tip, displayed inside the window. Vectoria finds that the relevant time is about 0.64 s. To generate an even more accurate answer, she can use the floating-point solve command, `fsolve`, to numerically solve the equation $vel = 5.0$ for the time, the result being labeled *t5*.

```
> t5:=fsolve(vel=5.0,t);
```

$$t5 := 0.6516884552$$

The badminton bird reached a speed of 5.0 m/s in about 0.65 seconds.

Experimentally, from Table 1.4, we see that the distance through which the bird falls in the first 1.501 s is 7.00 meters. From Figure 1.5, the theoretical value at this time must be very close to the experimental result. What is the theoretical value? This can be answered in two different ways. In her introductory physics course, Vectoria has learned that the area under the velocity curve for a certain time interval gives the distance traveled in that interval. To calculate the area, an integration must be performed. The distance (in meters) traveled will be equal to the definite integral from $t = 0$ to 1.501 seconds, viz.,

```
> distance:=Int(vel,t=0..1.501)=int(vel,t=0..1.501);
```

$$distance := \int_0^{1.501} 6.800000000 \frac{\sinh(1.442647059\, t)}{\cosh(1.442647059\, t)}\, dt = 7.001221152$$

Note that Maple is case sensitive here. Vectoria used the capitalized form `Int` to display the integral without evaluating it, while the lower-case form `int` allowed the integral to be calculated. Similarly, the capitalized form `Diff` will display the derivative without evaluating it. Another important example of case sensitivity is that the Maple command for the constant π is `Pi`, not `pi`.

The same distance should result if the time $t = 1.501$ is substituted into the analytic expression for y,

```
> subs(t=1.501,y);
```

$$4.713557594 \ln(\cosh(2.165413236))$$

and the previous line (referred to by the ditto operator %) numerically evaluated[9] with the floating-point evaluation command, `evalf`.

```
> y:=evalf(%);
```

$$y := 7.001221157$$

On rounding off to two decimal places, Vectoria observes that both of the analytically derived answers are in agreement with the experimental number in Table 1.4. So Newton's law of resistance does a very good job of explaining the dynamics of the falling badminton bird.

Vectoria is somewhat surprised that Peastrel, Lynch, and Armenti found that the more familiar Stokes's law of resistance ($F_{\rm res} = -k\, m\, v$), which is the drag relation cited in most elementary physics texts, does not prevail for the badminton bird. Can you offer any suggestions of why this is so?

PROBLEMS:

Problem 1-9: Limiting case

Determine the power law behavior of Equation (1.1) for small t. *Hint*: Use the analytic form for y given in the text and Taylor expand about $t = 0$ to third order. Remember to remove the "order of" sign.

[9] If Vectoria had used the evaluation command, `eval(y,t=1.501)`, instead of the substitute command, the numerical evaluation would have been automatically done. Try it and see.

Problem 1-10: Taylor expansion
Determine the Taylor expansion of $(1/x) - \cot x$ about $x = 0$ to the fourteenth order in x. Be sure to remove the "order of" term.

Problem 1-11: Stokes's law of resistance
If the drag force on the falling badminton bird were given by Stokes's law of resistance ($F_{\text{res}} = -k\,m\,v$), the theoretical distance y through which the bird falls from rest in time t would be given by

$$y = \frac{v_T^2}{g}(e^{-g\,t/v_T} - 1) + v_T\,t,$$

with the terminal velocity $v_T = g/k$. Taking $g = 9.81$ m/s^2 and $v_T = 6.80$ m/s:

(a) Plot the theoretical distance on the same graph as the experimental data. Discuss how well the formula agrees with the data.

(b) Analytically differentiate the distance formula to calculate the velocity.

(c) Plot the velocity as a function of time.

(d) About how long would it take the bird to reach its terminal speed according to this plot?

(e) Differentiate the velocity formula to obtain the acceleration.

(f) Plot the acceleration as a function of time.

Problem 1-12: Projectile motion
Table 1.5 shows the horizontal velocity v in meters/second as a function of time t in seconds for a shell fired from a 6-inch naval gun [Oha85].

Table 1.5: Projectile velocity as a function of time.

t	0	0.30	0.60	0.90	1.20	1.50	1.80	2.10	2.40	2.70	3.00
v	657	638	619	604	588	571	557	542	528	514	502

(a) Make a plot of velocity versus time.

(b) The velocity can be approximately represented by the formula

$$v = 655.9 - 61.4\,t + 3.26\,t^2.$$

Plot the velocity equation on the same graph as the experimental data.

(c) By integrating the area under the velocity curve, calculate the horizontal distance traveled by the shell in the first 3 seconds.

(d) Plot the acceleration versus time over the time interval 0 to 3.0 seconds by differentiating the velocity equation once with respect to time.

Problem 1-13: Hydrogen atom motion
The potential energy of one of the atoms in a hydrogen molecule is given by

$$U(x) = U_0(e^{-2\,(x-a)/b} - 2\,e^{-(x-a)/b})$$

with $U_0 = 2.36\,\text{eV}$ ($1\,\text{eV} \equiv 1.6 \times 10^{-19}$ J), $a = 0.37\,\text{Å}$(1 Å$\equiv 10^{-10}$ m), $b = 0.34\,\text{Å}$.

(a) Plot $U(x)$ in electron volts (eV) as a function of x in angstroms (Å).

(b) The force F on the atom is given by $F = -(dU/dx)$. Plot F versus x.

(c) Under the influence of this force, the atom moves back and forth along the x-axis between certain limits, called the *turning points*, determined by the total energy. If the total energy is $E = -1.15$ eV, find the turning points graphically. This can be done by plotting the total energy and potential energy versus x on the same graph and clicking the mouse with the cursor placed on the intersection points of the plot.

(d) Obtain more accurate turning points using the `fsolve` command.

Problem 1-14: Roots and minimum of a function
Consider the polynomial function $f(x) = x^4 + x - 3$.

(a) Plot the function over the range $x = -3$ to 3.

(b) By clicking on the plot, determine the approximate roots of the function.

(c) Determine the roots of the function using the `fsolve` command.

(d) Analytically, using the `diff` command, find the x-coordinate of the minimum in the curve and evaluate the function $f(x)$ at the minimum.

Problem 1-15: Definite integrals
Evaluate the following definite integrals, first obtaining the default output, and then simplifying where necessary:

$$\text{(a)} \int_0^\infty \frac{dx}{1+x^3}, \quad \text{(b)} \int_0^\infty \frac{\sqrt{x}}{1+x^2}\,dx, \quad \text{(c)} \int_{-1}^1 \frac{dx}{\sqrt{1-x^2}(1+x^2)},$$

$$\text{(d)} \int_0^\pi \frac{d\theta}{a+b\cos\theta}, \quad \text{(e)} \int_0^1 \frac{(\log x)^{16}}{\sqrt{1-x^2}}\,dx, \quad \text{(f)} \int_0^1 x^{20}\arcsin(x)\,dx.$$

Hint: For **(d)** assume that $a > 0$, $b > 0$, and $a > b$.

Problem 1-16: Integration and Differentiation
Consider the function $f(x) = \ln(1 + (x/100))\ln(6 - x)$.

(a) Determine the indefinite integral of $f(x)$. Does the answer involve any function with which you are unfamiliar? If so, use Maple's help facility to find out the meaning of the function.

(b) Determine the numerical value of the integral for the range $x = 0$ to 5.

(c) Calculate the first and second derivatives of $f(x)$.

Problem 1-17: Savings account

The accumulated amount A in a savings account earning interest at an annual rate of r_a% on an initial investment of P dollars is given by the compound interest formula $A = P\,(1 + 0.01\,r_a)^{\,y}$ after y years. Given an annual interest rate of 5% and $P =$ \$1000, plot A for a period of 10 years after the initial deposit. What is the accumulated amount at the end of 10 years?

Problem 1-18: Manhattan Island

In 1626, Peter Minuit, of the Dutch West Indies Company, bought the rights to Manhattan Island from the local residents for \$24. If this money had been invested by these residents at an interest rate of 6% compounded annually (i.e., once a year), how much would the \$24 investment have grown to by the year 2006? Plot the value of the investment over this 380-year time span.

Problem 1-19: Smoking is bad for your economic health

Although probably aware of the potential danger to one's physical health, the young smoker may not be quite as aware of the large cumulative drain that smoking will have on his or her economic health. Consider the following example, which is set up for the reader to complete.

(a) A 20-year-old stops smoking and then deposits the \$5 per day it costs to purchase cigarettes into a bank. Assuming the bank pays an interest of 5% annually, but compounded daily, what amount of money will this person have in the bank when he or she becomes 65? The exact amount A is given by the general compound interest formula $A = R\,\sum_{k=1}^{n\,y}(1 + r_a/n)^{\,k}$. Here, the principal investment $R =$ \$5.00 per day, r_a is the annual interest rate expressed as a decimal, y is the number of years elapsed, and n the number of compounding periods (365 here) per year. *Hint*: make use of the `add` command to evaluate the sum.

(b) Justify the formula given in part **(a)**.

(c) An approximate value for A can be obtained by using the integral expression $A = \int_0^T R(t) e^{(r_a/n)(T-t)}\,dt$ where $R(t)$ is the number of dollars invested in each time period and $T = n\,y$. By what percent does the approximate value of A differ from the exact value calculated in part **(a)**?

(d) Optional: Making use of the fundamental definition of the exponential function and assuming that the interest is compounded continuously, derive the integral expression used in part **(c)** to calculate A.

Problem 1-20: Investments

The owners of a small factory are making plans to purchase a much larger factory. The owners invest \$100,000 each month that they think will yield 24% annually after daily compounding. How much will these investments be worth 10 years from now? Use the integral approach of the previous problem.

Problem 1-21: Visitors to Erehwon

The number of foreign visitors to the planet Erehwon between 1984 and 1991 can be described by the functional form

$$n(t) = \frac{18000}{1 + 36.02\, e^{-0.8540\, t}} + 25000,$$

where t is the number of years since 1984. The fraction of these visitors who came from Earth is given by

$$f(t) = 1.429 \times 10^{-5}\, t^2 - 2.234 \times 10^{-3}\, t + 0.08955.$$

(a) Plot the number $N(t) = n(t)\, f(t)$ of visitors from Earth for the period 1984 to 1991.

(b) Calculate the analytic derivative of $N(t)$.

(c) Determine the inflection point of the curve $N(t)$. This can be done by calculating the second derivative of $N(t)$, which gives the curvature of the function. The inflection point corresponds to the t value at which the second derivative is zero. *Hint*: If you use the `fsolve` command, you may have to specify the t range.

(d) How many visitors came from Earth in 1989?

(e) How rapidly was this number changing at this time?

Problem 1-22: Milk sales

In a particular 31 day month, the selling price of milk $S(d)$ on day d, in dollars per gallon, and the number of gallons $G(d)$ sold were found to be given by

$$S(d) = 0.007\, d + 1.492, \qquad G(d) = 31 - 6.332\, (0.921)^d.$$

Plot the milk sales, i.e., $S(d)\, G(d)$, as a function of d. By integrating under the milk sales curve, determine the total milk sales in dollars for that month. Compare this result with the exact sum.

1.1.3 Minimizing the Travel Time

A traveler without knowledge is like a bird without wings.
Musharif-UD-DIN (1184–1291)

In addition to the free maps and tour books that the American Automobile Association (AAA) provides to its clients, it will design travel routes that follow scenic secondary highways or take the shortest routes or shortest traveling times between two cities. If you were to work for the AAA, you would want to avoid repeatedly working out the shortest traveling time or route each time a new customer came in and asked the question for a different pair of cities. Why not let the computer do the work for you. It might cost more effort initially, but if, say all the cities and towns in North America were entered with the travel times for all possible highway routes given, you could quickly tell your client the answer without having to laboriously work it out. To see how this can be done, we shall consider a simple example.

Our engineering friend Russell is transferred from Seattle to Phoenix and wants to take the inland route between the two cities that takes the shortest traveling time. Here is a computer algebra recipe that might be used.

The Maple networks package is accessed and cities[10] with longer names that lie on one of the various connecting routes are aliased, using the `alias` command, to save on subsequent typing. For example, typing in `Pen` will produce *Pendleton* in any displayed output.

```
> restart: with(networks): alias(Salt_Lake_City=SLC,
  Portland=Port,Yakima=Yak,Pendleton=Pen,Las_Vegas=LV,
  Winnemucca=Win,Seattle=S,Phoenix=Ph):
```

A new graph `G` is to be created.

```
> new(G):
```

All of the major cities on inland highway routes connecting Seattle (S) and Phoenix (Ph) are entered.

```
> cities:={S,Port,Yak,Pen,Bend,Reno,Win,Boise,SLC,LV,Ph}:
```

In the terminology of mathematical graph theory, the cities are added as vertices in the graph.

```
> addvertex(cities,G);
```

Salt_Lake_City, Yakima, Portland, Seattle, Pendleton, Las_Vegas, Winnemucca, Phoenix, Bend, Reno, Boise

The various cities are now connected. For example, in the next command line Seattle is connected to Portland, Oregon, and to Yakima, Washington. The average traveling times[11] for each connection (3.4 hours for Seattle to Portland and 2.7 hours for Seattle to Yakima) are included as "weights." The output gives names (equivalent to highway numbers) to the two routes.

```
> connect([S],[Port,Yak],weights=[3.4,2.7],G);
```

e1, e2

Here route *e1* (highway I-5 in real life) connects Seattle to Portland, while route *e2* (portions of highways I-90 and I-82) joins Seattle to Yakima. The remaining connections are entered in a similar manner. How many hours does it take to drive from Yakima to Bend? What is the corresponding route number (replace the colon with a semicolon)?

```
> connect([Yak],[Pen,Bend],weights=[2.6,4.5],G):
> connect([Port],[Bend,Reno],weights=[3.4,11.2],G):
> connect([Pen],[Win,Boise],weights=[10.0,4.2],G):
> connect([Bend],[Reno,Win],weights=[8.3,7.5],G):
> connect([Boise],[Win,SLC],weights=[5.2,6.6],G):
> connect([SLC,Win,Reno],[LV],weights=[8.3,9.9,9.0],G):
> connect([LV,SLC],[Ph],weights=[6.0,16.0],G):
```

[10] If any of these cities are unfamiliar, consult a highway map of the western United States.
[11] Obtained from the American Automobile Association travel books.

After all the connections have been finished, the number of different routes involved are displayed with the following `edges` command.

```
> routes:=edges(G);
```

$$routes := \{e1, e2, e3, e4, e5, e6, e7, e8, e9, e10, e11, e12, e13, e14, e15, e16, e17\}$$

According to the output, there are 17 of them.

If the colon is replaced with a semicolon in the following edge-weight command, the various times (in hours) between cities in the graph will be displayed.

```
> times:=eweight(G): #in hours
```

We will now create a graph made up of straight-line segments joining the vertices (cities) according to the connecting routes entered earlier. The graph, which is produced with the linear `draw` command, will not place the cities at their proper relative geographic positions, but instead will artistically group cities as follows. The first city in the argument is Seattle (S), which will be placed at the far left of the resulting picture. The next entry lists the cities of Portland (Port) and Yakima (Yak). These cities will be placed to the right of Seattle, and lined up vertically, one above the other. Each successive list of cities is placed further to the right and the members of the list organized vertically, until finally Phoenix (Ph) is placed on the far right of the graph G.

```
> draw(Linear([S],[Port,Yak],[Pen,Bend],[Reno,Win,Boise],
  [LV,SLC],[Ph]),G);
```

The resulting graph produced by the `draw` command is shown in Figure 1.6.

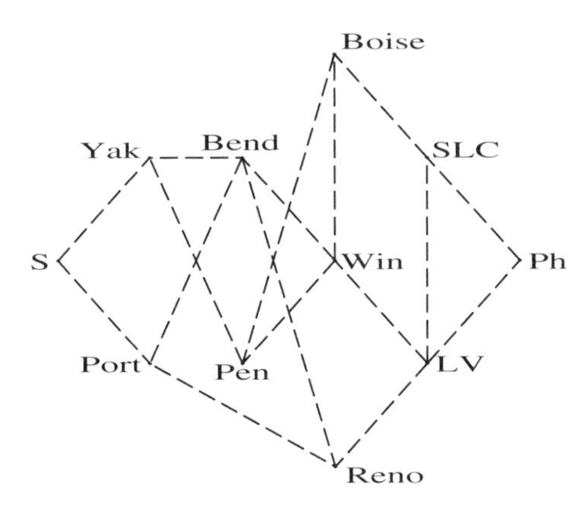

Figure 1.6: Various inland highway routes between Seattle and Phoenix.

The cities are correctly joined according to the connections that were entered earlier. The `allpairs` command is now used to calculate the shortest traveling time between any two cities in the graph. The output is suppressed for brevity.

```
> T:=allpairs(G): #traveling times between pairs of cities
```

The minimum traveling time between Seattle and Phoenix is desired, and is

```
> Minimum_Time:=T[S,Ph]; #between Seattle and Phoenix
```

$$Minimum\_Time := 29.6$$

found to be 29.6 hours. To determine the route that corresponds to this shortest time, we use the **shortpathtree** command, which implements *Dijkstra's algorithm*[12] for the shortest-path spanning tree. This produces Figure 1.7.

```
> SPT:=shortpathtree(G,S,Ph): draw(SPT);
```

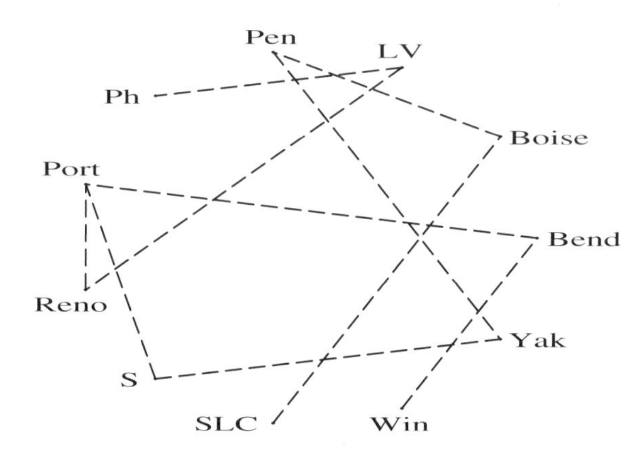

Figure 1.7: The shortest route between Seattle and Phoenix has no dead ends.

The shortest route between Seattle and Phoenix is the only one in Figure 1.7 that connects the two cities. To draw this route, remove those cities that clearly do not connect Seattle and Phoenix in the shortest path tree. For example, you can see that both Winnemucca (Win) and Salt Lake City (SLC) are dead ends in Figure 1.7 and can be deleted. Then Bend and Boise become dead ends, so delete them too. Finally, include Pendleton (Pen) and Yakima (Yak) in the following delete command. On applying the draw command to SPT,

```
> delete({Win,Bend,SLC,Boise,Pen,Yak},SPT): draw(SPT);
```

only the route that has the shortest traveling time between Seattle and Phoenix remains and is shown in Figure 1.8. The cities of Portland, Reno, and Las Vegas lie along this route. This is the route that would be recommended to our friend Russell. Again, remember that the cities in the picture are not oriented in their proper geographic positions.

Of course, for this example, the number of different routes was not so challenging that the calculation could not have been done fairly quickly with pen and paper. But imagine doing this for a graph with all the cities and towns in North America or in Europe. Further, once one has created the graph, the shortest path between any two towns or cities is easily obtained.

The scientifically inclined reader might wonder what the concept of mini-

[12]Named after its inventor, the Dutch computer scientist Edsger Dijkstra [CLR90].

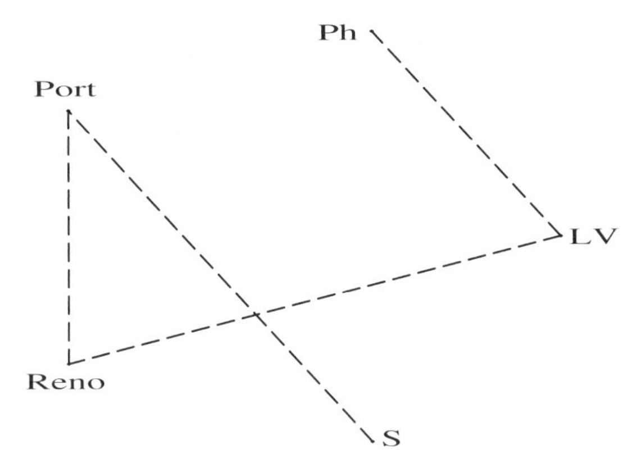

Figure 1.8: Route for the shortest traveling time.

mizing the time, which appeared in this example, has to do with engineering or physical phenomena. *Fermat's principle* in geometrical optics is based on a similar idea. In its simplest form [Tip91], it states that the path taken by light in traveling from one point to another is such that the time of travel is a minimum. Fermat's principle can be used to generate Snell's law in geometrical optics, as well as to account for the phenomena of mirages in a medium with a variable refractive index.

PROBLEMS:

Problem 1-23: Minimum distance

By consulting an appropriate set of maps, modify the provided recipe to determine the shortest distance from Seattle to Phoenix passing through the given cities. Be sure to draw all the appropriate graphs.

Problem 1-24: Boise to Reno

Use the provided recipe to determine the shortest travel time and the corresponding route to travel between Boise, Idaho, and Reno, Nevada. Show the appropriate graphs. Is this really the shortest route between these two cities? To answer this question, you might want to consult the appropriate road maps, add more connecting cities, and modify the recipe.

Problem 1-25: Planning your route

Choose two major American or European cities that are widely separated in distance and that have many possible connecting routes. Find the driving times and mileages between sizeable cities or towns along the various routes. To save on typing, do not include every village or hamlet that you would pass through. Repeat the procedure outlined in the text to determine the route between the two major cities that minimizes the total driving time. Determine the route that minimizes the total distance.

1.2 Log-Log (Power Law) Plots

1.2.1 Chimpanzee Brain Size

I believe that our Heavenly Father invented man because he was disappointed in the monkey.
Mark Twain (Samuel Langhorne Clemens), American humorist (1835–1910)

Heather is a premed student enrolled in the biological sciences program at MIT. In an introductory calculus text [AL79] that she is consulting, it is stated that within a given species of mammal, it is found that the brain volume V varies with the body mass m according to the power law $V = a\,m^b$. Here a and b are positive constants. Table 1.6, taken from the same text, shows the brain volumes (V in cm^3) as a function of body mass (m in kg) for a number of adult chimpanzees.

Table 1.6: Chimpanzee brain volumes V and body masses m.

m	31	36	38	41	42	45	47	48	50	53	55	57
V	365	380	382	395	397	410	410	415	420	427	437	440

The text further states that if these data satisfy the power law equation, the values of a and b can be determined by making a log-log plot of the data and determining b from the slope and a from the intercept of the best-fitting straight line. To confirm that this statement is true, Heather takes the log of both sides of the power law equation, yielding

$$\ln(V) = \ln(a\,m^b) = \ln(a) + b\,\ln(m) \tag{1.2}$$

which is a straight line of slope b and intercept $\ln(a)$ when $\ln(V)$ is plotted against $\ln(m)$. That is to say, if she sets $y \equiv \ln(V)$, $A \equiv \ln(a)$, and $x \equiv \ln(m)$, then the straight line equation $y = A + b\,x$ results. Once A is determined, then $a = e^A$. Heather decides to use the log-log plotting procedure to determine the parameters a and b in the power law formula for the chimpanzee data.

To make a log-log data plot, the plots library package must be entered.

```
> restart: with(plots):
```

The mass and volume values are inputted as separate Maple lists.

```
> mass:=[31,36,38,41,42,45,47,48,50,53,55,57]:
> volume:=[365,380,382,395,397,410,410,415,420,427,437,440]:
```

In the following command line, Heather uses the "arrow," or functional, operator, which is entered by typing a "hyphen" followed by a "greater than" sign, to group the mass and volume together as a list.

```
> pair:=(mass,volume)->[mass,volume];
```

$$pair := (mass, volume) \to [mass, volume]$$

The lists are then "zipped" together into a list of lists, ready for plotting, and assigned the name *points*.

```
> points:=zip(pair,mass,volume);
```

$$points := [[31, 365], [36, 380], [38, 382], [41, 395], [42, 397], [45, 410], [47, 410], [48, 415], [50, 420], [53, 427], [55, 437], [57, 440]]$$

A log-log plot of the data points (presented as size-12 blue boxes) is formed,

```
> pts:=loglogplot(points,style=point,symbol=box,
      color=blue,symbolsize=12):
```

and the display command used to produce Figure 1.9. The `display` command was utilized by Heather because it allows more control of the plot options than `loglogplot`. In particular, it allows use of the `view` command, which is often employed in this text to set the horizontal and vertical ranges in order to get a good picture. For complicated plot structures, it enables one to zoom in on any desired region without computing the graph again.

```
> display(pts,labels=["ln(m)","ln(V)"],tickmarks=[3,2],
  view=[1.5..1.8,2.5..2.65]);
```

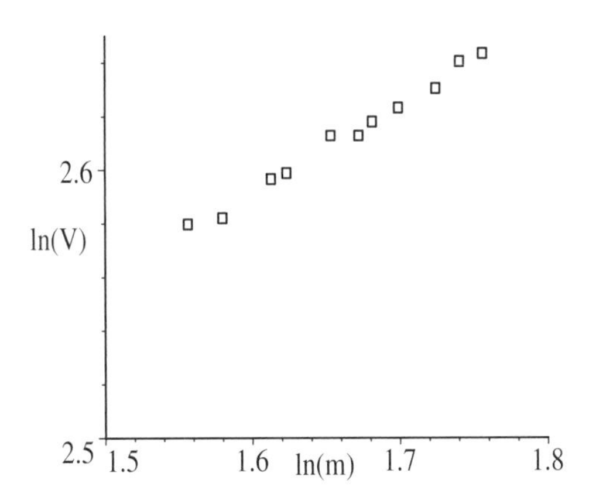

Figure 1.9: Log-log plot of brain volumes versus mass.

Heather notes that the data lie approximately along a straight line. From Table 1.6, she calculates that the slope is roughly

$$b \approx \frac{(\ln(440) - \ln(365))}{(\ln(57) - \ln(31))} = 0.31.$$

With this value as a starting point, she manages to establish by trial and error that $b = 1/3$. Similarly, she finds that $a \approx 114\frac{1}{2}$, so that the model equation for the chimpanzee data is of the power law form $V = 114.5\, m^{1/3}$.

To demonstrate how well this model formula fits the data, Heather makes a log-log plot of the power law equation for the mass range $m=30$ to 60 kg,

```
> eq:=loglogplot(114.5*m^(1/3),m=30..60):
```

and displays it and the data in the same graph, viz., the left plot of Figure 1.10.

```
> display({eq,pts},tickmarks=[3,2],labels=["ln(m)","ln(V)"]
  view=[1.5..1.8,2.5..2.65]);
```

The data and power law formula are now plotted "normally" (i.e., without taking logs) for the mass range $m=0$ to 60 kg,

```
> pts2:=plot(points,style=point,symbol=box,color=blue,
         symbolsize=12):
> v:=plot(114.5*m^(1/3),m=0..60):
```

and displayed together on the right-hand side of Figure 1.10.

```
> display({pts2,v},labels=["m","V"],tickmarks=[3,3]);
```

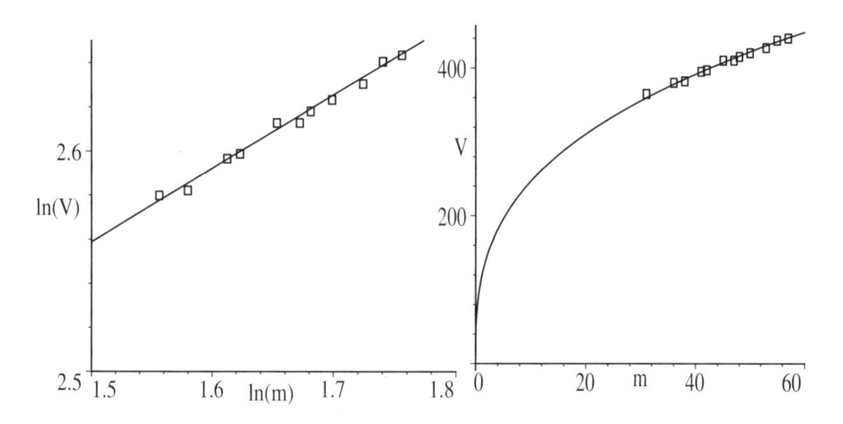

Figure 1.10: Log-log (left) and "normal" (right) plots of $V=114.5\,m^{\frac{1}{3}}$ and data.

Heather is pleased to note that the power law $V=114.5\,m^{1/3}$ fits the observational data quite well. However, she is bothered by the fact that she doesn't know why this should be the case. Perhaps her older sister Jennifer, who is an MIT applied mathematician, can provide an explanation? In the next story, we shall hear what Jennifer has to say about this example.

PROBLEMS:

Problem 1-26: Fir tree yield

The volume V of wood obtained from an average fir tree increases with age A as shown in Table 1.7. The age is in years and the volume in hundreds of board feet. A board foot is the volume of a board 1 foot square and 1 inch thick.

(a) Plot V versus A. Does the curve suggest a power law? Explain.

(b) Make a log-log data plot. Is the curve approximately a straight line?

(c) The straight-line data in the log-log plot can be fitted by the functional form $V = 5.6 \times 10^{-4}\, x^{5/2}$. Confirm that this is the case by plotting the log of the formula on the same graph as the log-log data.

(d) Plot the formula on the same graph as the original data.

Table 1.7: Volume V of wood as a function of age A.

A	50	75	100	125	150	175	200
V	11	28	56	98	158	225	330

Problem 1-27: Rowing times

Using dimensional scaling analysis (discussed in the next story) based on the idea that racing shells for different numbers n of oarsmen are geometrically similar, McMahon [PLA92] has shown that the time T (in minutes) to row 2000 meters is given by $T = k\, n^{-1/9}$ with k a positive constant. Table 1.8 gives times for I: 1964 Tokyo Olympics; II: 1968 Mexico Olympics; III: 1970 World Rowing Championships, St. Catherines, Ontario; and IV: 1970 International Championships, Lucerne, Switzerland.

Table 1.8: Rowing times T for different numbers n of oarsmen.

n	**T (min) I**	**T (min) II**	**T (min) III**	**T (min) IV**
8	5.87	5.92	5.82	5.73
4	6.33	6.42	6.48	6.13
2	6.87	6.92	6.95	6.77
1	7.16	7.25	7.28	7.17

Using a trial and error approach, find a value for the constant k that gives a good fit to the data when plotted on a log-log scale. Plot the theoretical formula and data in the same log-log graph and also produce a normal graph.

Problem 1-28: Newton's law of cooling

Russell, who has moved into his new lab in Phoenix, boils some water to make a cup of instant coffee. While sipping cautiously on his hot drink, he has placed a mercury-in-glass thermometer in the remaining boiling water for a few minutes and then removed the thermometer. He records the readings in degrees Celsius on the thermometer t seconds after removal, the results being shown in Table 1.9. The temperature of the room is a warm 26.0 °C. According to *Newton's law of cooling*, if ΔT_0 is the initial temperature difference between an object and its surroundings, the temperature difference t seconds later is given by the exponential law $\Delta T = \Delta T_0\, e^{-Kt}$, with K the cooling constant.

(a) Show that K is given by the slope of a straight line in a semilog plot.

Table 1.9: Temperature (T) readings t seconds after thermometer removal.

t	0	5	10	15	20	25	30	40
T	98.4	76.1	71.1	67.7	66.4	65.1	63.9	61.6
t	50	70	100	150	200	300	500	700
T	59.4	55.4	50.3	43.7	38.8	32.7	27.8	26.5

(b) Using the `logplot` command, show that most of the data in Table 1.9 lie on a straight line. What might give rise to data points that deviate from the straight line?

(c) By trial and error, plot a straight line that best fits the data and thus extract the approximate value of the cooling constant K.

1.2.2 Scaling Arguments and Gulliver's Travels

"What is the use of a book," thought Alice,
"without pictures or conversation?"
Lewis Carroll, *Alice's Adventures in Wonderland* (1865)

In the real world, the properties of many complex systems can, surprisingly, be described by simple power law curves. That such a curve might apply to a given set of data can be ascertained by checking to see whether the observation points lie on a straight line when a log-log plot of the data is made. This is the procedure that Heather applied in the last recipe for the brain volumes V of adult chimpanzees. In this case, the data were found to be consistent with a power law of the form $V = a\,M^{1/3}$, where M is the body mass of the chimpanzee and $a \approx 114\frac{1}{2}$. To understand how such a power law relation can arise, Heather has gone to see her sister Jennifer, who is a junior faculty member in the Institute of Applied Mathematics at MIT.

"Well, Heather," Jennifer begins, "To answer your question, I will have to tell you about the concept of scaling, which deals with how the properties and characteristics of various systems change with size. I will try to keep it simple, but if you want a more complete treatment I would refer you to an interesting paper entitled "Fundamentals of zoological scaling" written by Herbert Lin and published in the *American Journal of Physics* [Lin82]. As you are undoubtedly becoming aware, introductory science courses tend to deal with highly idealized models of the real world that are set up to give unique, well-defined answers. In reality, experimentalists are often confronted with complex systems for which the properties could depend on many factors. The appearance of power law behavior when one variable is plotted against another is particularly exciting to the experimentalist because it signals the possibility that the behavior could be explained on the basis of scaling arguments, i.e., on how the properties of a system change with a characteristic system size L. Let me give you a few simple examples. You're not in a hurry to get anywhere, are you?"

"No, I don't have any classes for the next couple of hours, so go ahead."

"OK, my first example, taken from Lin's article, is about the burning of wood of different diameters. A log of about 10 cm diameter will take around an hour to burn, while 1 cm diameter kindling will burn in several minutes, and a fuse of diameter 1 mm will burn in several seconds. To understand this, note that burning takes place only at the surface, so the rate of combustion is limited by the surface area S with which oxygen, necessary for burning, must make contact. The burning rate must be proportional to S, which in turn is proportional to L^2, where L is the diameter. On the other hand, the rate must be inversely proportional to the amount of material present, i.e., to the volume $V \propto L^3$. Combining these two aspects, the rate then should be proportional to $S/V \propto L^2/L^3 = 1/L$. Thus, according to this scaling argument, since the time to burn is inversely proportional to the rate of combustion, the log should burn about 10 times slower than the kindling, which in turn will burn about 10 times slower than the fuse. This is in rough agreement with the observations. Although the precise burning times clearly would depend on other factors, the gross observed behavior is dominated by changes in characteristic size L."

"I followed your argument, but what about the chimpanzee power law that I told you about? Can you also use scaling to explain it?"

"Oh, that's quite easy. Let's make the assumption that in order to maintain the same functional power, the adult chimpanzee brain volume V is proportional to the size L of the chimp. That is to say, the 'bigger' an animal of a given species is, the bigger is its brain. But the body mass M is equal to the density times the body volume. The density of all mammals is fairly constant, especially within a given species such as the chimps. So $M \propto L^3$, or conversely $L \propto M^{1/3}$. Thus, the brain volume satisfies $V \propto L \propto M^{1/3}$. Inserting the proportionality constant a, then the power law formula $V = a\, M^{1/3}$ results. And, as you verified, this power law is in very good agreement with the experimental data."

"That's interesting! Do you have any more simple examples of scaling?"

"Well, I have an example that I have been thinking about for a while. What got my mind going on it was all these movies that have appeared over the years featuring giant ants, apes, etc., which usually terrify humanity until some hero or heroine steps in to save the world. These movies are often not very good, but more importantly in the context of our discussion, they are flawed from a scaling viewpoint."

"What do you mean? Can you give me a concrete example?"

"Sure, but my example is from the world of classic fiction. Do you remember reading the novel *Gulliver's Travels* by Jonathan Swift? I am going to use scaling to punch some scientific holes in Swift's story. Recall that in the novel Gulliver travels to a number of strange lands. In two of these lands, Lilliput and

Brobdingnag, there exist inhabitants who look just like humans but are much smaller and larger than Gulliver. They are geometrically similar to humans, i.e., scaled-down and magnified replicas of humans. For calculation convenience, let's make the Lilliputians ten times as small and the Brobdingnagians (Brobs, for short) ten times as large. What effects might this have on their biological processes and on the scientific accuracy of Swift's tale?

Since body volume (V) and therefore weight W (proportional to mass) vary as L^3, the Lilliputians and Brobs would have weights 1000 times smaller and 1000 times larger, respectively, than Gulliver. On the other hand, as I pointed out in my first example, surface area S is proportional to L^2. So, the Lilliputians and Brobs would have surface areas 100 times smaller and 100 times larger than Gulliver. Now it is a well-known biological fact that warm-blooded animals lose heat through their skin. This means that the heat loss H satisfies $H \propto S \propto L^2$. Since heat is a form of energy, in equilibrium the heat energy loss must be balanced by energy intake in the form of food. Thus, it is reasonable to assume that the amount of food eaten is proportional to the area S. Now Swift's tale is in trouble! In the story, the Lilliputians live in a scaled-down version of human society, with all the trappings of civilization. Scaling will tell us that this is highly unlikely.

Suppose that Gulliver had a mass of 80 kg and that he ate 1/40 of his weight (2 kg) each day, which took him about 1 hour to consume. A Lilliputian is $10^3 = 1000$ times less massive, having a mass of 0.08 kg, but has a surface area only $10^2 = 100$ times smaller. The typical Lilliputian would have to eat $2/100 = 0.02$ kg each day, which amounts to one-quarter of his or her body weight. Assuming that scaling prevails, the time devoted to eating would be about 10 hours. When this time is combined with that required for acquiring and preparing the food, this would leave little time left over for the Lilliputians to develop a society similar to that of humans. If you think that this is pushing scaling too far, think of our world. Creatures that are much smaller than humans do indeed spend much of their time in gathering and eating food."

"Given your line of reasoning, Jennifer, surely everything would be fine in the land of Brobdingnag, the home of the mega-humans. They have weights 1000 times greater than Gulliver, but surface areas only 100 times larger. Thus, they would have to eat only about 1/400 of their body weight each day and probably could go for long stretches of time without eating. This is indeed the case for very large creatures in our own world. Therefore, it is possible that the Brobs would have sufficient time to develop an advanced civilization. So I see no problems with Swift's story in the land of Brobdingnag."

"Ah, but there is a different problem. It's in the bones! Let's look at the Brobs' leg bones, which must support their much greater weight. According to Lin's article, the simplest model assumes that the static compressive stress[13] σ sets the lower limit on the thickness d of the leg bone. If the leg bone has a cross-sectional area $A \propto d^2$ and supports the entire weight W, then by balancing

[13]Stress equals force per unit area.

forces, we have $\sigma A = W$. If d were to scale[14] with the characteristic size L, then $d^2 \propto L^2 \propto W^{2/3}$ and $\sigma \propto W^{1/3}$. Now, nature uses bone for skeletal frames, not titanium. As the weight W is increased sufficiently, a point would be reached where the bone could no longer sustain the compressive stress and would shatter. To prevent breaking, the thickness d must increase faster than L. If the compressive stress is kept constant, then $d^2 \propto W$, or $d \propto W^{1/2}$. Since the Brobs are 1000 times heavier, their legs would have to be about $\sqrt{1000} \approx 32$ times as thick. So, their legs would be disproportionately fatter, and they would not be simply perfectly magnified versions of humans."

"How good is the approximation $d \propto W^{1/2}$?"

"More refined scaling arguments presented in Lin's paper give a slight correction, so that $d \propto W^{5/12}$. Now, zoologists have measured the skeletal bone weight W_{sk} as a function of body weight W of various animals. Some values (expressed as kilogram weights) are given in Table 1.10.

Table 1.10: Body (W) and skeletal weights (W_{sk}) for some mammals.

Animal	Shrew	Mouse	Cat	Rabbit	Beaver	Human	Elephant
$\boldsymbol{W}$	0.0063	0.0295	0.845	2.0	22.7	67.3	6600
$\boldsymbol{W}_{sk}$	0.0003	0.0013	0.0436	0.181	1.15	12.2	1782

Assuming that $W_{sk} \propto L\,d^2$, then $W_{sk} \propto W^{1/3}\,(W^{5/12})^2 = W^{7/6}$. Let's check this power law formula against the empirical data. I will use a code very similar to the one that you used for the chimpanzee data, so you should be able to easily follow it without too much detailed explanation.

```
> restart: with(plots):
> weight:=[0.0063,0.0295,0.845,2.0,22.7,67.3,6600.0]:
> bone_weight:=[0.0003,0.0013,0.0436,0.181,1.15,12.2,1782]:
> pair:=(weight,bone_weight)->[weight,bone_weight]:
> points:=zip(pair,weight,bone_weight);
```

$$points := [[.0063, .0003], [.0295, .0013], [.845, .0436], [2.0, .181], [22.7, 1.15], [67.3, 12.2], [6600.0, 1782]]$$

```
> pts:=loglogplot(points,style=POINT,symbol=box,
      color=black,symbolsize=12):
```

Now I will create a log-log plot of the function $W_{sk} = a\,W^{7/6}$. Using a best-fit procedure, one finds that the constant a is equal to 0.065.

```
> a:=0.065:
> eq:=loglogplot(a*W^(7/6),W=.0002..7000,color=black):
```

[14]Since all other lengths, including leg length, are assumed to scale with L.

The two log-log plots, pts and eq, are then displayed together in Figure 1.11. Here b and w are the logs of the bone weight and body weight, respectively.

```
> display({pts,eq},labels=["w","b"],axes=framed,
  tickmarks=[3,3],view=[-2.5..4,-3.5..4]);
```

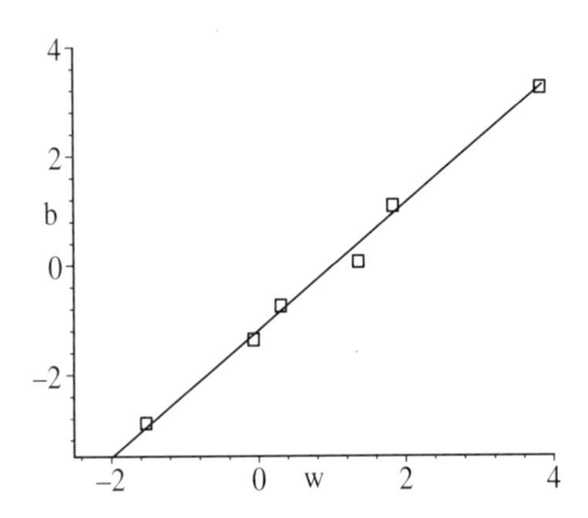

Figure 1.11: Log-log plot of the observational points and the model equation.

The agreement between the power law model equation, derived on the basis of scaling arguments, and the data is very good. Although scaling law arguments are somewhat empirical, they have proven very successful in understanding a wide variety of power law curves in the biological and physical sciences."

"Thanks for your help, Jennifer, but I have to be off to my next class. If I have any more questions on scaling, I will come to see you later."

PROBLEMS:

Problem 1-29: Human surface area

The surface area S of humans is related on average to their masses M and heights H by a formula of the structure $S = a\,M^b\,H^c$ where a, b, and c are constants. Table 1.11 gives the surface area in square meters and mass in kilograms for a group of people of the same height (1.80 m). For fixed H, then, the power law $S = A\,M^b$ should describe the data.

Table 1.11: Human surface area at a fixed height as a function of mass.

M	70	75	77	80	82	84	87	90	95	98
S	2.10	2.12	2.15	2.20	2.22	2.23	2.26	2.30	2.33	2.37

(a) Make a log-log plot of the data and show that the data lie approximately on a straight line.

(b) By trial and error fit the log of the power law to the log-log data to obtain approximate values of A and b.

(c) Noting that the height is fixed, use scaling arguments to plausibly account for the observed value of b.

Problem 1-30: Of lions & house cats
Vectoria has a house cat that weighs 13 lb (mass of 5.9 kg), is 0.85 m long (including the tail), and consumes about 150 g of food per day. Given that a female lion is, on average, about 8 ft long (including the tail), weighs 275 lb, and in the wild consumes about 30 lb of food every 6 days, use scaling arguments to estimate the expected weight and the expected daily food consumption of Vectoria's cat. Compare the estimated values with the actual values and give possible explanations for any discrepancies.

Problem 1-31: Attack of the giant killer ants?
For a certain species of ant, a typical ant has a length of 1.0 cm, a mass of 0.20 g, and is able to lift 100 times its body weight. A Hollywood movie director proposes to base a horror film on the theme of giant versions of these ants attacking and destroying downtown Metropolis. Each of the giant ants is supposed to have a length of 10 m. Using scaling arguments discuss what is wrong with the director's giant ant scenario. *Hint*: Can the giant ants lift their own weight?

Problem 1-32: Sequoias
According to Ohanian [Oha85], for tall trees the diameter at the base (or the diameter at any given point of the trunk, such as the midpoint) is roughly proportional to the 3/2 power of the distance x from the top of the tree. The tallest sequoia in Sequoia National Park, in California, has a length of 81 meters, a diameter of 7.6 meters at the base, and a mass of 6100 metric tons.

(a) Explain the observed scaling of diameter with distance x.

(b) Using the given data, derive the formula for the diameter d at a distance x from the top of a sequoia.

(c) Plot d over the range $x = 0$ to $x = 100$ meters.

(d) A petrified sequoia found in Nevada has a length of 90 meters. Estimate its diameter at the base when it was alive.

(e) Derive a formula for the volume V of a sequoia of length x meters.

(f) Plot V over the range $x = 0$ to $x = 100$ meters.

(g) Calculate the average density in metric tons per cubic meter of a sequoia using the data for the tallest tree. What was the mass of the petrified sequoia in metric tons when it was alive?

1.3 Contour and Gradient Plots

1.3.1 The Secret Message

Something deeply hidden had to be behind things.
Albert Einstein, theoretical physicist (1879–1955)

Mike, a mathematics student and amateur archaeologist, has been fortunate to obtain a summer job with an archaeological dig in the high Andes region near the Inca ruins of Machu Pichu. On the first day that Mike is there, the chief archaeologist takes him on a tour of the site and shows him how the area has been divided into squares with strings. Specifically, there are 121 squares, or "cells," in an 11×11 grid. Within each square, the type of artifact and the number found is recorded. At the moment, the number of artifacts discovered in each square is given in tabular form.

Mike's first task will be to plot these data in some suitable form. He is keen to get to work and asks for the data. Unbeknownst to him, he is not given the real numbers but some artificial data designed to carry a secret message of greetings from his new coworkers. He is also told that the rows and columns of the data array were inadvertently interchanged and should be transposed.

On the camp laptop computer, Mike loads Maple's plots and LinearAlgebra packages, the latter needed for the matrix representation of the data.

```
> restart: with(plots): with(LinearAlgebra):
```

To explicitly view the 11×11 data matrix (otherwise, the default maximum is 10×10), Mike includes the following `interface` command. Here `rtable` stands for "rectangular table," and although he could set `rtablesize` to 11, Mike sets the value to infinity in case the data matrix is later increased in size.

```
> interface(rtablesize=infinity):
```

He then enters the numbers given to him as the data matrix A.

```
> A:=Matrix([[1,0,1,0,1,1,2,1,0,2,1],[1,2,2,2,1,1,2,0,3,1,0],
  [1,1,2,0,1,1,3,1,0,1,0],[0,1,9,0,10,1,1,7,7,8,1],[1,2,8,2,10,
  3,2,1,9,2,0],[1,1,7,10,9,2,1,0,9,1,0],[1,1,7,1,9,2,1,2,8,0,1],
  [0,2,9,2,10,3,1,9,9,8,2],[2,1,2,1,0,0,3,2,0,1,0],
  [1,2,3,0,1,1,2,1,1,0,0],[0,1,2,1,0,3,1,0,2,1,0]]);
```

$$A := \begin{bmatrix} 1 & 0 & 1 & 0 & 1 & 1 & 2 & 1 & 0 & 2 & 1 \\ 1 & 2 & 2 & 2 & 1 & 1 & 2 & 0 & 3 & 1 & 0 \\ 1 & 1 & 2 & 0 & 1 & 1 & 3 & 1 & 0 & 1 & 0 \\ 0 & 1 & 9 & 0 & 10 & 1 & 1 & 7 & 7 & 8 & 1 \\ 1 & 2 & 8 & 2 & 10 & 3 & 2 & 1 & 9 & 2 & 0 \\ 1 & 1 & 7 & 10 & 9 & 2 & 1 & 0 & 9 & 1 & 0 \\ 1 & 1 & 7 & 1 & 9 & 2 & 1 & 2 & 8 & 0 & 1 \\ 0 & 2 & 9 & 2 & 10 & 3 & 1 & 9 & 9 & 8 & 2 \\ 2 & 1 & 2 & 1 & 0 & 0 & 3 & 2 & 0 & 1 & 0 \\ 1 & 2 & 3 & 0 & 1 & 1 & 2 & 1 & 1 & 0 & 0 \\ 0 & 1 & 2 & 1 & 0 & 3 & 1 & 0 & 2 & 1 & 0 \end{bmatrix}$$

As was suggested to him, Mike transposes the matrix A into a new matrix B, but suppresses the lengthy output.

```
> B:=Transpose(A):
```

Studying the Maple manual, Mike learns that there are various ways of plotting the matrix B. One approach is to use the matrixplot command and represent the data as three-dimensional, suitably colored, histograms, the height of each histogram corresponding to the number of artifacts in that particular square. The three-dimensional nature of the plot can be explored by clicking the left mouse button on the plot and dragging the mouse on the resulting three-dimensional box. If an interesting orientation occurs, the angular numbers θ and ϕ appearing at the top left of the computer screen can be permanently entered into the code with the orientation command so that viewpoint turns up the next time the code is run. Similarly, by clicking on the plot and the Color box at the top of the computer screen, different possible coloring schemes may be selected. In the following command line, Mike has chosen an orientation of $\theta = -60°$, $\phi = 15°$ and used shading=zhue to color the histogram boxes in the z-direction, i.e., the direction of increasing number N of artifacts.

```
> matrixplot(B,heights=histogram,style=patch,shading=zhue,
  axes=boxed,labels=["x","y","N"],orientation=[-60,15]);
```

On running the code,[15] Mike is surprised by the output, which seems to reveal a simple message that was not apparent to him in the tabulated data. To confirm the secret message, as well as to gain familiarity with other plotting styles, he tries three other graphical methods, the second approach being to make a list density plot of B. To reveal the message more clearly, Mike accepts Maple's default coloring in which each cell is assigned a gray level from white to black as the value of the cell increases. To assign colors other than shades of gray, he could have inserted the plots option colorstyle=HUE in the following command line. Try it and see what the message would then look like.

```
> listdensityplot(B,axes=boxed,labels=["x","y"]);
```

Did Mike make the right choice in accepting the default gray coloring scheme?

Converting the matrix B to an array C, the third method is to make a two-dimensional contour plot of C, choosing to take 20 contours (the default is eight contours). The contour plot creates lines corresponding to a fixed number of artifacts. Hikers often use contour maps that give lines of constant elevation.

```
> C:=convert(B,array):
> listcontplot(C,axes=boxed,labels=["x","y"],contours=20);
```

The final method is to create a three-dimensional contour plot of C, using filled=true to fill in the 20 contours with a color gradation. Mike takes the coloring and orientation of the figure to be the same as for the matrix plot.

```
> listcontplot3d(C,filled=true,axes=boxed,shading=zhue,
  labels=["x","y","N"],contours=20,orientation=[-60,15]);
```

At this point, Mike hears good-natured laughter outside the tent in which he

[15] The reader will have to do it in order to reproduce what Mike sees.

has been looking at the computer output. "Hi, Mike," someone yells, "we are going down to the village for some beer, or a Coke if you prefer. We will give you the real data tomorrow."

PROBLEMS:

Problem 1-33: Design your own secret message

Create your own data matrix that when plotted with the four methods of this subsection and the proper orientation and coloring scheme gives an interesting written message or pattern.

Problem 1-34: Electrostatic potential

The electrostatic potential Φ in the region $0 \leq x \leq a$, $y \geq 0$ with the boundary conditions $\Phi = 0$ along the sides $x = 0$, $x = a$, and for large y, and $\Phi = V$ for $y = 0$ and $0 < x < a$, is

$$\Phi = (2\,V/\pi)\,\arctan\left[\sin(\pi\,x/a)/\sinh(\pi\,y/a)\right].$$

Taking $a = 1$ meter, $V = 2$ volts, produce two-dimensional and color-coded three-dimensional contour plots of the potential. If necessary, play with the number of contours, the grid, and the view to produce nice plots.

Problem 1-35: Visual hallucination patterns

According to the text *Mathematical Biology*, by Jim Murray [Mur89], visual hallucination patterns can occur when an individual has a migraine headache, epileptic seizure, advanced syphilis, or as a result of taking drugs such as LSD or mescaline. From extensive experimental studies of drug-induced hallucinations, it appears that in the early stages, the test subject sees simple geometric patterns. A theoretical model for the underlying brain mechanism leads to a variety of visual patterns, one such pattern being described by the formula

$$V = \cos(a + k(\sqrt{3}y + x)/2)) + \cos(b + k(\sqrt{3}y - x)/2)) + \cos(c + kx).$$

(a) Taking $a = \pi/2$, $b = 0$, $c = 0$, and $k = 1$, form a two-dimensional, filled-in, color-coded, contour plot of the visual pattern V. Take the range $x = -10$ to 10, $y = -10$ to 10, and ten contours.

(b) What symmetry does the pattern have? Try some other values of the parameters and discuss your results.

Problem 1-36: Data storage matrices

As evident from the text recipe, matrices can be used to conveniently store data. For example, suppose that four varieties of wheat are treated with three different fertilizers and the outputs in bushels per acre are recorded, the data being stored in the 3×4 matrix A:

$$A = \begin{bmatrix} 30 & 39 & 33 & 24 \\ 36 & 42 & 33 & 27 \\ 33 & 33 & 39 & 36 \end{bmatrix}.$$

The element A_{ij}, with $i = 1, 2, 3$ denoting rows and $j = 1, 2, 3, 4$ indicating columns, gives the output of the jth wheat variety due to the ith fertilizer.

Now to the problem. Consider the following two different ecosystems, each consisting of three species, in which each species is the food source for the other:

(1) Each species consumes one each of the other two species.

(2) Species 1 consumes one of species 2 and none of species 3; species 2 consumes one-half each of species 1 and 3; species 3 consumes two of species 1 and none of species 2.

For each of the above ecosystems:

(a) Construct the consumption matrix C, where the element C_{ij} indicates the number of species j consumed daily by an individual of species i.

(b) Make a matrix plot for each consumption matrix.

(c) Make a `listcontplot3d` for each matrix.

Problem 1-37: Twin towers apartment complex

By using the binomial function to generate the binomial coefficients, the following code produces an architect's scale model of an apartment complex. Run the code and see what it looks like. Experiment with the command structure and see what apartment complexes you can design.

```
> restart: with(plots): with(LinearAlgebra):
> L:=[seq(i,i=0..8)]; #binomial(n,m)=n!m!/(n-m)!
> A:=[seq([seq(binomial(n,m),m=L)],n=L)]:
> B:=Matrix(A); C:=B+Transpose(B);
> matrixplot(C,heights=histogram,style=patch,
  orientation=[-135,50],shading=xyz,lightmodel=light3);
```

1.3.2 Designing a Ski Hill

Imagination is more important than knowledge.
Albert Einstein, theoretical physicist (1879–1955)

Rob, a young ski-hill designer and avid skier, has received a request to create a three-dimensional model of a mountainous area that allows ski runs of varying difficulties. He is asked not to build a physical model but instead to present the design in the form of a three-dimensional computer display that can be rotated so that the terrain can be viewed from different perspectives. It is also indicated that it is important to show the contours of constant elevation as well as the maximum slopes or gradients at various points on the ski hill.

In order to indicate the direction and magnitude of the gradients at various ski hill locations, it is necessary for Rob to load the `VectorCalculus` package.

```
> restart: with(plots): with(VectorCalculus):
```

Before attempting to design a realistic ski area, he first creates a simple mathematical model of a hill. A circularly symmetric mountain of peak height H with center at $x=a$ and $y=b$ can be produced using the exponential function

$$h(x,y) = H\,e^{-((x-a)^2+(y-b)^2)}, \tag{1.3}$$

where $h(x,y)$ is the elevation at the point (x,y). By experimenting with different functional forms, Rob realizes that more interesting terrain can be created by using additional exponential terms and by multiplying them by other functions such as simple polynomials. Adding slowly varying cosine terms can make the surrounding area have a rolling foothill appearance. Incorporating these aspects and using his imagination, Rob enters the following command line to produce an interesting mountainous terrain with slopes of different difficulties. All distances are in kilometers.

```
> h:=2*cos(0.4*x)*cos(0.4*y)+5*x*y*exp(-(x^2+y^2))
    +3*exp(-((x-2)^2+(y-2)^2));
```

$$h := 2\cos(0.4\,x)\cos(0.4\,y) + 5\,x\,y\,e^{(-x^2-y^2)} + 3\,e^{(-(x-2)^2-(y-2)^2)}$$

To generate a three-dimensional picture of the function h, the `plot3d` command is used with various plot options, the result being shown in Figure 1.12.

```
> plot3d(h,x=-3..4,y=-3..4,axes=framed,orientation=[-65,55],
  style=patchcontour,contours=20,scaling=constrained,
  shading=zgreyscale,lightmodel=light4);
```

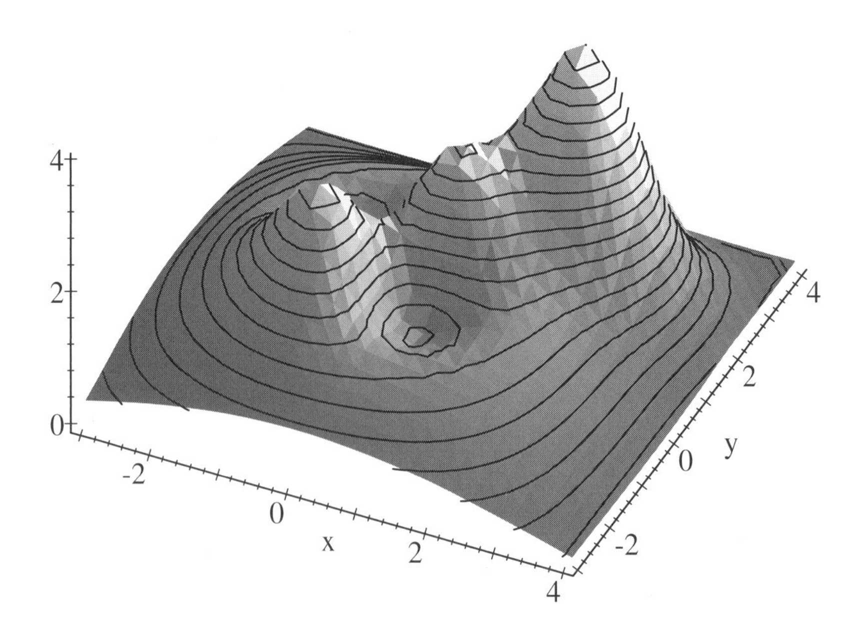

Figure 1.12: The ski hill produced by the input height function.

Rob has chosen a particular orientation in creating the figure and shown 20 elevation levels. For the purposes of this textbook he has, at our request, used a vertical gray scale with appropriate lighting. But he suggests that you experiment with different color and lighting schemes.

To fly around the ski hill designed by Rob and see it from different view points, click on the plot with the left mouse button and rotate the viewing box by dragging on the resulting box. Alternatively, you can rotate the plot in two-degree increments by clicking on the small up and down arrows adjacent to the two angular coordinate boxes near the top left of the computer screen.

To indicate the direction of maximum slope at a particular location, the gradient operator of mathematics must be used. If one has a function $f(x)$ of one independent variable, the slope of $f(x)$ at a particular point $x = c$ is obtained by calculating $(df(x)/dx)|_{x=c}$. A maximum or minimum of $f(x)$ is characterized by zero slope. For the ski hill, $h(x, y)$ is a function of two coordinates. If the coordinate y is held fixed, the slope in the x-direction is given by the partial derivative $\partial h(x, y)/\partial x$. Similarly, if x is held constant, the slope in the y-direction is given by $\partial h(x, y)/\partial y$. If $\hat{e}_x$ and $\hat{e}_y$ are unit vectors[16] in the x- and y-directions, then the gradient[17] of the function $h(x, y)$ is given by

$$\operatorname{grad} h(x,y) \equiv \nabla h(x,y) = \frac{\partial h(x,y)}{\partial x}\hat{e}_x + \frac{\partial h(x,y)}{\partial y}\hat{e}_y. \tag{1.4}$$

It can be shown that the gradient of any well-behaved $h(x, y)$ always points perpendicularly to the contours of constant elevation (or constant gravitational potential), i.e., $h(x, y)$=constant, and its magnitude at a point (x_0, y_0) is equal to the maximum slope at that location. The following command line produces the gradient of the function h in terms of the Cartesian coordinates x and y.[18] For other coordinate choices, the coordinate system must be specified in the argument, e.g., `'polar'[r,theta]`, for the plane polar coordinates r, θ.

```
> gradeq:=Gradient(h,[x,y]);
```

$$\begin{aligned} gradeq := (&-0.8\sin(0.4\,x)\cos(0.4\,y) + 5\,y\,\%1 - 10\,x^2\,y\,\%1 \\ &+ 3\,(-2\,x+4)\,e^{(-(x-2)^2-(y-2)^2)})\,\overline{e}_x + (-0.8\cos(0.4\,x)\sin(0.4\,y) \\ &+ 5\,x\,\%1 - 10\,x\,y^2\,\%1 + 3\,(-2\,y+4)\,e^{(-(x-2)^2-(y-2)^2)})\,\overline{e}_y \\ \%1 &:= e^{(-x^2-y^2)} \end{aligned}$$

Rob uses the `fieldplot` command to produce thick red arrows pointing in the direction of the positive gradient with the magnitude of the slope indicated by

```
> fp:=fieldplot(gradeq,x=-3..4,y=-3..4,arrows=THICK,color=red):
```

[16]Vectors of unit length.

[17]The standard mathematical symbol for the gradient operator is ∇.

[18]All Maple input/output appearing in this LaTeX-prepared text has been exported from the worksheet by clicking on File, Export As, and LaTeX. Lengthy Maple outputs will often appear here in terms of subexpressions, e.g., $\%1 := e^{(-x^2-y^2)}$ above, not seen on the computer screen. The overbars on the unit vectors in *gradeq* indicate that the gradient is a vector field, i.e., a vector-valued function of the coordinates.

the length of the arrow. Longer arrows correspond to a steeper maximum slope or gradient. The name fp is an acronym for field plot.

A two-dimensional contour plot, cp, is formed with 15 contours, the default being 8 contours. The smoothness of the contour lines is controlled through the grid option. The default grid is [25,25].

```
> cp:=contourplot(h,x=-3..4,y=-3..4,color=blue,
        grid=[30,30],contours=15,scaling=constrained ):
```

Figure 1.13 results from using the display command to superimpose fp and cp.

```
> display({fp,cp});
```

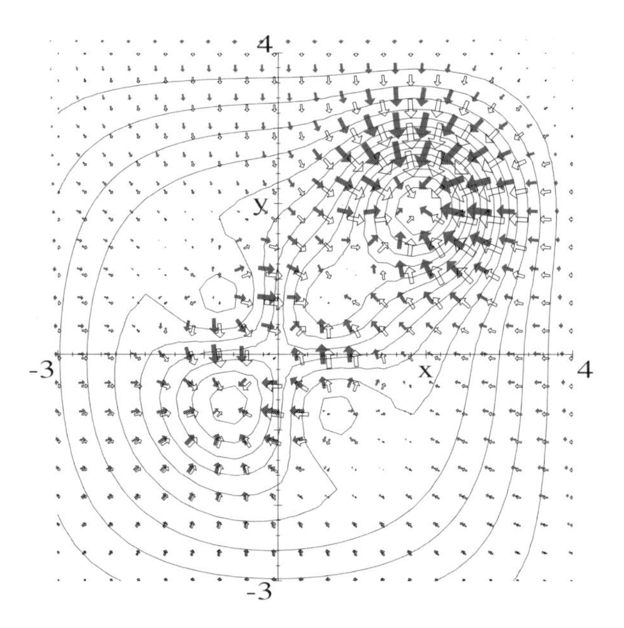

Figure 1.13: Contour plot of the ski hill with slope directions indicated.

From this plot, you can see confirmation that the gradient operator produces slope arrows perpendicular to the contours of constant elevation and that their length is an indication of hill steepness in the direction of the arrows.

The two maxima and two minima are easy to spot, and their locations can be approximately found by clicking the mouse with the cursor arrow placed on the relevant points or more accurately by using the fsolve command. At the maxima and minima, the slope is zero. In the following command line, Rob locates the coordinates of the top of the tallest peak in Figure 1.13, which lies in the quadrant $x=0$ to 4, $y=0$ to 4. The entries gradeq[1] and gradeq[2] in the Maple set refer to the first and second (x and y) components in gradeq. Note that if Rob hadn't set these entries equal to zero, Maple would have done so, unless Rob had otherwise indicated.

```
> coords:=fsolve({gradeq[1]=0,gradeq[2]=0},{x=0..4,y=0..4});
```

$$coords := \{y = 1.926405613,\ x = 1.926405613\}$$

The tallest peak is located at $x = y \approx 1.93$ km. Substituting these coordinates into `h` and numerically evaluating the result to 3 digits

```
> subs(coords,h): height:=evalf(%,3);
```

$$height := 4.01$$

yields a peak height of about 4 km.

According to Rob, the reader should have no difficulty in choosing a ski run of suitable steepness for his or her ability. The maximum slope at a given point, for example, at $x = 0$, $y = 2$ km, is easily evaluated. Substituting the specified coordinates into `gradeq[1]` and numerically evaluating the result yields the x-component, s_x, of the maximum slope s.

```
> subs({x=0,y=2},gradeq[1]): s[x]:=evalf(%);
```

$$s_x := 0.4029440556$$

The y-component, s_y, of the maximum slope at $x = 0$, $y = 2$ is similarly obtained.

```
> subs({x=0,y=2},gradeq[2]): s[y]:=evalf(%);
```

$$s_y := -0.5738848727$$

The maximum slope is then given by $s = \sqrt{s_x^2 + s_y^2}$.

```
> max_slope:=sqrt(s[x]^2+s[y]^2);
```

$$max\_slope := 0.7012187669$$

The maximum slope at $x = 0$, $y = 2$ is about 0.7, i.e., a vertical rise of 7 meters for every 10 meters horizontally. To translate this result into degrees, Rob first calculates the angle in radians,

```
> angle:=arctan(max_slope); #in radians
```

$$angle := 0.6115434605$$

and converts it to degrees (to 3 figures), using the fact that π radians $\equiv 180\,°$.

```
> angle:=evalf(angle*180/Pi,3); #in degrees
```

$$angle := 35.0$$

At $x = 0$, $y = 2$, the maximum slope corresponds to an angle of 35° with the horizontal. It is left as an exercise for the reader to determine what direction the maximum slope points in at this location.

PROBLEMS:

Problem 1-38: Ski-hill design

Design your own unique ski hill and repeat the steps in the text.

Problem 1-39: A different hill

Consider the height function

$$h(x, y) = 50\,(2\,x\,y - 3\,x^2 + 14\,x - y^2 - 2\,y + 10)$$

over the range $x = -3$ to 9 km , $y = -3$ to 9 km with the height in meters.

(a) Produce a three-dimensional plot of this function, using the Maple options `orientation=[-20,70]`, `axes=framed`, and `style=patchcontour`.

(b) Produce a two-dimensional contour plot with 15 contours and gradient arrows included.

(c) Where is the top of the hill located and what is the height there?

(d) If a person is standing at $x = 1$ km, $y = 1$ km, in what direction is the steepest slope? What is the slope at this location?

Problem 1-40: Enon on the hill

The height h of a certain hill (in meters) on the planet Erehwon is given by

$$h(x, y) = 2\,x\,y - 3\,x^2 - 4\,y^2 + 14\,x + 10\,y + 12,$$

where x and y are the easterly and northerly coordinates in kilometers.

(a) Produce a three-dimensional contour map of the hill.

(b) Where is the top of the hill located?

(c) How high is the hill?

(d) The small town of Enon is located on the hill at $x=2$, $y=1$ km. When it rains what is the natural direction in which the water drains out of town?

(e) Make a two-dimensional plot that shows the contours of constant elevation, the direction arrows for water drainage, and the location of Enon. Consider the range $x=1.5$ to 2.5, $y=0.5$ to 1.5 km and take 16 contours.

(f) Assuming that the draining water is frictionless and $g \simeq 9.8$ m/s^2, use Newton's second law to determine its acceleration down the hill at Enon.

Problem 1-41: Climbing a hill

Suppose that you are climbing a hill whose height is given by

$$h = 1000 - 0.01\,x^2 - 0.02\,y^2 \text{ meters}$$

and you are standing at a point with coordinates $(60, 100, 764)$ meters.

(a) Produce a three-dimensional contour map of the hill.

(b) In which direction should you proceed initially in order to reach the top of the hill in the shortest distance.

(c) Make a two-dimensional plot that shows the contours of constant elevation, the direction arrows pointing in the direction of positive gradient, and your present location indicated by a suitably sized circle.

(d) If you climb in the direction found in part **(b)**, at what angle above the horizontal will you be climbing initially?

Problem 1-42: Follow that mosquito

The temperature in a warehouse is given by $T = x^2 + y^2 - z$. A mosquito located at $(1, 1, 2)$ in the warehouse desires to fly in such a direction that it will get warm as soon as possible. In what direction must it fly? Express your answer as a three-dimensional vector.

Problem 1-43: Electric dipole potential

An electric dipole potential in the $z=0$ plane containing two charges of equal but opposite sign is given by

$$V = -\frac{0.4}{\sqrt{(x-1)^2+y^2}} + \frac{0.4}{\sqrt{(x+1)^2+y^2}}.$$

(a) Use the `plot3d` command to make a three-dimensional plot of V. As your plotting options, take $x=-3$ to 3, $y=-3$ to 3, `zhue` color, 20 contours, `patchcontour` style, `lightmodel = light3`, and constrained scaling.

(b) The electric field $\vec{E}$ is equal to $-\nabla V$. Calculate $\vec{E}$ for the dipole.

(c) Use the `fieldplot` and `contourplot` commands to plot the electric field, represented by arrows, and the equipotentials in the same graph. Take the range to be $x=-1.5$ to 1.5, $y=-1.5$ to 1.5, and use suitable arrows. You will probably wish to set the zoom magnification to 200% (the largest magnifying glass in the tool bar) to view the arrows.

Problem 1-44: Electric quadrupole potential

An electric quadrupole potential in the $z=0$ plane containing four charges of equal but alternating sign placed at the four corners of a square is given by

$$V = \frac{0.4}{\sqrt{(x+1)^2+y^2}} - \frac{0.4}{\sqrt{(x-1)^2+y^2}} + \frac{0.4}{\sqrt{x^2+(y+1)^2}} - \frac{0.4}{\sqrt{x^2+(y-1)^2}}.$$

For this quadrupole potential, carry out the same steps as in the electric dipole problem, but choose whatever options that you think give the best plots.

Problem 1-45: Temperature variation

Consider the temperature function $T(x, y, z) = 80/(1 + x^2 + 2\,y^2 + 3\,z^2)$, where T is measured in °C and x, y, and z in meters. In which direction does the temperature increase the fastest at the point $(1, 1, -2)$? What is the maximum rate of increase?

Problem 1-46: Van der Waals equation of state

The suitably normalized Van der Waals equation of state can be written as

$$\left(P + 3/V^2\right)(V - 1/3) = (8/3)\,T.$$

Here P, V, and T are the normalized pressure, volume, and temperature. Use `plot3d` to plot the following *isotherms* over the range $V=0.1$ to 3, $P=0$ to 3:

$$T = 0.85, 0.9, 0.95, 1.0, 1.05, 1.10, 1.15, 1.20, 1.25, 1.5, 2.0, 2.5$$

Take `grid=[150,150]`, `numpoints=5000`, `style=patchcontour`, `shading=zhue`, `view=[0..3,0..3,0..3]`, and unconstrained scaling. Choose an orientation that shows the isotherms in the P-V plane.

1.4 Animated Plots

1.4.1 Waves Are Dynamic

Life is a wave, which in no two consecutive moments of its existence is composed of the same particles.
John Tyndall, British physicist (1820–1893)

In most introductory physics courses, science students learn about wave motion and in particular how to add different wave forms to study such wave phenomena as standing waves, beats, and interference patterns. The associated physics text will usually have static pictures of the resulting waves at some instant in time. In the real world, waves tend to move and evolve with time. In other words, they are dynamic. The dynamic behavior of wave forms can be studied by animating them, that is to say, creating a time sequence of frames that when displayed rapidly give the illusion of continuous wave motion.

Two illustrative examples, which can be easily altered to investigate other wave phenomena, are now animated. To use the `animate` command, the plots package must be loaded.

```
> restart: with(plots):
```

The displacement U of a sinusoidal wave of amplitude A and wavelength λ traveling in the x-direction with velocity v may be represented by

$$U = A \sin\left(\frac{2\,\pi}{\lambda}(x - v\,t) + \delta\right). \tag{1.5}$$

If $v > 0$ the wave travels in the positive x-direction, while if $v < 0$ it travels in the negative x-direction. Here t is the time and δ the phase angle, which shifts the location of the maxima and minima. Longitudinal waves such as sound have their displacement in the direction of propagation, while transverse waves such as light have their displacement perpendicular to the direction of motion.

In the next command line, two different sinusoidal waves with, in general, different amplitudes, wavelengths, velocities, etc., are added together.

```
> U[1]:=A[1]*sin((2*Pi/lambda[1])*(x-v[1]*t)+delta[1])+
        A[2]*sin((2*Pi/lambda[2])*(x-v[2]*t)+delta[2]);
```

$$U_1 := A_1 \sin\left(\frac{2\,\pi\,(x - v_1\,t)}{\lambda_1} + \delta_1\right) + A_2 \sin\left(\frac{2\,\pi\,(x - v_2\,t)}{\lambda_2} + \delta_2\right)$$

Depending on the values assigned to the parameters, various wave phenomena can occur. For example, let's consider two waves of the same amplitude ($A_1 = A_2 = 1$), wavelength ($\lambda_1 = \lambda_2 = 2\pi$), and phase angle ($\delta_1 = \delta_2 = 0$), traveling with the same speeds ($|v_1| = |v_2| = 0.5$) but in opposite directions. Other possibilities can be explored in the problems at the end of this subsection.

```
> A[1]:=1: A[2]:=1: v[1]:=0.5: v[2]:=-0.5: lambda[1]:=2*Pi:
  lambda[2]:=2*Pi: delta[1]:=0: delta[2]:=0:
```

The resulting waveform with the parameter values substituted is now displayed,

```
> Wave[1]:=U[1];
```

$$Wave_1 := \sin(x - 0.5000000000\,t) + \sin(x + 0.5000000000\,t)$$

and animated.

```
> animate(Wave[1],x=-20..20,t=0..4*Pi,frames=40,numpoints=250);
```

The spatial range has been taken to be from $x = -20$ to $+20$ and the time interval from $t = 0$ to $4\,\pi$. Depending on what input parameters you choose, you might have to alter these ranges. The default setting of the animation command is to give 25 equally spaced time frames. Here we have chosen 40 frames as a compromise between obtaining a visually smooth time sequence and the increase in computing time that results when more frames are specified. A sufficient number of plotting points (minimum default number is 50) must be selected to give a smooth spatial profile that is fine enough not to miss any important structural details. Specifying more points slows the calculation down, so once again a compromise usually must be achieved.

When the `animate` command line is executed, the initial frame of the wave animation appears on the screen. Clicking on the picture with the left mouse button places the picture in a viewing box and opens up an animation bar at the top of the screen. The animation is started by clicking on the arrowhead (▷) and stopped by clicking on the square (□). If you want the sequence to be continuous, click on the loop arrow. In this particular animation, the resultant wave travels neither to the right nor to the left, but simply oscillates up and down. This is an example of a *standing wave.* Such a wave can be achieved experimentally, for example, by suitably displacing a string, fixed at both ends, in the transverse direction.

Our second example involves a traveling wave made up of a superposition of many sinusoidal waves,

```
> restart: with(plots):
```

the nth wave having a velocity $c = 1/(1 + a\,n)$ with $a \geq 0$. Let us suppose that $N = 5$ waves are present so that, since $n = 1, 2, \ldots, N$, each wave in general has a different velocity.

```
> c:=1/(1+a*n); N:=5:
```

$$c := \frac{1}{1 + a\,n}$$

We now add waves described by the following mathematical structure. This particular series, in the limit $N \to \infty$, is an example of what mathematicians call a *Fourier series.*

```
> U[2]:=(4/Pi)*(1+add(sin((2*n+1)*(x-c*t))/(2*n+1),n=0..N));
```

$$U_2 := 4\left(1 + \sin(x - t) + \frac{1}{3}\sin\left(3\,x - \frac{3\,t}{1+a}\right) + \frac{1}{5}\sin\left(5\,x - \frac{5\,t}{1+2\,a}\right)\right.$$
$$\left. + \frac{1}{7}\sin\left(7\,x - \frac{7\,t}{1+3\,a}\right) + \frac{1}{9}\sin\left(9\,x - \frac{9\,t}{1+4\,a}\right) + \frac{1}{11}\sin\left(11\,x - \frac{11\,t}{1+5\,a}\right)\right)/\pi$$

First, let's take $a = 0$, so that $c = 1$ for each sinusoidal wave in the sum.

```
> a:=0: c:=c;
```

$$c := 1$$

Then the resultant wave is given by the output of the following command line.

```
> Wave[2]:=U[2];
```

$$Wave_2 := 4\left(1+\sin(x-t)+\frac{1}{3}\sin(3\,x-3\,t)+\frac{1}{5}\sin(5\,x-5\,t)+\frac{1}{7}\sin(7\,x-7\,t)\right.$$
$$\left.+\frac{1}{9}\sin(9\,x-9\,t)+\frac{1}{11}\sin(11\,x-11\,t)\right)/\pi$$

The wave is animated, an alternative syntax to that given above being used. The syntax is animate(plotcommand,[plotarguments],t=a..b,options). As shown in the next recipe, this form can be readily generalized to 3-dimensional animations by replacing plot with plot3d and including the range of the second spatial argument in plotarguments. Stylistic options can also be included.

```
> animate(plot,[Wave[2],x=-20..20,numpoints=500],t=0..10,
  frames=50);
```

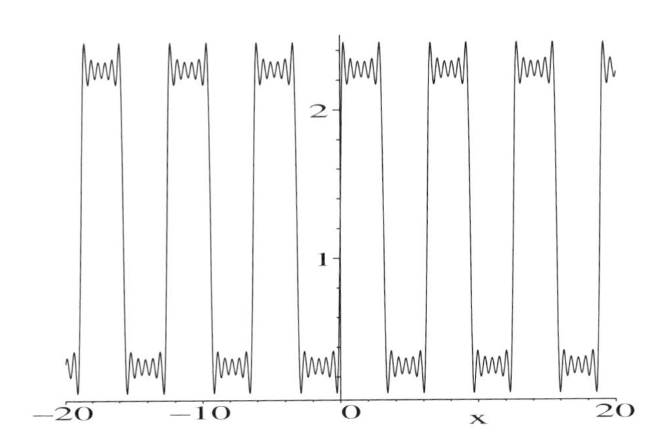

Figure 1.14: A traveling wave of approximately rectangular pulses.

The opening picture on the computer screen is shown in Figure 1.14. Except for the wiggles, the profile is a periodic sequence of rectangularly shaped pulses. The finite Fourier series for $N=5$ is an approximation to a rectangular pulse train. On being animated, the rectangular pulses travel to the right without changing shape. By default, the numerical value of the animation parameter t is displayed to 5 significant digits for each frame in the animation in the title region of the plot. If desired, the optional argument paraminfo=false can be included to turn this information off.

Now the reader should set $a=1$, so that the speed c is equal to $1/(1+n)$. On running the animation you should see that the initial shape alters dramatically because the individual sinusoidal waves are traveling at different velocities. The wave is said to be *dispersive*, whereas for $a=0$ the wave was *nondispersive*.

PROBLEMS:

Problem 1-47: Beats

Consider the addition of two waves with $A_1 = A_2 = 1$, $v_1 = v_2 = 1$, $\delta_1 = \delta_2 = 0$, $\lambda_1 = 2\pi$, and $\lambda_2 = 2.05\pi$.

(a) Animate the resultant wave, taking an x range from -200 to $+200$, $t = 0$ to 500, 25 frames, and 600 points.

(b) Describe the envelope of the resultant wave. If these were sound waves what intensity variation would an observer hear at a fixed spatial point as the resultant wave passed by? This is the phenomenon of *beats*.

Problem 1-48: Triangular wave train

Making use of the animation command with $c=1$ and $N=20$, show that

$$y = \sum_{n=1}^{N} (-1)^{n+1} \frac{2}{n} \sin(n\,(x - c\,t))$$

is a traveling wave consisting of approximately triangular pulses.

Problem 1-49: Wave pulse

Animate and discuss the behavior of the following wave profile:

$$y = 0.03\,(x - v\,t)/(1 + (x - v\,t)^4),$$

with $v = 2$ m/s. Take $x = -5$ to 25 meters, $t = 0$ to 10 seconds, 250 points, 100 frames, and use the color magenta. Discuss what happens if the v in the denominator is replaced with v^2. You may wish to increase the x and t ranges.

1.4.2 The Sands of Time

Thou seest the mountains and thou deemest them affixed,
(verily) they are as fleeting as the clouds.
The Koran

In an earlier tale, Rob, a young ski-hill designer and gung-ho skier, was asked to create a three-dimensional computer model of a mountainous area suitable for ski runs of varying difficulties. Although we were able to "fly" around the ski hill by rotating the viewing perspective in the computer file, the model itself was static, not displaying the temporal evolution that real mountains would exhibit over sufficiently long times due to the competing geological forces of uplift and water and wind erosion. Although this feature would be of no concern to the ski-hill designer, it is of interest to geologists concerned with the evolution of real mountain ranges over the eons.

In this upcoming recipe, we shall show how to build geological erosion into the earlier ski-hill model and illustrate the time evolution of the mountain range by producing a three-dimensional animation. The relevant command structure is very useful in animating the transverse vibrations of elastic membranes and other dynamic models in science and engineering. A call is first made to the plots package, which is needed in order to use the `animate` command.

```
> restart: with(plots):
```

Two decay coefficients, $\alpha_1=0.5$ and $\alpha_2=0.005$, are introduced,

```
> alpha[1]:=0.5: alpha[2]:=0.005:
```

and exponentially decaying time-dependent functions formed.

```
> d[1]:=exp(-alpha[1]*t); d[2]:=exp(-alpha[2]*t);
```

$$d_1 := e^{(-0.5\,t)} \qquad d_2 := e^{(-0.005\,t)}$$

Geological uplift might be simulated, for example, by changing the sign of one of the decay coefficients and executing the recipe over a limited time so as to avoid "exponential overflow." Of course, one need not use only exponential functions for the time dependence.

Now the ski hill in recipe **01-3-2** is modified by inserting the time-dependent functions d_1 and d_2 into the height function. The new height function h given below was chosen, using a trial and error approach, in such a way that the erosion looks somewhat realistic, displaying asymmetric spatial variation.

```
> h:=2*d[2]*cos(0.4*x)*cos(0.4*y)
   +5*x*y*d[1]*exp(-(x^2+d[1]*y^2))
   +3*d[1]*exp(-(d[1]*(x-2)^2+(y-2)^2));
```

$$h := 2\,e^{(-0.005\,t)}\cos(0.4\,x)\cos(0.4\,y) + 5\,x\,y\,e^{(-0.5\,t)}\,e^{(-x^2-e^{(-0.5\,t)}\,y^2)}$$
$$+\,3\,e^{(-0.5\,t)}\,e^{(-e^{(-0.5\,t)}\,(x-2)^2-(y-2)^2)}$$

By running the following `animate` command,[19] the reader can view the erosion of the ski-hill mountain range. The opening argument is the Maple plotting command, `plot3d`.

```
> animate(plot3d,[h,x=-3..4,y=-3..4],t=0..15,frames=40,axes=
  framed,orientation=[-65,55],style=patchcontour,contours=20,
  scaling=constrained,shading=xyz,lightmodel=light2);
```

The plot arguments, h and the two spatial ranges, were entered as a list. The range of the animation parameter, time t, was then given, followed by the number of frames and a large number of style and color options. We have retained the original 20 contours appearing in the earlier recipe, so that the temporal evolution of the hill can be more easily visualized. Instead of the `zgreyscale` coloring that Rob was instructed to use, here we take `shading=xyz` to color the hill in all three directions. The lighting has also been changed from that used in the original ski-hill recipe.

You can experiment with controlling the erosion by inserting different temporally decaying functions and playing with the decay coefficients. Also feel free to change the coloring and lighting.

[19]This command syntax supersedes the `animate3d` command used in earlier releases.

PROBLEMS:

Problem 1-50: Uplift

Modify the erosion recipe to illustrate the geological phenomenon of uplift of a portion of the mountain chain. Be careful with your choice of parameters and the total running time, or the initial profile will look extremely small compared to the final profile.

Problem 1-51: Normal modes of a rectangular membrane

The transverse displacement ψ of a light, uniform, horizontal, rectangular membrane fixed along its four edges at $x=0$, $x=a$, $y=0$, and $y=b$ will in general be described by a linear combination of its characteristic functions, or *normal modes*. The (m,n)th normal mode ($m=1,2,3,\ldots$, $n=1,2,3,\ldots$) with amplitude $A_{m,n}$ is given at time t by [Mor48]

$$\psi_{m,n} = A_{m,n}\,\sin(m\,\pi\,x/a)\,\sin(n\,\pi\,y/b)\,\cos(2\,\pi\,\nu_{m,n}\,t)$$

with characteristic (*eigen*) frequencies $\nu_{m,n} = (1/2)\sqrt{T/\sigma}\sqrt{(m/a)^2+(n/b)^2}$. Here T is the tension in the stretched membrane and σ the mass per unit area of the membrane. Take $T=1$, $\sigma=1$, $a=3/2$, and $b=1$.

(a) Animate the membrane for the normal mode $m=2$, $n=2$ with $A_{2,2}=1$. Take the time sufficiently long to show several complete vibrations.

(b) Animate the membrane for several other values of m and n.

(c) Animate different combinations of two or more normal modes, choosing amplitudes $A_{m,n}$ that give interesting vibrational patterns.

1.4.3 These Arrows Are Useful

If the Third World War is fought with nuclear weapons,
the fourth will be fought with bows and arrows.
Lord Louis Mountbatten, British admiral, member of royal family (1900–1979)

The "arrow" or functional operator, which appeared in some earlier recipes, is much more useful than its debut may have indicated. Here are two examples, a simple one to start with and then a more advanced illustration of its use.

Suppose that young Justine's personal weekly income consists of $p=2\,d^2$ pennies put into her piggy bank by her parents each day d that she is good. At the beginning of each new week, d is reset to 1, so that her parents don't ultimately go broke. On the first day ($d=1$) of the week, she receives 2 pennies, on the second ($d=2$) day she receives $2\times 2^2=8$ pennies, and so on. If she is bad, the income ceases for the rest of the week. Let's consider a (rare?) week when Justine was good every day.

The plots library package is loaded because it contains the `pointplot` command, which will be used to plot Justine's daily income in a point format.

Alternatively, one could use the `plot` command, with the option `style=point`. However, the `display` command will also be needed in order to superimpose plots in our second example, so the plots library package is required in any case.

```
> restart: with(plots):
```

Using the arrow operator, Justine's income formula can be entered as follows:

```
> p:=d->2*d^2;
```

$$p := d \rightarrow 2\,d^2$$

When the day d of the week is supplied to `p` as an argument, Justine's income for that day will be calculated. For example, the number of pennies deposited in her piggy bank on day 3 is obtained by entering `p(3)`.

```
> p(3);
```

$$18$$

So, Justine receives 18 pennies on the third day. Using the `pointplot` command, Justine's daily income can be plotted as seven size-16 circles for the seven days of the week. The resulting power law behavior is shown in Figure 1.15.

```
> pointplot([seq([d,p(d)],d=1..7)],symbol=circle,
  symbolsize=16,labels=["day","income"]);
```

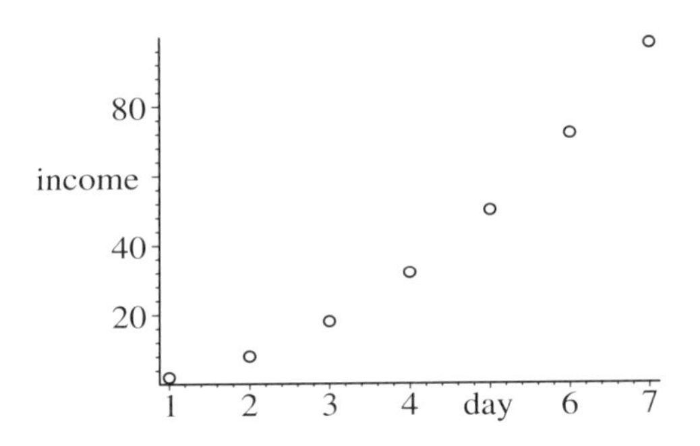

Figure 1.15: Justine's daily income versus day of the week.

Her total allowance for the week is obtained by adding up the daily totals.

```
> total:=add(p(d),d=1..7);
```

$$total := 280$$

Thus, Justine would receive \$2.80 in a week that she was good every day.

In this simple example, there was only one input variable in the functional operator, namely the number d of the day. The extension to more input variables and a more complex situation is easily handled, as is now demonstrated.

Recall that in the introductory recipe, **Bridge Design 101**, Russell the engineer found that in order to minimize the cost of the bridge it should have eight piers including one at each end. We have asked Russell to return and create a recipe that schematically draws the bridge, with a river below. Then he is to place a square box on the horizontal bridge deck and make the box undergo simple harmonic motion about the center of the bridge, with turning points at the ends of the bridge.

Russell decides that he will use solidly colored rectangles to depict the river, the bridge deck, and the piers. The necessary Maple command, `rectangle`, is in the plottools library package. This package contains many other geometrical shapes that can be plotted. Replace the colon with a semicolon in the following command line to see what shapes are available.

```
> with(plottools):
```

The command `rectangle([x1,y1],[x2,y2],color=c)` creates a plot data object, which when displayed is a two-dimensional rectangle whose top left corner is located at `x1, y1` and bottom right corner at `x2, y2`. The color `c` must be specified. An arrow operator `r` (`r` for rectangle) is formed, with the five quantities `x1, y1, x2, y2, c` required as input variables.

```
> r:=(x1,y1,x2,y2,c)->rectangle([x1,y1],[x2,y2],color=c):
```

Using this operator, a green bridge deck, a blue river, and eight equally spaced magenta-colored piers are created. A surrealistically colored scene indeed!

```
> deck:=r(1,2.2,8.2,2,green):
> river:=r(1,0,8.2,-0.5,blue):
> piers:=seq(r(i,2,i+0.2,0,magenta),i=1..8):
```

Russell uses the `textplot` command to add the red-colored words "pier," "deck," and "river" to the plot. Each word is entered as a Maple string, and the associated numbers specify the horizontal and vertical positions for that word. The word positions are determined initially by examining the parameter values used in the above three commands, and then fine-tuned by trial and error. Each word grouping is put into a list format and then a list of lists formed.

```
> tp:=textplot([[5.7,1,"pier"],[2.6,1.85,"deck"],
      [4.6,0.15,"river"]],color=red):
```

The four plots are superimposed with the `display` command, and assigned the acronym `bg` standing for "background." The annotated bridge and river will provide the stationary background in the animation of the moving box.

```
> bg:=display({deck,tp,piers,river}):
```

Using the following arrow operator, a red box is drawn whose position at time t is determined when t is given as input. The horizontal input coordinates (first and third arguments of `r`) are such that the box will undergo simple harmonic motion along the bridge deck surface as t increases, the motion being about the center of the bridge, with the turning points at the bridge ends.

```
> box:=t->PLOT(r(4.4+3.4*sin(t),2.6,4.8+3.4*sin(t),2.2,red)):
```

Simple harmonic motion of the box on the bridge deck is produced with the `animate` command over the time interval $t = 0$ to $4\,\pi$.

```
> animate(box,[t],t=0..4*Pi,frames=100,background=bg,
  scaling=constrained,axes=NONE,paraminfo=false);
```

For a reasonably smooth animation, one hundred frames are used. In the `background` option, Russell specifies `bg`. The scaling is constrained and all axes are removed from the resulting figure by including `axes=NONE`. The time parameter will not appear in the title region of the animated plot, because the option `paraminfo=false` is included. The opening frame of the animation is shown in Figure 1.16.

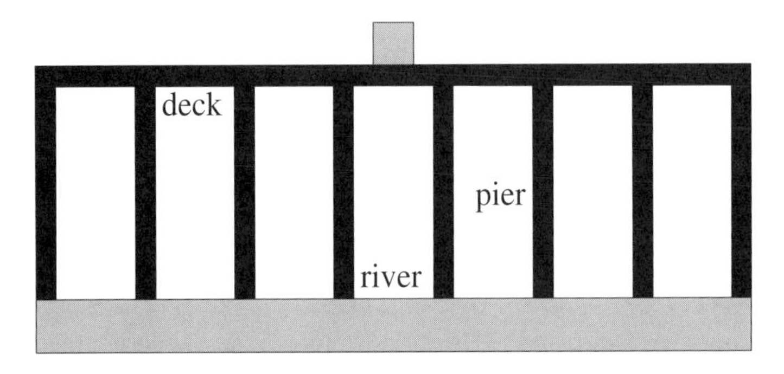

Figure 1.16: Box undergoing simple harmonic motion on the bridge deck.

To initiate the animation, click on the computer plot with the left mouse button and then on the start arrow in the tool bar.

PROBLEMS:

Problem 1-52: Mathematical example

Use the arrow operator to apply the operation $5\,t^5 - 2\,t^4 + t^2$ to the sequence of integers 3 to 9. Use the `add` command to total the sequence.

Problem 1-53: Another math example

For the function $f(x,y) = 2\,x^5 + 4\,y^4$, use the arrow operator to evaluate $f(5,9)$. Create a 3-dimensional plot of $f(x,y)$ over the range $x = -2$ to 2, $y = -2$ to 2.

Problem 1-54: Altering the text recipe

Change the shape of the sliding box in the text recipe into a circular disk. Change the shapes of the piers into triangles with the flat sides down. Alter the colors of the background components and allow the time parameter to appear in the animation.

Chapter 2

Deriving Model Equations

Aristotle could have avoided the mistake of thinking that women have fewer teeth than men by... asking Mrs. Aristotle to open her mouth.
Bertrand Russell, British philosopher and mathematician (1872–1970)

In Chapter 1 you learned how to plot observational and experimental data and functional forms in a variety of ways. Still other important types of graphs and other plotting commands will be encountered in ensuing chapters. The main purpose of first creating pictures in science is to gain a qualitative idea of the overall behavior of the data or the often complicated mathematical equations used to describe physical phenomena. When one is presented with a graph of the data, the next step is to derive a model equation or mathematical form that best describes the data. In this chapter, we shall illustrate how this is done using the *method of least squares*. The least squares method, which will be explained shortly, can be easily implemented by accessing the Statistics library package. This package, which was briefly encountered in the previous chapter, contains all of the relevant Maple statistical analysis and graphing commands. It supersedes the stats and statplots library package of earlier Maple releases.

Our specific goal is to further our understanding of which types of data structures lead to the model equations of Table 2.1 and, more importantly, to learn how to derive these equations using simple Maple command structures.

Table 2.1: Types of model equations encountered in this chapter.

Linear model $y = a + b\,x$	**Power law model** $y = a\,x^n$	**Polynomial model** $y = a + b\,x - c\,x^2 + \cdots$
Exponential model $y = a\,e^{-b\,x}$	**Logistic model** $y = \dfrac{a}{1 + b\,e^{-c\,x}}$	**Functional model** $y = f(x)$

The qualitative shapes of the curves associated with the linear, power law, polynomial, exponential, and logistic models were qualitatively drawn in Figure 1.1 of Chapter 1. The reader might wish to review these shapes before proceeding. The label "functional models" will be used in this chapter to refer to model equations involving functions $f(x)$ other than powers and exponentials.

Once a particular model equation has been determined for a specified set of data, it can then be used either to interpolate between data points or to extrapolate outside the range of the given data and make predictions of the anticipated behavior of the physical phenomena. Of course, the latter can prove dangerous if the underlying assumptions of the model equation no longer hold. Before plunging into the least squares derivation of model equations from a wide variety of interesting data, we begin by introducing the linear correlation coefficient r. If you suspect that the plotted data can be fitted, at least approximately, with a straight line, the calculation of r can quickly confirm whether your intuition is correct.

2.1 Linear Correlation

If a set of N paired quantities (x_i, y_i), $i = 1, 2, \ldots, N$, appear to lie along a straight line it is useful to first calculate the *linear correlation coefficient* r. If

$$\bar{x} = \frac{1}{N}\sum_{i=1}^{N} x_i, \qquad \bar{y} = \frac{1}{N}\sum_{i=1}^{N} y_i, \tag{2.1}$$

denote the mean values of the x and y data, respectively, the linear correlation coefficient is defined by

$$r = \frac{\displaystyle\sum_{i=1}^{N}(x_i - \bar{x})(y_i - \bar{y})}{\sqrt{\displaystyle\sum_{i=1}^{N}(x_i - \bar{x})^2}\sqrt{\displaystyle\sum_{i=1}^{N}(y_i - \bar{y})^2}}. \tag{2.2}$$

The value of r lies between -1 and 1. It takes on the value of 1 when the data points lie on a perfectly straight line with positive slope, x and y increasing together. *Complete positive correlation* is said to exist. To see that this is true, let the data points lie on the straight line $y_i = a\,x_i + b$, with the slope a positive. Then, $y_i - \bar{y} = a\,(x_i - \bar{x})$ and $r = a\,\sum_i (x_i - \bar{x})^2 / a\,\sum_i (x_i - \bar{x})^2 = 1$. Thus, since a cancels out, the value 1 holds no matter what the magnitude of the slope.

Complete negative correlation exists when $r = -1$. In this case the data points lie on a perfectly straight line with negative slope, y decreasing as x increases. If r is close to zero, the variables x and y are said to be *uncorrelated*, and a straight-line relation between them would not be expected.

Since the mathematical steps involved in Equation (2.2) are easily accomplished with a single simple command, the following "corny" example should suffice to show how the correlation coefficient is calculated with Maple.

2.1.1 The Corn Palace

"Who did you pass on the road?" the King went on....
"Nobody," said the Messenger.
"Quite right," said the King, "this young lad saw him too.
So of course Nobody walks slower than you."
Lewis Carroll, *Through the Looking Glass* (1872)

Driving through South Dakota on I-90 without stopping can be a somewhat monotonous drive, unless one takes time to visit one or more of the attractions that are off the Interstate. The main natural feature along this stretch of freeway is Badlands National Park, where spectacular examples of weathering and erosion can be viewed. There are also several man-made attractions that dot the highway. From west to east, these are Mount Rushmore National Memorial with its colossal sculpted heads of George Washington, Thomas Jefferson, Abraham Lincoln, and Theodore Roosevelt; then, further to the east, the block-long "world famous" Wall Drugstore in Wall, South Dakota; and finally, still further to the east, the Corn Palace in Mitchell, South Dakota. Mitchell is a trade center for locally produced corn, grain, and cattle. Its local events are centered around corn, and the Corn Palace is perhaps the ultimate artistic tribute to corn. Quoting from the American Automobile Association tour book, "The Corn Palace is of Moorish architecture with minarets and kiosks. Portions of the exterior and interior are covered with designs of corn outlined with grasses and grains. Each year 2000–3000 bushels of various shades of corn and grasses are used to redecorate the building, which is illuminated at night."

In the spirit of the Corn Palace, let's consider an example involving corn.

Table 2.2: United States corn yield for the period 1950–1971.

Year	**Corn output (bushels/acre)**	**Year**	**Corn output (bushels/acre)**
1950	38	1961	63
1951	36	1962	65
1952	41	1963	67
1953	40	1964	70
1954	38	1965	73
1955	40	1966	72
1956	45	1967	80
1957	46	1968	79
1958	52	1969	87
1959	53	1970	83
1960	54	1971	88

Table 2.2 shows the corn output [AL79], expressed in bushels per acre, for the United States for the period 1950–1971.

The call with(Statistics) is needed in order to calculate the linear correlation coefficient and to make a scatter plot of the corn data.

```
> restart: with(Statistics):
```

Taking the year 1950 to be 0, 1951 to be 1, etc., the sequence command is used to enter the years as a list of numbers from 0 to 21.

```
> year:=[seq(n,n=0..21)]:
```

The corn output given in Table 2.2 is also entered as a list.

```
> output:=[38,36,41,40,38,40,45,46,52,53,54,63,65,67,70,
          73,72,80,79,87,83,88]:
```

Using the ScatterPlot command, the data are plotted in Figure 2.1 and appear to lie approximately along a straight line.

```
> ScatterPlot(year,output,view=[0..21,0..90],symbol=circle,
  labels=["year","output"],tickmarks=[3,3]);
```

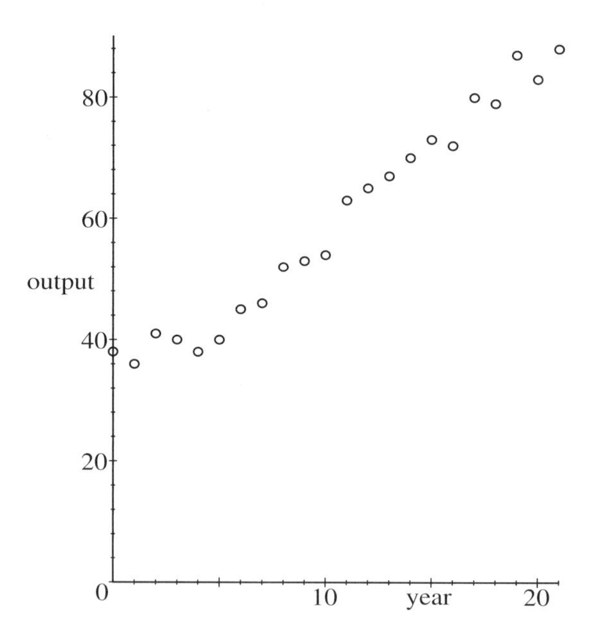

Figure 2.1: United States corn output (bushels/acre) for the period 1950–1971.

The linear correlation coefficient between output and year is calculated.

```
> r:=Correlation(year,output);
```

$$r = 0.9842870603$$

The correlation coefficient is 0.984, which is near +1, so a straight-line relation with positive slope is indicated. The best-fitting straight line can be found with the linear least squares procedure discussed in the next section.

PROBLEMS:

Problem 2-1: Deer antler weights

Table 2.3 shows the weights W, expressed in 0.01 kg, as a function of age A in months for a number of different deer antlers [AL79]. Plot the data and calculate the linear correlation coefficient. Is there a linear relationship between deer antler weights and age?

Table 2.3: Deer antler weights W as a function of age A.

A	20	22	30	34	42	43	46	54	56	68	70
W	8	10	15	20	27	26	31	36	40	49	49

Problem 2-2: Purchasing power of the dollar

In Table 1.1, the purchasing power of the U.S. dollar, as measured by consumer prices, was given for the period 1983 to 1992. The purchasing power in the base year 1983 was assigned the value 1.00. By calculating the linear correlation coefficient and showing that it is quite close to the value -1, demonstrate that the data can be well fitted by a straight line with negative slope.

2.2 Least Squares Derivations

A standard procedure in fitting observational data that appear to lie along a straight line is to find the best-fitting straight line using the method of least squares. Statisticians refer to this procedure as *regression analysis*. Let the equation of the proposed straight line be $Y = a + b\,x$, where the intercept a and the slope b are to be determined. Label the N data points (x_i, y_i) with $i = 1, 2, 3, \ldots, N$. If only two data points ($N = 2$) are present, a straight line can be found immediately that passes exactly through the two points. Unless all the points lie precisely on a straight line, for $N > 2$ there will in general be a difference between the Y-coordinate of the straight line and the value of y_i at the same value of x (i.e., at the x_i). Depending on whether the y_i lie above or below the straight line, the differences will be either positive or negative. To eliminate the possibility of producing an average error of zero, the differences are squared. Adding up the squares of the differences, a "total error" is formed.

$$\text{Total error} = \sum_{i=1}^{N} (y_i - Y_i)^2 = \sum_{i=1}^{N} (y_i - a - b\,x_i)^2 \tag{2.3}$$

If the x_i are exactly known but there is an uncertainty σ_i for each y_i, e.g., due to measurement error, this result is slightly modified to yield the *chi-square* merit function of statistics, viz.,

$$\chi^2(a, b) = \sum_{i=1}^{N} \left(\frac{y_i - a - b\,x_i}{\sigma_i} \right)^2 . \tag{2.4}$$

Points with smaller uncertainty σ_i are given more weight in the chi-square function. If the individual uncertainties σ_i are not known, then the σ_i are taken to be all the same and arbitrarily assigned the value 1, thus reducing chi-square to the total error defined above. In the examples that follow, we shall take all the $\sigma_i = 1$ and use the phrases chi-square and total error interchangeably. The best-fitting straight line is the one that minimizes chi-square or the total error. At the minimum, the derivatives of chi-square with respect to a, b vanish, yielding the two equations

$$S_y = a\,N + b\,S_x, \quad S_{xy} = a\,S_x + b\,S_{xx}, \tag{2.5}$$

with $S_y \equiv \sum_i y_i$, $S_x \equiv \sum_i x_i$, $S_{xx} \equiv \sum_i x_i^2$, and $S_{xy} \equiv \sum_i x_i\,y_i$. This pair of equations is easily solved for the intercept a and slope b of the best-fitting straight line,

$$b = (N\,S_{xy} - S_x\,S_y)/\Delta, \quad a = \bar{y} - b\,\bar{x}, \tag{2.6}$$

where $\bar{y} = S_y/N$ and $\bar{x} = S_x/N$ are the mean values and $\Delta = N\,S_{xx} - (S_x)^2$. Maple has a built-in least squares procedure that takes care of finding the best-fit coefficients a and b for us.

As a measure of the accuracy of the least squares fit, statisticians define the standard error or *standard deviation* as

$$\sigma = \sqrt{\frac{\chi^2}{N-2}} = \sqrt{\frac{\sum_i (y_i - a - b\,x_i)^2}{N-2}} = \sqrt{\frac{S_{yy} - a\,S_y - b\,S_{xy}}{N-2}}, \tag{2.7}$$

where use has been made of the least squares equations (2.5) to obtain the last form. The factor $N-2$ is inserted in the denominator rather than N to reflect the fact that only $N-2$ data points are really independent of the fitting procedure. The straight-line fit involves two unknowns, a and b, which require two data points to determine. Assuming a normal distribution of data points about the mean, there is a 68.3% probability of a data point being within one standard deviation of the mean, 95.4% of being within $2\,\sigma$ of the mean, 99.7% of being within $3\,\sigma$ of the mean, and so on.

It can be shown [PFTV89] that the standard deviations, σ_a and σ_b, in the least squares estimates of a and b, respectively, are given by

$$\sigma_a = \sqrt{(S_{xx}/\Delta)}\,\sigma, \quad \sigma_b = \sqrt{(N/\Delta)}\,\sigma. \tag{2.8}$$

If the observational data do not lie along a straight line but instead along one of the other curves of Table 2.1, the least squares method can be generalized to handle these situations as well. You will see several such examples as you progress through the chapter.

PROBLEMS:

Problem 2-3: Verification

Derive Equations (2.5) and (2.6).

2.2.1 Will You Be Better Off Than Your Parents?

He was a self-made man who owed his lack of success to nobody.
Joseph Heller, *Catch 22* (1961)

One of the great worries of the younger generation is that they won't be as economically well-off as their parents. To see whether this concern is backed by recent historical trends, the disposable (per capita) personal income over a period of time can be tracked and a model equation formulated. The validity of the model can then be checked by comparing its predictions with the actual disposable personal income at some later date. Table 2.4 shows the disposable

Table 2.4: United States per capita disposable income and GNP.

Year	Income	GNP	Year	Income	GNP
1960	8660	12585	1973	13539	18572
1961	8794	12651	1974	13310	18360
1962	9077	13215	1975	13404	18032
1963	9274	13587	1976	13793	18878
1964	9805	14184	1977	14095	19611
1965	10292	14897	1978	14662	20367
1966	10715	15661	1979	14899	20794
1967	11061	15896	1980	14813	20497
1968	11448	16485	1981	15009	20756
1969	11708	16809	1982	14999	20090
1970	12022	16616	1983	15277	20702
1971	12345	16959	1984	16252	21896
1972	12770	17694	1985	16597	22443

personal income (in dollars) for citizens of the United States for the period 1960 to 1985. To account for inflation, the dollar amounts have been "chained" to 1992 dollars. The per capita GNP is also shown.[1] Separate lists are formed for the year (using the sequence command) and for the disposable income. We take the year 1960 to be 0, the year 1961 to be 1, and so on. A comment to this effect has been added to the command line by using the sharp symbol #.

```
> restart: with(Statistics): with(plots):
> year:=[seq(n,n=0..25)]: #year since 1960
> income:=[8660,8794,9077,9274,9805,10292,10715,11061,11448,
        11708,12022,12345,12770,13539,13310,13404,13793,14095,
        14662,14899,14813,15009,14999,15277,16252,16597];
```

[1] GNP stands for gross national product, the annual sum of all goods and services produced by a country.

The year and income lists are then zipped together to form a single list of lists for the plotting points.

```
> pair:=(year,income)->[year,income]:
> points:=zip(pair,year,income);
```

$$points := [[0, 8660], [1, 8794], [2, 9077], [3, 9274], [4, 9805], [5, 10292], [6, 10715], [7, 11061], [8, 11448], [9, 11708], [10, 12022], [11, 12345], [12, 12770], [13, 13539], [14, 13310], [15, 13404], [16, 13793], [17, 14095], [18, 14662], [19, 14899], [20, 14813], [21, 15009], [22, 14999], [23, 15277], [24, 16252], [25, 16597]]$$

A call is made to the plot command, with its stylistic options, to generate a graph (labeled `pts`) of the plotting points, which are then shown in Figure 2.2. The `view` option is included so as to force the vertical axis to begin at zero, which it would not otherwise do.

```
> pts:=plot(points,style=point,symbol=circle,symbolsize=12,
  color=blue,tickmarks=[3,2],labels=["year","income"],
  view=[0..25,0..16000]): pts;
```

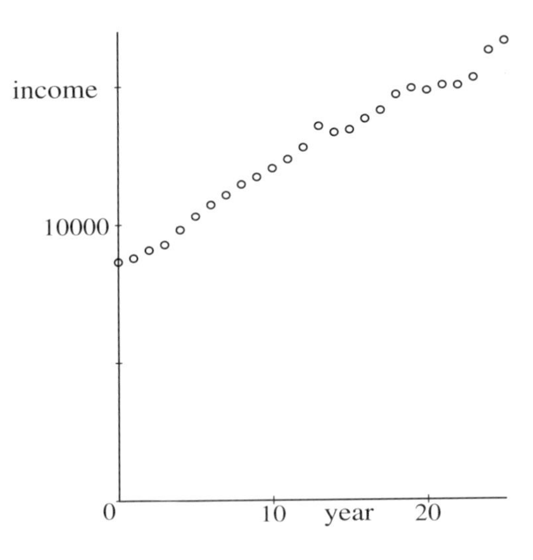

Figure 2.2: Personal income from Table 2.4 versus year after 1960.

Aside from some small oscillations, it appears that the points can be fitted by a straight line. This can be confirmed by calculating the linear correlation coefficient r.

```
> r:=Correlation(year,income);
```

$$r := 0.9922275441$$

Since the correlation coefficient r is close to 1, it is reasonable to seek a straight-line fit to the data using the least squares fitting procedure. This can be accomplished with two different Maple command structures. First, the `Fit` command is used with the linear form $a + b\,x$ given. The independent variable x is the number of years after 1960. The income formula is assigned the name y.

```
> y:=Fit(a+b*x,year,income,x);
```

$$y := 8757.98290598290806 + 310.499829059828869\,x$$

Alternatively, the same result can be obtained using the `LinearFit` command.

```
> y2:=LinearFit([1,x],year,income,x);
```

$$y2 := 8757.98290598290806 + 310.499829059828869\,x$$

In either case, the default answer is given to higher floating-point accuracy than the "normal" ten digits. On rounding off to 4 digits with the `evalf` command, the best-fitting straight line relating the personal income (p) to the year (x) since 1960 is

```
> p:=evalf(y,4);
```

$$p := 8758. + 310.5\,x$$

The default "straight-line model" equation, y, based on the data for 1960–1985, will now be checked for its predictive accuracy by including "future" data points for the period 1986–1995 taken from Table 2.5.

Table 2.5: Income and GNP data for 1986–1995.

Year	Income	GNP	Year	Income	GNP
1986	16981	22866	1991	17756	24119
1987	17106	23296	1992	18062	24490
1988	17621	23979	1993	18075	24767
1989	17801	24553	1994	18320	25305
1990	17941	24642	1995	18757	25588

To carry out the check, a single picture will be created, containing the future data points, the data points from 1960 to 1985, and the best-fitting straight line y. The latter is now plotted in `Gr` over the range $x = 0$ to 35.

```
> Gr:=plot(y,x=0..35,color=red):
```

Next a plot of the future points (shown as size 12 green boxes) is created.

```
> futurepts:=plot([[26,16981],[27,17106],[28,17621],
      [29,17801],[30,17942],[31,17756],[32,18062],[33,18075],
      [34,18320],[35,18757]],color=green,style=point,
      symbol=box,symbolsize=12):
```

Finally, all three plots are superimposed in a single graph by the use of the by-now-familiar `display` command.

```
> display({Gr,pts,futurepts},labels=["year","income"],
  tickmarks=[4,2],view=[0..35,0..20000]);
```

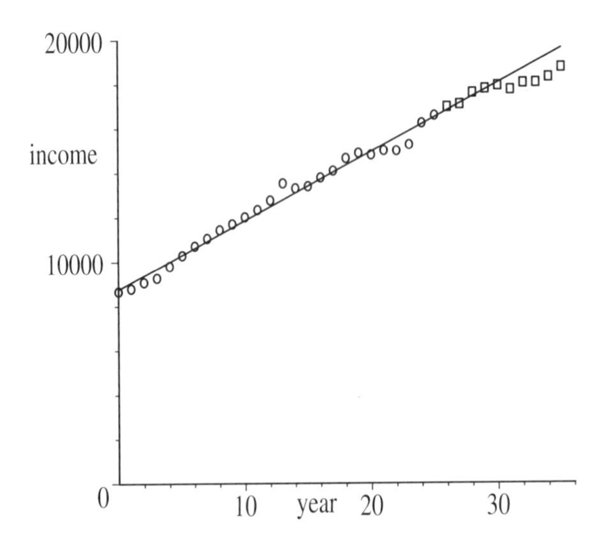

Figure 2.3: Best-fitting straight line for personal income data.

Figure 2.3 shows the results of applying Maple's best-fitting straight-line routine. The first five future data points (boxes) for the period 1986 to 1990 (x = 26 to 30) lie on the model equation's extrapolated straight line. Not surprisingly, the linear (straight-line) model does a good job in predicting the short-term future. The last five future data points for 1991 to 1995 lie below the straight-line prediction. Is this a trend or part of the small oscillations seen for the earlier data? You might try searching the literature or, perhaps, going to the Internet for some more data points since 1995.

Assuming that the linear model displays the correct trend, one could use it to predict the per capita disposable income in, say, the year 2005. Obviously, one could obtain the answer by substituting (use `subs`) these numbers into the linear model equation y, or by using the evaluation (`eval`) command. Alternatively, we can turn y into a functional operator f by using the `unapply` command:

```
> f:=unapply(y,x);
```

$$f := x \to 8757.98290598290806 + 310.499829059828869\, x$$

The appearance of the arrow in the output signals the presence of a functional operator. It indicates that if you specify a value of x, then the operation on the right-hand side of the arrow will be applied to x. Thus, since the year 2005 corresponds to $x = 45$, the command

```
> f(45);
```

$$22730.47522$$

yields \$22,730.48 as the predicted per capita income in that year. Again, to check on the model equation, the reader could compare the predicted income values with the actual numbers as they become available.

PROBLEMS:

Problem 2-4: Predicting the GNP

Using the data of Table 2.4 for the period 1960–1985, find the best fitting straight line for the GNP of the United States. Then include the 1986–1995 data of Table 2.5 to see how well the straight line predicts the "future" GNP. What GNP does the linear model predict for the years 2000, 2005, and 2010?

Problem 2-5: Deer antlers

Table 2.3 shows the weights W, expressed in 0.01 kg, as a function of age A in months for a number of different deer. Find the best-fitting straight line for these data. What is the predicted weight in kilograms for deer antlers belonging to a deer of age 60 months?

Problem 2-6: Corn yield

Table 2.2 shows the corn output, expressed in bushels per acre, for the United States for the period 1950–1971. Determine the best-fitting straight line for these data. Assuming that the linear trend prevailed, what should have been the corn output for the year 1975? 1985? 1995? Do an Internet or literature search and find out what the actual corn output was for these years and compare with the predicted values.

Problem 2-7: Heating-oil consumption

Table 2.6 gives the number of gallons of oil in a tank used for heating a condo complex d days after January 1, the day that the tank was last filled.

Table 2.6: Number of gallons in the tank on day d.

d	0	1	2	3	4	5	6
gallons	30,000	29,525	29,250	28,775	28,300	27,800	27,300

(a) Determine the linear correlation coefficient for these data. Does your result suggest a straight-line fit?

(b) Determine the best-fitting straight line to the data.

(c) Predict the number of gallons of oil in the tank on the last day of January. What assumptions are you making in using the best-fit equation for your prediction?

(d) If the pattern of fuel consumption remains unchanged, when would the tank be empty? Express your answer in terms of the day of the month.

(e) If the tank is to be refilled when 25% remains, on what day of which month should the oil tanker return to fill the tank?

Problem 2-8: T-shirts

Colleen, the MBA graduate, was employed one summer at a tourist shop, which among other things sold printed T-shirts with the city logo on the front. At the time, the owner of the shop wanted to order 650 more T-shirts, but the catalogue prices, which are reproduced in Table 2.7, only went up to 350 T-shirts. The owner tried to contact her supplier to see what the total cost would be for 650 T-shirts but the supplier was away for a few days. In the meantime,

Table 2.7: Total cost of T-shirts as a function of the number ordered.

Number purchased	50	100	150	200	250	300	350
Total cost	250	375	500	600	700	825	950

she asked Colleen to answer the following questions, which you can also try:

(a) What is the linear correlation coefficient for the data? Does this suggest a linear model relating the total cost to the number purchased?

(b) What is the equation of the straight line that best fits the tabulated data?

(c) Using the model equation, how much should 650 T-shirts cost the store owner?

(d) If 650 shirts are bought, what is the average cost per shirt?

(e) If the store owner plans to mark the T-shirts up 300% over the cost, what should a T-shirt be sold for?

(f) At the marked-up price, how many T-shirts would have to be sold to just cover the total cost of the 650 shirts? That is, what is the break-even number of shirts that must be sold?

Problem 2-9: Enon Revisited
The hill on which the small town of Enon is located is described by

$$h(x, y) = 2\,x\,y - 3\,x^2 - 4\,y^2 + 14\,x + 10\,y + 12 \quad \text{meters}$$

where x and y are the easterly and northerly coordinates in kilometers and h is the height. Convert h into a functional operator. If Enon's coordinates are $x = 2$, $y = 1$, use the functional operator to determine Enon's elevation? What is the elevation difference between Enon and the top of the hill?

2.2.2 What Was the Heart Rate of a Brachiosaur?

The progress of science is strewn, like an ancient desert trail, with the bleached skeleton of discarded theories which once seemed to possess eternal life.
Arthur Koestler, British writer (1905–1983)

Previously, the general scientific concensus was that dinosaurs were cold-blooded animals (reptiles), but opinions have shifted, and many scientists now believe that they were actually warm-blooded. Learning of this possibility in one of her zoology courses, Heather, the premed student, asks her mathematician sister Jennifer the following question:

"Even though dinosaurs have been extinct for millions of years, can the heart rate of, say, the brachiosaur be determined? This creature was believed to be the largest dinosaur that ever lived, with an estimated body mass of about 75 thousand kilograms."

To which Jennifer replies, "Assuming that the brachiosaurs were warm-blooded, the answer is yes. Given the heart rate data as a function of body mass for present-day warm-blooded animals, we can formulate a model equation to estimate the heart rate for a brachiosaur. Not only that, the model equation can be understood by using scaling arguments similar to those that we discussed earlier."

"OK, Jennifer, but where are we going to get the necessary data and how are we going to use it to extract the model equation?"

"Remember Herbert Lin's paper [Lin82] on zoological scaling. Using data from this paper, we can construct Table 2.8 showing the approximate (average

Table 2.8: Heart rate and body mass for some warm-blooded animals.

	mouse	rabbit	dog	human	tiger	donkey	elephant	whale
Mass	0.015	2.0	15	63	99	407	3000	50000
Rate	624	210	76	72	55	46	37	17

values for a species) body mass in kilograms and heart rate in beats per minute for a selection of present-day warm-blooded animals. Notice that the heart rate and body mass are inversely related, the heart rate decreasing as the body mass increases. Further, look at the wide span of the data. The ratio of largest heart rate to smallest is about 37 (624/17), while the ratio of largest body mass to the smallest is about 3,000,000 (50,000/0.0150). As you recall from our earlier scaling examples, since the data span many orders of magnitude, this suggests that we create a log-log plot and look for a power law of the structure $y = k\,x^b$."

"I understand what you're saying, Jennifer. If the points can be fitted by such a *power law model*, the log-log graph will be a straight line of slope b and intercept $\ln(k)$ if $\ln(y)$ is plotted against $\ln(x)$. So if we set $Y \equiv \ln(y)$, $X \equiv \ln(x)$, and $a \equiv \ln(k)$, then the straight-line equation is $Y = a + b\,X$. So by using a best-fit procedure, we should be able to extract the model equation parameters a and b."

"That's precisely, what we are going to do. Let's begin by forming a log-log plot of the data in Table 2.8. For our purposes it will suffice to work to five digits accuracy. The number of entries in each data list will be `N=8`.

```
> restart: with(Statistics): with(plots):
> Digits:=5: N:=8:
```

We enter the heart rate and body mass data as separate lists. I prefer to work with floating-point numbers, so let's add decimal zero to the numbers.

```
> rate:=[624.0,210.0,76.0,72.0,55.0,46.0,37.0,17.0]:
> mass:=[0.015,2.0,15.0,63.0,99.0,407.0,3000.0,50000.0]:
```

Next we need to take the log of each data entry in both lists. This can be accomplished using the `ln` command on the ith entry of each list and the sequence (`seq`) command to include all `N=8` entries. The names *lograte* and *logmass* are then attached to the new lists.

```
> lograte:=[seq(ln(rate[i]),i=1..N)];
```

$$lograte := [6.4362, 5.3471, 4.3307, 4.2767, 4.0073, 3.8286, 3.6109, 2.8332]$$

```
> logmass:=[seq(ln(mass[i]),i=1..N)];
```

$$logmass := [-4.1997, .69315, 2.7081, 4.1431, 4.5951, 6.0088, 8.0064, 10.820]$$

Making use of the arrow operator and the `zip` command, the two new lists are then "zipped" together into a list of lists named *coords*.

```
> pair:=(lograte,logmass)->[lograte,logmass];
```

$$pair := (lograte, logmass) \to [lograte, logmass]$$

```
> coords:=zip(pair,logmass,lograte);
```

$$coords := [[-4.1997, 6.4362], [.69315, 5.3471], [2.7081, 4.3307], [4.1431, 4.2767], [4.5951, 4.0073], [6.0088, 3.8286], [8.0064, 3.6109], [10.820, 2.8332]]$$

The log-log plot for the data is created but not displayed.

```
> Gr1:=plot(coords,x=-5..11,view=[-5..11,0..10],style=point,
      symbol=circle,symbolsize=12,color=black):
```

The best-fitting straight line $y = a + b\,x$ to the log-log data is then found using the `Fit` command. The output is evaluated to 5 digits.

```
> y:=evalf(Fit(a+b*x,logmass,lograte,x),5);
```

$$y := 5.3092 - 0.23807\,x$$

The best-fitting straight line is of the form $y = 5.31 - 0.24\,x$, from which we can identify $b \approx -0.24$ and $a \approx 5.31$. The log-log plot `Gr1` for the data points and the plot `Gr2` of the best-fitting straight line are displayed together in Figure 2.4.

```
> Gr2:=plot(y,x=-5..11):
> display({Gr1,Gr2},tickmarks=[3,3],labels=["x","y"]);
```

The straight line clearly fits the data reasonably well. To find the model equation for the original data, we need to extract b and a,

```
> b:=coeff(y,x); a:=coeff(y,x,0);
```

$$b := -0.23807 \qquad a := 5.3092$$

so that the model equation is of the form $r = e^a\, m^b$, where r is the heart rate and m the body mass.

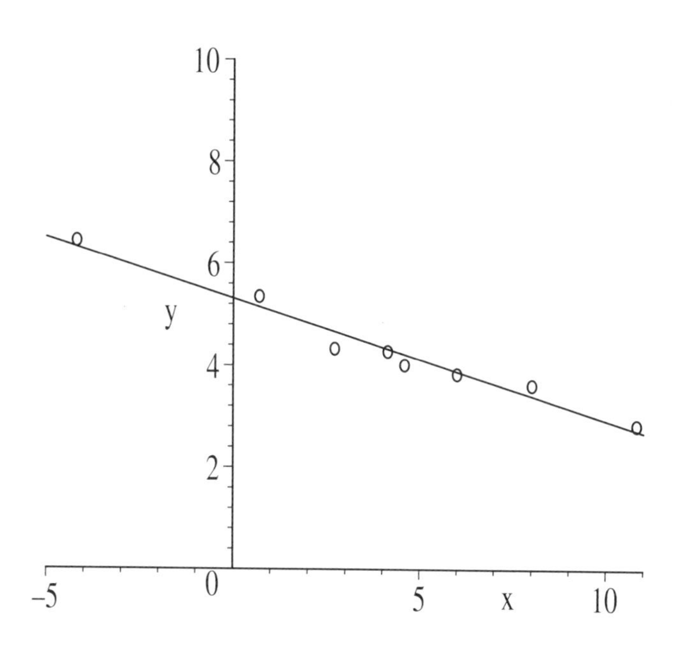

Figure 2.4: Best-fitting straight line for log-log data.

The relation between heart rate and body mass

```
> heartrate:=exp(a)*bodymass^b;
```

$$heartrate := \frac{202.19}{bodymass^{0.23807}}$$

is given by the inverse power law $r = 202/\, m^{0.24}$, which can be plotted.

```
> Gr3:=plot(heartrate,bodymass=0.01..55000,
        view=[0..55000,0..100]):
```

The original two lists for the heart rates and body mass are zipped together, plotted, and displayed in Figure 2.5 along with the empirical formula."

```
> pair:=(rate,mass)->[rate,mass];
```

$$pair := (rate,\, mass) \rightarrow [rate,\, mass]$$

```
> coords2:=zip(pair,mass,rate);
```

$$coords2 := [[.015,\, 624.0],\, [2.0,\, 210.0],\, [15.0,\, 76.0],\, [63.0,\, 72.0],\, [99.0,\, 55.0],\, [407.0,\, 46.0],\, [3000.0,\, 37.0],\, [50000.0,\, 17.0]]$$

```
> Gr4:=plot(coords2,view=[0..55000,0..100],style=point,
        symbol=box,symbolsize=12,color=blue):

> display({Gr3,Gr4},tickmarks=[3,4],labels=["m","r"]);
```

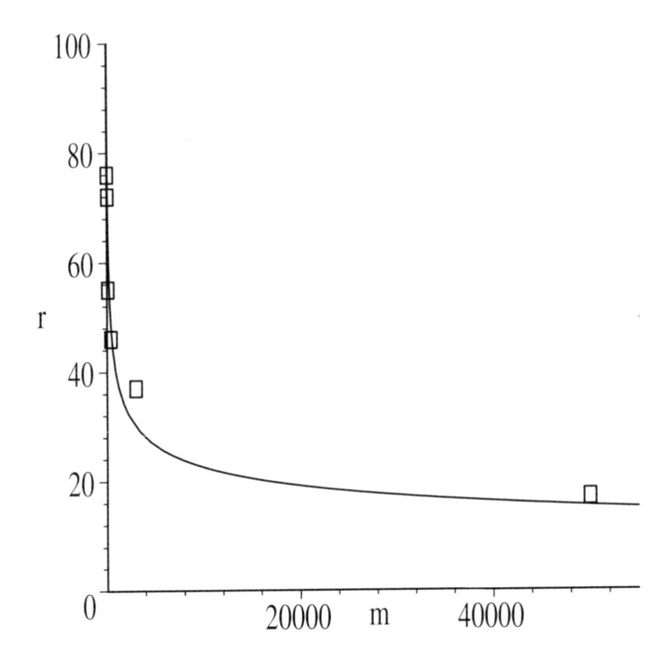

Figure 2.5: Best-fitting curve for the data of Table 2.8.

"I hate to interrupt, Jennifer, but your curve doesn't do a bad job of fitting the data, but you have used only eight data points, and there is clearly a large region in the figure where there are no data points at all. How good is this power law formula, given the limited data?"

"Actually, in Lin's paper 54 data points are used, and I could certainly get some more points for you. However, his quoted value for b is $b = -0.24 \pm 0.01$, so we haven't done too badly with only eight points."

"How did Lin calculate the error in the slope?"

"Did you read the general discussion on estimating error at the beginning of this section? You did. Good. We can calculate the standard deviation σ_b for the slope b in the best-fit straight line using

$$\sigma_b = \sqrt{N/\Delta}\,\sigma,$$

where

$$\Delta = N\,S_{xx} - (S_x)^2, \quad \sigma = \sqrt{(S_{yy} - a\,S_y - b\,S_{xy})/(N-2)},$$

with $S_x = \sum_i x_i$, $S_{xx} = \sum_i x_i^2$, etc.

In the present problem, the x_i are the `logmass[i]`, while the y_i are the `lograte[i]`. Evaluating all the relevant sums in the next few command lines,

```
> Sx:=sum(logmass[i],i=1..N): Sy:=sum(lograte[i],i=1..N):
```

```
> Sxx:=sum(logmass[i]^2,i=1..N): Syy:=sum(lograte[i]^2,i=1..N):
> Sxy:=sum(logmass[i]*lograte[i],i=1..N8): Delta:=N*Sxx-(Sx)^2:
> sigma_b:=sqrt(N/Delta)*sqrt((Syy-a*Sy-b*Sxy)/(N-2));
```

$$sigma\_b := 0.017511$$

we see that the standard deviation for b is about 0.018. Not surprisingly, this is slightly higher than Lin's calculated standard deviation, since we have used a lot fewer points. The standard deviation scales as $1/\sqrt{N}$. Using our result, we have $b = -0.238 \pm 0.018$."

"Jennifer, you mentioned earlier that we could use a scaling argument to support our model equation. How does the argument go for this example?"

"A simple scaling argument presented in Lin's paper predicts that $b = -1/4 = -0.25$, which is within one standard deviation of our b value. The scaling argument goes like this. Suppose that an animal has a body size characterized by the dimension or "length" L. The weight W of the animal then is proportional to the volume, i.e., $W \propto L^3$, since the density of all warm-blooded animals is about the same. To determine an animal's energy requirements, note that its metabolic heat production P must be balanced by heat loss through its surface. Since surface area is proportional to L^2, then $P \propto L^2 \propto W^{2/3}$. A more detailed argument due to McMahon [McM73] alters this mathematical relationship slightly, yielding $P \propto W^{3/4}$. This latter scaling relationship has been found to hold empirically over several decades of weight.

The heart acts as a pump, delivering with each heartbeat a given amount of blood into the circulatory system. The oxygen carried by the blood reaches the cell level, where it is used in the metabolic process. The rate R at which oxygen is delivered must be proportional to $P \propto W^{3/4}$. But $R \propto A/T_{\rm hb}$, where A is the amount of oxygen delivered in one heartbeat of duration $T_{\rm hb}$. Now, A is proportional to the volume of blood delivered in one heartbeat. The volume of blood delivered must be proportional to the volume of the heart. If the heart's volume is assumed to be proportional to body weight W, then

$$R \propto \frac{A}{T_{\rm hb}} \propto \frac{W}{T_{\rm hb}} \propto W^{3/4},$$

or solving for $T_{\rm hb}$, we have $T_{\rm hb} \propto W^{1/4}$. The heart rate is the reciprocal of $T_{\rm hb}$, so the heart rate satisfies $r \propto 1/W^{1/4} \propto m^{-1/4}$, where m is the mass."

"OK," Heather interjects, "now that we are fairly confident that our model equation is reasonable, I can see how to answer my original question. If we enter the estimated body mass of the brachiosaur,

```
> bodymass:=75000; #brachiosaur mass in kg
```

$$bodymass := 75000$$

its heart rate is readily calculated,

```
> heartrate;
```

$$13.969$$

and must have been about 14 beats per minute."

PROBLEMS:

Problem 2-10: Lifetime

As a rule of thumb, the larger the animal, the longer it lives. Table 2.9 gives the

Table 2.9: Lifetime L and body mass m for some warm-blooded animals.

	Mouse	Guinea Pig	Fox	Goat	Human	Gorilla	Elephant
m	0.02	0.26	3	34	63	190	3500
L	3.5	7.5	14	18	70	36	70

body mass (m in kilograms) and lifetime (L in years) of various animals [Lin82]. Making use of the best-fitting straight line to a log-log plot of the data, determine the empirical dependence of L on m. It is known that cardiac (heart) muscle can tolerate only a fixed number (about 100 million to 1 billion) of contractions for all warm-blooded animals. Assuming that heart failure is the major limiting factor for longevity, the lifetime = number of heart beats (fixed) $\times$ period of one heart beat. So, theoretically, $L \propto m^{1/4}$. How does the exponent in this theoretical prediction compare with the exponent obtained from the data? Discuss reasons for any discrepancy.

Problem 2-11: Kepler's third law

Table 2.10 shows the semimajor axis a, expressed in millions of kilometers, and the period T, in years, for the planets of our solar system [Oha85]. Use these data to verify Kepler's third law, which states that the square of the period is proportional to the cube of the semimajor axis of the planetary orbit. Plot the given data and the best-fitting curve in the same graph.

Table 2.10: Semimajor axis a and period T for the planets of our solar system.

Planet	$a(10^6\,\text{km})$	T**(years)**	**Planet**	$a(10^6\,\text{km})$	T**(years)**
Mercury	57.9	0.241	Saturn	1430	29.5
Venus	108	0.615	Uranus	2870	84.0
Earth	150	1.00	Neptune	4500	165
Mars	228	1.88	Pluto	5890	248
Jupiter	778	11.9			

Problem 2-12: Chimpanzee brain volumes

Within a given species of mammal, it is found that the brain volume V varies with the body mass m according to a power law $V = a\,m^b$, where a and b are constants. Table 1.6 shows the brain volumes for a number of adult chimpanzees. Using the best-fit procedure, determine a and b and determine the standard deviations for each parameter.

Problem 2-13: Olympic 100-meter times

Table 2.11 shows the winning times in seconds for the 100-meter run at the Olympic games, (S.A. = South Africa; G.B. = Great Britain; Tr. = Trinidad).

Table 2.11: Olympic 100 meter times.

Year	**Winner**	**Time**	**Year**	**Winner**	**Time**
1896	T. Burke, U.S.	12.0	1948	H. Dillard, U.S.	10.3
1900	F. Jarvis, U.S.	10.8	1952	L. Remigino, U.S.	10.4
1904	A. Hahn, U.S.	11.0	1956	B. Morrow, U.S.	10.5
1908	R. Walker, S.A.	10.8	1960	A. Hary, Germany	10.2
1912	R. Craig, U.S.	10.8	1964	B. Hayes, U.S.	10.0
1920	C. Paddock, U.S.	10.8	1968	J. Hines, U.S.	9.9
1924	H. Abrahams, G.B.	10.6	1972	V. Borzov, U.S.S.R.	10.14
1928	P. Williams, Canada	10.8	1976	H. Crawford, Tr.	10.06
1932	E. Tolan, U.S.	10.3	1980	A. Wells, G.B.	10.25
1936	J. Owens, U.S.	10.3	1984	C. Lewis, U.S.	9.99

(a) Make a log-log plot of the data and find the best-fitting straight line.

(b) Use this result to obtain a model equation for the winning time as a function of Olympic year.

(c) What is the predicted winning time for the 100 meter run at the 1996 Atlanta Olympic games? How does your prediction compare with the actual winning time of 9.84 seconds posted by Donovan Bailey of Canada?

Problem 2-14: Long-distance running

Table 2.12 shows the world record times T, expressed in hours (h), minutes (m), and seconds (s), for various long-distance runs (D km) as of 1983. Calculate the

Table 2.12: World record times for long-distance runs.

D	2	3	5	10	20	25	30	42.195
h						1	1	2
m	4	7	13	27	57	13	29	08
s	51.40	32.10	00.41	22.40	24.20	55.80	18.80	13.00

average velocity V in meters/second for each run. According to Strnad [Str85], the average velocity should be given by a formula of the structure $V = V_1/T^n$, where T is in seconds and V_1, n are positive constants. Make a log-log plot and use the best-fit procedure to determine the values of the two constants. Calculate the standard deviations for each constant.

2.2.3 Senate Renewal

Practical politics consists of ignoring facts.
Henry Adams, *The Education of Henry Adams* (1838–1918)

In 1954, Strom Thurmond, of South Carolina, was elected for the first time to the U.S. Senate and began serving his six-year term in 1955. There were 96 senators in this congress, including future presidents John Kennedy and Lyndon Johnson, and presidential candidates Hubert Humphrey and Barry Goldwater. Of those 96, the number still serving (i.e., those who were reelected) one, two, and three terms later is given in Table 2.13. Can one use this admittedly limited data to estimate how many of the original 96 senators served in subsequent congresses, for example, in the congress beginning in 1997?

Table 2.13: Number of 1955 senators serving in subsequent congresses.

Year	1955	1961	1967	1973
Number	96	64	42	25

Political scientists tend to favor an *exponential model* to describe legislative turnover [LFHC95]. If there are N senators after an election, then it is assumed that there are $S = N\,e^{-k\,t}$ of those senators still serving t terms later, with k a positive constant. For the congress beginning in 1997, $t = 7$.

Although we could set $N = 96$ and determine k by trial and error using the remaining three data points, a better way is to use a semilog plot for the data and use all four data points to determine the values of both N and k that give the best fit. To see why a semilog plot is used, take the natural log of S so that $y = A - k\,t$, where $y \equiv \ln S$ and $A \equiv \ln N$. Thus, once again we are looking for the best-fitting straight line to determine the coefficients A and k. Once these parameters are found, then $S = e^{A}\,e^{-k\,t}$.

As in the previous examples, the Statistics and plots packages are loaded.

```
> restart: with(Statistics): with(plots):
```

Separate Maple lists are entered for the year and number and a new list created for the log of the number. All of the number entries are converted to floating-point numbers, by adding decimal zero to them, so that the logs are evaluated.

```
> year:=[0,1,2,3]:
> number:=[96.0,64.0,42.0,25.0]:
> number_log:=[seq(ln(number[i]),i=1..4)];
```

$$number\_log := [4.564348191,\ 4.158883083,\ 3.737669618,\ 3.218875825]$$

The year and number_log are zipped together into a list of lists.

```
> pair:=(year,number_log)->[year,number_log]:
> coordinates:=zip(pair,year,number_log):
```

A graph of the `coordinates` is formed but not displayed.

```
> Gr1:=plot(coordinates,view=[0..4,3..5],style=point,
        symbol=circle,color=blue,symbolsize=16):
```

A linear least squares fit is sought to the *year* and *number_log* data.

```
> y:=Fit(A-k*t,year,number_log,t);
```

$$y := 4.58858876370000068 - 0.445763056300000404\, t$$

A plot of the best-fitting straight line, y, is created

```
> Gr2:=plot(y,t=0..4):
```

and displayed in Figure 2.6 along with a semilog plot of the input data.

```
> display({Gr1,Gr2},view=[0..4,2.5..5],
  tickmarks=[2,3],labels=["year","log(n)"]);
```

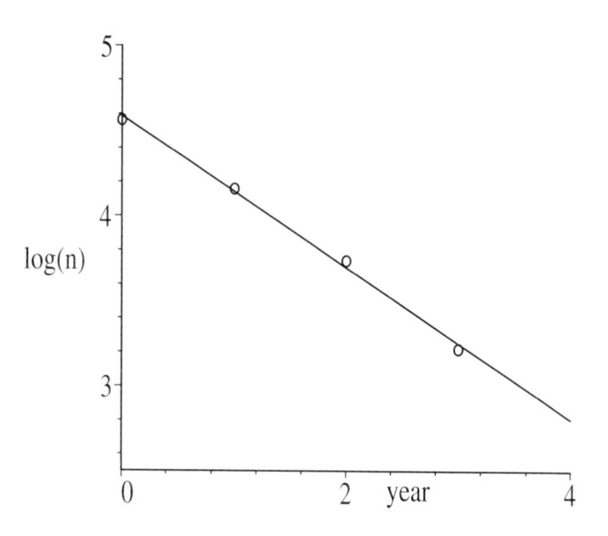

Figure 2.6: Best-fitting straight line for a semilog plot of the input data.

The coefficients A and k are extracted from y.

```
> A:=coeff(y,t,0); k:=-coeff(y,t);
```

$$A := 4.58858876370000068 \qquad k := 0.445763056300000404$$

Forming $S = e^{A}\, e^{-k\,t}$, we see that the number of 1955 senators still serving t terms later is given by

```
> S:=exp(A)*exp(-k*t);
```

$$S := 98.35552931\, e^{(-0.445763056300000404\, t)}$$

Actually, Maple has a built-in algorithm (`ExponentialFit`) for determining the best exponential fit to the original year and number data, viz.,

```
> S2:=ExponentialFit(year,number,t);
```

$$S2 := 98.355529332525\, e^{(-0.445763056487510190\, t)}$$

which is the same as obtained in our "first principles" calculation. Of the original 96 senators, the number still serving in the 1997 Congress is obtained by setting $t = 7$ in either S or $S2$ and rounding off to the nearest whole senator.

```
> t:=7: Number[1997]:=round(S);
```

$$Number_{1997} := 4$$

So, the exponential model equation predicts that approximately four of the original 1955 senators should be in the 1997 Congress. In actuality, only one of the original 96 remained as of 1998, namely Strom Thurmond. The exponential model overestimated the number somewhat, but considering the sparsity of the data and the fact that 42 years had elapsed, the estimate was not too bad.

PROBLEMS:

Problem 2-15: Nikita Khrushchev's Secret Purge

In 1956 a total of 133 members were elected by the Party Congress to the Central Committee of the Communist Party of the U.S.S.R. The number of these who were reelected in 1961, 1966, and 1971 are given [LFHC95] in Table 2.14.

Table 2.14: Number of 1956 Central Committee members reelected.

Year	1956	1961	1966	1971
Number	133	66	54	35

However, in 1957 Nikita Khrushchev secretly purged some of his opponents from the Central Committee. The number who were purged is not known. Assuming that an exponential model approximately fits the last three data points, estimate the number who were purged. State any assumptions.

Problem 2-16: Blood alcohol level

A college student, celebrating the end of final exams, consumes a substantial amount of whiskey. His blood alcohol level rises to 0.22 mg/ml and then slowly decreases as indicated [AL79] in Table 2.15. The elapsed time is in hours.

Table 2.15: Blood alcohol level as a function of time.

Time	0	0.5	0.75	1.0	1.5	2.0	2.5	3.0
Alcohol level	0.22	0.18	0.15	0.13	0.10	0.08	0.06	0.05

Assuming that the data can be approximated by an exponential model, find the best-fitting curve and plot it along with the original data. What is the blood alcohol level after 4 hours have elapsed?

Problem 2-17: Effect of a bactericide

A bactericide is added to a solution containing 10^7 bacteria. The number of bacteria remaining at various elapsed times is given in Table 2.16. Assuming that the data can be modeled with an exponential curve, find the best-fitting function and plot it along with the original data. How many minutes after administering the bactericide is the bacterial count down to 70?

Table 2.16: Bacterial count (in thousands) as a function of time (in minutes).

Time	0	10	20	30	40	50	60	70	80
Count	10000	3200	1000	320	100	32	10	3.2	1

Problem 2-18: Intel processor chips
In 1965, Intel cofounder Gordon Moore predicted (known as Moore's law) that the number of transistors on a chip would double about every two years. The number of transistors in various Intel processor chips and the date of their appearance is given in Table 2.17. (U.S.A. TODAY, Feb. 17, 1995.)

Table 2.17: Number of transistors (in millions) in Intel chips.

Year	1971	1986	1989	1993	1995
Processor chip	4004	386DX	486DX	Pentium	P6
Number	0.0023	0.275	1.2	3.3	5.5

Determine the best-fitting exponential curve and then use the model equation to determine the annual percentage increase in the number of transistors used in an Intel chip. Comment on the applicability of Moore's law.

2.2.4 Bikini Sales and the Logistic Curve

Everything you see, I owe to spaghetti.
Sophia Loren, Italian film actress (1934–)

Colleen, the manager of the ladies' leisure section of the Glitz department store, ordered 800 bikini swimsuits for the current year. At the end of September, she looks over her records of cumulative sales numbers for the first 9 months of the year, which are reproduced in Table 2.18. She notes that only four bikinis were sold by the end of the first month (by January 31), 12 bikinis sold by the end of February (second month), and so on. Observing that a total of 769 bikinis

Table 2.18: Number of bikinis sold in the first nine months of the year.

Month	1	2	3	4	5	6	7	8	9
Number	4	12	25	58	230	439	648	748	769

had been sold by the end of September, Colleen wonders whether she will sell all of this year's model of bikini or will have some swimsuits left over at the end of the year. Having taken a statistics course in college, Colleen decides to develop a model equation that can be used to predict the total sales by the end of December (month 12).

Using the `ScatterPlot` command, Colleen graphs the sales data,

```
> restart: with(Statistics):
> month:=[1,2,3,4,5,6,7,8,9];
> number:=[4,12,25,58,230,439,648,748,769];
> graph1:=ScatterPlot(month,number,symbol=box,symbolsize=14,
          labels=["month","number"],tickmarks=[6,2]):
> graph1;
```

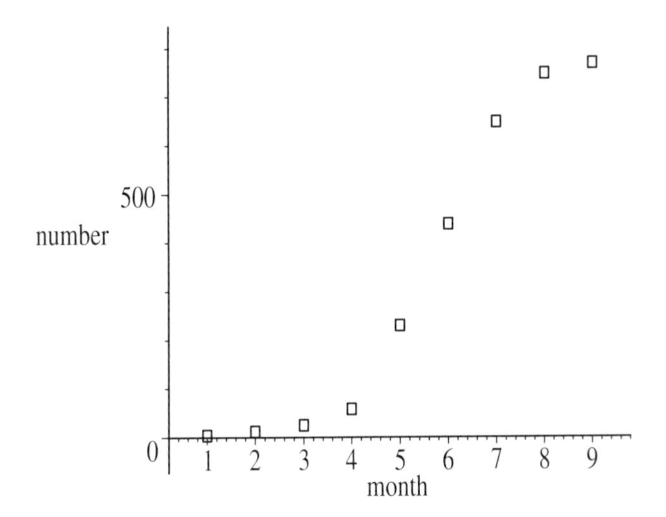

Figure 2.7: Cumulative number of bikinis sold at the end of months 1 to 9.

which is shown in Figure 2.7. The number of sales rises slowly in January and February, takes off in the summer months, and begins to plateau in the fall. Colleen recalls that such data can be modeled by the *logistic* curve, which is described by the mathematical form

$$n(m) = \frac{a}{1 + b\,e^{-c\,m}}, \tag{2.9}$$

where $n(m)$ is the number of bikinis sold by the end of month m and the parameters a, b, and c have to be adjusted to give the best fit to the data. As m increases, the number n will become closer and closer to the constant a. Since a fixes the height of the plateau, Colleen anticipates that a will have a value greater than the total sales number of 769 for the first 9 months.

She could use a linear least squares approach similar to that in the earlier recipes. To this end, she could rewrite the logistic equation (2.9) in the form

$$b\,e^{-c\,m} = (a/n(m)) - 1. \qquad (2.10)$$

Then, taking the natural logarithm of the above equation yields

$$\ln(b) - c\,m = \ln\left((a/n(m)) - 1\right) \equiv N(a, m). \qquad (2.11)$$

The lhs of Equation (2.11) is a straight line with slope $-c$ and intercept $\ln(b)$. Thus, if $N(a, m)$ is plotted versus m for a given value of a, the linear least squares method will yield c and b. The parameter a is chosen to minimize the total error.

Guided by the Senate example, Colleen wonders whether she could bypass the above linearized approach and find the a, b, and c values simultaneously that minimize the total error $\sum_{i=1}^{N}(y_i - n_i)^2$, where N is the number of data points, y_i the y coordinate of the ith data point, and $n_i \equiv n(m_i)$, where m_i is the ith month. Since n is a nonlinear function of the parameters a, b, and c, a nonlinear fitting algorithm is required. Consulting the Statistics library package, Colleen finds that the `NonlinearFit` command is available. Colleen now uses this command, supplying search ranges for the parameters.

```
> f:=NonlinearFit(a/(1+b*exp(-c*m)),month,number,m,
    parameterranges=[a=769..2000,b=500..2000,c=1..2]);
```

$$f := \frac{786.704443551278814}{1 + 1464.70166905573888\,e^{(-1.26109661752213742\,m)}}$$

The best-fitting function f is plotted and superimposed on the data,

```
> graph2:=plot(f,m=0..12,thickness=2):
> plots[display]({graph1,graph2});
```

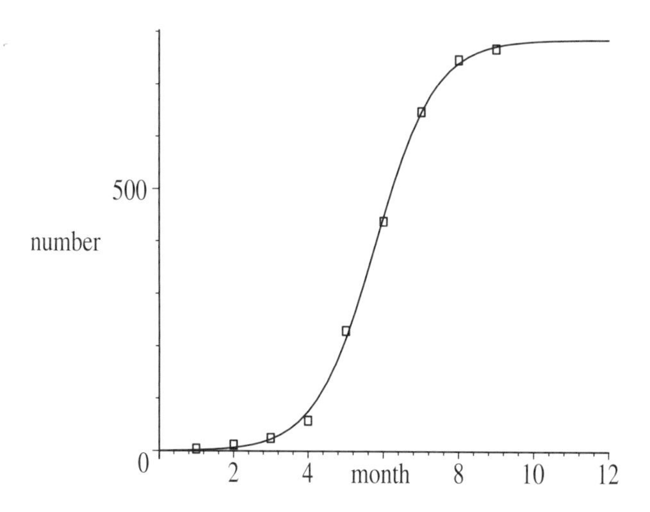

Figure 2.8: Best-fit logistic curve and bikini sales numbers.

the resulting picture being shown in Figure 2.8. It should be noted that if the search ranges had not been specified in the `NonlinearFit` command, the fit of the logistic curve to the bikini sales numbers would be quite poor.

Now that Colleen has her model equation, she can apply it to estimating how many swimsuits will be left over at the end of December. The projected number (rounded to the nearest whole swimsuit)

```
> projected_number:=round(eval(f,m=12));
```

$$projected\_number := 786$$

is approximately 786 bikinis sold for the year. Since Colleen originally purchased 800 bikinis, there will be about 14 swimsuits unsold at the end of the year.

PROBLEMS:

Problem 2-19: Linearized approach

Carry out the linearized approach discussed in the text and compare the result with that obtained using the `NonlinearFit` command. Discuss the goodness of fit and the projected number of unsold swimsuits.

Problem 2-20: Polio epidemic

Table 2.19 shows the cumulative number (N) of polio cases diagnosed each month (M) in the U.S. polio epidemic of 1949, the second worst in that country's history (National Foundation for Infantile Paralysis, *12th Annual Report*, 1949).

Table 2.19: Cumulative number of polio cases in 1949.

M	January	February	March	April	May	June
N	494	759	1016	1215	1619	2964
M	July	August	September	October	November	December
N	8489	22377	32618	38153	41462	42375

(a) Plot the data to see whether they suggest trying a logistic curve fit.

(b) Determine the best-fitting logistic curve and plot it along with the data.

(c) Comment on the goodness of fit and offer some plausible reasons for any deviations of the data from the curve.

Problem 2-21: The great flu epidemic, revisited

Table 1.3 gives death statistics for the great flu epidemic of 1918:

(a) Find the best-fitting logistic equation for each of the navy, army, and civilian deaths due to the flu.

(b) Plot each best-fitting logistic curve on the same graph as the data.

(c) Which set of data appears to be best-fitted by a logistic curve? Which set has the worst fit? Can you offer any plausible explanation for the difference in goodness of fit?

2.2.5 Following the Dow Jones Index

Money is better than poverty, if only for financial reasons.
Woody Allen, American writer, comedian, actor, and film director (1935–)

The Dow Jones index is of great interest to stock market investors. This industrial average is quoted in the business section of most daily newspapers. The "Dow," as it is usually called, is one of the most widely recognized stock market indicators in the world. The Dow tracks the performance of 30 very large "blue-chip" companies (e.g., Boeing, Disney, G.E., IBM, Microsoft), and it is felt that changes in the value of the Dow mirror the general state of the United States, and, to some degree, the global economies. The Dow has been in existence for over 100 years, and the daily Dow Jones value is readily available from the Internet for the period 1900 onward. To obtain a model curve, we shall work with the Dow at year's end, i.e., on the last trading day of the year, for the period 1900 to 1990. The year 1900 will be labeled as 0, the year 1901 as 1, and so on.

```
> restart: with(plots): with(Statistics):
```

The sequence command is used to enter the 91 years,

```
> year:=[seq(n,n=0..90)]:
```

and the Dow index values (obtained from the Internet) inputted.

```
> Dow:=[70.71,64.56,64.29,49.11,69.61,96.20,94.35,
    58.75,86.15,99.05,81.36,81.68,87.87,78.78,54.58,
    99.15,95.00,74.38,82.20,107.23,71.95,81.10,98.73,
    95.52,120.5,156.66,157.20,202.40,300.01,248.48,164.58,
    77.90,59.93,99.90,104.64,144.13,179.90,120.85,154.76,
    150.24,131.13,110.96,119.40,135.89,152.32,192.91,177.20,
    181.16,177.30,200.13,235.41,269.23,291.90,280.90,404.39,
    488.40,499.47,435.69,583.65,679.36,615.89,731.14,652.10,
    762.95,874.13,969.26,785.69,905.11,943.75,800.36,838.92,
    890.20,1020.02,850.86,616.24,852.41,1004.65,831.17,
    805.01,838.74,963.99,875.00,1046.54,1258.64,1211.57,
    1546.67,1895.95,1938.83,2168.57,2753.20,2633.66]:
```

It is important to realize that the Dow values given above are not corrected for inflation. Inflation-corrected values are available in graphical form from the Internet. Trusting that the numbers have been inputted correctly, the number of operands (`nops`) command checks that there are 91 entries in the Dow list.

```
> N:=nops(Dow);
```

$$N := 91$$

A graph of the Dow index as a function of year is created using the `ScatterPlot` command, the data points being represented by size-12 circles.

```
> pts:=ScatterPlot(year,Dow,symbol=circle,symbolsize=12,
    view=[0..91,0..3500],labels=["year","Dow"],
    tickmarks=[3,3]):
```

The data points are colored blue and displayed in Figure 2.9.

```
> display(pts,color=blue);
```

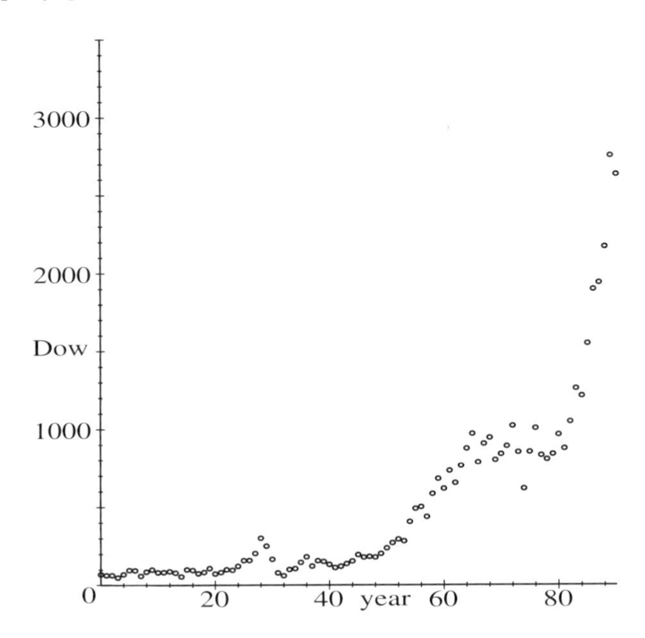

Figure 2.9: Dow Jones index at year's end for the period 1900–1990.

From the figure, it is clear that the data cannot be fitted by a single straight line, but has an overall trend characterized by a curve with low slope (on the scale of the plot) for the first 30–40 years and then rising much more quickly in recent times. Superimposed on the general trend are fluctuations that are of great importance to short-term investors or those who cannot leave their money in stocks for a long period of time. Can you spot the infamous stock market crash of 1929?

The mechanisms giving rise to the overall trend and the fluctuations are the subject of many so-called theories about stock market behavior. Delving into these theories is beyond the scope of this text or the expertise of the authors. However, in the spirit of this chapter, it is possible to find a best-fitting polynomial model equation that does a good job of fitting the 1900–1990 data trend and successfully predicting the Dow for several following years.

A functional operator, labeled `eq`, is now formed to obtain the best-fitting nth-order polynomial equation to the Dow Jones data. The nth-order polynomial $\sum_{i=0}^{n} a_i\, y^i$ is created using the `add` command.

```
> eq:=n->Fit(add(a[i]*y^i,i=0..n),year,Dow,y):
```

Using `eq` in the following sequence command, best-fitting polynomial equations are generated for $n = 2$, 4, and 6.

```
> seq(p[2*i]=eq(2*i),i=1..3);
```

$$p_2 = 239.691790299461274 - 18.7401940528387492\,y + .404515215547016750y^2,$$
$$p_4 = 178.942577528342412 - 28.2568098375523143\,y + 1.65050110494074832y^2$$
$$- 0.0309520539296583228\,y^3 + 0.000210617039324588283\,y^4,$$
$$p_6 = 189.147571067008556 - 81.5765465285548858\,y + 11.5240779259858410\,y^2$$
$$- 0.603890924929016193\,y^3 + 0.0143426153049926175\,y^4$$
$$- 0.000155412644461251410\,y^5 + 0.627688870349042276\,10^{-6}\,y^6$$

To see how well the second-, fourth-, and sixth-order polynomials fit the data, we will create graphs of the above polynomial equations. A functional operator, assigned the name `Graph`, is formed to plot the nth-order polynomial for the period 1900–2000. Entering `Graph(4)`, for example, will produce a plot of the quartic polynomial.

```
> Graph:=n->plot(eq(n),y=0..100,thickness=2):
```

To test the predictive power of the various polynomials, the Dow Jones index values for the years 1991 to 1999 are plotted as size-12 green boxes using the `pointplot` command. The graph is assigned the name `futurepts`.

```
> futurepts:=pointplot([[91,3168.83],[92,3301.11],[93,3754.09],
      [94,3834.44],[95,5117.12],[96,6448.27],[97,7908.25],
      [98,9181.43],[99,11497.12]],color=green,symbol=box,
      symbolsize=12):
```

A "do loop" will be used to display the superposition of the nth-order polynomial on the "past" and "future" data points for $n = 2$, 4, and 6.

The general syntax for a do loop is

for <name> from <expression> by <expression> to <expression>
while <expression> do <statement sequence> end do

where the <*statement sequence*> is the main body of the do loop and the loop ends with *end do*. In the following do loop, <*name*> is the index n, the first <*expression*> is 2, the second <*expression*> is also 2, the third <*expression*> is 6, and there is no conditional *while* <*expression*> present. On executing the do loop, pictures are generated for $n = 2$, 4, 6.

```
> for n from 2 by 2 to 6 do
> display({Graph(n),pts,futurepts},
  view=[0..100,0...12000],labels=["year","Dow"]);
> end do;
```

And the winner is? Of the three curves, the sixth-order polynomial shown in Figure 2.10 does the best job of fitting the 1900–1990 data (circles) and predicting the "future" points (square boxes) for 1991 to 1999. The $n = 2$ and 4 curves can be viewed on the computer screen. The do loop can be altered to observe the polynomial fits for other values of n. Note that if the *by* <*expression*> is missing in the do loop, the default is to increment n by 1 each time the <*statement sequence*> is executed.

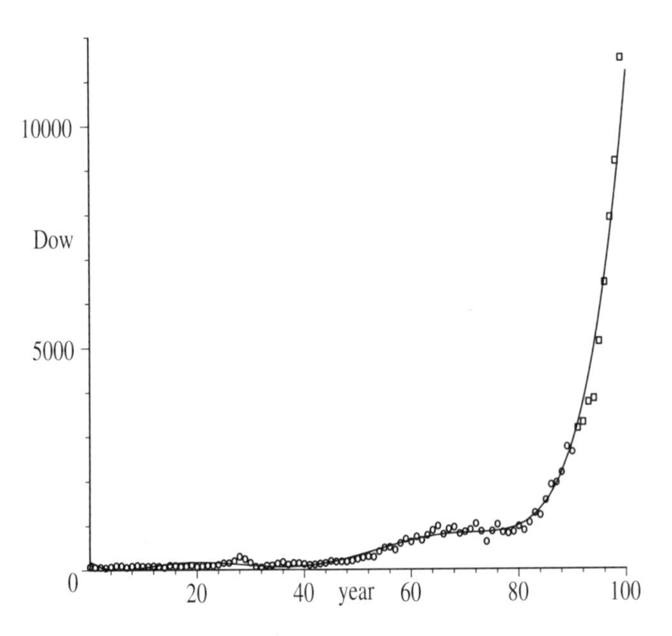

Figure 2.10: Best-fitting polynomial curve for $n = 6$.

A standard approach to deciding which polynomial fits best, other than simply looking at the plot, is to generalize the formula for calculating the standard deviation σ. Suppose that there are N data points, and the proposed best-fitting curve is the nth order polynomial $Y = a_0 + a_1 x + a_2 x^2 + \cdots + a_n x^n$. The polynomial will then have $n+1$ fitting parameters, or coefficients, $a_0, a_1, \ldots, a_n$. If y_i and Y_i are the y coordinates of the data point and the polynomial curve at the same x_i, respectively, the standard deviation becomes

$$\sigma = \sum_{i=1}^{N} \sqrt{\frac{\chi^2}{N-(n+1)}} = \sum_{i=1}^{N} \sqrt{\frac{(y_i - Y_i)^2}{N-n-1}}. \tag{2.12}$$

If $Y = a_0 + a_1 x$, i.e., a straight line, then $n + 1 = 2$ and σ reduces to the form introduced at the beginning of this section. The reasoning behind dividing through by the factor $N - (n + 1)$ in the square root is the same as before. In fitting a polynomial with $n + 1$ coefficients, only $N - (n + 1)$ data points are really independent of the fitting procedure. Another way to look at it is to note that the higher the order of the polynomial, the easier it is to fit a fixed number of data points. Dividing by $N - n - 1$ is a way of evening out this advantage.

A functional operator s is formed to calculate the standard deviation σ given by Equation (2.12) for the best-fitting polynomial of order n.

```
> s:=n->evalf(add(sqrt((Dow[i]-subs(y=year[i],eq(n)))^2),
        i=1..N)/sqrt(N-n-1)):
```

Employing s, the sequence command is used to generate σ_2, σ_4, and σ_6.

```
> seq(sigma[2*i]=s(2*i),i=1..3);
```

$$\sigma_2 = 1296.991659, \; \sigma_4 = 1267.886913, \; \sigma_6 = 678.7838616$$

As expected, the sixth-order polynomial has the lowest value of σ of the three polynomial curves. It also does the best job of fitting the "future" points for the period 1991–1999.

However, a strong word of caution is in order! Using model equations to predict the future growth of the stock market could be dangerous to your economic health. Extrapolation is a dangerous game, particularly if one uses a high-order polynomial to make a long-range prediction. As you can easily check by running the Maple file, the sixth-order polynomial just keeps on increasing as time evolves. As has been demonstrated over and over again (e.g., the stock market crash of 1929), unchecked growth is not the way the world works. Indeed, more recently the "bubble" burst in the year 2000, the year-end Dow value dropping to 8341.63 by 2002, before beginning a slow upward trend. The drop in the value of the tech stocks was particularly bad, with some sectors dropping by 90 percent, with very little improvement by the year 2004.

"Aha," the reader might say. "If I can wait out the short-term fluctuations, the trend in the Dow according to Figure 2.9 has always been historically upward." An important ingredient has been left out of the analysis, namely the effect of inflation. Go to the Internet and look at the inflation-adjusted Dow graph and you will see that there are actually large-amplitude cycles of about 30 years duration, a long time to wait if you invest at the beginning of a downturn.

PROBLEMS:

Problem 2-22: Black Friday

Take the years 1900 to 1925 as the "present" points and the years 1926 to 1932 as the "future" points. Fit various polynomial curves to the present data and see whether any curves predict the precipitous stock market crash of 1929 to 1932. Do a historical search as to the meaning of the phrase Black Friday in connection with the crash.

Problem 2-23: Are you going to bet all your money on this horse?

At year's end, the Dow was 10786.85, 10021.50, 8341.63, 10453.92, 10783.01 for the years 2000 to 2004. Add these data points to the future points in the text recipe and execute the code. How well does the sixth-order polynomial model account for these data points? Experiment with other polynomial models.

Problem 2-24: Cleveland's population

Table 2.20 gives the population statistics for the city of Cleveland, Ohio, for the period 1900 to 1980.

(a) Taking 1900 to be year zero, determine the best-fitting quadratic (parabolic) curve and plot it on the same graph as the data points.

(b) What is the projected population for Cleveland in 1990? Go to the Internet and find out what Cleveland's actual population was in 1990. Compare the two numbers and comment on the accuracy of the prediction.

Table 2.20: Population of Cleveland from 1900 to 1980.

Year	1900	1910	1920	1930	1940
Population	381,768	560,663	796,841	900,429	878,336
Year	1950	1960	1970	1980	1990
Population	914,808	876,050	750,879	573,822	?

Problem 2-25: Natural gas prices
Table 2.21 shows the average price in dollars per thousand cubic feet of natural gas for household use in the United States from 1980 to 1990.

Table 2.21: Natural gas prices.

Year	1980	1981	1982	1983	1984	1985
Price	3.68	4.29	5.17	6.06	6.12	6.12
Year	1986	1987	1988	1989	1990	
Price	5.83	5.54	5.47	5.64	5.77	

(a) Taking 1980 as year zero, find the best-fitting cubic equation to the data and plot it on the same graph as the data.

(b) Compare the fit with that of a fifth-order polynomial. In particular, how do the sigma values compare?

Problem 2-26: Imported cars
Table 2.22 gives the number (in thousands) of imported cars sold in the United States from 1984 to 1992.

Table 2.22: Number of imported cars sold in the United States.

Year	1984	1985	1986	1987	1988	1989	1990	1991	1992
Number	2439	2838	3245	3196	3004	2699	2403	2038	1938

(a) Plot the data and qualitatively decide which curve fits best: linear, quadratic, or cubic. Take 1984 to be year zero.

(b) Confirm your guess by finding the best-fitting linear, quadratic, and cubic equations and calculating the corresponding sigma values.

(c) Plot all three curves along with the data points.

(d) For the best model equation, how many imported cars are predicted to be sold in 1994?

Problem 2-27: Lung cancer death rates
Table 2.23 shows the death rates (deaths per 100,000 males) due to lung cancer every decade from 1930 to 1990.

Table 2.23: Death rates due to lung cancer. [SAU94]

Year	1930	1940	1950	1960	1970	1980	1990
Death rate	5	11	21	39	59	66	67

(a) By plotting the data, decide on qualitative grounds which model, linear, polynomial, exponential, or logistic, is most appropriate for the data.

(b) Based on your choice, obtain the best-fitting model equation.

(c) Use your model equation to predict the death rate in the year 2000.

(d) If possible, compare your prediction with the actual death rate. Comment on how good the prediction is.

Problem 2-28: World mile records
In the time interval since Roger Bannister of Great Britain broke the 4-minute barrier for the mile run with a time of 3 minutes, 59.4 seconds, on May 6, 1954, the world mile record has been lowered progressively as indicated in Table 2.24.

Table 2.24: Progressive lowering of world record times for the mile run.

Year	**Runner**	**Time**	**Year**	**Runner**	**Time**
1954	J. Landy (N.Z.)	3:58.0	1975	J. Walker (N.Z.)	3:49.4
1957	D. Ibbotson (G.B.)	3:57.2	1979	S. Coe (G.B.)	3:49.0
1958	H. Elliot (A.)	3:54.5	1980	S. Ovett (G.B.)	3:48.8
1962	P. Snell (N.Z.)	3:54.4	1981	S. Coe (G..B.)	3:48.53
1964	P. Snell (N.Z.)	3:54.1	1981	S. Ovett (G.B.)	3:48.40
1965	M. Jazy (F.)	3:53.6	1981	S. Coe (G.B.)	3:47.33
1966	J. Ryan (U.S.A.)	3:51.3	1985	S. Cram (G.B.)	3:46.32
1967	J. Ryan (U.S.A.)	3:51.1	1993	N. Morcelli (Al.)	3:44.39
1975	F. Bayi (T.)	3:51.0	1999	H. El Guerrouj (M.)	3:43.13

Develop a model equation from these data and estimate in what year a runner will break the 3:40.00 minute barrier. Note that the times after 1980 were measured to one-hundredths of a second. The country of each runner is indicated: N.Z. = New Zealand, G.B = Britain, A = Australia, F = France, T = Tanzania, Al = Algeria, M = Morocco.

Problem 2-29: Boiling point temperature

Table 2.25 shows the variation of the boiling-point temperature (in degrees Celsius) of water as the atmospheric pressure (in 10^{-2} atmospheres) changes. Derive the best-fit equation and plot it on the same graph as the data.

Table 2.25: Boiling-point data.

Temperature	0	10	20	30	40	50	60
Pressure	.605	1.21	2.30	4.30	7.30	12.2	19.7
Temperature	70	80	90	100	110	120	130
Pressure	31.2	46.7	69.2	100	143	196	267

Problem 2-30: Brain weight

Table 2.26 lists the average percentage weight of the brain in human males as a percentage of body weight from birth to age 16.

Table 2.26: Brain weight data.

Age	0	2	4	6	8	10	12	14	16
% Brain Weight	11	8	7	6	5	4.5	4	3.5	3.25

(a) Fit quadratic and exponential models to the data. Which model yields the better fit?

(b) Use the better-fitting model to determine the age in months at which the brain weight is 10% of the body weight.

(c) What is the predicted percentage brain weight in an average 18-year-old male?

2.2.6 Variation of "g" with Latitude

Where in this small-talking world can I find
A latitude* with no platitude?
Christopher Fry, English dramatist (1907–). *longitude in original quote.

In this section, Vectoria will look at an elementary example from the world of physics and show a new wrinkle to finding the best-fitting curve to the data. Instead of assuming a straight line or a polynomial to fit the data, she will consider a functional form guided by simple theoretical analysis.

```
> restart: with(Statistics): with(plots):
```

Neglecting the earth's rotational motion, the acceleration a_1 of an object of mass m near the earth's (mass M_e) surface due to gravitational attraction is determined by inserting the gravitational force law into Newton's second law:

$$m\,a_1 = \frac{G\,m\,M_e}{R_e^2}. \tag{2.13}$$

Here $G = 6.67 \times 10^{-11}$ N $\cdot$ m^2/kg^2 is the gravitational constant and R_e is the earth's radius. The earth's mass is $M_e = 5.98 \times 10^{24}$ kg, and Vectoria assumes that the earth is nearly spherical with a mean radius $R_e = 6.37 \times 10^6$ meters.

```
> G:=6.67*10^(-11): M[e]:=5.98*10^(24): R[e]:=6.37*10^6:
```

Canceling out m in Eq. (2.13), the acceleration in m/s^2 due to gravity then is

```
> a[1]:=G*M[e]/R[e]^2;
```

$$a_1 := 9.829878576$$

Vectoria now includes the effect of the earth's rotation, noting that the earth rotates about its axis with a period of approximately 23 hours, 56 minutes, or

```
> T:=(23+56/60)*60*60;
```

$$T := 86160$$

86,160 seconds. Letting θ be the latitude with $\theta = 0$ at the equator and 90° or $\pi/2$ radians at the North Pole, a point at the equator moves in a circle of radius R_e, while at latitude θ the radius (see Fig. 2.11) of the circle is $r = R_e \cos(\theta)$.

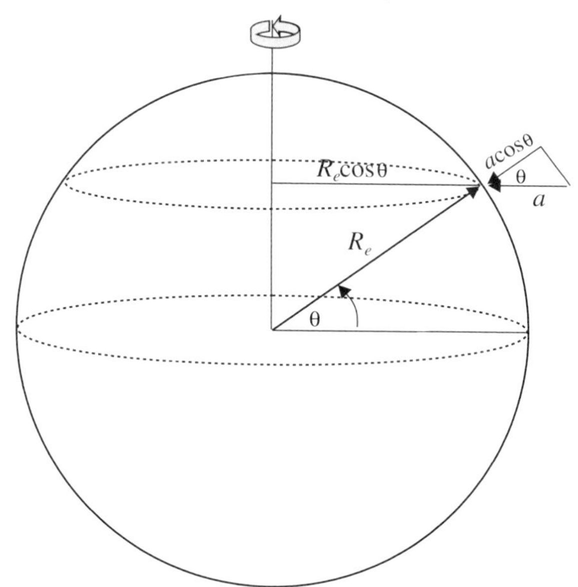

Figure 2.11: Contribution to "g" due to the earth's rotation.

Since $v = 2\,\pi\,r/T$ is the speed, the corresponding acceleration is

$$a = \frac{v^2}{r} = \frac{4\,\pi^2\,R_e \cos(\theta)}{T^2}. \tag{2.14}$$

Thus, referring to Figure 2.11, the rotational contribution to the gravitational acceleration is given by $a_2 = a\cos(\theta)$,

```
> a[2]:=evalf(4*Pi^2*R[e]*(cos(theta))^2/(T^2));
```

$$a_2 := 0.03387566182\cos(\theta)^2$$

or, on introducing the angle y in degrees, by the output of the following line.

```
> a[2]:=subs(theta=y*evalf(Pi)/180,%);
```

$$a_2 := 0.03387566182\cos(0.01745329252\,y)^2$$

The net acceleration, or net "g," is the difference[2] between a_1 and a_2.

```
> net_accel:=a[1]-a[2];
```

$$net\_accel := 9.829878576 - 0.03387566182\cos(0.01745329252\,y)^2$$

Vectoria now creates a plot of the theoretical formula for the net acceleration.

```
> Gr1:=plot(net_accel,y=0..90,color=black):
```

From the introductory physics text by Hans C. Ohanian [Oha85], Vectoria extracts Table 2.27, which gives the acceleration g (in m/s^2) due to gravity at

Table 2.27: Variation of "g" with latitude.

Location	**lat**	g	**Location**	**lat**	g
Quito, Ecuador	0	9.780	London, England	51	9.811
Madras, India	13	9.783	Oslo, Norway	60	9.819
Hong Kong	22	9.788	Murmansk, U.S.S.R.	69	9.825
Cairo, Egypt	30	9.793	Spitsbergen, Norway	80	9.831
New York City	41	9.803	North Pole	90	9.832

several locations at different latitudes (in degrees north of the equator) on the earth's surface. Vectoria intends to compare the theoretical curve derived above with the data and also obtain a best-fitting curve for the data. Data lists are formed for the latitude (**lat**) and observed $\boldsymbol{g}$ values.

```
> lat:=[0,13,22,30,41,51,60,69,80,90]:
> g:=[9.780,9.783,9.788,9.793,9.803,9.811,9.819,9.825,
      9.831,9.832]:
```

The `ScatterPlot` command is used to plot the data as size-12 circles.

```
> pts:=ScatterPlot(lat,g,symbol=circle,symbolsize=12):
```

Guided by the theoretical analysis, Vectoria fits the data with a curve of the structure $a - b\cos(\pi\,y/180)^2$, with the two parameters a and b determined.

```
> eq:=Fit(a-b*(cos(y*evalf(Pi)/180))^2,lat,g,y);
```

$$eq := 9.83201711513076404 - 0.0517244526832718946\cos(0.01745329252\,y)^2$$

[2]Do you see why it is the difference?

The equation *eq* is the best-fitting curve of the postulated form. To compare the curves and data, *eq* is plotted using a thick red dashed line style,

```
> Gr2:=plot(eq,y=0..90,color=red,linestyle=DASH,thickness=2):
```

and all three plots displayed in Figure 2.12.

```
> display({Gr1,pts,Gr2},labels=["latitude","g"],
  tickmarks=[4,3],view=[0..90,9.77..9.85],
  title="Variation of g with latitude");
```

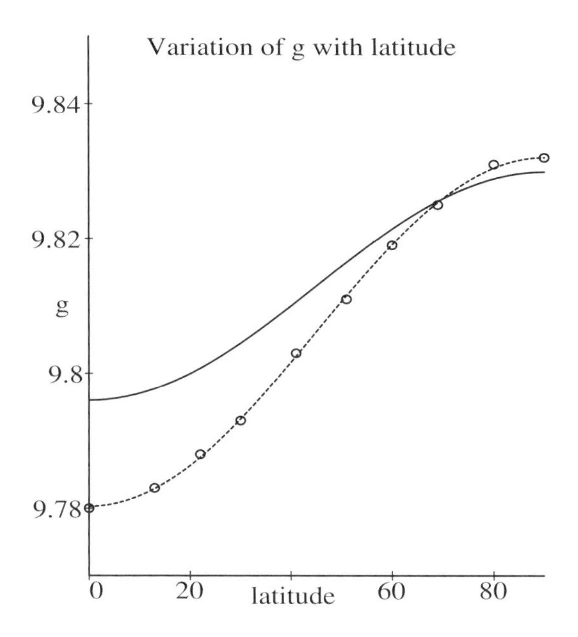

Figure 2.12: Data (circles) and theoretical (solid) and best-fit (dashed) curves.

In Figure 2.12 the best-fit curve passes right through the observational data. Vectoria notes that the theoretical curve has the right shape but is not in quantitative agreement. (Note that the differences are very small, the vertical scale being a little misleading.) She thinks this is due to the fact that the earth was modeled as being spherical and the mean radius used.

Due to rotation, the earth flattens slightly inward at the poles and bulges slightly outward at the equator. A point at the equator is at a radius greater than the mean radius, so g, which decreases with larger radius, will be slightly decreased at the equator. Conversely, the poles are at a radius smaller than the mean so g is slightly increased at the poles.

Vectoria believes that taking this slight distortion of the spherical shape into account will bring the theoretical curve into good agreement with the data. As an excercise you might estimate how big this distortion would have to be.

PROBLEMS:

Problem 2-31: Differential cross section

For a certain nuclear reaction, measurements of the differential scattering cross section $\sigma(\theta)$ (units are 10^{-30} cm^2/steradian) as a function of angle θ (in degrees) yield the values listed in Table 2.28.

Table 2.28: Differential scattering cross section data.

$\boldsymbol{\theta}$	30	45	90	120	150
$\boldsymbol{\sigma(\theta)}$	11	13	17	17	14

(a) Make a least squares fit to $\sigma(\theta)$ of the form

$$\sigma = a + b\cos\left(\frac{\pi\,\theta}{180}\right) + c\cos^2\left(\frac{\pi\,\theta}{180}\right).$$

(b) Plot the data points and the least squares curve on the same graph over the range 0° to 180°.

(c) What is the predicted differential cross section at 0°?

(d) Using the best-fit equation, calculate the total scattering cross section by performing the integration

$$\sigma = 2\,\pi \int_0^{\pi} \sigma(\theta)\, \sin\theta\, d\theta.$$

Problem 2-32: Swimsuit sales

Table 2.29 shows the cumulative number of men's swimsuits that the Glitz department store had sold by the end of January (month 1), by the end of February, and so on, for the first three-quarters of last year.

Table 2.29: Cumulative number of men's swimsuits sold.

Month	1	2	3	4	5	6	7	8	9
Number	10	16	25	58	230	439	648	748	795

Letting x label the month and y the cumulative number of swimsuits sold, an equation of the form $y = a + b\tanh(0.60\,(6.0 - x))$ is believed to model the data.

Find the coefficients a and b that give a best fit. Then plot the least squares curve and the data points on the same graph. If the manager had ordered 900 swimsuits at the beginning of the previous year, how many swimsuits are predicted to remain unsold at year's end?

2.2.7 Finding Romeo a Juliet

You have so many computers,
why don't you use them in the search for love?
Lech Walesa, Polish president, Interview in Paris, on his first journey outside the Soviet bloc (1988)

A certain lonely bachelor, Mat, decides to try to find his Juliet through the Happy Hearts dating service. As part of her interviewing process, the manager of the service explains how initial matches are achieved. First, the client is asked to fill in a questionnaire answering questions about income, hobbies, religious affiliation, and so on. The answer to each question is entered in numerical form so that a computer search can be done, comparing the client's responses to those of thousands of possible dates. A least squares process is then used to select a possible Juliet for the lonely bachelor. A similar procedure is used to find a possible Romeo for a lonely, single, female.

Mat, a professional who is interested in computing, asks the sales manager for a specific example of how the least squares matches are carried out. To protect the identity of her actual clients, the manager decides to create a list of 10 fictitious possible dates using a random matrix generator to create the numerical responses that they might have given to the questions on the questionnaire. In her program, she loads the necessary `LinearAlgebra` package.

```
> restart: with(LinearAlgebra):
```

To create a random matrix, the command `randomize()` is first entered. With no argument specified, this command will set the random number seed to a number based on the computer system clock. The number of possible dates, $N=10$, is entered and each date is asked $q=5$ questions. For possible matches involving real people, the values of N and q are, of course, much larger.

```
> N:=10: q:=5: randomize():
```

The potential Romeo's responses to the five questions are entered as a matrix[3] along with the client's name, in this case Mat. For each question, the client is asked to give a numerical answer between 1 and 5. In this example, Mat has given a response of 2 to the first question, 3 to the second one, and so on. The form of the resulting matrix, which has 1 row and 6 columns, is displayed.

```
> Romeo:=Matrix([[Mat,2,3,1,2,2]]);
```

$$Romeo := \begin{bmatrix} Mat & 2 & 3 & 1 & 2 & 2 \end{bmatrix}$$

The names of 10 possible dates are entered as a "column matrix"[4] (output not displayed here), having N rows and only 1 column.

```
> dates:=Matrix([[Ann],[Lynda],[Karen],[Mary],[Judy],[Sue],
        [Betty],[Lara],[Rose],[Cindy]]);
```

The responses of each of the $N = 10$ possible dates to the $q = 5$ questions is

[3]This is the "long" form. A "short" form for entering the row matrix is `<<Mat|2|3|1|2|2>>`.
[4]The "short" form is `<<Ann,Lynda,Karen,Mary,Judy,Sue,Betty,Lara,Rose,Cindy>>`.

simulated using the RandomMatrix command with the entries determined by generator=rand(1..5), which generates random integers between 1 and 5.

```
> A:=RandomMatrix(N,q,generator=rand(1..5)):
```

The numerical responses embodied in matrix A, which were suppressed, are joined to the names of the dates and the new A matrix displayed.

```
> A:=<dates|A>;
```

$$A := \begin{bmatrix} Ann & 5 & 5 & 5 & 5 & 3 \\ Lynda & 1 & 5 & 2 & 4 & 3 \\ Karen & 2 & 5 & 4 & 4 & 2 \\ Mary & 5 & 2 & 1 & 3 & 1 \\ Judy & 2 & 3 & 1 & 4 & 3 \\ Sue & 3 & 3 & 3 & 3 & 1 \\ Betty & 3 & 4 & 2 & 2 & 2 \\ Lara & 4 & 4 & 1 & 5 & 4 \\ Rose & 3 & 1 & 2 & 3 & 3 \\ Cindy & 5 & 2 & 1 & 1 & 1 \end{bmatrix}$$

So Ann has given a response 5 to the first question, 5 to the second, etc. Of course, in this example, Ann's responses will differ the next time the program is run, since the random number seed changes with computer clock time.

Now a comparison of each of the $N = 10$ date's answers to Romeo's (Mat's) responses is made using a repetitive do loop. As a measure of compatibility, the response on each question for each potential Juliet is subtracted from Mat's response to the same question. The difference is then squared, to accentuate large differences in personalities or traits, and a sum of squares over all questions formed. The date with the lowest least squares total, or score, will then be the computer's choice for Mat. (If a tie occurs, the first lowest score is selected.)

```
> for j from 1 to N do:
> B[j]:=add((Romeo[1,i]-A[j,i])^2,i=2..q+1);
> end do:
```

The totals for the 10 possible dates are put into a column matrix form B,

```
> B:=<<seq(B[j],j=1..N)>>;
```

which is joined to the A matrix. The least squares totals are in the last column.

```
> A:=<A|B>;
```

$$A := \begin{bmatrix} Ann & 5 & 5 & 5 & 5 & 3 & 39 \\ Lynda & 1 & 5 & 2 & 4 & 3 & 11 \\ Karen & 2 & 5 & 4 & 4 & 2 & 17 \\ Mary & 5 & 2 & 1 & 3 & 1 & 12 \\ Judy & 2 & 3 & 1 & 4 & 3 & 5 \\ Sue & 3 & 3 & 3 & 3 & 1 & 7 \\ Betty & 3 & 4 & 2 & 2 & 2 & 3 \\ Lara & 4 & 4 & 1 & 5 & 4 & 18 \\ Rose & 3 & 1 & 2 & 3 & 3 & 8 \\ Cindy & 5 & 2 & 1 & 1 & 1 & 12 \end{bmatrix}$$

In this case, the number of possible dates is so small that one glance at the last column of the output will tell the sales manager who the best match is. However, remember that in practice she is working with thousands of possible dates, and it would be tedious to skim through the data for each new client. So, the identification of the best match is automated. In the next line, the score (least squares total) of the first date (Ann) listed in the matrix A is recorded.

```
> score:= A[1,q+2];
```

$$score := 39$$

A do loop containing a conditional "if...then" statement is used to compare the scores of the remaining possible Juliets in the matrix A to Ann's score. The loop runs over the rows $i = 2$ to N of the matrix.

```
> for i from 2 to N do;
```

If the last $(q + 2)$ entry in the ith row of A is less than *score*, then this value becomes the new score. This process continues until the lowest score is achieved. The corresponding row value, k, is recorded and the "if...then" statement ended.

```
> if (A[i,q+2]<score) then score:=A[i,q+2]; k:=i; end if;
> end do:
```

In some runs it may turn out that Ann has the lowest score, so no numeric value is generated for k in the do loop. In the following command line, the `type` command is used to check whether k has a numeric value. If it does, then the value of k is chosen. Otherwise k must have the value 1.

```
> if type(k,numeric) then k:=k; else k:=1; end if:
```

For the illustrative example, the name of the potential Juliet is obtained by extracting the matrix element from A corresponding to the kth row and first column. And the potential Juliet is

```
> Juliets_name:=A[k,1]; Her_score:=score;
```

$$Juliets\_name := Betty \qquad Her\_score := 3$$

Betty, with a winning low score of 3. If it is desired to compare her responses to individual questions with those of Romeo, her complete record can be obtained by extracting the entire kth row from A. This might be important to a client who weights the response to certain questions more highly than to others.

```
> Juliets_responses:=Row(A,k);
```

$$Juliets\_responses := [Betty, 3, 4, 2, 2, 2, 3]$$

The sales manager, having finished her example, asks Mat whether he is interested in using their dating service. On learning of the cost of the service, Mat hems and haws and says that he will think about it.

PROBLEMS:

Problem 2-33: More selection

Modify the Romeo and Juliet file to include the five responses given by 25 Juliets and write your code so that it gives the names of the three leading candidates with the three lowest scores.

Problem 2-34: The Microhard training procedure
The Microhard Computer Company has a training procedure in which its new employees are assigned to work with experienced and proven group leaders. Kevin, a newcomer to the company, is to be assigned to a group leader who best matches his skills and personality, using a computer matching process similar to that in the Romeo and Juliet recipe. The relevant data can be obtained by running "readdata.mws," which reads the two Maple data files "hiredname.m" and "trainernames.m." Use these data and modify **02-2-7** to see which group leader should train Kevin.

2.3 Multiple Regression Equations

In the previous section, almost all of the examples involved only two variables: the independent variable x and the dependent variable y. If the data appeared to lie along a straight line, a straight-line model equation $Y = a + b\,x$ was sought. As mentioned earlier, statisticians refer to the least squares procedure that produces the "best" values of a and b as *regression analysis*. Suppose that y is thought to depend linearly on two independent variables, x_1 and x_2. For example, medical school admissions are generally based on both the student's college GPA (x_1) and the student's MCAT score (x_2). In this case, the proposed model equation would be assumed to be of the form $Y = a + b_1\,x_1 + b_2\,x_2$. In geometrical terms, the three variables x_1, x_2, and y form a three-dimensional space, and the model equation defines a plane in this three-dimensional space. The least squares procedure in this case corresponds to minimizing the sums of the squares of the distances of the data points from the plane. Determining the coefficients by the generalization of the least squares method is referred to as *multiple regression analysis*, and the best-fitting equation is called the *multiple regression equation*. To carry out the multiple regression analysis, we again introduce χ^2, which now takes the form

$$\chi^2 = \sum_{i=1}^{N} (y_i - a - b_1\,x_1 - b_2\,x_2)^2. \tag{2.15}$$

Differentiating χ^2 with respect to a, b_1, and b_2 yields the following three equations, which are an obvious generalization of Equations (2.5), for the three unknowns a, b_1, and b_2:

$$\begin{aligned} S_y &= a\,N + b_1\,S_{x1} + b_2\,S_{x2}, \\ S_{x1\,y} &= a\,S_{x1} + b_1\,S_{x1\,x1} + b_2\,S_{x1\,x2}, \\ S_{x2\,y} &= a\,S_{x2} + b_1\,S_{x1\,x2} + b_2\,S_{x2\,x2}. \end{aligned} \tag{2.16}$$

Here, $S_{x1} \equiv \sum_i (x_1)_i$, $S_{x1\,y} \equiv \sum_i (x_1\,y)_i$, etc. Once again, we can make use of Maple's built-in least squares procedure, which solves these equations and produces the best-fit coefficients a, b_1, and b_2. Noting that there are now three fitting coefficients, the standard deviation for N data points is given by $\sigma = \sqrt{\chi^2/(N-3)}$. The multiple regression analysis can be easily generalized to more than two independent variables. For a more complete discussion of deriving multiple regression equations and the validity of such models, we refer the reader to the statistics text by Anderson et al. [ASW87].

2.3.1 Real Estate Appraisals

Everyone lives by selling something.
Robert Louis Stevenson, Scottish writer (1850–1894)

Syd Boffo, chief appraiser and co-owner of the Boffo Brothers Real Estate Company, wishes to develop a linear mathematical formula that predicts the selling price of a house in his town, given the square footage of the house and the square footage of the lot. The company accountant gives him the data, which are displayed in Table 2.30. The house size is in hundreds of square feet, the lot size in thousands of square feet, and the selling price in thousands of dollars.

Table 2.30: Data to develop real estate formula.

Price	135	111	88	96	102	147	159	195	223	264
House size	21	16	17	14	19	18	23	22	24	26
Lot size	11	13	7	9	11	25	12	13	15	22

Syd enters the lists for the house size (`hs`), lot size (`ls`), and selling price (`pr`). The number of entries N should be 10 in each list, and this is checked for the house sizes with the `nops` command.

```
> restart: with(Statistics): with(plots): with(LinearAlgebra):
> hs:=[21,16,17,14,19,18,23,22,24,26]: #house size
> N:=nops(hs);
```

$$N := 10$$

```
> ls:=[11,13,7,9,11,25,12,13,15,22]: #lot size
> pr:=[135,111,88,96,102,147,159,195,223,264]: #price
```

Syd decides to first check how good his assumption is of a linear relationship between price and house size and between price and lot size. He calculates the linear correlation coefficient for price versus house size,

```
> lincoeff1:=Correlation(hs,pr);
```

$$lincoeff1 := 0.8858411229$$

obtaining a value of 0.89. There appears to be quite a reasonable correlation between these two variables. Similarly, he finds that the linear correlation coefficient for price versus lot size,

```
> lincoeff2:=Correlation(ls,pr);
```

$$lincoeff2 := 0.6134269770$$

is about 0.61. The correlation between price and lot size is not quite as good as that between price and house size. Nevertheless, he decides to use both the house size and lot size data. To apply the `Fit` command, he forms a matrix of the house size and lot size data and transposes the matrix.

```
> X:=Transpose(Matrix([hs,ls]));
```

$$X := \begin{bmatrix} 21 & 11 \\ 16 & 13 \\ 17 & 7 \\ 14 & 9 \\ 19 & 11 \\ 18 & 25 \\ 23 & 12 \\ 22 & 13 \\ 24 & 15 \\ 26 & 22 \end{bmatrix}$$

Using X in the `Fit` command, a linear model equation with two independent variables $x1$ (corresponding to house size) and $x2$ (lot size) is obtained for the selling price sp.

```
> sp:=Fit(a+b1*x1+b2*x2,X,pr,[x1,x2]);
```

$$sp := -127.361560581726636 + 11.7242891875249864\,x1 \\ +3.25186788632078994\,x2$$

By first using the selling price formula to calculate the predicted prices ppr,

```
> ppr:=seq(subs(x1=hs[i],x2=ls[i],sp),i=1..N);
```

$$ppr := 154.6190592,\ 102.5013489,\ 94.71443080,\ 66.04529907, \\ 131.1704808, 164.9723420,\ 181.3195054,\ 172.8470841, \\ 202.7993983,\ 249.0110518$$

the standard deviation, or error, σ in the selling price is easily obtained.

```
> sigma:=sqrt(sum((pr[i]-ppr[i])^2,i=1..N)/(N-3));
```

$$\sigma := 24.48051456$$

The standard deviation is about $24,000. Syd decides to check graphically on how well the model equation fits the data. He first creates a three-dimensional plot of the selling price sp, choosing a style option that colors the planar surface with no grid.

```
> graph1:=plot3d(sp,x1=10..30,x2=5..30,axes=boxed,
          style=patchnogrid):
```

He then combines the ith entry of each list into triplets of numbers that give the coordinates of the ith plotting point,

```
> points:=seq([hs[i],ls[i],pr[i]],i=1..N);
```

$$points := [21, 11, 135], [16, 13, 111], [17, 7, 88], [14, 9, 96], [19, 11, 102], [18, 25, 147], [23, 12, 159], [22, 13, 195], [24, 15, 223], [26, 22, 264]$$

and creates a three-dimensional graph of the points.

```
> graph2:=pointplot3d({points},style=point,symbol=circle,
        color=blue):
```

The two graphs are superimposed with the display command and a suitable orientation chosen. The result is shown in Figure 2.13.

```
> display({graph1,graph2},orientation=[-20,60],tickmarks=[3,3,3],
  labels=["x1","x2","sp"],view=[10..30,5..30,75..325]);
```

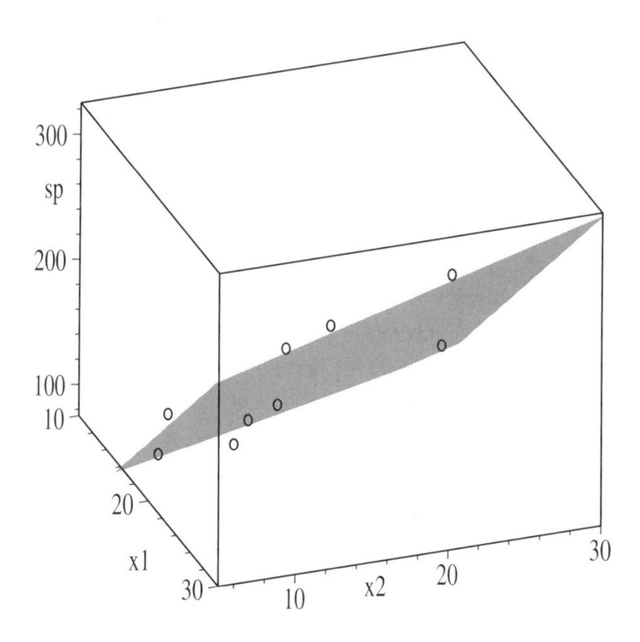

Figure 2.13: Data points and planar surface corresponding to sp formula.

The data points seem to be moderately well described by the linear relationship, although the standard error is a bit bigger than Syd would like. This is undoubtedly due to the somewhat weak linear correlation between price and lot size. Syd's brother and co-owner, Benny, is skeptical that any such formula will indeed be useful. Benny gives Syd a test on the accuracy of the predicted selling price formula.

"OK, Syd, we have a new house to put on the market. It is 2000 ft^2 in size and the lot is 90 feet by 200 feet. What should we set the selling price at?"

Syd replies, "We input $x1 = 20$ and, since $90 \times 200 = 18{,}000$ square feet, $x2 = 18$ into our formula for the selling price,

```
> x1:=20; x2:=18; selling_price:=sp;
```

$$x1 := 20 \quad x2 := 18 \quad selling\_price := 165.6578452$$

which tells us that the house should sell for about $166,000."

Benny snorts and says, "Would you like to bet a filet mignon dinner at the Hungry Heifer Steak House on your prediction?"

"Well," Syd answers, "I do not like the sizeable standard error of $24,000 for my answer, which undoubtedly is associated with the fact that the lot size data didn't linearly correlate that well with the price. Do we have any other additional data?"

"Sure," Benny replies, "we have the median neighborhood selling price in thousands of dollars for the homes in your sample. I will read them off and you can input them."

So then Syd proceeds to enter the new data in the following median price list `mpr`.

```
> mpr:=[150,120,69,96,117,135,147,210,198,273]: #median price
```

Again, as a preliminary check, he calculates the linear correlation coefficient for the price versus the median selling price,

```
> lincoeff3:=Correlation(mpr,pr);
```

$$lincoeff3 := 0.9672802592$$

obtaining 0.97. "Ah, this is better," Syd remarks, "I should get an improved model equation using these new data."

He then assumes a relationship of the form $y = A + B_1\, X1 + B_2\, X2 + B_3\, X3$, with $X3$ referring to the median selling price. In this case, Syd would have had to solve the four equations

$$\begin{aligned}
S_y &= A\,N + B_1\,S_{x1} + B_2\,S_{x2} + B_3\,S_{x3},\\
S_{x1\,y} &= A\,S_{x1} + B_1\,S_{x1\,x1} + B_2\,S_{x1\,x2} + B_3\,S_{x1\,x3},\\
S_{x2\,y} &= A\,S_{x2} + B_1\,S_{x1\,x2} + B_2\,S_{x2\,x2} + B_3\,S_{x2\,x3},\\
S_{x\,3y} &= A\,X_{x3} + B_1\,S_{x1\,x3} + B_2\,S_{x2\,x3} + B_3\,S_{x3\,x3},
\end{aligned} \tag{2.17}$$

for the four unknowns A, B_1, B_2, B_3, if Maple's least squares fitting procedure were not available to do the work for him. The new selling price (nsp) formula is given by the output of the following command line.

```
> nsp:=Fit(A+B1*X1+B2*X2+B3*X3,Transpose(Matrix([hs,ls,mpr])),
     pr,[X1,X2,X3]);
```

$$nsp := -42.5535998199035106 + 3.83297084662808540\,X1$$
$$+1.18029719945884404\,X2 + 0.670667204850229459\,X3$$

Syd extends the formula for the standard error to the new situation and calculates a new standard error,

```
> new_ppr:=seq(subs(X1=hs[i],X2=ls[i],X3=mpr[i],nsp),i=1..N);
```

$$new\_ppr := 151.5221379, 114.5978619, 77.14502211, 86.11471850, 121.7241784, 146.4873781, 158.3563752, 197.9557354, 199.9342651, 266.1623275$$

```
> sigma[new]:=sqrt(sum((pr[i]-new_ppr[i])^2,i=1..N)/(N-4));
```

$$\sigma_{new} := 15.47382409$$

of about \$15,000, considerably reduced from his previous calculated value. Benny tells him that the median selling price for his calculation can be taken as $X3$ = \$156,000. Syd enters the values of $X1$, $X2$, and $X3$ into his formula,

```
> X1:=20: X2:=18: X3:=156: new_selling_price:=nsp;
```

$$new\_selling\_price := 159.9752507$$

and finds that the new predicted selling price of \$160,000 is slightly lower than the original estimate.

"Benny," Syd says, "If you will allow me one standard deviation on either side of this value, the bet is on."

"Come on," Benny replies, "That's a spread from about \$145,000 to \$175,000. I will buy the dinner if the house sells for between \$152,000 and \$167,000!"

"You have always been a hard bargainer, oh skeptical brother of mine, but I will take your bet. My mouth is already drooling in anticipation of a medium rare filet mignon. I understand that they also serve a delicious chocolate dessert that is almost worth dying for. If you're going to pick up the tab, as I anticipate will happen, I am going to splurge at your expense!"

In the spirit of the typical soap opera, the reader will have to wait until the next exciting episode, when Syd and Benny reappear to tell us who has won the bet.

PROBLEMS:

Problem 2-35: Effect of Advertising

In Table 2.31 the owners of a movie theater chain [ASW87] have recorded their gross weekly revenue (GR, in thousands of dollars) and the amounts (also in thousands) spent on advertising on TV and in the newspapers (NP).

Table 2.31: Weekly revenue and advertising expenditures.

GR	96	90	95	92	95	94	94	94
TV	5.0	2.0	4.0	2.5	3.0	3.5	2.5	3.0
NP	1.5	2.0	1.5	2.5	3.3	2.3	4.2	2.5

(a) Derive a linear best-fit formula for the weekly gross revenue in terms of the two types of advertising.

(b) If both the weekly television and newspaper advertising are \$5000, what is the estimated weekly gross revenue?

(c) What is the standard deviation for your estimate?

(d) Produce a plot of the data points and the planar surface corresponding to the best fit.

Problem 2-36: Car prices

Table 2.32 gives the list price (P in thousands of dollars), masses (M in thousands of kg), interior volumes (V in m^3), the gas consumption (G in liters per 100 km), the sales volume (S in thousands of cars), and country of origin (C: U.S. = United States, J. = Japan, G.B. = Great Britain, G. = Germany) for 10 different models of cars sold in North America.

Table 2.32: Car data.

P	13.5	9.6	15.3	24.5	35.5	6.9	14.5	7.5	9.5	17.4
M	1.3	1.2	1.0	1.4	1.3	1.0	1.2	1.1	1.2	1.5
V	3.4	3.4	2.9	1.9	3.0	2.7	3.0	2.8	3.3	3.5
G	8.5	9.0	7.4	10.0	8.0	7.0	9.5	7.2	8.5	9.9
S	350	170	85	10	25	140	210	160	100	55
C	U.S.	U.S.	J.	G.B.	G.	U.S.	U.S.	J.	U.S.	U.S.

(a) Calculate the correlation coefficient between list price and each of the possible factors (e.g., weight) that could influence the list price. Also produce a scatter plot in each case. Is there a strong dependence of the list price on any of the factors?

(b) Remove the non-U.S. cars from the list and repeat part **(a)**. Do any of the correlation coefficients improve?

(c) Choose the two factors that have the highest correlation coefficient and derive a best-fit formula for the list price in terms of these two factors.

(d) Produce a plot of the data points and the planar surface corresponding to the best fit.

(e) Calculate the standard deviation in list price for your best fit formula.

2.3.2 And the Winner Is?

There's no such thing as a free lunch.
Milton Friedman, American economist (1912–)

It's two weeks later and Syd and Benny are comfortably seated in the main dining room of the Hungry Heifer Steak House. This is the night of their wives' biweekly ladies bridge session, and the Boffo brothers are combining a business meeting with a night out. They have both consumed sizeable filet mignons done to perfection and quaffed a bottle of vintage red Cabernet wine produced from grapes grown in the Columbia River basin of eastern Washington. The real estate business has been good recently, and the house for which Syd predicted the selling price has been sold to a nice young family. Looking over the dessert menu, Syd orders a chocolate pecan torte followed by a cup of Sumatran coffee. On finishing, he asks the waiter to bring the bill.

With a gentle belch, Syd put his hand on the bill, turns it over, and looks at the total. "This wasn't a cheap meal," he remarks. He then pushes the bill across the table to Benny. "I believe this is yours tonight. That nice young couple got the house for \$165,000, which is within the range of \$152,000 to \$167,000 that you allowed me."

"Ah, you were lucky with your prediction, Syd. If you feel so confident, we can make a similar bet with regard to our associated trucking company. I will talk to Cousin Brenda, who manages that side of our operations, and see if she has any recent numbers that you can use. I will give you the data in the office tomorrow."

"OK, Syd, but I will have to see how good a model equation I can develop before I make any new bet with you."

The next day, Syd is given the information [ASW87] displayed in Table 2.33. The table gives the miles traveled, the number of deliveries, and the travel time

Table 2.33: Data for the Boffo Trucking Company.

Day	Miles traveled	Delivery number	Travel time
1	100	4	9.3
2	50	3	4.8
3	100	4	8.9
4	100	2	5.8
5	50	2	4.2
6	80	1	6.8
7	75	3	6.6
8	80	2	5.9
9	90	3	7.6
10	90	2	6.1

in hours for 10 consecutive days for the Boffo Trucking Company. The objective is to develop a linear model equation that predicts the travel time, given the number of miles traveled and the number of deliveries. The problem is similar to the one Syd solved for the housing prices. Although he could simply use a cut and paste approach, replacing the old data and labels in his earlier recipe with the new data and labels, Syd decides to create a generic Maple "procedure" instead. This has the advantage that he has then only to recall the procedure, type in the new data where indicated in the worksheet, and execute the worksheet.

```
> restart: with(Statistics): with(LinearAlgebra):
```

The linear least squares fitting procedure is assigned the name `fitlin`. In the argument of the procedure command, `proc`, Syd inserts the generic names of the three data lists, which will have to be entered in order for any explicit calculations to be done.

```
> fitlin:=proc(x1list,x2list,ylist)
```

For the example at hand, the miles traveled will be entered as the `x1list`, the number of deliveries as the `x2list`, and the travel time as the `ylist`.

The next two command lines indicate which Maple names in the fitting procedure are taken to be as "global" and which are to be regarded as "local" to this procedure. Maple regards local variables in different procedures to be different variables, even if they have the same name. Global variables hold outside procedures. If a variable is not specified as global or local, Maple will automatically assume that it is local if it appears on the left-hand side of an assignment (name) statement. Since they are all quantities that Syd wishes to evaluate outside the procedure, the Maple assignments `lincoeff1`, `lincoeff2`, `eq`, `sigma`, and `Y`, whose meaning will be made clear shortly, are indicated to be global.

```
> global lincoeff1,lincoeff2,eq,sigma,Y;
```

The remaining Maple names in the procedure are taken as being local.

```
> local N,predy,x1,x2,X;
```

Using the `nops` command, the number of entries in the `x1list` is determined.

```
> N:=nops(x1list):
```

The linear correlation coefficients between the $x1$ and y data,

```
> lincoeff1:=Correlation(x1list,ylist);
```

and the $x2$ and y data are calculated.

```
> lincoeff2:=Correlation(x2list,ylist);
```

The `x1list` and `ylist` are formed into a tranposed matrix.

```
> X:=Transpose(Matrix([x1list,x2list]));
```

The least squares `Fit` command is used to derive the linear model equation,

$$y = a + b1\,x1 + b2\,x2, \tag{2.18}$$

that best fits the three data lists.

```
> eq:=Fit(a+b1*x1+b2*x2,X,ylist,[x1,x2]);
```

The next two command lines determine the standard deviation σ.

```
> predy:=seq(subs(x1=x1list[i],x2=x2list[i],eq),i=1..N);
> sigma:=sqrt(sum((ylist[i]-predy[i])^2,i=1..N)/(N-3));
```

Given specific input values of *X1* (miles traveled) and *X2* (number of deliveries), the predicted travel time in hours, Y, is evaluated.

```
> Y:=subs(x1=X1,x2=X2,eq);
```

The procedure syntax is then terminated with the `end proc` command.

```
> end proc:
```

In the next three command lines, Syd enters the relevant trucking data lists. For other problems in which it is desired to use the same fitting procedure, the new data can be entered where indicated by the comments. If fewer or more dependent variables were used than in this example, the procedure could be easily modified and saved under a new procedure name.

```
> x1list:=[100,50,100,100,50,80,75,80,90,90]: #enter data
> x2list:=[4,3,4,2,2,1,3,2,3,2]: #enter data
> ylist:=[9.3,4.8,8.9,5.8,4.2,6.8,6.6,5.9,7.6,6.1]: #enter data
```

As an example, Syd wishes to use the model equation to estimate the travel time for a day on which the number of miles traveled is 90 with 3 deliveries.

```
> X1:=90: X2:=3: #enter data
```

After entering the various data, a call is now made for the `fitlin` procedure.

```
> fitlin(x1list,x2list,ylist):
```

The global quantities are now explicitly evaluated for the specified data. The linear correlation coefficients LC_{x1y} (between miles traveled and travel time) and LC_{x2y} (between delivery number and travel time)

```
> LC[x1y]:= lincoeff1; LC[x2y]:= lincoeff2;
```

$$LC_{x1y} := 0.7755115490 \qquad LC_{x2y} := 0.6338656910$$

are 0.78 and 0.63, respectively. The best-fitting linear model equation is given by the output of the following command line,

```
> y:=eq;
```

$$y := 0.03665523502118317 46 + 0.05616300181561423 99\, x1 + 0.76386927577163676 8\, x2$$

while the standard deviation is 0.85.

```
> standard_deviation:=sigma;
```

$$standard\_deviation := 0.8493805413$$

Corresponding to the input values of 90 miles traveled and three deliveries,

```
> travel_time:=Y;
```

$$travel\_time := 7.382933226$$

the model equation predicts a travel time of 7.38 hours

After doing the calculation, Syd is not certain as to whether he should take Benny up on his bet or once again ask for more information in order to develop a better model equation. What do you recommend that Syd do?

PROBLEMS:

Problem 2-37: Generalizing the text procedure

Generalize the Maple procedure of this section to deal with a list of y values depending linearly on lists of x_1, x_2, and x_3 values.

Apply this generalized procedure to the housing example considered by Syd Boffo where the selling price depended on house size, lot size, and median neighborhood selling price.

Problem 2-38: New home cost

According to *U.S. News and World Report* (April 6, 1992), the median cost C of a new house, the number of new housing starts S in 1991–92, and the average household income I are as given in Table 2.34. All entries are in thousands. Making use of the Maple procedure in the text:

Table 2.34: Housing market data.

City	C	S	I	City	C	S	I
Atlanta	100.6	24.2	54.7	Mobile, Ala.	68.5	1.0	41.0
Baltimore	121.8	11.1	62.8	Oklahoma City	68.9	3.3	53.2
Chicago	181.8	12.9	61.0	Pittsburgh	79.5	4.9	49.2
Cleveland	122.9	5.4	54.0	Richmond, Va.	102.8	6.0	64.5
Columbia, S.C.	90.3	3.1	57.4	San Antonio	72.5	1.5	57.0
Dayton, Oh.	107.8	3.8	48.4	Scranton, Pa.	81.6	2.5	44.8
Gary, Ind.	98.2	3.2	45.7	Tacoma	96.1	4.1	51.1
Jacksonville	82.1	8.0	47.5	W. Palm Beach	130.4	8.9	58.1

(a) Develop a linear least squares formula relating the cost to the number of new housing starts and the household income.

(b) Estimate the median cost for a new house for a city with 8000 new housing starts and average household income of $50,000.

(c) Determine the standard deviation.

(d) Calculate the relevant linear correlation coefficients and comment on how good you think the model equation is.

Problem 2-39: Commercial office buildings

Generalize the Maple procedure of this section to deal with a list of y values depending linearly on lists of x_1, x_2, x_3, and x_4 values. Apply this generalized procedure to the following problem. The Boffo brothers are considering buying a group of small office buildings in the downtown Metropolis business district. Syd has obtained the data shown in Table 2.35 for 11 representative office buildings.

Table 2.35: Office building data.

Floor space (x_1)	**Offices** (x_2)	**Entrances** (x_3)	**Age** (x_4)	**Value** (y)
2310	2	2	20	142
2333	2	2	12	144
2356	3	1.5	33	151
2379	3	2	43	150
2402	2	3	53	139
2425	4	2	23	169
2448	2	1.5	99	126
2471	2	2	34	142.9
2494	3	3	23	163
2517	4	4	55	169
2540	2	3	22	149

The variables are floor space in square feet (x_1), number of offices (x_2), number of entrances (x_3), age of the office building in years (x_4), and the assessed value (y) in thousands of dollars of the office building. The two values "1.5" in the x_3 column indicate that for these two buildings, one of the two entrances is a delivery entrance only and has been counted as half an entrance.

(a) Help Syd develop a linear multiple regression equation relating the assessed value to the four independent x values.

(b) Use this formula to estimate the assessed value of an office building in the same business district that has 2500 square feet, three offices, two entrances, and is 25 years old.

(c) Which x variable has the highest linear correlation coefficient? Is this coefficient positive or negative. What does this indicate?

(d) Which x variable has a negative correlation coefficient? What does this indicate?

Part II

THE ENTREES

Science is the knowledge of many,
orderly and methodically digested and arranged,
so as to become attainable by one.
John F. W. Herschel, English astronomer (1792–1871)

It isn't so much what's on the table that matters
as what's on the chairs.
William S. Gilbert, English librettist (1836–1911)

"Take some more tea," the March Hare said to Alice, ...
"I've had nothing yet," Alice replied
in an offended tone: "so I can't take more."
"You mean you can't take less," said the Hatter:
"it's very easy to take more than nothing."
Lewis Carroll, English writer and mathematician (1832–1898)

Chapter 3

Algebraic Models. Part I

For the sake of... different types, scientific truth should be presented in different forms, and should be regarded as equally scientific, whether it appears in the robust form and... vivid coloring of a physical illustration, or in the tenuity and paleness of a symbolic expression.
James Clerk Maxwell, Scottish physicist (1831–1879)

Because of the typical mathematical background of the students involved, introductory college science courses tend to concentrate on simple algebraic models that can be solved analytically. Usually, the concepts of derivatives and integrals are introduced as well as the associated idea of finding the maximum or minimum of a function. Also, the student learns about dot and cross products as well as vector operators, and is expected to solve simultaneous equations.

In contrast to conventional programming languages, which require numerical values for all variables, computer algebra systems have the additional advantage of allowing the user to symbolically derive, manipulate, and solve a wide variety of interesting scientific model equations. This introductory chapter of the **Entrees** illustrates the application of the Maple CAS to a wide variety of intellectually stimulating scalar algebraic models. Vector and matrix models are covered in the following chapter, followed by chapters dealing with linear ODE and difference equation models. We are confident that you will not find the symbolic expressions appearing in our recipes and stories to be tenuous and pale, for they will generally be embedded in robust and colorful physical illustrations.

3.1 Scalar Models

In this chapter, we concentrate on scalar algebraic models and show how Maple's symbolic manipulative ability can make one's mathematical life much easier and the learning process more fun. However, remember that using the computer to do the algebra is no substitute for thinking. The computer cannot derive the basic equations. It's up to you to correctly enter the relevant expressions. This is the hard part of the task, which you will master only by carefully studying the recipes and trying the provided problems.

3.1.1 Bombs Versus Schools

I have never let my schooling interfere with my education.
Mark Twain, American author (1835–1910)

In 1872, Samuel Butler wrote a satiric novel about an adventurer who stumbles across an unknown civilization where everything seems to be done backward. The novel was called *Erewhon*, which is an anagram for *nowhere*. On the planet Erehwon (note the different spelling), which is featured periodically in this text, the civilization is not particularly backward, but the inhabitants have an idiosyncratic habit of often spelling names backward. Keep this in mind whenever a strange name is encountered in a story involving this planet.

Planet Erehwon has been spending an enormous amount of money (100 trillion ehs[1]) on bombs (defense) and much less (1 trillion ehs) on schools (education). It has been decided to increase the base budget of 101 trillion ehs next year by 5 trillion ehs, an approximate 5% increase. The 5 trillion increase is to be divided between the two groups in a way that maximizes the effective use of the money.

Now, as it happens, Trebla Nietsnie, the administrator deciding how the division is to occur, is a former scientist who wishes to rationalize how the money is to be split by constructing a reasonable mathematical model. Trebla believes that a group's effective use of the extra funds is linear for small percentage increases but grows less rapidly for large percentage increases. A phenomenological model that mimics this behavior is needed and created. As a measure of the effectiveness, letting x be the increase in funding and the constant C represent the base budget, Trebla defines an "effectiveness index" i,

```
> restart: with(plots):
> i:=ln(C+x)-ln(C);
```

$$i := \ln(C+x) - \ln(C)$$

and then combines the ln terms in the previous output into a simpler expression.

```
> i:=combine(%,ln,symbolic);
```

$$i := \ln\left(\frac{C+x}{C}\right)$$

The second argument in the `combine` command instructs Maple to use known logarithmic transformations. However, since the parameter C could, from a general mathematical viewpoint, be negative, the argument `symbolic` must be included here. This instructs Maple to assume that all parameters are positive.

The behavior of the index for small x can be obtained by Taylor expanding i about $x=0$ out to second order in x.

```
> taylor(i,x=0,2);
```

$$\frac{1}{C}\,x + \mathrm{O}(x^2)$$

[1]This unusual name for their currency arises from the fact that the planet's citizens end every sentence with an "eh"—pronounced to rhyme with "play."

The term of order x^2 is removed from the Taylor series output of the previous command line by converting it to polynomial form.

```
> small_x_behavior:=convert(%,polynom);
```

$$small\_x\_behavior := \frac{x}{C}$$

Thus, for small x the index is linear in x, with slope $1/C$. The complete behavior of the index can be examined by substituting a specific C value, e.g., $C=1$, in units of 1 trillion ehs.

```
> i:=subs(C=1,i);
```

$$i := \ln(1 + x)$$

The index i and the linear form x (valid for small x, with $C=1$) are now plotted together over the range $x = 0$ to 5. The two functions are entered as a Maple list here because order is important. Trebla wants the curve for i to be colored blue and the curve for x to be colored red, so the colors are also entered as a list. Because colors do not show up in a black-and-white rendering (such as this text), different line styles are chosen for the two curves, the two line styles also entered as a list. The linestyle 1 produces a solid curve, the linestyle 3 a dashed line. Equivalently, Trebla could have entered `linestyle=[SOLID,DASH]`. The other plot options should be familiar to the reader by now.

```
> plot([i,x],x=0..5,color=[blue,red],linestyle=[1,3],thickness
  =2,view=[0..5,0..2],tickmarks=[4,2],labels=["x","i"]);
```

The resulting picture is shown in Figure 3.1.

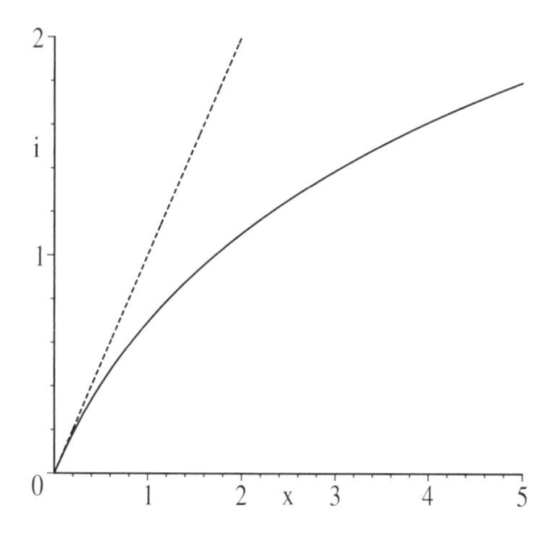

Figure 3.1: Variation of effectiveness index i with increase x in funding.

The deviation of i away from linear behavior as x increases is clearly seen. For large x, the behavior of the effectiveness index can be obtained by expanding around $x=\infty$, to second order, and removing the "order of" term.

```
> taylor(i,x=infinity,2);
```

$$\ln(x) + \frac{1}{x} + \mathrm{O}\left(\frac{1}{x^2}\right)$$

```
> large_x_behavior:=convert(%,polynom);
```

$$large\_x\_behavior := \ln(x) + \frac{1}{x}$$

For very large x, the index behavior is dominated by the log term, with the first-order correction being $1/x$.

In the bombs versus schools situation, Trebla lets x refer to the amount of extra funds allocated to building bombs and $y = 5 - x$ to building schools. He labels the effectiveness index for building bombs B and for schools S. Then,

```
> B:=(ln(1+x/C[b]));
```

$$B := \ln\left(1 + \frac{x}{C_b}\right)$$

```
> y:=5-x;
```

$$y := 5 - x$$

```
> S:=ln(1+y/C[s]);
```

$$S := \ln\left(1 + \frac{5 - x}{C_s}\right)$$

With the base budgets $C_b = 100$ trillion ehs and $C_s = 1$ trillion ehs, the `spacecurve` command is used to produce a three-dimensional plot of B versus S versus x over the range $x = 0$ to 5.

```
> C[b]:=100: C[s]:=1:
> spacecurve([B,S,x],x=0..5,labels=["B","S","x"],axes=normal,
  tickmarks=[3,3,3],thickness=2,orientation=[-90,0]);
```

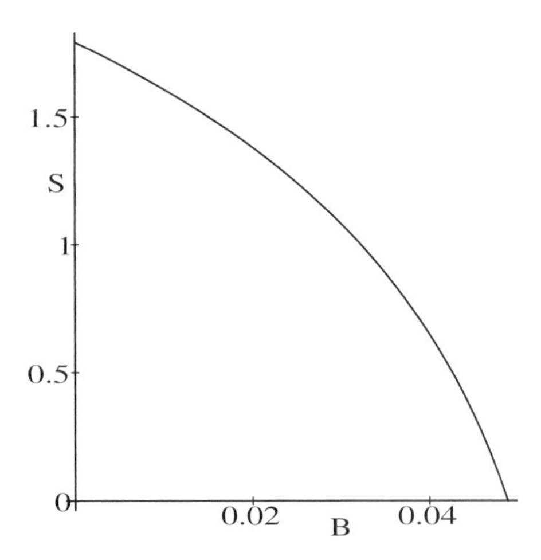

Figure 3.2: Relationship between the bomb index B and the school index S.

The orientation was chosen so as to produce Figure 3.2, showing the school index S versus the bomb index B. The three-dimensional character of the picture can be observed by clicking on the computer plot and dragging the mouse to change the orientation. Once the expressions for the individual indices have been written down, an overall index f that is the product of the two indices, $f = B\,S$, is formed and plotted, the result being shown in Figure 3.3. With tongue in cheek, Trebla refers to this as the BS model.

```
> f:=B*S;
```

$$f := \ln\left(1 + \frac{x}{100}\right)\ln(6 - x)$$

```
> plot(f,x=0..5,tickmarks=[4,3],labels=["x","f"]);
```

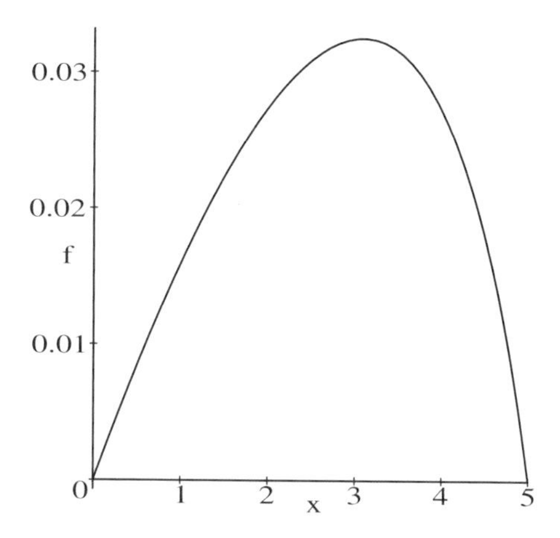

Figure 3.3: Overall effectiveness index f as a function of x.

Many problems dealing with decision theory and bargaining models generate curves similar to that shown in Figure 3.3.

The approximate location of the maximum can be found by clicking the mouse on the curve's peak and reading the coordinate values that appear in the small window in the upper left corner of the computer screen. The value of x at the maximum can also be found by first analytically differentiating f with respect to x and setting the result equal to zero.

```
> diff(f,x)=0;
```

$$\frac{1}{100}\,\frac{\ln(6-x)}{1+\dfrac{x}{100}} - \frac{\ln\left(1+\dfrac{x}{100}\right)}{6-x} = 0$$

The output is a complicated transcendental equation that must be solved by numerical means. The command `fsolve`, which is based on Newton's method, is used to produce a floating-point solution of the last output equation.

```
> x:=fsolve(%,x);
```

$$x := 3.080667922$$

Thus, the value of x that maximizes the overall index is $x \approx 3.08$ trillion ehs,

```
> y:=y; #in trillion of ehs
```

$$y := 1.919332078$$

and therefore $y \approx 1.92$ trillion ehs. Notice that the percentage increase of the bomb group's budget is small (3.08%) even though the actual eh [sic] increase is more than that allotted for schools. However, in this model the school group benefits proportionally more because it gains 1.92 trillion ehs, which is a percentage increase of 192%. This conclusion is supported by calculating the separate effectiveness indices, B and S.

```
> B:=B; S:=S; #effectiveness indices for bombs and schools
```

$$B := 0.03034167921 \qquad S := 1.071354850$$

The effectiveness index S for schools is much larger than that for bombs.

PROBLEMS:

Problem 3-1: Drug concentration

If A units of a drug are injected into a patient, the concentration c (in mg/ml) of the drug in the bloodstream after t hours is given by $c(t) = A\,t\,e^{-t/3}$. The maximum safe concentration of the drug is 1 mg/ml.

(a) What amount A should be injected to reach the maximum safe concentration and when does this maximum occur? Plot the concentration over the time interval $t=0$ to 15 hours.

(b) By clicking on the plot, determine the approximate time when the concentration has dropped to 0.25 mg/ml.

(c) An additional amount of drug is to be administered when this concentration is reached. Use the `fsolve` command to determine the time to the nearest minute when this second injection is to be given.

Problem 3-2: Newton's method

Newton's method makes use of the first two terms of the Taylor series expansion of a function $f(x)$ about a point $x=a$, i.e., $f(x)=f(a)+(x-a)\,f'(a)+\cdots$, where the prime denotes a derivative with respect to x. The root of $f(x)$ corresponds to setting $f(x)=0$. Then $x \approx a - f(a)/f'(a)$. If x_1 is the first guess for the root, one repeatedly iterates the relation $x_{i+1}=x_i - f(x_i)/f'(x_i)$ for $i=1,2,\ldots$, until the solution has converged to an answer of acceptable accuracy.

In the text example, the root of the function

$$g(x) = \frac{\ln(6-x)}{100+x} - \frac{\ln(1+x/100)}{6-x} = 0$$

is found to be $x = 3.080667922$. Using Newton's method and the initial guess $x_1=2$, determine x_2, x_3, x_4, and x_5. How many iterations of Newton's method does it take to get the above value for x?

Problem 3-3: Tug of war

In a tug of war contest, each team consists of five persons. The average mass m of each person is 60 kilograms. Each individual on Team 1 exerts an average force $F_1 = 900\,e^{-0.1\,t}$ newtons as a function of time t in seconds, while each person on Team 2 exerts a force $F_2 = 950\,e^{-0.13\,t}$ newtons. The negative exponents indicate a tiring effect on the part of both teams. Neglecting any dissipation and assuming that the mass of the connecting rope can be neglected and that the two teams are initially at rest, answer the following questions:

(a) What is the acceleration, velocity, and displacement of the two-team system at arbitrary time $t > 0$?

(b) Plot the displacement up to $t = 7$ seconds.

(c) Which team is the first to pull the other team a distance of 0.75 meters?

(d) Using `fsolve`, find how long it takes to move this distance.

(e) If the object of the game is to pull the other team a distance of 1 meter, which team wins and at what time does it win? Use the `fsolve` command.

Problem 3-4: Richter earthquake magnitude scale

Large earthquakes may generate seismic waves with energies E as much as 10^{18} joules. The Richter magnitude scale has been historically used to express the magnitude of an earthquake as a much smaller number. The Richter magnitude M is related to E by the relation

$$M = 0.67 \log E - 2.9,$$

where the logarithm is to the base 10. The great San Francisco earthquake of 1906 had an energy of about 10^{17} joules.

(a) What was its Richter magnitude? (Use the `log10` command.)

(b) What energy corresponds to $M = 0$?

(c) Analytically solve for the energy E in terms of M.

(d) Plot the resulting formula for the range $M = 7$ to 9. Put the data point for the San Francisco earthquake on the same graph.

Problem 3-5: Income versus study time

In economics, the "utility function" U is a measure of how valuable or desirable alternative products or options are considered to be. Alice, an economics student, has a part-time job at the Hungry Heifer Steak House which pays $8 an hour. She deems her utility function for earning i dollars and spending s hours studying to be $U = i^{1/4}\,s^{3/4}$. The total amount of time she spends each week working in the restaurant and studying is 100 hours. Interpret Alice's utility function. How should she divide her time up in order to maximize her utility? What would her weekly income be? Arrive at the answer by carrying out the following steps. Express the utility function entirely in terms of s. Plot the utility function over the appropriate range. Use the mouse to determine the value of s that makes U a maximum. Analytically determine this value of s.

Problem 3-6: Alice's sweet tooth

Alice loves fruit and chocolates. Her utility function for f units of fruit and c units of chocolate is $U = f^{5/6}\, c^{1/3}$. She has \$49 to spend each month on fruit and chocolates. Fruit costs \$1 per unit and chocolate \$2 per unit.

(a) Make a three-dimensional plot of Alice's utility function.

(b) Make a plot of her utility function as a function of c alone.

(c) Graphically and analytically, determine how many units of fruit and of chocolate Alice should buy each month to maximize her utility function.

Problem 3-7: Ground-state energy of an electron

For an electron of mass m confined to a rectangular potential well of height V_0 and width a, the lowest or "ground-state" energy E must be found by solving the transcendental equation [Wie73],

$$X \tan X = \sqrt{R^2 - X^2}, \quad \text{where} \quad R \equiv \sqrt{\frac{2\,\pi^2\, m\, V_0\, a^2}{h^2}}, \quad X \equiv \sqrt{\frac{2\,\pi^2\, m\, E\, a^2}{h^2}},$$

and h is Planck's constant. Taking $m = 9 \times 10^{-31}$ kg, $V_0 = 1.6 \times 10^{-18}$ joules, $a = 10^{-10}$ meters, and $h = 6.63 \times 10^{-34}$ joule · seconds, what is the numerical value of the normalized ground-state energy E/V_0?

Problem 3-8: Pressure to sink a plate in mud [BF89]

The pressure (force per unit area) required to sink a large object in muddy soil lying above a hard soil base can be predicted by the pressure required to sink smaller objects in the same mud. According to M. G. Becker [Bec69], the pressure p in psi (pounds per square inch) required to sink a circular plate of radius r inches a distance d inches in the mud, where the hard soil base lies a distance $D > d$ below the mud surface, is given approximately by the formula

$$p = k_1\, e^{k_2\, r} + k_3\, r,$$

with $k_2 > 0$. The parameters k_1, k_2, k_3, depend on d and the consistency of the mud, but not on r.

(a) It is observed that to sink circular plates of radii 1, 2, and 3 inches to a depth of 1 foot in a certain muddy field requires pressures of 10, 12, and 15 psi, respectively. Determine the parameters k_1, k_2, and k_3.

(b) For the above field, what is the minimal radius of circular plate that is required to sustain a load of 500 lb without sinking more than 1 ft?

Problem 3-9: Taylor series expansion

Expand the following functions in a Taylor series as indicated:

(a) $f(x) = 1/\sqrt{1 + 3\,x^2}$ about $x = 0$ to order 10;

(b) $f(x) = x/(e^x - 1)$ about $x = 0$ to order 10;

(c) $f(x) = e^{\arctan(x)}$ about $x = \pi/4$ to order 6;

(d) $f(x) = \ln(1 + \sqrt{1 + x^2})$ about $x = 0$ to order 10;

(e) $f(x) = \ln(\sin x)$ about $x = 0$ to order 10.

Problem 3-10: The rule of 72, the measure of money growing
Investors soon learn that the number 72 is a magical number in financial circles. If 72 is divided by the annual investment return in percent, the resulting number is the number of years it will take to (approximately) double the original amount of money. For example, if one earns 6% annually on a savings bond, it will take $72/6 = 12$ years to double the money. Many investors are apparently not aware of why this rule works and how good it is for normal ranges of interest. It has to do with the way compound interest works. If the return is, say, 6%, then \$1 produces \$1.06 the next year, $1.06 \times 1.06 = (1.06)^2$ the second year, and so on. Let the number of years to double, according to the rule of 72, be $0.72/r$, where r is the annual rate of interest expressed as a decimal, e.g., $r = 0.06$ for the example above. Then, according to the rule of 72, $(1 + r)^{0.72/r} \approx 2$.

(a) Calculate the Taylor expansion of this expression about $r = 0$ out to r^4.

(b) Plot the difference between the Taylor expansion and 2 for $r=0$ to 0.12, i.e., 0 to 12%. At what percent annual return does the rule of 72 exactly agree with the Taylor expansion?

3.1.2 Kirchhoff Rules the Electrical World

"The rule is, jam tomorrow, and jam yesterday–
but never jam today."
"It must come sometimes to 'jam today,'" Alice objected.
"No, it can't," said the Queen. "It's jam every other day:
today isn't any other day, you know."
Lewis Carroll, *Through the Looking Glass* (1872)

In the world of electrical circuit theory, the application of two rules established by the German physicist Gustav Kirchhoff (1824–1887) to mathematical models of actual circuits helps in the design of complex useful electrical systems. In the simplest circuits the energy sources, batteries, are idealized to have constant potentials that push the current in only one direction (a direct current, DC). The resistance of the wires is neglected in comparison to that of the resistors. Figure 3.4 shows a simple DC circuit containing three batteries with voltages V_1, V_2, V_3, and five resistors with resistances R, $2\,R$, $3\,R$, $4\,R$, $6\,R$. The objective is to find the current through each resistor in terms of the given battery voltages and resistances. To achieve this, Kirchhoff's rules will be applied.

The first rule states that the algebraic sum of the potential drops around any closed loop is zero. The convention is that if one goes through a battery from the negative (labeled with a minus sign) terminal to the positive (plus sign) terminal, the potential change is positive. If one goes through the battery from plus to minus the potential change is negative. For each loop chosen, one assumes a direction for the current[2] i. Traveling through a resistor R ohms in

[2]Note that Maple reserves capital I, a common symbol for current, to stand for $\sqrt{-1}$.

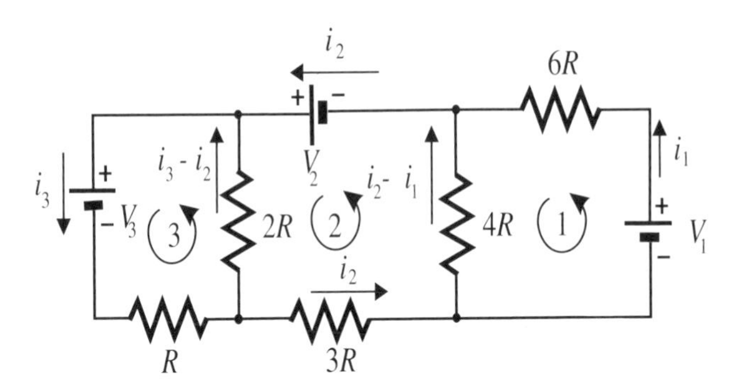

Figure 3.4: A simple DC circuit.

the sense of the assumed current produces a potential drop written as $-i\,R$. Going through the resistor in the opposite sense produces a $+i\,R$ contribution. The product $i\,R$ for the potential difference is just a statement of Ohm's law,[3] a well-known result for electrical circuits.

Kirchhoff's second rule states that the sum of the currents flowing out of a junction point (a dot in the figure) is equal to the sum of the currents flowing into the junction point. For example, if the current flowing through the battery V_1 is i_1 and assumed to be counterclockwise, and the current through the battery V_2 (also assumed counterclockwise) is i_2, then there is a current $i_2 - i_1$ flowing up through the $4\,R$ resistor. The sum of the currents into the junction between the battery V_2 and the $6\,R$ resistor is $i_1 + (i_2 - i_1) = i_2$, which matches the current out of the junction.

With this brief review behind us, our task is to solve for the currents through each resistor. We go through each of the loops in a counterclockwise sense and assume that the currents are flowing in that direction.

```
> restart:
```

Loop 1 is the loop furthest to the right in the circuit with the current i_1 through the battery labeled V_1. Loop 2 is the middle loop with the current through V_2 taken to be i_2. Finally, Loop 3 is the loop furthest to the left with the current through V_3 labeled as i_3. Applying Kirchhoff's potential drop rule to each loop yields the following system of equations for the three loops.

```
> Loop_1:=V[1]-6*R*i[1]+4*R*(i[2]-i[1])=0;
```

$$Loop\_1 := V_1 - 6\,R\,i_1 + 4\,R\,(i_2 - i_1) = 0$$

```
> Loop_2:=V[2]+2*R*(i[3]-i[2])-3*R*i[2]-4*R*(i[2]-i[1])=0;
```

$$Loop\_2 := V_2 + 2\,R\,(i_3 - i_2) - 3\,R\,i_2 - 4\,R\,(i_2 - i_1) = 0$$

```
> Loop_3:=-V[3]-R*i[3]-2*R*(i[3]-i[2])=0;
```

$$Loop\_3 := -V_3 - R\,i_3 - 2\,R\,(i_3 - i_2) = 0$$

[3]Discovered by another German physicist, George Ohm, in 1827.

The system of loop equations is analytically solved for the currents i_1, i_2, i_3.

```
> Current:=solve({Loop_1,Loop_2,Loop_3},{i[1],i[2],i[3]});
```

$$Current := \left\{ i_1 = \frac{1}{182}\,\frac{12\,V_2 - 8\,V_3 + 23\,V_1}{R},\ i_2 = \frac{1}{91}\,\frac{6\,V_1 + 15\,V_2 - 10\,V_3}{R},\right.$$
$$\left. i_3 = \frac{1}{91}\,\frac{+4\,V_1 + 10\,V_2 - 37\,V_3}{R} \right\}$$

In order to extract the individual currents, e.g., i_1, *Current* must be assigned. Otherwise, entering `i[1]` would generate the symbol i_1, not the analytic solution for this current.

```
> assign(Current):
```

As a check, the assigned values of the currents are substituted into a list of the three loop equations and expanded.

```
> expand([Loop_1,Loop_2,Loop_3]);
```

$$[0 = 0, 0 = 0, 0 = 0]$$

For each loop equation, we obtain $0 = 0$, confirming that each loop equation is satisfied by the analytically calculated currents. Then the current through each resistor is calculated for some specific values of the voltages and resistances. If, for example, $R=10$ ohms, $V_1=27$ volts, $V_2=10.5$ volts, and $V_3=5$ volts,

```
> R:=10: V[1]:=27: V[2]:=10.5: V[3]:=5:
```

the currents, in amperes, are as follows:

```
> Current_through_R_resistor:=i[3];
```

$$Current\_through\_R\_resistor := 0.03076923077$$

```
> Current_through_2R_resistor:=simplify(i[3]-i[2]);
```

$$Current\_through\_2R\_resistor := -0.2653846154$$

```
> Current_through_3R_resistor:=i[2];
```

$$Current\_through\_3R\_resistor := 0.2961538462$$

```
> Current_through_4R_resistor:=simplify(i[2]-i[1]);
```

$$Current\_through\_4R\_resistor := -0.09230769230$$

```
> Current_through_6R_resistor:=i[1];
```

$$Current\_through\_6R\_resistor := 0.3884615385$$

The appearance of a minus sign for a current indicates that the current is in the opposite direction to that assumed.

As a second, more advanced, example, we consider the circuit shown in Figure 3.5, which has an alternating current (AC) source, with real voltage amplitude V and frequency ω, three identical resistors R, a capacitor with capacitance C, and an inductor with inductance L. Kirchhoff's rules still apply at each instant of time t, but the concept of resistance must be generalized [FLS64] to that of impedance Z. Using the complex representation[4] $V\,e^{I\,\omega\,t}$ for the AC

[4]The real AC voltage is the real part of $V\,e^{I\,\omega\,t}$, namely, $V\cos(\omega\,t)$.

voltage, with $I = \sqrt{-1}$, the impedances of a resistor, inductor, and capacitor are $Z = R$ (as in the DC case), $Z = I\,\omega\,L$, and $Z = -I/(\omega\,C)$, respectively. The potential change across an impedance through which an AC current i is passing is equal to $-i\,Z$ in the direction of the assumed current, an obvious generalization of Ohm's law.

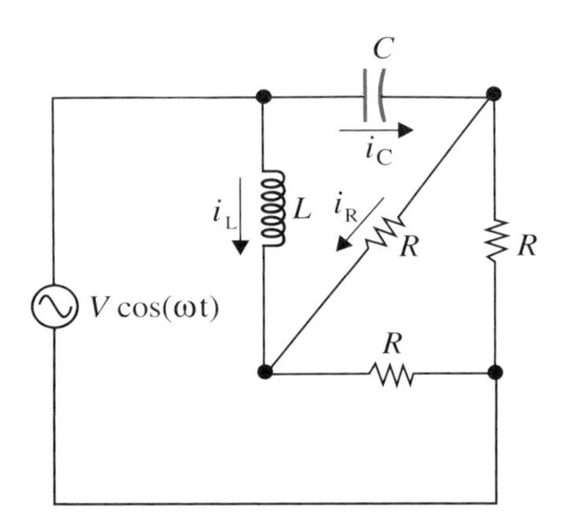

Figure 3.5: An AC circuit.

Suppose that our goal is to determine the steady-state current through the diagonal resistor in the AC circuit of Figure 3.5. Let i_C, i_L, and i_R be the amplitudes of the currents through the capacitor, the inductor, and the diagonal resistor. The instantaneous assumed directions of the currents are as indicated.

```
> restart:
```

The first loop equation is for the outermost loop of the circuit passing clockwise through the AC source. The loop equations are applied to the amplitudes. The voltage amplitude V is assumed to be real, but the various current amplitudes will in general be complex. (Note the Maple entry I for $\sqrt{-1}$.)

```
> Loop_1:=V+I/(omega*C)*i[C]-(i[C]-i[R])*R=0;
```

$$Loop\_1 := V + \frac{I\,i_C}{\omega\,C} - (i_C - i_R)\,R = 0$$

The second loop equation is taken counterclockwise through the inductor L, the diagonal resistor R, and the capacitor C.

```
> Loop_2:=-I*omega*L*i[L]+i[R]*R-I/(omega*C)*i[C]=0;
```

$$Loop\_2 := -I\,\omega\,L\,i_L + i_R\,R - \frac{I\,i_C}{\omega\,C} = 0$$

The third loop is taken counterclockwise through the diagonal resistor, the horizontal resistor, and the vertical resistor.

```
> Loop_3:=-i[R]*R-(i[L]+i[R])*R+(i[C]-i[R])*R=0;
```

$$Loop\_3 := -i_R\,R - (i_L + i_R)\,R + (i_C - i_R)\,R = 0$$

The three loop equations are solved for the three current amplitudes,

```
> solution:=solve({Loop_1,Loop_2,Loop_3},{i[C],i[R],i[L]});
```

$$solution := \left\{ i_R = \frac{I\,V\,(\omega^2\,L\,C+1)}{3\,\omega\,L + 2\,I\,\omega^2\,L\,C\,R - 2\,I\,R + R^2\,\omega\,C}, \right.$$
$$i_L = \frac{V\,(R\,\omega\,C - 3\,I)}{3\,\omega\,L + 2\,I\,\omega^2\,L\,C\,R - 2\,I\,R + R^2\,\omega\,C},$$
$$\left. i_C = \frac{V\,\omega\,C\,(3\,I\,\omega\,L + R)}{3\,\omega\,L + 2\,I\,\omega^2\,L\,C\,R - 2\,I\,R + R^2\,\omega\,C} \right\}$$

and the *solution* is assigned.

```
> assign(solution):
```

We are interested in the current i_R through the diagonal resistor. As it stands, the answer is expressed as the ratio of two complex quantities, both denominator and numerator involving $I = \sqrt{-1}$. To put i_R in the form (Real Part) + I (Imaginary Part), the complex evaluation command, evalc, is applied.

```
> current:=evalc(i[R]);
```

$$current := \frac{V\,(\omega^2\,L\,C+1)\,(2\,\omega^2\,L\,C\,R - 2\,R)}{(3\,\omega\,L + R^2\,\omega\,C)^2 + (2\,\omega^2\,L\,C\,R - 2\,R)^2} + \frac{I\,V\,(\omega^2\,L\,C+1)\,(3\,\omega\,L + R^2\,\omega\,C)}{(3\,\omega\,L + R^2\,\omega\,C)^2 + (2\,\omega^2\,L\,C\,R - 2\,R)^2}$$

The *current* expression is pretty formidable, so let's remove the real term by assuming that the frequency of the AC source is given by $\omega = 1/\sqrt{LC}$.

```
> omega:=1/sqrt(L*C):
```

With ω automatically substituted, the current is simplified with the radnormal command. This command normalizes expressions containing radical numbers.

```
> Diagonal_current:=radnormal(current);
```

$$Diagonal\_current := \frac{2\,I\,V\,\sqrt{L\,C}}{3\,L + R^2\,C}$$

The current amplitude is still expressed as a complex number. To relate it to the real voltage amplitude V, we shall recast the current amplitude in polar form. A complex number z can be rewritten as $z = re^{I\theta}$, where r is the modulus and θ is the phase angle. Assuming that all circuit parameters are positive,

```
> assume(V>0,L>0,C>0,R>0,omega>0,t>0);
```

the current amplitude through the diagonal resistor is converted to polar form.

```
> Diagonal_current:=polar(%);
```

$$Diagonal\_current := \mathrm{polar}\left(2\,\frac{V\,\sqrt{L\,C}}{3\,L + R^2\,C},\ \frac{\pi}{2}\right)$$

Note that the ditto sign allowed the polar command to be applied to the last output, not to the assume command line. The first argument in the polar command output is the modulus (magnitude), while the second is the phase angle of the current relative to that of the voltage source. The two arguments of *Diagonal_current* can be extracted with the operand (op) command.

```
> r:=op(1,Diagonal_current); theta:=op(2,Diagonal_current);
```

The real current through the diagonal resistor is the real part of $r\,e^{I(\omega\,t+\theta)}$, viz.,

```
> Re(r*exp(I*(omega*t+theta)));
```

$$-\frac{2\,V\,\sqrt{L\,C}\sin\left(\frac{t}{\sqrt{L\,C}}\right)}{3\,L+R^2\,C}$$

This result tells us that the maximum in the diagonal AC current occurs 1/4 of a cycle earlier than the maximum in the AC voltage source.

PROBLEMS:

Problem 3-11: Double square configuration

A piece of uniform (constant resistance) wire is made up into two connected squares arranged as a horizontal figure 8 as in Figure 3.6.

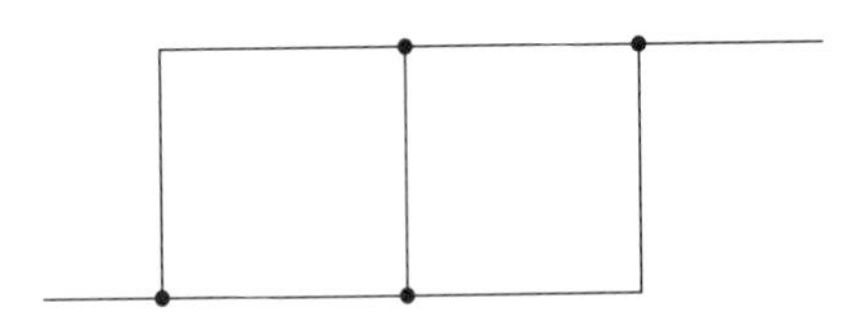

Figure 3.6: Horizontal figure 8 configuration.

A current enters the lower left-hand corner of the figure 8 and leaves at the upper right-hand corner. What fraction of the entering current passes through the common side of the two squares?

Problem 3-12: DC circuit

Five resistors, of resistance $R_1=2$, $R_2=4$, $R_3=6$, $R_4=2$, and $R_5=3$ ohms, are connected to a 12-volt battery as shown in the left circuit of Figure 3.7. What is the current in each resistor? What is the potential difference between the points A and B?

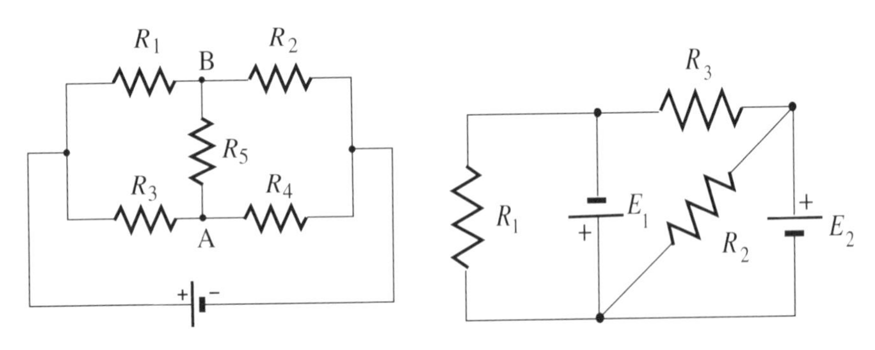

Figure 3.7: Left: circuit for Problem 3-12. Right: circuit for Problem 3-13.

Problem 3-13: Another DC circuit

Two batteries with $E_1 = 6$ and $E_2 = 3$ volts are connected to three resistors with $R_1 = 6$, $R_2 = 4$, and $R_3 = 2$ ohms as shown in the right circuit of Figure 3.7. Find the current in each resistor and the current in each battery.

Problem 3-14: Warning! Do not attempt by hand.

The circuit diagram in Figure 3.8 shows 14 resistors with two batteries present. Analytically determine the currents through each resistor.

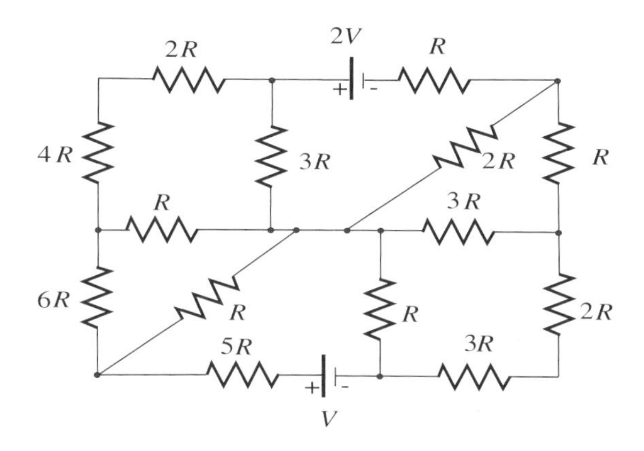

Figure 3.8: Circuit with 14 resistors.

Problem 3-15: An AC circuit

Consider the circuit [LC90] shown in Figure 3.9.

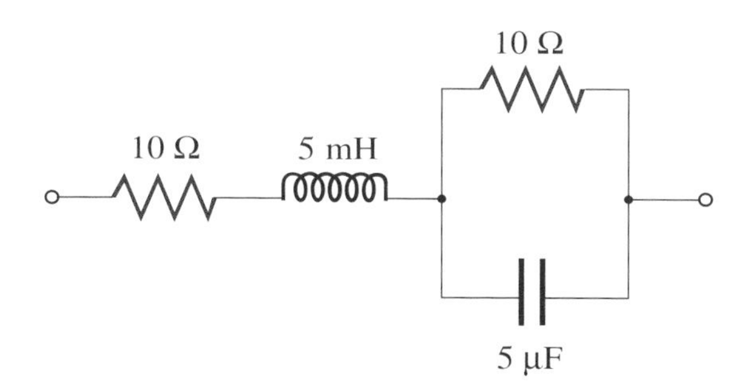

Figure 3.9: AC circuit.

(a) Calculate the impedance Z of the circuit for an arbitrary frequency ω.

(b) What are the magnitude and phase angle of Z at 1 kHz?

(c) Can the real part of the impedance become negative?

(d) For what frequency range is the circuit equivalent to (i) a resistor in series with an inductor; (ii) a resistor in series with a capacitor?

(e) At what frequency is the circuit equivalent to a pure resistance?

Problem 3-16: Phase shifter

It is often necessary to shift the phase of a signal. A simple circuit [LC90] for doing this is shown in Figure 3.10. The resistances are adjustable, but are kept equal. Use the polarities shown. Here V_i is the voltage of the top terminal with respect to the bottom one, and V_0 is the voltage of the right terminal with respect to the left-hand one. The connection through V_0 carries no current.

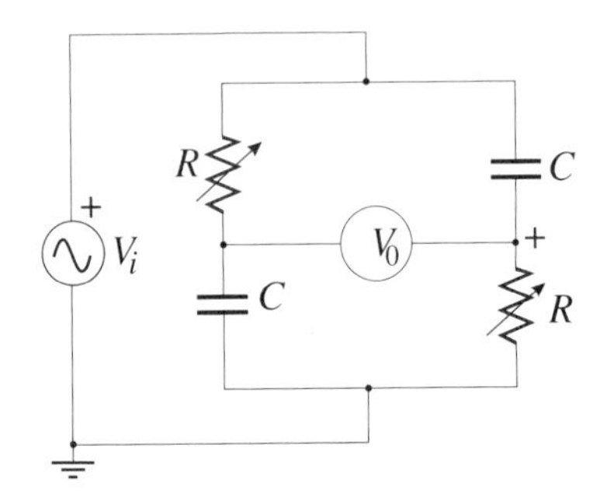

Figure 3.10: Phase shifter.

(a) Show that $V_0/V_i = \exp[2\,I\,\arctan(1/(R\,\omega\,C))]$.

(b) Draw a graph of the phase of V_0 with respect to V_i, in the range $R\,\omega\,C = 0.1$ to 10. Use a log scale for $R\,\omega\,C$.

3.1.3 The Window Washer's Secret

A professor can never better distinguish himself in his work than by encouraging a clever pupil, for the true discoverers are among them, as comets among the stars.
Carl Linnaeus, Swedish botanist (1707–1778)

A retired engineering professor is getting too old to climb a tall ladder to wash the windows on the second floor of his house. So he calls on the services of the Dirty Bird window washing service, which sends out a young lady by the name of Heather. Striking up a conversation with her, the elderly professor notes that Heather seems to be too well educated to be permanently employed at the window washing occupation. As the professor surreptitiously watches, she does some apparently odd things. The surface on which the ladder is standing is fairly smooth and slippery, so Heather experiments with leaning the ladder at different angles to the vertical and concludes that the ladder will begin to slip when the angle is about 45°. After some deep thought and furious scribbling, she is heard to exclaim, "Aha! Since the ladder has a mass of about 25 kg and my mass[5] is 75 kg, and since I want to go three-quarters of the way up the

[5]Heather rows in the varsity eights and has "bulked up" for that purpose.

ladder, I should keep the angle that the ladder makes with the vertical smaller than 33°." The professor is intrigued by what he observes. He confronts the window washer and says, "You must have taken an introductory college physics course and done well in the subject." Heather grins and confesses to the fact of actually being a premed science student who is trying to earn money to pay for next semester's tuition. We shall now reproduce her reasoning.

The ladder problem is modeled by the two-dimensional picture shown in Figure 3.11. Assuming the ladder to be of length L and of uniform composition,

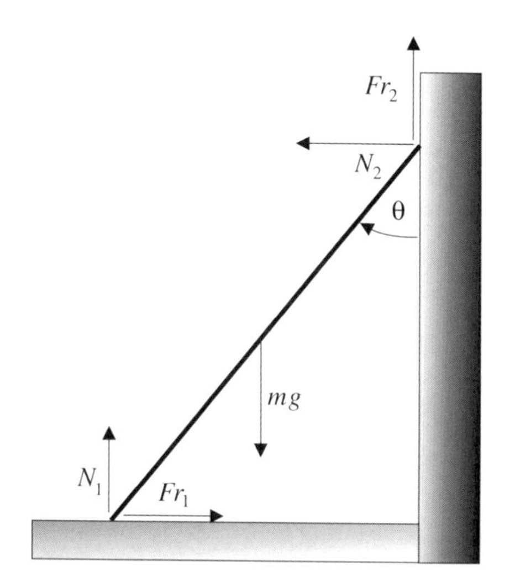

Figure 3.11: Free body diagram for window washer's problem.

all of its weight is considered for purposes of calculation to be located at a single point (the *center of gravity*) halfway along the ladder. If the ladder's mass is m, its weight is $w = m\,g$, where g is the acceleration due to gravity. Heather makes a *free-body diagram* indicating all of the external forces acting on the ladder. The ground and the wall exert normal (perpendicular) forces N_1 and N_2 on the bottom and top ends of the ladder, respectively. If the ladder is not to slip, there also must be frictional forces Fr_1 and Fr_2 acting on the ends of the ladder as shown. If μ is the coefficient of static (not moving) friction, the maximum values of the frictional forces are $Fr_1 = \mu\,N_1$ and $Fr_2 = \mu\,N_2$. If Heather's weight is $W = M\,g$, then the following are entered into the calculation.

```
> restart:
> Fr[1]:=mu*N[1]: Fr[2]:=mu*N[2]: w:=m*g: W:=M*g:
```

In order to calculate μ, Heather first assumes that she is not on the ladder. For static equilibrium, the horizontal and vertical components (in the x- and y-directions, respectively) of the forces must balance or, equivalently, algebraically sum to zero.

```
> x_eq:=N[2]-Fr[1]=0;
```

$$x\_eq := N_2 - \mu\, N_1 = 0$$

```
> y_eq:=N[1]+Fr[2]-w =0;
```

$$y\_eq := N_1 + \mu\, N_2 - m\, g = 0$$

A third condition, to prevent rotation of the ladder, is that the algebraic sum of the moments[6] (torques) of the forces must add to zero. If moments are taken about the bottom end of the ladder, the forces N_1 and Fr_1 have zero moment arms and do not contribute to the rotational equilibrium equation. Including the moments of the remaining forces, the rotational equilibrium equation is

```
> rot_eq:=N[2]*(L*cos(theta))+(Fr[2])*(L*sin(theta))
      -w*((L/2)*sin(theta))=0;
```

$$rot\_eq := N_2\, L\cos(\theta) + \mu\, N_2\, L\sin(\theta) - \frac{1}{2}\, m\, g\, L\sin(\theta) = 0$$

Since the length L is a common factor, it is divided out of the equation.

```
> rot_eq:=simplify(rot_eq/L);
```

$$rot\_eq := N_2\cos(\theta) + \mu\, N_2\sin(\theta) - \frac{1}{2}\, m\, g\sin(\theta) = 0$$

Assuming that the angle θ at which slipping occurs is known, the three equations (x_eq, y_eq,rot_eq) are solved symbolically for the unknowns N_1, N_2, and μ.

```
> sol:=solve({x_eq,y_eq,rot_eq},{N[1],N[2],mu});
```

$$sol := \left\{\mu = \%1,\ N_1 = \frac{1}{2}\,\frac{m\, g\,(\%1\tan(\theta) + 2)}{1 + \%1\tan(\theta)},\ N_2 = \frac{1}{2}\,\frac{\%1\, m\, g\,(\%1\tan(\theta) + 2)}{1 + \%1\tan(\theta)}\right\}$$

$$\%1 := \mathrm{RootOf}(-\tan(\theta) + \_Z^2\tan(\theta) + 2\,\_Z)$$

In the subexpression %1, RootOf is a placeholder for all the roots of the quadratic equation in $\_Z$. Heather assigns the solution, *sol*, and selects μ, whose value she first wants to calculate.

```
> assign(sol): mu:=mu;
```

$$\mu := \mathrm{RootOf}(-\tan(\theta) + \_Z^2\tan(\theta) + 2\,\_Z)$$

To find the actual roots of the quadratic equation contained in RootOf, the `allvalues` command is applied to the previous line.

```
> sol2:=allvalues(mu);
```

$$sol2 := \frac{-1 + \sqrt{1 + \tan(\theta)^2}}{\tan(\theta)},\ -\frac{1 + \sqrt{1 + \tan(\theta)^2}}{\tan(\theta)}$$

Since the negative square root answer will yield a negative value for μ, which is physically unacceptable, the positive square root answer must be selected.

```
> mu:=sol2[1]; #choose positive square root
```

$$\mu := \frac{-1 + \sqrt{1 + \tan(\theta)^2}}{\tan(\theta)}$$

[6] The moment of a force is the product of the magnitude of the force and the perpendicular distance from the line of action of the force to the axis about which the moment is calculated.

Heather, the window washer (science major), observed that the ladder slipped at 45° or $45\times\pi/180=\pi/4$ radians, so the numerical value of μ can be evaluated.

```
> theta:=evalf(Pi/4): mu:=mu;
```

$$\mu := 0.4142135620$$

Thus, the coefficient of static friction μ is about 0.41. This value will be used in the remainder of the calculation.

Now Heather redoes the calculation with her weight W concentrated at a point three-quarters of the way up the ladder. So that Maple will not remember the value $\theta=\pi/4$ in the first part of the calculation or the values obtained for N_1 and N_2 in *sol*, the variables θ, N_1, and N_2 must be unassigned.

```
> unassign('theta','N[1]','N[2]');
```

The horizontal force equation remains unchanged, but with the numerical value of μ now appearing.

```
> x_eq2:=N[2]-Fr[1]=0;
```

$$x\_eq2 := N_2 - 0.4142135620\,N_1 = 0$$

Heather's weight W is included in the vertical force equation.

```
> y_eq2:=N[1]+Fr[2]-w-W=0;
```

$$y\_eq2 := N_1 + 0.4142135620\,N_2 - m\,g - M\,g = 0$$

The rotational equilibrium equation is also modified, the contribution of Heather's moment being included (last term in the following command line).

```
> rot_eq2:=N[2]*(L*cos(theta))+Fr[2]*(L*sin(theta))
        -w*((L/2)*sin(theta))-W*(3/4)*L*sin(theta)=0;
```

$$rot\_eq2 := N_2\,L\cos(\theta) + 0.4142135620\,N_2\,L\sin(\theta) - \frac{1}{2}\,m\,g\,L\sin(\theta) - \frac{3}{4}\,M\,g\,L\sin(\theta) = 0$$

```
> rot_eq2:=simplify(rot_eq2/L);
```

$$rot\_eq2 := N_2\cos(\theta) + 0.4142135620\,N_2\sin(\theta) - 0.5000000000\,m\,g\sin(\theta) - 0.7500000000\,M\,g\sin(\theta) = 0$$

The three equilibrium equations are solved for the unknowns N_1, N_2, and θ (output not shown here) and the solution *sol3* assigned.

```
> sol3:=solve({x_eq2,y_eq2,rot_eq2},{N[1],N[2],theta});
> assign(sol3):
```

The values for the masses are entered and the angle θ in radians is determined,

```
> m:=25: M:=75: theta:=theta;
```

$$\theta := 0.5787993638$$

or, on using the following `convert` command, expressed in degrees.

```
> max_angle:=evalf(convert(theta,degrees)); #angle in degrees
```

$$max\_angle := 33.16276073\,degrees$$

Thus Heather concludes that the angle with the vertical should be kept below about 33°.

When Heather is done washing the professor's windows, the professor's wife invites her in for a drink of lemonade while her husband searches for his wallet. The professor's wife is pleased with the job that Heather has done and would like to learn a little more about her and her career aspirations. Both the professor and his wife like to encourage promising students and have missed the contact with eager, bright, young people since his retirement.

PROBLEMS:

Problem 3-17: Coulomb's law

The force between two point charges Q_1 and Q_2 separated by a distance r is given by Coulomb's law, which in SI units states that the magnitude F_e of the electrical force between the charges is given by $F_e = Q_1 Q_2/(4\pi\epsilon_0 r^2)$, where ϵ_0 is the permittivity of free space. The force acts along the line between the two charges and is repulsive (attractive) for charges of the same (opposite) sign.

Two point charges of equal mass m and charge Q are suspended from a common point by two threads of negligible mass and each of length L. Show that at equilibrium the inclination angle α of each thread with the vertical is given by $\sin^3\alpha/\cos\alpha = Q^2/(16\pi\epsilon_0 m g L^2)$.

Problem 3-18: Suspension bridge

A suspension bridge in Rainbow County is to span a deep river gorge 54 m wide as shown in Figure 3.12. The "floor" of the bridge is a steel truss of 48,000 kg.

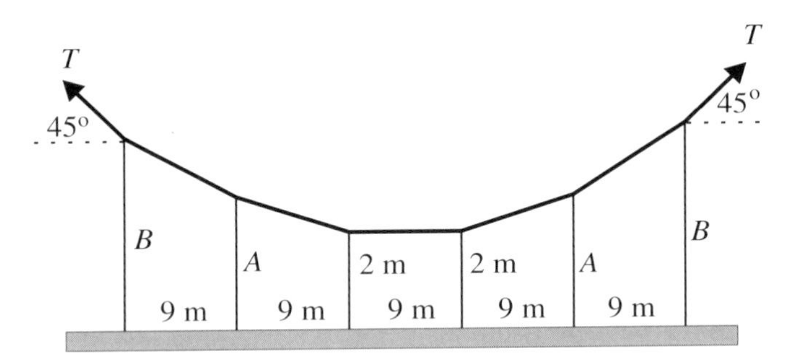

Figure 3.12: Suspension bridge.

The six pairs of vertical cables are spaced 9 m apart and are to carry an equal amount of the weight. The two central pairs of vertical cables are 2 m in length. The end cables of the suspension arc make an angle of 45° with the horizontal. Determine the lengths A and B of the remaining vertical cables and the tension T in the end cables of the arc. Neglect the weights of all cables.

Problem 3-19: Traffic lights

On one of Metropolis's main streets, three traffic lights, each of mass 20 kg, hang from a wire stretched between two telephone poles 15 m apart. The horizontal spacing of the traffic lights is uniform. At each pole, the wire makes a downward angle of 10° with the horizontal. Determine the tensions in all the segments of wire as well as the distance of each lamp below the horizontal line.

Problem 3-20: Pushing a box

A cubical wooden box of mass 75 kg and 0.5 m in length on each side sits on a concrete floor in Rob's basement. The box is of uniform density and the coefficient of static friction between the floor and the box is $\mu_s = 0.80$. If Rob exerts a sufficiently strong horizontal push against one side of the box, it will either tip over or start sliding without tipping over, depending on how high above the floor level he pushes. What is the maximum height at which Rob can push if he wants the box to slide? What is the magnitude of the force that he must exert to start the box sliding?

Problem 3-21: Playing cards

While playing poker with his engineering colleagues, Russell leans two cards against each other to form an "A-frame" roof. If the frictional coefficient between the bottom of the cards and the table is μ_s, what is the maximum angle that the cards can make with the vertical without slipping?

Problem 3-22: A challenging inclined plane problem

The inclined plane problem that follows, with certain numerical values supplied to make the problem easier, was once used on one of the author's first-year physics exams. After making up the exam, the author gave it to his colleague, Professor X, to produce the solution key. When Professor X encountered the problem, he tried to solve it first symbolically before substituting in the numerical values of the masses, coefficients of friction, etc. This is the standard method of attack favored by physics professionals, but not by the majority of first-year students. Professor X was able to easily set up the relevant general equations, but struggled to solve them correctly analytically. It turned out that substituting in the numerical values first and evaluating the coefficients in the algebraic equations made the problem considerably easier to solve. However, if Professor X had used a CAS, he would have had no difficulty in solving the general symbolic problem. Here is that challenging inclined plane problem for you to try. A symbolic solution is sought, so do not substitute numbers until

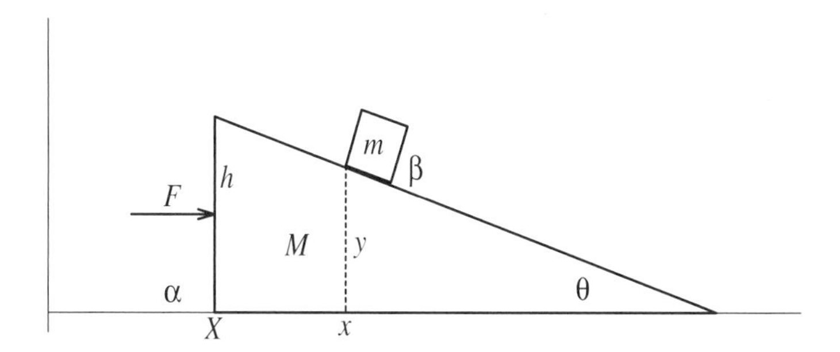

Figure 3.13: Geometry of the inclined plane problem.

you are told to do so! Referring to Figure 3.13, a horizontal force F is applied to the inclined plane whose mass is M and angle with the horizontal is θ. The coefficient of kinetic friction between the bottom of the inclined plane and the level surface is α. Resting on the inclined plane is a block of mass m. The coefficient of kinetic friction between the block and the upper surface of the inclined plane is β.

(a) Make separate free body diagrams for the block and for the inclined plane, indicating all of the external forces acting on each. Clearly label your forces. Be careful here, because this is where the heart of the physics is.

(b) Choose the horizontal direction to the right as the positive x-axis and the upward vertical direction as the positive y-axis. Resolve the forces and accelerations into their x- and y-components.

(c) Using Newton's second law, write down the equations for the x- and y-components of the accelerations of the block and of the inclined plane.

(d) From the geometry of the figure, write down an equation that relates the acceleration components to each other in terms of the angle θ. *Hint*: Referring to the figure, note that $\tan\theta = (h - y)/(x - X)$.

(e) Express the frictional forces in terms of the normal forces.

(f) Write down the normal force that is exerted by the level surface on the bottom of the inclined plane.

(g) Solve the system of equations for the three acceleration components and the normal force on the block in terms of θ, F, α, β, M, m, and g (the acceleration due to gravity). The form of the answers is quite formidable! You will appreciate the sweat raised on Professor X's brow in trying to carry out this step by hand without any mistakes.

(h) Given $M = 8\,\text{kg}$, $m = 2\,\text{kg}$, $\alpha = 0.2$, $\beta = 0.6$, $F = 37\,\text{N}$, $\theta = 35°$, and $g = 9.8\,\text{m/s}^2$, determine the numerical values of the acceleration components and the normal forces.

3.1.4 The Science Student's Summer Job Interview

I evidently knew more about economics than my examiners.
John Maynard Keynes, English economist (1883–1946),
explaining why he performed badly in the Civil Service examinations

Wanting a job experience more directly related to science than the window washing job held last summer, our premed student Heather goes for an interview with an engineering firm that has a student work-term job available. After some preliminary chitchat, the interviewer gets down to business and says, "I notice that you have done quite well in your courses and that you have taken an economics course as one of your options. Let's see whether you remember what you have learned. Here are a couple of little problems for you taken from my old economics text by Riggs, Rentz, Kahl, and West [RRKW86]. I would like to know what you would recommend or conclude in each case. Although you could do the problems by hand, we like our students to have some computing skills. From your résumé it appears that you have used the computer algebra system Maple. Therefore, I would like you to sit at this computer and do the problems with Maple. Be sure to clearly explain your reasoning and show me that you know how to plot functions and obtain graphical and algebraic solutions. Here's the first problem. Which alternative is the best choice?"

An engineering firm has a contract to supervise construction of a sewage treatment plant in an isolated town. It is estimated that the installation phase will probably last at most 2 years and two of the firm's engineers will be required to supervise the construction. Separate living accommodations for the two engineers and an office will be needed. The following alternatives are available:

(1) A building can be rented with furnished living quarters and an office for \$3000 a month including utilities and upkeep.

(2) The firm could rent office space at \$800 a month and buy two furnished trailers at a cost of \$24,000 each. The trailer company has agreed to buy back the used trailers for 40% of the purchase price at any time up to 2 years. The water and electricity, site rental, and upkeep of the trailers is \$200 per trailer per month.

(3) Finally, the firm could buy the two house trailers as in the second alternative and a third smaller one for the office for \$16,000. The same buyback conditions apply, since it is the same dealer as in alternative **(2)**. The \$200 per month outlay for each trailer also still holds.

"OK," says Heather, as she sits down at the computer, "we are going to need some specialized plots, so let's start by including the plots library package.

```
> restart: with(plots):
```

The total cost, TC_1, of alternative 1 after n months will be as follows:

```
> TC[1]:=3000*n;
```

$$TC_1 := 3000\,n$$

Since it is anticipated that the trailers will be sold back within 2 years, they cost 60% of the original purchase price. Thus, the total cost of alternative (2) after n months is given by the following input line.

```
> TC[2]:=(2*24000*0.6)+((2*200)+800)*n;
```

$$TC_2 := 28800.0 + 1200\,n$$

For the third alternative, the total cost is given by TC_3.

```
> TC[3]:=(2*24000+16000)*0.6+(3*200)*n;
```

$$TC_3 := 38400.0 + 600\,n$$

We now create a plot for the three total cost formulas and put labels and a title on the graph.

```
> P1:=plot({TC[1],TC[2],TC[3]},n=0..24,labels=["n","D"],color
      =black,title="Total costs in dollars(D) vs months(n)"):
```

So that the three alternative curves can be identified, the `textplot` command will be used to add text names for the alternatives to the above plot. The locations of the names are determined by trial and error.

```
> P2:=textplot([[10,21000,"TC[1]"],[5,30000,"TC[2]"],
      [5,45000,"TC[3]"]]):
> display({P1,P2},tickmarks=[3,3]);
```

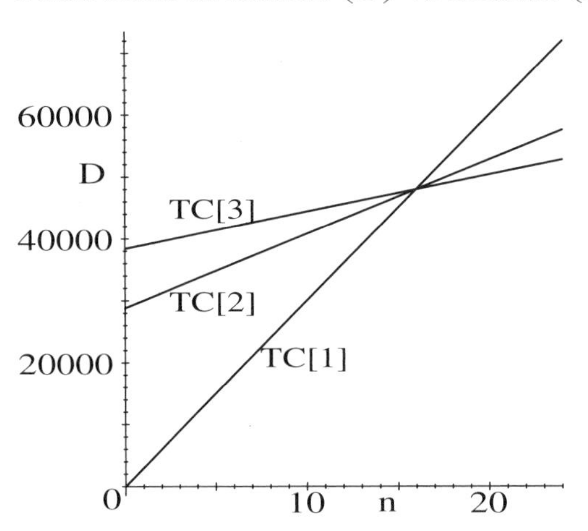

Figure 3.14: Graph of the three alternatives.

As you can see from the figure (Figure 3.14), there is apparently a single break-even point (BEP) at which the three lines cross. By clicking on the plot and then on the BEP, one can see that the crossing point occurs for n about 16 months. Better yet, one can equate the total costs, e.g. for the first and second alternatives, and algebraically solve for n.

```
> eq:=TC[1]=TC[2];
```

$$eq := 3000\,n = 28800.0 + 1200\,n$$

```
> BEP:=solve(eq,n);
```

$$BEP := 16.$$

Equating the total costs for the first and third alternatives and solving for n,

```
> BEP:=solve(TC[1]=TC[3],n);
```

$$BEP := 16.$$

yields exactly the same value for `BEP`. So 16 months is indeed the BEP for the three alternatives. It appears that the choice must be between alternatives 1 and 3. Up to 16 months, the straight rent alternative (1) is the cheapest way to go. After 16 months, alternative 3 (buying three trailers) is less costly. Since it is not certain that the project will last 24 months, and to avoid possible loan expenses in buying the trailers and the hassle with trade-in of the used trailers, it might be best to choose the first alternative."

"Very good! I like your approach and your ability to articulate your reasoning and conclusion," replies the interviewer.

"Now here's a second hypothetical case for you to examine."

A manufacturing plant keeps monthly records of operating expenses and revenues in turning out and selling a new product. By making use of a least squares polynomial fit, it is determined that if n is the number of units made and sold per month, the total cost (TC) in dollars to produce n units is given by the formula

$$TC = 200{,}000 + 4\,n + 0.005\,n^2, \tag{3.1}$$

while the total revenue (TR) is given by the sales price (SP) per unit,

$$SP = 100 - 0.001\,n, \tag{3.2}$$

multiplied by the number of units. The plant is designed to produce 12,000 units per month. Based on these formulas, determine the BEPs at which the profit is zero and the levels of output that produce the largest profit and the least average unit cost.

"To get a feeling for the total cost and revenue, I will plot the formulas that you have given to me," states Heather.

```
> restart: with(plots):
```

"The total cost, TC, for n units is entered,

```
> TC:=200000+4*n+0.005*n^2;
```

$$TC := 200000 + 4\,n + 0.005\,n^2$$

along with the sales price, SP, per unit.

```
> SP:=100-0.001*n;
```

$$SP := 100 - 0.001\,n$$

The total revenue, TR, for n units is equal to $SP \times n$.

```
> TR:=SP*n;
```

$$TR := (100 - 0.001\,n)\,n$$

As in the first problem, it is convenient to put both graphs in the same plot with the BEPs indicated.

```
> P1:=plot(TC,n=0..15000,color=blue):
> P2:=plot(TR,n=0..15000,color=red):
> P3:=textplot([[3000,194000,"BEP"],[12500,1180000,"BEP"],
   [7300,900000,"Total Revenue"],[7700,370000,"Total Cost"]]):
> display({P1,P2,P3},view=[0..15000,0..1500000],
  tickmarks=[2,2],labels=["n","D"]);
```

The profit is the difference between the total revenue and total cost curves. As can be seen from the figure (Figure 3.15), for small n the total cost lies above the total revenue, and the profit is negative, i.e., losses occur.

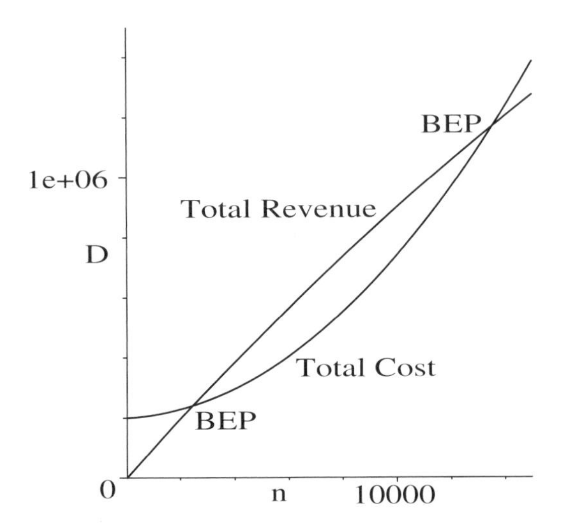

Figure 3.15: Total cost and total revenue curves.

As n is increased, there is a BEP at which the profit passes through zero and becomes positive, i.e., the company makes money. At much larger n there is a second BEP at which losses again occur. Although we could again click on the graph to find approximate values for the two BEPs, the profit formula is easily derived,

```
> Profit:=expand(TR-TC);
```

$$Profit := 96\,n - 0.006\,n^2 - 200000.$$

and the BEPs found by solving the quadratic equation in n:

```
> BEP:=fsolve(Profit=0);
```

$$BEP := 2462.250758,\ 13537.74924$$

The BEPs are at $n \approx 2462$ and $n \approx 13{,}538$ units. As long as the plant operates between the two BEPs it will make a profit. To find the maximum profit, simply differentiate the profit formula with respect to n and set the result equal to zero.

```
> eq:=diff(Profit,n)=0;
```

$$eq := 96 - 0.012\,n = 0$$

```
> N[max]:=solve(eq,n);
```

$$N_{max} := 8000.$$

The maximum profit occurs for an output of 8000 units per month. The average unit cost, AUC, is given by dividing the total cost by n.

```
> AUC:=expand(TC/n);
```

$$AUC := \frac{200000}{n} + 4 + 0.005\,n$$

The least average unit cost is found by differentiating AUC with respect to n and setting the result equal to zero."

```
> eq2:=diff(AUC,n)=0;
```

$$eq2 := -\frac{200000}{n^2} + 0.005 = 0$$

```
> Sol:=solve(eq2,n);
```

$$Sol := 6324.555320, \ -6324.555320$$

Heather concludes with, "Thus, selecting the positive answer,

```
> N[min]:=Sol[1];
```

$$N_{min} := 6324.555320$$

and rounding it off to the nearest integer,

```
> N[min]:=round(%);
```

$$N_{min} := 6325$$

the least average unit cost occurs for an output of 6325 units per month. Would you like me to say any more on this case, or solve any more problems?"

To which the interviewer replies, "No, it is clear that you know some of the basic concepts of engineering economics and how to apply them effectively. Our engineering firm would be only too happy to offer you a summer job that makes use of the skills you have demonstrated!"

PROBLEMS:

Problem 3-23: Widget production

The selling price (SP) in dollars per widget is given by $SP = 21{,}000/\sqrt{n}$, where n is the number of widgets produced per day. The total cost (TC) in dollars per day is given by $TC = 100{,}000 + 1000\,n$.

(a) Plot the total revenue (TR) formula and TC in the same graph.

(b) Derive the profit formula and plot it.

(c) Determine the break-even values of n. First obtain approximate values by clicking on the graph and then more precise values using `fsolve`.

(d) At what n value is the profit a maximum? First obtain an approximate value and then use an analytic approach.

Problem 3-24: Another revenue problem

A small manufacturing plant can sell all n items it produces per hour at a selling price of $750 per item. The total cost function, in thousands of dollars, is given by $TC = (n^3 - 8\,n^2 + 25\,n + 30)/25$.

(a) Construct a graph of the cost and total revenue curves with the scale in thousands of dollars and such that the BEPs are clearly seen. Add text to the graph indicating the two curves and the BEPs.

(b) Derive the profit formula and plot it.

(c) Determine accurate numerical values for the BEPs.

(d) Derive an analytic formula for n that corresponds to the maximum profit. Numerically solve for this value of n. How many items are produced per 8 hour shift?

3.1.5 Envelope of Safety

Mechanics is the paradise of the mathematical sciences because by means of it one comes to the fruits of mathematics.
Leonardo da Vinci, Italian painter, sculptor, architect, scientist, musician, and natural philosopher (1452–1519)

A previously dormant volcanic mountain in the North Cascades range of the Pacific Northwest has erupted and is throwing rocks into the atmosphere at speeds of up to $Vo=700$ m/s. Colleen's sister Sheelo, who is a part-time *National Geographic* photographer, has been assigned to film the spectacle from the air. Hiring an aircraft in Seattle, Sheelo instructs the pilot to get as close to the erupting mountain as safety will allow. Neglecting the volcano's height and air resistance, and assuming that rocks are thrown out uniformly in all directions, what is the envelope of safety that the hired plane should stay outside?

```
> restart: with(plots):
```

If Vo is the initial speed of a rock and its initial angle with the horizontal is θ, the distance that the rock will travel horizontally in time t seconds is given by

```
> xeq:=x=Vo*cos(theta)*t;
```

$$xeq := x = Vo\cos(\theta)\,t$$

In the same time, the rock will rise through a vertical distance,

```
> yeq:=y=Vo*sin(theta)*t-(1/2)*g*t^2;
```

$$yeq := y = Vo\sin(\theta)\,t - \frac{g\,t^2}{2}$$

where g is the acceleration due to gravity. Then xeq is solved for the time,

```
> t:=solve(xeq,t);
```

$$t := \frac{x}{Vo\cos(\theta)}$$

and the expression for t is automatically substituted into yeq.

```
> yeq;
```

$$y = \frac{\sin(\theta)\,x}{\cos(\theta)} - \frac{1}{2}\,\frac{g\,x^2}{Vo^2\cos(\theta)^2}$$

For a given value of θ, the last result is a parabolic equation relating x and y, i.e., neglecting air resistance, the rocks travel along parabolic trajectories. This equation can be cast into a simpler form by substituting the trigonometric identities $\sin\theta = \cos\theta \times \tan\theta$ and $1/\cos^2\theta = 1 + \tan^2\theta$.

```
> yeq2:=subs({sin(theta)=cos(theta)*tan(theta),
        1/cos(theta)^2=1+tan(theta)^2},yeq);
```

$$yeq2 := y = \tan(\theta)\,x - \frac{1}{2}\,\frac{g\,x^2\,(1+\tan(\theta)^2)}{Vo^2}$$

To find the envelope of safety, we must determine the upper bounding curve that is tangent to all possible parabolas. This can be accomplished as follows. Since $y = y(\theta, x)$, then $dy = (\partial y/\partial\theta)\,d\theta + (\partial y/\partial x)\,dx$. At a fixed value of x, the upper bounding curve is determined by setting dy equal to zero, which implies that $(\partial y/\partial\theta) = 0$. Thus, the curve is determined by setting the derivative of *yeq2* with respect to θ equal to zero,

```
> diff(yeq2,theta);
```

$$0 = (1+\tan(\theta)^2)\,x - \frac{g\,x^2\tan(\theta)\,(1+\tan(\theta)^2)}{Vo^2}$$

solving for $\tan\theta$,

```
> sol:=solve(%,tan(theta));
```

$$sol := I,\ -I,\ \frac{Vo^2}{g\,x}$$

and selecting the real solution (the third answer in *sol*).

```
> tan(theta):=sol[3];
```

$$\tan(\theta) := \frac{Vo^2}{g\,x}$$

With the above assignment for $\tan\theta$, *yeq2* becomes

```
> yeq2;
```

$$y = \frac{Vo^2}{g} - \frac{g\,x^2\left(1+\dfrac{Vo^4}{g^2\,x^2}\right)}{2\,Vo^2}$$

and expanding yields the equation *yeq3* for the envelope of safety,

```
> yeq3:=expand(yeq2);
```

$$yeq3 := y = \frac{Vo^2}{2\,g} - \frac{g\,x^2}{2\,Vo^2}$$

which is also parabolic in shape. Taking $g \approx 10$ m/s^2 to make the coefficients somewhat nicer, and inserting the upper limit $Vo = 700$ m/s^2 on the initial speed of the rocks, *yeq3* for the envelope of safety becomes

```
> Vo:=700: g:=10: yeq3;
```

$$y = 24500 - \frac{x^2}{98000}$$

Unassigning $\tan\theta$, *yeq2* for the parabolic trajectory for a given initial angle θ is given by

```
> unassign('tan(theta)'): yeq2;
```

$$y = \tan(\theta)\,x - \frac{1}{98000}\,x^2\,(1+\tan(\theta)^2)$$

The following do loop creates a set of plots of 41 parabolas for different values of θ ranging from $\theta = 1.51 - 20 \times 0.075 = 0.01$ radians up to $1.51 + 20 \times 0.075 = 3.01$ radians (just under 180° or 3.14 radians).

```
> for i from -20 to 20 do
> theta:=1.51+(i*.075):
> pl[i]:=plot(rhs(yeq2),x=-50000..50000,numpoints=500);
> end do:
```

Since there are 1000 meters in 1 kilometer, the x range is from -50 to 50 km. To superimpose the 41 parabolas, the `display` command is used in `d`.

```
> d:=display(seq(pl[i],i=-20..20));
```

The envelope of safety is now plotted, but not displayed.

```
> p:=plot(rhs(yeq3),x=-50000..50000,y=0..25000,
      color=black,thickness=2):
```

Finally, the envelope of safety is superimposed on the family of parabolas, the result being shown in Figure 3.16. Since all directions are uniformly possible, this plot should be mentally rotated around the vertical axis. Note that the scaling is not constrained here.

```
> display({d,p});
```

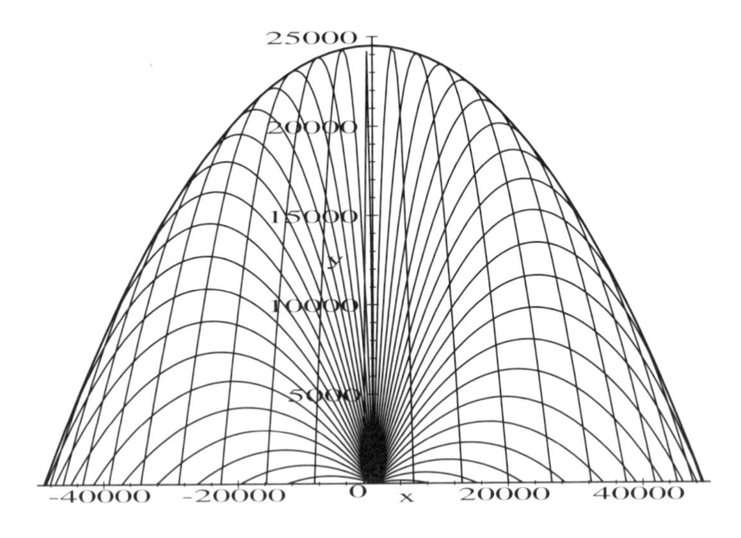

Figure 3.16: Parabolic trajectories and envelope of safety curve.

Thus, neglecting air resistance and the height of the mountain, the aircraft carrying Sheelo, the *National Geographic* photographer, should fly outside the envelope of safety shown in Figure 3.16. Taking air resistance into account will alter the parabolic trajectories and, not surprisingly, shrink the envelope of safety, thus allowing Sheelo to fly somewhat closer to the erupting mountain.

PROBLEMS:

Problem 3-25: Throwing rocks at a tree

Justine is whiling away her time by throwing rocks at trees. The maximum speed with which she can throw a rock is about 25 m/s. Assuming that she can throw accurately, can Justine hit a tall tree 50 m away and 13 m up from the point where the stone leaves her hand? What is the maximum height that she can hit on this tree? Make a plot of the rock's trajectory in the latter case. Neglect air resistance and take $g = 9.8$ m/s^2.

Problem 3-26: Is it a homer?

At a recent baseball game that Russell attended, Boomer Bailey hit a ball with a velocity of 132 ft/s at an angle of 26° above the horizontal. The ball was 3 ft above home plate when hit towards an 8 ft high bleacher wall located 386 ft from home plate. Assuming that the outfielder was unable to reach the ball, did Boomer hit a home run or did the ball hit the wall? How long did it take the ball to reach the wall? Plot the trajectory of the ball over this time inerval. Take $g = 32$ ft/s^2.

Problem 3-27: Invaders beware

A gun on the shore of a beleaguered town fires a shell at an enemy ship that is heading directly toward the gun at a constant speed of 40 km/hr. At the instant of firing, the ship is 15 km away. The muzzle velocity of the shell is 700 m/s and air resistance is to be neglected. What is the required angle of elevation for the gun in order for the shell to hit the ship? How much time elapses between the firing of the shell and its impact with the ship? Animate the motion of the shell and the ship up to the moment of impact. ($g = 9.8$ m/s^2)

Problem 3-28: A military problem

At the Erehwon Military Academy, the army cadets are presented with the following hypothetical problem. Referring to Figure 3.17, an enemy gun em-

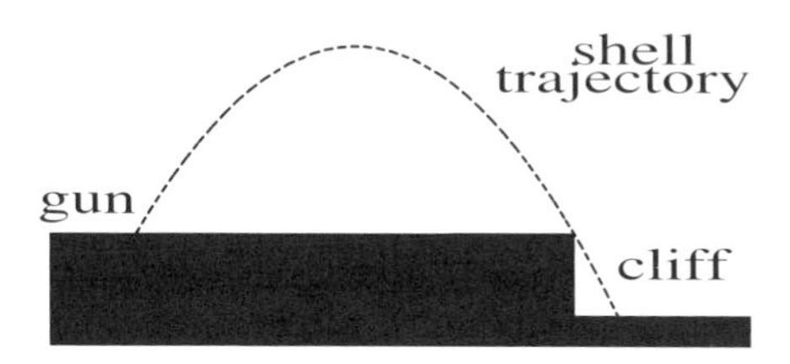

Figure 3.17: Schematic representation of military problem.

placement is set 8230 m horizontally from the edge of a vertical cliff that drops 107 m down from the level of the gun emplacement to a flat plain. How close to the bottom edge of the cliff should the invading cadets remain in order to guarantee that they will not be directly hit by an incoming shell? The muzzle speed of the shells is 305 m/s. Take $g = 9.8$ m/s^2.

3.1.6 Rainbow County

Climb ev'ry mountain, ford ev'ry stream
Follow ev'ry rainbow, till you find your dream!
Oscar Hammerstein II, American songwriter (1895–1960)

The county in which Russell, our engineering friend, spent his early childhood years is called Rainbow County, because of the spectacular rainbows that often occur there. Russell has studied the origin of rainbows in his introductory college physics class and cannot resist setting up the theoretical development as a computer algebra exercise. He knows from frequent observation that the primary rainbow has red at the top and violet at the bottom. Occasionally a secondary rainbow is seen above the primary one with the color order reversed.

Doing some background reading, Russell finds that the Greek philosopher Aristotle recorded only four colors, red, yellow, green, and blue, and suggested that the rainbow was due to reflection by the raindrops. The role of the individual drops was recognized by Roger Bacon, who noted in the year 1267 that the primary rainbow subtends an angle of about 42°. An explanation of the origin of rainbows quite close to our modern view was developed by Theodoric of Freiburg in 1304 and a nearly complete explanation given by the French philosopher and mathematician René Descartes in 1635. The coup de grace to the rainbow problem came in Isaac Newton's famous prism experiments. Clearly the origin of rainbows has intrigued great thinkers for centuries. On finishing his reading, Russell is ready to begin his calculation of how a rainbow is formed.

Russell models a typical raindrop as a sphere of water whose refractive index $n \approx 1.33$ is greater than that of the surrounding air, which has a refractive index $n \approx 1$. He draws a picture as illustrated in Figure 3.18. A light ray from the sun enters the raindrop at an angle θ to the normal to the spherical surface.

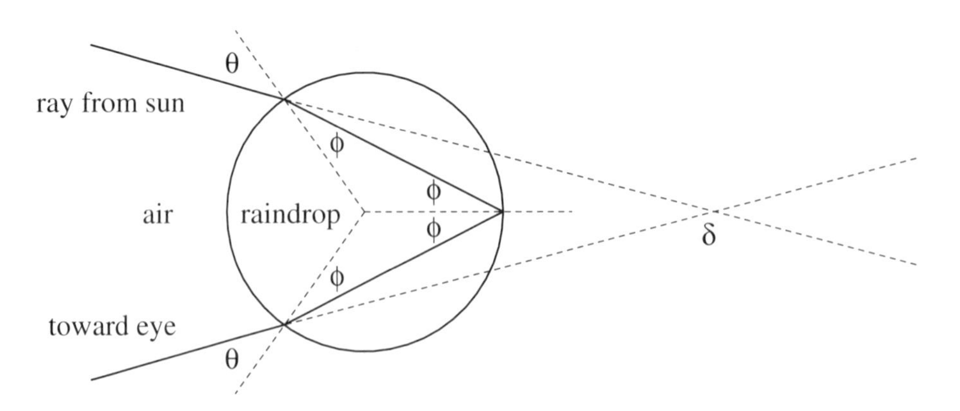

Figure 3.18: Ray diagram for a spherical raindrop in air.

Snell's law states that a light ray passing from a medium of refractive index n_1 into a medium with index n_2 will have its angle of refraction ϕ with the normal determined by the relation $n_1 \sin\theta = n_2 \sin\phi$. Since the refractive index of water is approximately 1.33 and thus greater than that for the surrounding air, one has $\phi < \theta$ and the refracted ray is bent toward the normal. Inside the raindrop, the ray is reflected off the "back side" of the sphere. Since the geometry in the figure is that of an isosceles triangle, the light ray is incident on the back side at an angle ϕ to the normal. But making use of another law of geometrical optics, the angle of reflection is equal to the angle of incidence, and thus equal to ϕ as well. Finally, the light ray is incident on the inner front surface of the drop at an angle ϕ and is then refracted back into air at an angle θ to the normal toward the observer's eye.

Russell begins his calculation by feeding in the refractive indices of air and water, leaving the latter value unspecified for the moment, since its precise value n depends on the wavelength component of the incident light.

```
> restart: n[1]:=1: n[2]:=n:
```

Next, Snell's law is stated.

```
> Snells_law:=n[1]*sin(theta)=n[2]*sin(phi(theta));
```

$$Snells\_law := \sin(\theta) = n\sin(\phi(\theta))$$

The angle ϕ depends on θ. How ϕ changes with θ is determined by differentiating the previous line.

```
> diff_eq:=diff(Snells_law,theta);
```

$$diff\_eq := \cos(\theta) = n\cos(\phi(\theta))\left(\frac{d}{d\theta}\,\phi(\theta)\right)$$

Another relationship is needed to eliminate the quantity $d\phi(\theta)/d\theta$. From the figure, the total angular deflection δ of the incident ray is given by the sum of $\theta - \phi$ at the air–water interface plus $\pi - 2\phi$ at the back side plus $\theta - \phi$ at the water–air interface.

```
> delta:=2*(theta-phi(theta))+(Pi-2*phi(theta));
```

$$\delta := 2\,\theta - 4\,\phi(\theta) + \pi$$

Descartes had experimentally discovered that the primary rainbow occurred when the angle of deflection δ was a minimum and that the angle subtended by the continuation of the incident and outgoing rays, i.e., $180 - \delta$, ranged from about 40° to 42°. So, Russell differentiates δ with respect to θ and sets the result equal to zero,

```
> diff(delta,theta)=0;
```

$$2 - 4\left(\frac{d}{d\theta}\,\phi(\theta)\right) = 0$$

which is easily solved for $d\phi(\theta)/d\theta$,

```
> diff(phi(theta),theta)=solve(%,diff(phi(theta),theta));
```

$$\frac{d}{d\theta}\,\phi(\theta) = \frac{1}{2}$$

yielding the value $\frac{1}{2}$. This value for $d\phi(\theta)/d\theta$ is substituted into *diff_eq*.

```
> eq1:=subs(%,diff_eq);
```

$$eq1 := \cos(\theta) = \frac{1}{2}\, n \cos(\phi(\theta))$$

Both sides of *eq1* are squared,

```
> eq2:=lhs(eq1)^2=rhs(eq1)^2;
```

$$eq2 := \cos(\theta)^2 = \frac{1}{4}\, n^2 \cos(\phi(\theta))^2$$

and the well-known trigonometric substitution $\cos^2(\phi) = 1 - \sin^2(\phi)$ made.

```
> eq3:=subs(cos(phi(theta))^2=1-sin(phi(theta))^2,eq2);
```

$$eq3 := \cos(\theta)^2 = \frac{1}{4}\, n^2\, (1 - \sin(\phi(\theta))^2)$$

Next, Snell's law is solved for $\sin(\phi)$,

```
> sin(phi(theta)):=solve(Snells_law,sin(phi(theta)));
```

$$\sin(\phi(\theta)) := \frac{\sin(\theta)}{n}$$

so that *eq3* becomes

```
> eq3;
```

$$\cos(\theta)^2 = \frac{1}{4}\, n^2 \left(1 - \frac{\sin(\theta)^2}{n^2}\right)$$

Substituting the trig identity $\sin^2(\theta) = 1 - \cos^2(\theta)$,

```
> subs(sin(theta)^2=1-cos(theta)^2,eq3);
```

$$\cos(\theta)^2 = \frac{1}{4}\, n^2 \left(1 - \frac{1 - \cos(\theta)^2}{n^2}\right)$$

and isolating $\cos^2(\theta)$ to the lhs of the equation yields the following expression for the critical angle of incidence, which is then factored.

```
> eq4:=isolate(%,cos(theta)^2);
```

$$eq4 := \cos(\theta)^2 = \frac{n^2}{3} - \frac{1}{3}$$

```
> critical_angle_eq:=factor(eq4);
```

$$critical\_angle\_eq := \cos(\theta)^2 = \frac{(n-1)\,(n+1)}{3}$$

The visible spectrum of light ranges from red to violet. The index of refraction for red light in water is $n = 1.3311$, whereas $n = 1.3435$ for violet light. The critical angle of incidence for red light that produces a minimum in δ is evaluated in the next few command lines,

```
> n:=1.3311; critical_angle_eq;
```

$$n := 1.3311 \qquad \cos(\theta)^2 = 0.2572757367$$

```
> theta[c]:=fsolve(%,theta);
```

$$\theta_c := 1.038836227$$

and found to be about 1.04 radians,

```
> theta[c]:=evalf(%*180/Pi);
```

$$\theta_c := 59.52093141$$

or 59.5°. The angle ϕ is evaluated,

```
> phi(theta):=evalf(arcsin(sin(%%)/n)*180/Pi);
```

$$\phi(\theta) := 40.34926868$$

and found to be 40.3°. The minimum angle of deflection δ,

```
> delta:=180+2*theta[c]-4*phi(theta);
```

$$\delta := 137.6447881$$

is 137.6°. Finally, the angle between the incoming and outgoing rays is obtained by calculating $180 - \delta$,

```
> Angle:=180-delta;
```

$$Angle := 42.3552119$$

which yields 42.4° for red light. By changing the value of n to 1.3435, the reader can check that the angle for violet light is 40.6°. These results are in agreement with Descartes's observations.

It should be noted that only one color reaches the eye from any particular drop. By considering a number of raindrops, one above the other, it follows that a viewer's eye would observe the primary rainbow with red light at the top and violet light at the bottom. Because δ is less for red light than violet light, red light must come from a higher raindrop than for violet light.

The secondary rainbow, which is fainter, involves two internal reflections inside the raindrop. Considerations similar to those above lead to the conclusion that there is an inversion of the color order compared to the primary rainbow.

PROBLEMS:

Problem 3-29: Applying Fermat's principle

Suppose that a light ray in going from point A to point B traverses distances d_1, d_2,..., d_N in media of refractive indices n_1, n_2,..., n_N, respectively. The total time of flight is then

$$t = \frac{1}{c}\sum_{i=1}^{N} n_i d_i,$$

where c is the vacuum speed of light. The summation is referred to as the optical path length. In its simplest form, *Fermat's principle* states that in traveling from A to B, a light ray travels a path that minimizes the time, or, equivalently, minimizes the optical path length. By minimizing the relevant optical path lengths, prove the following laws of geometrical optics: (i) the angle of reflection is equal to the angle of incidence; (ii) Snell's law.

Problem 3-30: Applying Snell's law

A wide glass container (refractive index $n_g = 1.51$) is 8 cm tall with a glass bottom and top lid, each of a uniform thickness of 1 cm. The interior of the container is filled with two liquids, the bottom half with water ($n_w = 1.33$), the top half with carbon disulphide ($n_c = 1.63$). The glass container is in air ($n_a = 1$). A light ray enters the container through the top glass lid making an angle of 50° with the vertical to the lid and exits through the glass bottom. Determine the angles that the beam makes with the vertical as it passes through glass, carbon disulphide, water, glass, and air. Make a labeled plot showing the path traversed by a representative light ray. By how many centimeters is the exit point displaced from the entry point of the light ray?

3.2 Integral Examples

In the next two recipes, scalar algebraic models are presented, involving one-, two- and three-dimensional integrations.

3.2.1 The Great Pyramid of Cheops

Who shall doubt "the secret hid under Cheops' pyramid"
Was that the contractor did Cheops out of several millions?
Rudyard Kipling, British Nobel laureate in literature (1907)

Mike, the mathematics student and amateur archaeologist whom we met on an archaeological dig at Machu Pichu, is impressed by the ingenuity and effort that must have gone into the crafting and assembling of the massive, precisely cut stone blocks in the Inca ruins. On the recommendation of the chief archaeologist, when he returns to his university campus at the end of the summer he goes to the library and searches for some books dealing with a similar impressive achievement by the Egyptian pharaohs in the building of their pyramids.[7]

Mike comes across some factual information as well as accounts by the fifth century B.C. Greek historian Herodotus, who was called the Father of History. The Great Pyramid near present-day Gizeh, Egypt, was built by a pharaoh whom Herodotus referred to as Cheops. Cheops, also known as Khufu, ruled Egypt about 2600 B.C. The Great Pyramid was originally about 481 ft (nearly 150 m) high, but has lost nearly 7 m of its height due to the stripping of its marble casing. At its base, each of the four sides is about 230 m in length. The pyramid is solid, with only a few blocks having been omitted to leave a secret passageway for depositing Cheops's earthly remains. There are about $2\frac{1}{2}$ million limestone blocks in the pyramid, some weighing as much as 15 tons, but the average being about $2\frac{1}{2}$ tons. The rocks were quarried in the Arabian mountains, dragged to the Nile, and taken in boats to a site near the pyramid. According to Herodotus, 100,000 workers, which was about one-tenth of Egypt's

[7] *Pyramid* derives from the Egyptian word pi-re-mus, which translates as altitude [Dur54].

population, toiled for 20 years in building the road along which the rocks were moved and in constructing the pyramid.

Being a mathematics student, Mike cannot resist playing around with the above numbers involved in this historical event. He first calculates the number of worker-years involved in the building of the Great Pyramid of Cheops.

```
> restart:
> Herodotus_estimate:=100000*20*worker*years;
```

$$Herodotus\_estimate := 2000000\, worker\ years$$

The project involved about 2 million worker-years. Next Mike decides to construct a hypothetical model of the number of workers involved as a function of time. He assumes that the number of workers can be modeled as two logistic curves placed back to back. He creates the following piecewise function in *eq1* to accomplish this.

```
> eq1:=piecewise(t<=10,100000/(1+100*exp(-3*t))+5000,
      t>=10,105000/(1+0.00015*exp(1.25*(t-10))));
```

$$eq1 := \begin{cases} \dfrac{100000}{1+100\,e^{(-3\,t)}} + 5000 & t \leq 10 \\ \dfrac{105000}{1+0.00015\,e^{(1.25\,t-12.50)}} & 10 \leq t \end{cases}$$

Note that the two logistic curves do not precisely match at $t = 10$. The parameters of the model have been adjusted so that the area under the curve yields approximately the correct worker-year total. The initial number is not zero but reflects the permanent population already present in the nearby settlement. Mike plots the piecewise function *eq1* for the time span 0 to 20 years.

```
> plot(eq1,t=0..20,tickmarks=[4,2],labels=["years","workers"]);
```

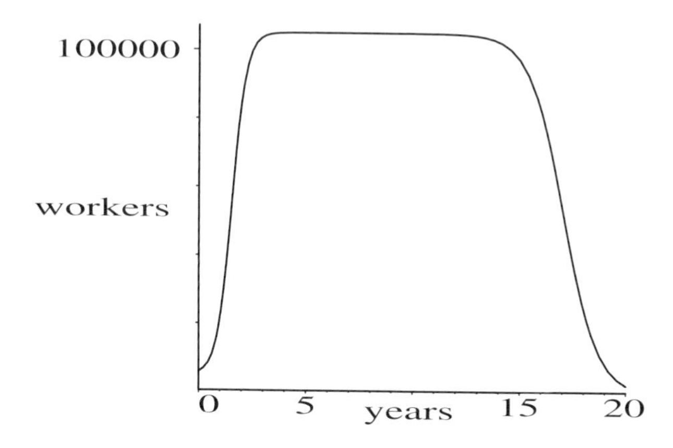

Figure 3.19: Plot of the piecewise model equation.

Given the shape of the curve, he decides to see how easy it is to extract various pieces of information. Mike determines the location of the two inflection points at which the curvature changes by differentiating *eq1* twice with respect to time and labeling the result as *eq2*. Note that Mike uses the alternative syntax `t$2` for the second argument in the `diff` command, instead of `t,t`.

```
> eq2:=diff(eq1,t$2);
```

Looking at the output (which has been artificially suppressed here because of its length), Mike notes that because the two logistic curves are not precisely joined at $t=10$, the slope (first derivative) has an artificial discontinuity and hence the second derivative is undefined at this value. To determine the inflection points, he applies the floating-point solve command to $eq2 = 0$, specifying the ranges of t to be searched for each solution.

```
> inflection_point1:=fsolve(eq2=0,t=0..9);
```

$$inflection\_point1 := 1.535056729$$

```
> inflection_point2:=fsolve(eq2=0,t=11..20);
```

$$inflection\_point2 := 17.04390021$$

The first inflection point is at about 1.54 years, the second at 17.04 years. Mike calculates the number of workers at the second inflection point by applying the evaluation (`eval`) command to *eq1*.

```
> workers:=eval(eq1,t=inflection_point2);
```

$$workers := 52500.00010$$

There are about 52,500 workers at the time of the second inflection point. The total number of worker-years is obtained by integrating *eq1* with respect to time to find the area under the curve for the period $t=0$ to 20.

```
> area_1:=Int(eq1,t=0..20) =int(eq1,t=0..20.0)* worker*years;
```

$$area\_1 := \int_0^{20} \begin{cases} \dfrac{100000}{1 + 100\, e^{(-3\, t)}} + 5000 & t \leq 10 \\ \dfrac{105000}{1 + 0.00015\, e^{(1.25\, t - 12.50)}} & 10 \leq t \end{cases} dt$$
$$= 0.1633723348\; 10^7 \; worker\; years$$

Mike's model gives about 1.63 million worker-years, which is not too far from the Herodotus estimate of 2 million worker-years. The number of worker-years from the second inflection point at 17.04 years to the twentieth year is obtained by a similar integration. Mike uses three ditto operators in series to pick up the inflection point that was calculated three command lines earlier.

```
> area_2:=Int(eq1,t=%%%..20)=int(eq1,t=%%%..20)*worker*years;
```

$$area\_2 := \int_{17.04390021}^{20} \begin{cases} \dfrac{100000}{1 + 100\, e^{(-3\, t)}} + 5000 & t \leq 10 \\ \dfrac{105000}{1 + 0.00015\, e^{(1.25\, t - 12.50)}} & 10 \leq t \end{cases} dt$$
$$= 56162.94008 \; worker\; years$$

About 56,000 worker-years were involved for this period. Now Mike decides to see what other deductions he can make from the numbers that he has gleaned from the literature. He wants to calculate how much work it took to assemble the pyramid, i.e., to lift the blocks into place.

Mike idealizes the pyramid by considering four straight lines from the four basal corners to the apex of the pyramid as shown on the left-hand side of Figure 3.20. He takes the basal length to be $2\,L$ and lets each horizontal coordinate range from $x=-L$ to $+L$ as shown in the two-dimensional view on the right-hand side of the same figure.

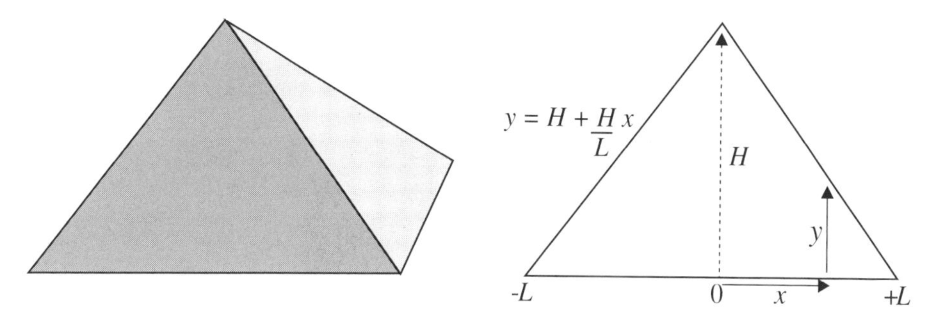

Figure 3.20: Diagram of the pyramid.

Assuming the height at the apex to be H, the mathematical equation describing the slanted surface between $x=-L$ and $x=0$ is $y=H+H(x/L)$.

```
> y:=H+H*x/L;
```

$$y := H + \frac{H\,x}{L}$$

At $x=0$, $y=H$ and at $x=-L$, $y=0$, so the entered form of y checks out. If the density (assumed to be constant) of the stones in the pyramid is ρ, the total mass can be calculated by multiplying the volume by ρ. The mass dM of a thin square volume element with sides of length $2\,x$ and thickness dy at a height y will be

$$dM = \rho\,(2\,x)\,(2\,x)\,dy = 4\,\rho\,x^2\left(\frac{dy}{dx}\right)\,dx \equiv \left(\frac{dM}{dx}\right)\,dx. \qquad (3.3)$$

By integrating (3.3) from $x=-L$ to 0, the formula for the total mass can be determined. The integrand dM/dx is entered, with left quotes being attached to the assigned name to prevent Maple from thinking that a mathematical division is intended.

```
> `dM/dx`:= 4*rho*x^2*diff(y,x);
```

$$dM/dx := \frac{4\,\rho\,x^2\,H}{L}$$

The mass of the pyramid is determined by integrating the previous command line output from $x=-L$ to $x=0$.

```
> Mass:=Int(%,x=-L..0)=int(%,x=-L..0);
```

$$Mass := \int_{-L}^{0} \frac{4\,\rho\,x^2\,H}{L}\,dx = \frac{4\,\rho\,H\,L^2}{3}$$

The total mass of the pyramid is $4\,\rho\,H\,L^2/3$. The amount of work to lift a small mass dM through a vertical distance y from the ground is

$$dW = dM\,g\,y = \left(\frac{dM}{dx}\right) g\,y\,dx \equiv \left(\frac{dW}{dx}\right) dx, \tag{3.4}$$

where g is the acceleration due to gravity. Mike enters the quantity dW/dx, again using left quotes to enclose the name.

```
> `dW/dx`:=`dM/dx`*g*y;
```

$$dW/dx := \frac{4\,\rho\,x^2\,H\,g\left(H + \frac{H\,x}{L}\right)}{L}$$

Carrying out the integration of the previous line from $x = -L$ to $x = 0$,

```
> W:= Int(%,x=-L..0) =int(%,x=-L..0);
```

$$W := \int_{-L}^{0} \frac{4\,\rho\,x^2\,H\,g\,(H + \frac{H\,x}{L})}{L}\,dx = \frac{\rho\,H^2\,L^2\,g}{3}$$

Mike finds that the work to lift all the blocks in the pyramid is $W = \rho\,g\,H^2\,L^2/3$. Although he could look up the density of limestone and calculate the work using this formula, Mike decides to use the information given at the beginning of this story. To accomplish this, he solves for the density in terms of the total mass M and pyramid dimensions,

```
> density:=solve(rhs(Mass)=M,rho);
```

$$density := \frac{3\,M}{4\,H\,L^2}$$

and substitutes the density expression into the right-hand side of W.

```
> Work:=subs(rho=density,rhs(W));
```

$$Work := \frac{M\,H\,g}{4}$$

This yields the expression $W = M\,g\,H/4$ for the work. The mass M of the pyramid is obtained by multiplying the number (about 2.5×10^6) of blocks by the average weight (about 2.5 tons) in tons of each block and converting the result into kilograms. To accomplish the latter, Mike notes that a weight of 1 ton corresponds to a mass of 907.2 kg.

```
> M:=2.5*10^6*2.5*907.2; #kilograms
```

$$M := 0.5670000000\,10^{10}$$

The pyramid has a mass $M = 0.567 \times 10^{10}$ kg. With this number automatically entered, Mike takes $H = 150$ m and $g = 9.81$ m/s^2 and evaluates *Work*.

```
> Work:=eval(Work,{H=150,g=9.81})*joules;
```

$$Work := 0.2085851250\,10^{13}\,joules$$

Thus, the total amount of work to lift the stones into place is 0.21×10^{13} J. Mike is curious about whether Herodotus's estimate of the number of worker-years needed to build the pyramid is reasonable. He decides to calculate how much work energy was available from the workers. He assumes that the average food consumption was about 2000 food calories per day and converts this number into joules by recalling that there are 4186 joules in a food calorie.

```
> Food_energy:=(2000*calories/(worker*day))
                *(4186*joules/calories);
```

$$Food\_energy := \frac{8372000\, joules}{worker\ day}$$

Thus, the average food consumption per worker is about 8 million joules per day. Now only a small fraction, perhaps 1%, of this ends up as useful work, the rest being lost in sweat, etc.

```
> Useful_work_energy:=0.01*%;
```

$$Useful\_work\_energy := \frac{83720.00\, joules}{worker\ day}$$

Over the 20 year period, taking 365 days in a year, the total useful work is,

```
> Total_useful_work_energy:=%*Herodotus_estimate*365*day/years;
```

$$Total\_useful\_work\_energy := 0.6111560000\, 10^{14}\, joules$$

about 6.1×10^{13} J. This is about 30 times the amount of energy needed to build the pyramid. Of course, the energy needed to build the roads along which the limestone blocks were hauled and the ramps up the pyramid has not been included.

After doing this calculation, Mike can't wait to get another opportunity to combine his archaeological and mathematical interests. His contact with the chief archaeologist at Machu Pichu has given him the inside track on a dig in a remote region of Asia not that far from the Great Wall of China, which he hopes to join in his next summer semester.

PROBLEMS:

Problem 3-31: A bizarre proposal

Suppose that some modern Egyptian developers propose to build a pyramid whose height would have been H but has the top 20% removed so that the pyramid has a flat (horizontal) top. This pyramid would be located in the Valley of the Pyramids and the roof would accommodate an upscale restaurant where prominent socialites could feast and enjoy the view of the surrounding monuments. Using a first principles integration approach, derive the work formula for assembling this pyramid in terms of the total mass M, height H, and the acceleration due to gravity g. Check your result by using a more clever approach.

Problem 3-32: A cone-shaped "pyramid"

Suppose that the Egyptians had built a "pyramid" out of the same limestone blocks but in the shape of a right circular cone of height 150 m and radius 115 m. How much work would have to be done in order to lift the blocks into place? How does your answer compare with the work done to build the Cheops pyramid discussed in the text? Does your answer make sense? Explain.

Problem 3-33: Rocket flight
A rocket fired vertically has a constant upward acceleration of $2g$ during the burning of the rocket motor, which lasted for 50 seconds.

(a) Neglecting air resistance and the variation of g with altitude, express the acceleration as a piecewise function for the two time intervals before and after 50 seconds.

(b) By carrying out the indefinite integral of the acceleration, obtain an analytical expression for the velocity as a function of time.

(c) Plot the velocity expression for the time interval that it takes the rocket to reach maximum altitude. Remember that the velocity is zero at this point.

(d) Carry out the definite integral of the velocity expression with respect to time to find the maximum altitude. Express your answer in kilometers.

(e) Calculate the total time that the rocket is in the air.

(f) Plot the altitude versus time for the entire flight of the rocket.

3.2.2 Noah's Ark

If Noah had been truly wise, he would have swatted those two flies.
Attributed to Helen Castle in the 1999 Merrill Lynch daily planner.

Noah, a meteorologist by training and TV weatherman by profession, has enrolled in a course on computer-assisted sailing-ship design at the local community college. Because it has been unusually rainy lately, he has taken some good-natured ribbing about the weather from his TV viewers and friends. On learning of his enrollment in the ship-design course, one of his neighbors, Jerry, was heard to chortle, "Hey, Noah, when are you going to build your ark?"

Fortunately, Noah is used to this syndrome of blaming the messenger for bad news, and is able to take such comments in his stride. "Very funny, Jerry. No, as you are aware, I am into sailing and have participated in transoceanic yacht races such as the one from Victoria, British Columbia, to Maui in the Hawaiian Island chain. I simply want to learn more about designing a high-speed racing hull and I have always been interested in applying the computer to solving technical problems. Why don't you take the course with me? You like to come out sailing with me in the summer, and at the least, the course might help both of us to improve our computer skills."

"OK, Noah, it sounds like it might be interesting. I will take the course with you, if you promise to let me crew in one of your next yacht races."

Having agreed to Jerry's request, Noah takes Jerry along with him to the first evening of the computer-assisted ship-design course. To get his students up to speed on the computers, the instructor, Buzz, has given them a simple warm-up exercise designed to show them some aspects of the Maple software system that they will be using. Buzz asks them to create a colored plot, suitably annotated, of the planar cross section of a ship's hull that is mathematically bounded on the bottom by the parabola $Y = x^2$ and on the top by the horizontal deck line $Y = 1$. For this planar hull cross section, they are then to use integral calculus and Maple's integral command to determine its area, its mass, and the location of the center of mass assuming that the mass density function is $\rho = a\,y^n\,e^{-b\,y}$ with the choice of parameter values, a, n, and b, left up to Noah and Jerry. The location of the center of mass (the point at which all the mass can be thought of as being concentrated) is important in determining the stability of the ship's hull against tipping.

Noah and Jerry begin by making a call to the following library packages. The plottools package is needed in order to place an arrow indicating an integration direction on the planar hull plot.

```
> restart: with(plots): with(plottools):
```

Next, Noah and Jerry enter the parabolic equation $Y = x^2$.

```
> Y:=x^2;
```

$$Y := x^2$$

Buzz hasn't specified any units of length, so Jerry jocularly says "let's take distances to be in units of 5 cubits. I am sure that Buzz doesn't know how big a cubit is."

"I don't know, either," retorts Noah.

"A cubit is an ancient measure of length, about 18 to 22 inches," Jerry replies. "It was originally the length of the arm from the end of the middle finger to the elbow. So 5 cubits is about $7\frac{1}{2}$ to 9 feet, or $2\frac{1}{4}$ to 3 meters. Since Buzz has said that the deck line corresponds to $Y = 1$, then equating this to $Y = x^2$ gives us a deck width whose coordinates range from $x = -1$ to $+1$. So, the deck width for the planar cross section would be about 15 to 18 feet."

Noah and Jerry form a plot of Y. "Let's make an artistic plot," says Noah, "and fill the region below the parabolic hull bottom with an aquamarine color to represent tropical sea water."

```
> plot||1:=plot(Y,x=-1..1,filled=true,color=aquamarine):
```

Noah and Jerry have used the concatenation operator `||` to attach the number `1` to the name `plot`. The sequence command can be applied to sets of concatenated names that include the number in this manner.

"Remember that we are going to have to calculate the cross-sectional area A of the planar hull," points out Jerry. "We can make use of a double integral, writing the area in the form $A = \iint dA = \iint dx\,dy$ and inserting the correct limits for the hull cross section. Let's integrate vertically in the y-direction first, then horizontally in the x-direction. To let Buzz know what we are doing, let's create a thick red arrow placed at $x = 0.4$ with its tail on the parabola $Y = x^2$ and its tip at $Y = 1$. The numbers 0.035, 0.1, 0.1 in the following command line refer to the width of the arrow, the width of the arrowhead, and the ratio of the arrowhead length to the overall length of the arrow.

```
> plot||2:=arrow([0.4,0.4^2],[0.4,1],0.035,0.1,0.1,color=red):
```

While we're at it, let's fill the interior of the hull with a grey color,

```
> plot||3:=plot(1,x=-1..1,filled=true,color=grey):
```

and use the `textplot` command to add some black labeling to the plot."

```
> plot||4:=textplot([[0.48,0.18,"Y=x^2"],[0.48,0.96,"Y=1"]],
            align=RIGHT,color=black):
```

"OK, Jerry, if we are going on a coloring binge, let's outline the hull in red for easy visualization on the computer screen."

```
> plot||5:=plot([[-1,1],[1,1]],color=red,thickness=2):
> plot||6:=plot(Y,x=-1..1,color=red,thickness=2):
```

"We should have enough for our plot," notes Noah, "so let's superimpose the six plots and see what we have created. Note that the plots are put into a list format in the `display` command and the order of the plots matters. If `plot||3` preceded `plot||1` in the list, the picture would be completely grey. In `plot||3`, everthing below $y = 1$ is colored grey, but it does not color over a previously plotted fill."

```
> display([seq(plot||k,k=1..6)],tickmarks=[2,4],
  labels=["x","y"]);
```

Figure 3.21 shows a black-and-white rendition of the colorful and informative picture that Noah and Jerry have created.

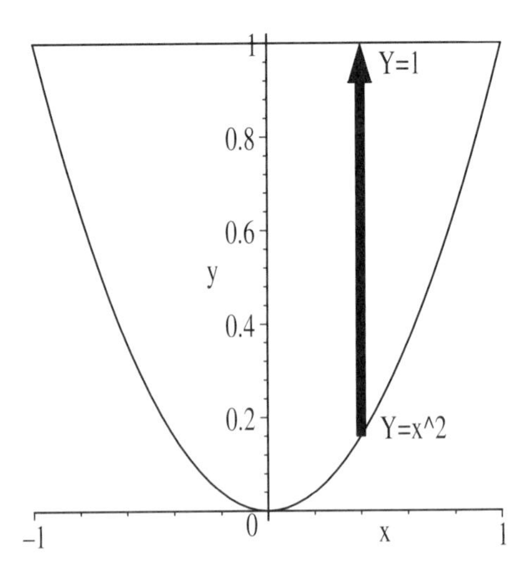

Figure 3.21: Geometry of the planar parabolic hull.

Having shown Buzz their artistic creation, Noah and Jerry next evaluate the area of the hull enclosed by the curves $Y = x^2$ and $Y = 1$. From Figure 3.21, the double integral, expressed in Cartesian coordinates, takes the specific form

$$A = \int_{-1}^{1} \int_{x^2}^{1} dy\, dx. \tag{3.5}$$

The vertical arrow in Figure 3.21 indicates the range of the y integration.

"We don't need to use a computer to do this simple integral," Noah remarks. "In fact, we didn't even have to express the area as a double integral but could have chosen to take vertical strips of thickness dx and written the area integral as the difference between the area under the curve $Y = 1$ and the parabolic curve $Y = x^2$. This would yield $A = \int_{-1}^{1} dx - \int_{-1}^{1} x^2\, dx = 2 - 2/3 = 4/3$ units of area. As a check, let's use the double integral form (3.5) to evaluate the area. The following command line will produce the inert form of the area integral,

```
> Area:=Int(Int(1,y=Y..1),x=-1..1);
```

$$Area := \int_{-1}^{1} \int_{x^2}^{1} 1\, dy\, dx$$

which can be evaluated with the `value` command.

```
> Area:=value(%);
```

$$Area := \frac{4}{3}$$

The cross-sectional area of the hull is 4/3 units, which is what we expected. You can multiply this by $5^2 = 25$, Jerry, to express the answer in terms of cubits squared. Let's now calculate the mass of this planar hull cross section by first entering the density profile that Buzz gave us.

```
> rho:=a*y^n*exp(-b*y);
```

$$\rho := a\, y^n\, e^{(-b\, y)}$$

Buzz didn't specify the parameter values or the units, so I am going to choose $a = 3$, $n = 1$, and $b = 2$ and see what the density profile looks like, by plotting ρ as a thick blue curve.

```
> a:=3: n:=1: b:=2:
> plot(rho,y=0..1,tickmarks=[3,3],labels=["y","density"],
  color=blue,thickness=2);
```

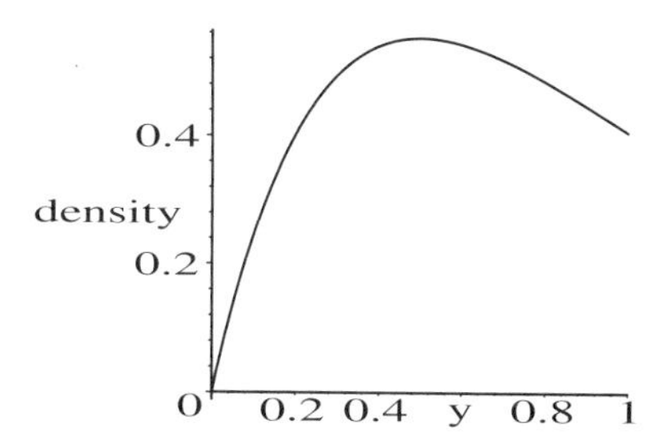

Figure 3.22: Variation of the density ρ with the vertical coordinate y.

For my choice of parameters, the density rises to a maximum roughly midway between the bottom of the hull and the top and then begins to decrease with increasing y. The mass of the planar hull cross section will be given by"

$$M = \int_{-1}^{1} \int_{x^2}^{1} \rho \, dy \, dx. \tag{3.6}$$

Noah and Jerry enter this mass integral, and calculate its value.

```
> Mass_int:=Int(Int(rho,y=Y..1),x=-1..1);
```

$$Mass\_int := \int_{-1}^{1} \int_{x^2}^{1} 3\, y\, e^{(-2\,y)} \, dy \, dx$$

```
> Mass:=value(%);
```

$$Mass := -\frac{21}{4}\, e^{(-2)} + \frac{9}{16}\sqrt{2}\,\sqrt{\pi}\,\mathrm{erf}(\sqrt{2})$$

They note that the answer is specified in terms of a "special" function, called the *error function*, which is defined as

$$\mathrm{erf}(u) = \frac{2}{\sqrt{\pi}} \int_{0}^{u} e^{-t^2} \, dt. \tag{3.7}$$

The error function and therefore the mass can be numerically evaluated,

```
> Mass:=evalf(%);
```

$$Mass := 0.6353137782$$

so Jerry and Noah find that the mass is about 0.64 mass units.

"Let's now calculate the location of the center of mass," Noah comments. "I recall from my freshman physics text [Oha85] that the y- and x-coordinates of the center of mass are defined by the expressions"

$$y_{\rm cm} = \frac{1}{M} \int_{-1}^{1} \int_{x^2}^{1} y\, \rho \, dy \, dx, \quad x_{\rm cm} = \frac{1}{M} \int_{-1}^{1} \int_{x^2}^{1} x\, \rho \, dy \, dx. \tag{3.8}$$

Entering the y center of mass coordinate,

```
> y_cm:=Int(Int(y*rho,y=Y..1),x=-1..1)/Mass_int;
```

$$y\_cm := \frac{\displaystyle\int_{-1}^{1} \int_{x^2}^{1} 3\, y^2\, e^{(-2\,y)} \, dy \, dx}{\displaystyle\int_{-1}^{1} \int_{x^2}^{1} 3\, y\, e^{(-2\,y)} \, dy \, dx}$$

and evaluating it,

```
> y_cm:=value(y_cm);
```

$$y\_cm := \frac{-\frac{153}{16}\, e^{(-2)} + \frac{45}{64}\sqrt{2}\,\sqrt{\pi}\,\mathrm{erf}(\sqrt{2})}{-\frac{21}{4}\, e^{(-2)} + \frac{9}{16}\sqrt{2}\,\sqrt{\pi}\,\mathrm{erf}(\sqrt{2})}$$

```
> y_cm:=evalf(%);
```

$$y\_cm := 0.6109364936$$

Noah and Jerry find that $y_{\rm cm} \approx 0.61$, i.e., about 3 cubits above the hull bottom.

"By symmetry," Jerry points out, "the x-coordinate of the center of mass must clearly be equal to zero."

"You're right," Noah replies, "but let's check it anyway." So Noah enters the integral expression for x_cm,

```
> x_cm:=Int(Int(x*rho,y=Y..1),x=-1..1)/Mass_int;
```

$$x\_cm := \frac{\int_{-1}^{1}\int_{x^2}^{1} 3\,x\,y\,e^{(-2\,y)}\,dy\,dx}{\int_{-1}^{1}\int_{x^2}^{1} 3\,y\,e^{(-2\,y)}\,dy\,dx}$$

and evaluates the double integral,

```
> x_cm:=value(x_cm);
```

$$x\_cm := 0$$

obtaining zero as required.

"Noah, I think that we should locate the center of mass on the figure of the planar hull that we created earlier. Let's place a red circle at x_cm, y_cm,

```
> plot||7:=plot([[x_cm,y_cm]],style=point,symbol=circle,
            symbolsize=16,color=red):
```

and place the phrase `center of mass` adjacent to the circle.

```
> plot||8:=textplot([0.05,y_cm,"center of mass"],
            align={ABOVE,RIGHT},color=red):
```

We don't need the arrow on the diagram anymore so let's remove `plot||2` and superimpose all the other plots, i.e., `plot||1` and `plot||3` to `plot||8`, to create a new figure." The resulting picture is shown in Figure 3.23.

```
> display([plot||1,seq(plot||k,k=3..8)],
  tickmarks=[2,2],labels=["x","y"]);
```

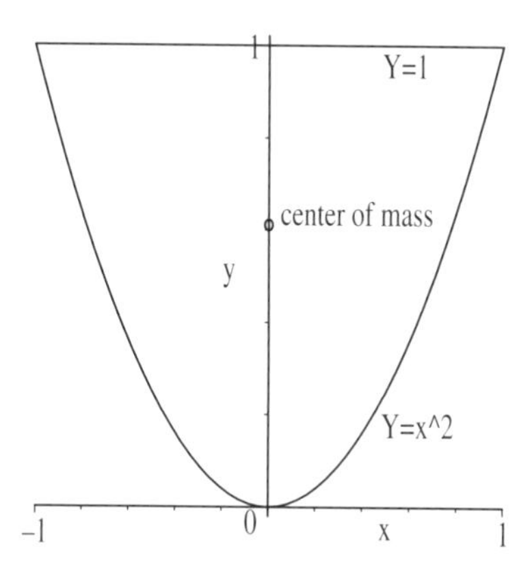

Figure 3.23: Center of mass for planar parabolic hull.

At this point, Buzz takes a look at their picture and comments, "A nice picture guys, but I think that your center of mass is quite high in relation to the hull's vertical dimension and as a consequence your hull might not be very stable. You might want to play around with the parameters and lower $y_{\rm cm}$. You can do this later, however, as I have another exercise for you. Keeping the density profile the same, I would like you to create a three-dimensional paraboloid of revolution, by rotating the parabola about its symmetry axis, $x = 0$, and then calculate the volume, mass, and center of mass of this paraboloid."

To create the paraboloid, Noah and Jerry apply the `contourplot3d` command to the function $y = x^2 + z^2$, which corresponds to rotating the parabola around the y-axis.

```
> plot||9:=contourplot3d(((x^2+z^2)),x=-1..1,z=-1..1,
              contours=14,filled=true,lightmodel=light2):
```

They also add labels to the three-dimensional figure with `textplot3d`,

```
> plot||10:=textplot3d([[-0.6,0.3,0.1,"Y=x^2+z^2"],
              [-0.9,0.7,1,"Y=1"]],align=RIGHT,color=black):
```

and use the `display` command to show the completed plot, which is similar to that in Figure 3.24. (The colored and shaded version on the computer screen displays substantially fewer contours than shown in the text figure.)

```
> display([seq(plot||k,k=9..10)],axes=framed,
  view=[-1..1,-1..1,0..1],orientation=[54,50],tickmarks=
  [2,2,2],labels=["x","z","y"],shading=Z);
```

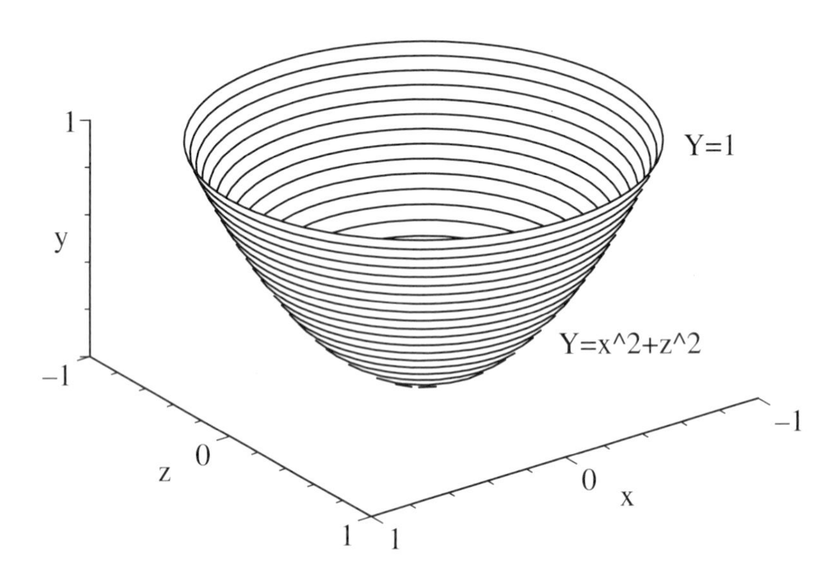

Figure 3.24: Paraboloidal hull shape.

"Excuse my language, Noah, but from a sailing ship viewpoint, that's a H_LL of a hull. The drag would be tremendous. I guess that this is just a warm-up exercise and we will be looking at more realistic designs later in the course. Let's determine the volume, $V = \iiint dV$, for the paraboloid of revolution by carrying out the triple integration. If we take the z integration to range from $z = -\sqrt{y - x^2}$ to $z = \sqrt{y - x^2}$, the remaining two integrals will be identical to those used in the planar case. So the volume is symbolically expressed as

```
> Volume:=Int(Int(Int(1,z=-sqrt(y-Y)..sqrt(y-Y)),y=Y..1),
        x=-1..1);
```

$$Volume := \int_{-1}^{1}\int_{x^2}^{1}\int_{-\sqrt{y-x^2}}^{\sqrt{y-x^2}} 1\,dz\,dy\,dx$$

which is easily evaluated.

```
> Volume:=value(%);
```

$$Volume := \frac{\pi}{2}$$

The volume is equal to $\pi/2$ in normalized units. To express the volume in cubic cubits, we can multiply this result by $5^3 = 125$. Let's evaluate the mass integral, which is now of the form $M = \iiint \rho\,dV$." Noah types in the mass integral,

```
> Mass2_int:=Int(Int(Int(rho,z=-sqrt(y-Y)..sqrt(y-Y)),
        y=Y..1),x=-1..1);
```

$$Mass2\_int := \int_{-1}^{1}\int_{x^2}^{1}\int_{-\sqrt{y-x^2}}^{\sqrt{y-x^2}} 3\,y\,e^{(-2\,y)}\,dz\,dy\,dx$$

and attempts to evaluate it.

```
> Mass2:=value(%);
```

$$Mass2 := \int_{-1}^{1} -\frac{3}{16}(16\,(1-x^2)^{(3/2)}\,e^{(-2+2\,x^2)} + 12\,\sqrt{1-x^2}\,e^{(-2+2\,x^2)}$$
$$- 3\,\sqrt{2}\,\sqrt{\pi}\,\mathrm{erf}(\sqrt{1-x^2}\,\sqrt{2}) + 16\,x^2\,\sqrt{1-x^2}\,e^{(-2+2\,x^2)}$$
$$- 4\,x^2\,\sqrt{2}\,\sqrt{\pi}\,\mathrm{erf}(\sqrt{1-x^2}\,\sqrt{2}))e^{(-2\,x^2)}dx$$

"That's interesting. Maple isn't able to perform the x integration."

"Who cares," Jerry snorts. "let's evaluate the result numerically."

```
> Mass2:=evalf(%);
```

$$Mass2 := 0.7618132467$$

That's better. The mass is 0.76 mass units. Now, why don't you type in the integral form for the y-coordinate of the center of mass,

```
> y2_cm:=Int(Int(Int(y*rho,z=-sqrt(y-Y)..sqrt(y-Y)),
        y=Y..1),x=-1..1)/Mass2_int;
```

$$y2\_cm := \frac{\int_{-1}^{1}\int_{x^2}^{1}\int_{-\sqrt{y-x^2}}^{\sqrt{y-x^2}} 3\,y^2\,e^{(-2\,y)}\,dz\,dy\,dx}{\int_{-1}^{1}\int_{x^2}^{1}\int_{-\sqrt{y-x^2}}^{\sqrt{y-x^2}} 3\,y\,e^{(-2\,y)}\,dz\,dy\,dx}$$

and then evaluate it.

```
> y2_cm:=evalf(value(%));
```

$$y2\_cm := 0.6628492940$$

So, $y_{\rm cm} = 0.66$, which is even higher than the 0.61 value that we obtained for the planar case. We had better play around with the parameters and lower the center of mass before Buzz comes back. As far as the values of $x_{\rm cm}$ and $z_{\rm cm}$ are concerned, they must clearly be equal to zero by symmetry. You might check one of these, for example $x_{\rm cm}$."

As instructed, Noah enters the integral expression for $x_{\rm cm}$,

```
> x2_cm:=Int(Int(Int(x*rho,z=-sqrt(y-Y)..sqrt(y-Y)),
          y=Y..1),x=-1..1)/Mass2_int;
```

$$x2\_cm := \frac{\int_{-1}^{1}\int_{x^2}^{1}\int_{-\sqrt{y-x^2}}^{\sqrt{y-x^2}} 3\,x\,y\,e^{(-2\,y)}\,dz\,dy\,dx}{\int_{-1}^{1}\int_{x^2}^{1}\int_{-\sqrt{y-x^2}}^{\sqrt{y-x^2}} 3\,y\,e^{(-2\,y)}\,dz\,dy\,dx}$$

```
> x2_cm:=evalf(value(%));
```

$$x2\_cm := 0.$$

and finds that it is indeed zero.

With the exercise completed, Buzz informs Noah and Jerry that their first class is over and he will see them next week. However, he doesn't let them off easily, assigning the following set of problems to be completed before the next class. Perhaps you should try them as well.

PROBLEMS:

Problem 3-34: Lowering the center of mass

Noah and Jerry's choice of parameters led to a $y_{\rm cm}$ for both the planar and three-dimensional calculations that was substantially above the bottom of the hull. Help them out by lowering the center of mass. Change the parameter b, holding all other parameter values unchanged, to a value that lowers $y_{\rm cm}$ just below 0.4. What is this critical value of b for each case?

Problem 3-35: Confirmation that $z_{\rm cm} = 0$

For the three-dimensional case, confirm that $z_{\rm cm} = 0$.

Problem 3-36: Varying n in the density function

Keeping all other parameters unchanged, calculate $y_{\rm cm}$ for the planar hull case for $n = 2, 3, 4, \ldots$ and plot this center of mass coordinate as a function of n.

Problem 3-37: A more realistic hull

A somewhat more realistically shaped hull can be defined by

$$\left(\frac{x}{a}\right)^2 + \left(\frac{z}{b}\right)^2 + \left(\frac{y-1}{c}\right)^2 \leq 1,$$

with $y \leq 1$, where y points down into the water. Taking $a=1$, $b=6$, and $c=1$:

- **(a)** Make a three-dimensional plot of this hull, using both constrained and unconstrained scaling.
- **(b)** Calculate the volume, mass, and center of mass of the hull if the density is given by $\rho = A\, y^n\, e^{-B\,y}$, with $A = 3$, $n = 1$, and $B = 2$.
- **(c)** Plot the center of mass along with the outline of the hull in both the x–y and z–y planes, using constrained scaling.

Problem 3-38: Triangular plate

A thin triangular plate with vertices $(0, 0)$, $(1, 0)$, and $(0, 2)$ has density $\rho(x, y) = 1 + 3\,x + y$. Using double integrals, calculate:

- **(a)** the mass of the plate;
- **(b)** the center of mass coordinates;
- **(c)** the moment of inertia $I_x = \iint y^2 \rho(x, y)\, dx\, dy$, about the x-axis;
- **(d)** the moment of inertia $I_y = \iint x^2 \rho(x, y)\, dx\, dy$, about the y-axis.

Create a plot of the triangular plate, suitably colored, and superimpose the center of mass in the form of a colored circle on the plate.

Problem 3-39: Semicircular plate

The density at any point on a thin semicircular plate of radius a is proportional to the distance from the center of the circle. Making use of double integrals and appropriate plotting tools:

- **(a)** calculate the mass of the plate;
- **(b)** calculate the coordinates of the center of mass;
- **(c)** plot the center of mass, appropriately labeled, on a colored picture of the plate.

Hint: Use polar coordinates (r, θ) and note that the area element is $dA = (r\, d\theta)\, dr$.

Problem 3-40: Charge distribution

Charge of surface charge density $\sigma(x, y) = x\,y$ C/m^2 is distributed over the triangular region defined by the lines $x = 1$, $y = 1$, $y = 1 - x$. Using double integrals and appropriate plotting tools:

- **(a)** calculate the total charge;
- **(b)** calculate the center of mass coordinates of the charge distribution;
- **(c)** plot the charge density distribution over the triangular plate;

(d) plot the center of mass, appropriately labeled, on the plate.

Problem 3-41: Mass of a solid tetrahedron

A solid tetrahedron of mass density $\rho = z$ is bounded by the four planes $x = 0$, $y = 0$, $z = 0$, and $x + y + z = 1$. Making use of triple integrals, calculate:

(a) the volume of the tetrahedron;

(b) the mass of the tetrahedron;

(c) the center of mass coordinates.

Chapter 4

Algebraic Models. Part II

It is written in the language of mathematics, and its characters are triangles, circles, and other geometrical figures, without which it is impossible to understand a single word of it; without these, one is wandering about in a dark labyrinth.
Galileo Galilei, Italian astronomer and physicist (1564–1642)

In the first section of this chapter, the recipes illustrate how Maple can be used to formulate and explore vector models in Cartesian as well as other orthogonal[1] curvilinear coordinate systems. The key library package for entering and manipulating vectors is the VectorCalculus package.

The second section features matrix models, the LinearAlgebra library package being of central importance for dealing with matrices.

4.1 Vector Models

If $\hat{e}_x$, $\hat{e}_y$, and $\hat{e}_z$ are Cartesian unit vectors (vectors of unit length) pointing along the x, y, and z axes, respectively, the sum of two vectors $\vec{A} = A_x\,\hat{e}_x + A_y\,\hat{e}_y + A_z\,\hat{e}_z$ and $\vec{B} = B_x\,\hat{e}_x + B_y\,\hat{e}_y + B_z\,\hat{e}_z$ is given by

$$\vec{A} + \vec{B} = (A_x + B_x)\,\hat{e}_x + (A_y + B_y)\,\hat{e}_y + (A_z + B_z)\,\hat{e}_z.$$

The dot or scalar product between two vectors $\vec{A}$ and $\vec{B}$ is defined by

$$\vec{A} \cdot \vec{B} = A_x\,B_x + A_y\,B_y + A_z\,B_z = A\,B\,\cos\theta,$$

where $A = \sqrt{A_x^2 + A_y^2 + A_z^2}$ and $B = \sqrt{B_x^2 + B_y^2 + B_z^2}$ are the magnitudes of $\vec{A}$ and $\vec{B}$, and θ is the angle between them.

The cross or vector product of $\vec{A}$ and $\vec{B}$, written as $\vec{A} \times \vec{B}$, is another vector whose magnitude $|\vec{A} \times \vec{B}|$ is equal to $A\,B\,\sin\theta$. The direction of $\vec{A} \times \vec{B}$ is given by the right-hand rule. Put the fingers of the right hand along $\vec{A}$ and curl them toward $\vec{B}$ in the direction of the smaller angle between $\vec{A}$ and $\vec{B}$. The thumb then points in the direction of the new vector.

[1]The angle between unit vectors is 90°.

4.1.1 Vectoria's Mathematical Heritage

There are some things which cannot be learned quickly, and time, which is all we have, must be paid heavily for their acquiring. They are the very simplest things and because it takes a man's life to know them the little new that each man gets from life is very costly and the only heritage he has to leave.
Ernest Hemingway, American writer, *Death in the Afternoon (1932)*

Vectoria, a physics student whom we have already met on a number of occasions, is in the process of learning about the magical kingdom of vectors. At least this field of mathematics is somewhat magical to her, since it inspired her mathematically inclined mother, Dorothy (Dot, for short) Product, to give Vectoria her unique first name. Taking after her mother, who is an avid computer algebra fan, Vectoria soon grows tired of doing messy vector manipulation problems by hand and says, "Let's see what we can do with Maple."

To plot vectors and mathematically manipulate them, she finds that a call must be made to the plots, plottools, and VectorCalculus packages. To suppress four warning messages that would otherwise appear on "loading" these packages, the warning level is first set to zero in the `interface` command.

```
> restart: interface(warnlevel=0):
> with(plots): with(plottools): with(VectorCalculus):
```

Vectoria decides to first carry out various standard vector operations, such as the analytical and graphical addition of vectors, finding the angle between two vectors, and determining dot and cross products for combinations of vectors.

She decides to work with some representative three-dimensional vectors in Cartesian coordinates, viz.,

$$\vec{A} = 2\,a\,\hat{\mathrm{e}}_{\mathrm{x}} + a\,\hat{\mathrm{e}}_{\mathrm{y}}, \qquad \vec{B} = b\,\hat{\mathrm{e}}_{\mathrm{x}} + 5\,b\,\hat{\mathrm{e}}_{\mathrm{z}}, \qquad \vec{C} = c\,\hat{\mathrm{e}}_{\mathrm{x}} + 4\,c\,\hat{\mathrm{e}}_{\mathrm{y}} + 3\,c\,\hat{\mathrm{e}}_{\mathrm{z}}, \tag{4.1}$$

where a, b, and c are real constants. The three vectors are entered, the "long" syntax form being used for $\vec{A}$, the "short" forms for $\vec{B}$ and $\vec{C}$.

```
> A:=Vector([2*a,a,0]); B:=<b,0,5*b>; C:=<c,4*c,3*c>;
```

$$A := 2\,a\,\mathrm{e}_{\mathrm{x}} + a\,\mathrm{e}_{\mathrm{y}} \quad B := b\,\mathrm{e}_{\mathrm{x}} + 5\,b\,\mathrm{e}_{\mathrm{z}} \quad C := c\,\mathrm{e}_{\mathrm{x}} + 4\,c\,\mathrm{e}_{\mathrm{y}} + 3\,c\,\mathrm{e}_{\mathrm{z}}$$

Vectoria observes that the hat symbol doesn't appear on the unit vectors in the output, which is notationally consistent with the fact that vector arrows do not appear on A, B, and C.

For the specific vectors that she has chosen, Vectoria can easily calculate the resultant vector $\vec{R} = \vec{A} + \vec{B} + \vec{C}$ in her head. However, she checks to see whether Maple will add the vector components properly.

```
> R:=A+B+C;
```

$$R := (2\,a + b + c)\,\mathrm{e}_{\mathrm{x}} + (a + 4\,c)\,\mathrm{e}_{\mathrm{y}} + (5\,b + 3\,c)\,\mathrm{e}_{\mathrm{z}}$$

Vectoria can see that the answer is correct, and undoubtedly you can too!

Another standard problem encountered by beginning science students is to find the angle θ between any two vectors, say $\vec{A}$ and $\vec{B}$. It follows from the definition of the dot product that the angle between $\vec{A}$ and $\vec{B}$ is given by

$$\theta = \arccos((\vec{A} \cdot \vec{B})/(A\,B)). \tag{4.2}$$

Vectoria notes that the dot product of two vectors $\vec{A}$ and $\vec{B}$ can be accomplished with either the long form of the command, viz., `DotProduct(A,B)`, or the short form `A . B`. She chooses to use the short form in evaluating the angle θ between the two vectors $\vec{A}$ and $\vec{B}$ entered earlier.

```
> theta:=arccos((A . B)/sqrt((A . A)*(B . B)));
```

$$\theta := \arccos\left(\frac{1}{65}\,\frac{a\,b\,\sqrt{130}}{\sqrt{a^2\,b^2}}\right)$$

In the last output line, the parameters a and b should have canceled but didn't because the term $\sqrt{a^2b^2}$ could yield either a positive or negative result. Since the angle should be positive, Vectoria applies the `simplify` command with the `symbolic` option (which assumes that a and b are positive).

```
> theta:=simplify(theta,symbolic);
```

$$\theta := \arccos\left(\frac{\sqrt{130}}{65}\right)$$

The angle between $\vec{A}$ and $\vec{B}$ is numerically evaluated,

```
> theta:=evalf(theta);
```

$$\theta := 1.394472488$$

yielding 1.39 radians. The angle can be converted from radians to degrees,

```
> theta:=convert(theta,units,radians,degrees));
```

$$\theta := 79.89738818$$

so the angle between $\vec{A}$ and $\vec{B}$ is $\theta = 79.9°$.

Vectoria has heard that the cross, or vector, product (the latter being the inspiration for her own name) tends to give many beginning students trouble, so she reviews the definition given at the beginning of this chapter. She also notes that it is straightforward to prove that

$$\vec{A} \times \vec{B} = (A_y B_z - A_z B_y)\,\hat{e}_x + (A_z B_x - A_x B_z)\,\hat{e}_y + (A_x B_y - A_y B_x)\,\hat{e}_z. \tag{4.3}$$

It may also be shown that the magnitude $|\vec{A} \times \vec{B}|$ is equal to the area of the parallelogram having $\vec{A}$ and $\vec{B}$ along two of the sides.

Vectoria uses Maple to calculate the cross products $\vec{A} \times \vec{B}$ and $\vec{B} \times \vec{C}$ for the previously entered vectors, using the long form of the cross product command for the former, and the short form for the latter.

```
> AcrossB:=CrossProduct(A,B); BcrossC:=B &x C;
```

$$AcrossB := 5\,a\,b\,e_x - 10\,a\,b\,e_y - a\,b\,e_z$$

$$BcrossC := -20\,b\,c\,e_x + 2\,b\,c\,e_y + 4\,b\,c\,e_z$$

She calculates the areas, *Area1* and *Area2*, of the two parallelograms having $\vec{A}$, $\vec{B}$ and $\vec{B}$, $\vec{C}$, respectively, as adjacent edges. For *Area1*, she employs the `simplify(symbolic)` command once again, while for *Area2* she uses the `assuming` command. This command differs from the `assume` command in that it applies the assumption only to the command line in which it appears, while the latter applies the assumption to the entire worksheet after being entered.

```
> Area1:=simplify(sqrt(AcrossB . AcrossB),symbolic);
```

$$Area1 := 3\sqrt{14}\, a\, b$$

```
> Area2:=sqrt(BcrossC . BcrossC) assuming b>0,c>0;
```

$$Area2 := 2\sqrt{105}\, b\, c$$

The volume of the parallelepiped having $\vec{A}$, $\vec{B}$, and $\vec{C}$ as adjacent edges can be calculated by forming the "mixed" product $\vec{A}\cdot(\vec{B}\times\vec{C})$ and taking the absolute value. Vectoria labels the result *Volume1*. She also calculates the volume $|(\vec{A}\times\vec{B})\cdot\vec{C}|$, calling this result *Volume2*.

```
> Volume1:=abs(A . (B &x C)); Volume2:=abs((A &x B) . C);
```

$$Volume1 := 38\, |a\, b\, c| \qquad Volume2 := 38\, |a\, b\, c|$$

The two volumes are identical. If the absolute values are not taken, Vectoria is able to confirm the "vector identity"

$$\vec{A}\cdot(\vec{B}\times\vec{C}) = (\vec{A}\times\vec{B})\cdot\vec{C}, \tag{4.4}$$

by subtracting the right-hand side from the left-hand-side,

```
> `Adot(BcrossC)-(AcrossB)dotC`:=A . (B &x C) - (A &x B) . C;
```

$$Adot(BcrossC) - (AcrossB)dotC := 0$$

and obtaining zero. Vectoria has had to use "left quotes" on the assigned name because it contains mathematical operations.

In the next two command lines, she finds, however, that $\vec{A}\times(\vec{B}\times\vec{C})$ is not the same as $(\vec{A}\times\vec{B})\times\vec{C}$. In mathematical language, the "associative law" does not hold for the cross product.

```
> `Across(BcrossC)`:=A &x (B &x C);
```

$$Across(BcrossC) := 4\, a\, b\, c\, e_x - 8\, a\, b\, c\, e_y + 24\, a\, b\, c\, e_z$$

```
> `(AcrossB)crossC`:=(A &x B) &x C;
```

$$(AcrossB)crossC := -26\, a\, b\, c\, e_x - 16\, a\, b\, c\, e_y + 30\, a\, b\, c\, e_z$$

To plot the four vectors $\vec{A}$, $\vec{B}$, $\vec{C}$, and $\vec{R}$, Vectoria chooses some specific values for the parameters, viz., $a = 3$, $b = 2$, and $c = 1$. She also creates a functional operator `F` to evaluate an arbitrary quantity V with these parameter values.

```
> params:={a=3,b=2,c=1}: F:=V->eval(V,params):
```

Then, using `F`, the forms of $\vec{A}$, $\vec{B}$, $\vec{C}$, and $\vec{R}$ are explicitly displayed with the parameter values substituted.

```
> A:=F(A); B:=F(B); C:=F(C); R:=F(R);
```

$$A := 6\, e_x + 3\, e_y \qquad B := 2\, e_x + 10\, e_z \qquad C := e_x + 4\, e_y + 3\, e_z$$

$$R := 9\, e_x + 7\, e_y + 13\, e_z$$

The `arrow` command is used to produce a red arrow connecting the tail (which is placed at the origin, (0,0,0)) of the vector $\vec{A}$ to its tip. The numbers `0.2,1,0.2` refer to the width of the arrow's body, the width of the arrow's head, and the arrowhead length expressed as a ratio of the arrow body length, respectively. The arrow plot is suppressed temporarily and given the label AA.

```
> AA:=arrow([0,0,0], A,0.2,1,0.2,color=red):
```

To carry out the vector addition of $\vec{A}$ and $\vec{B}$ graphically, the vector $\vec{B}$ is placed with its tail at the tip of the vector $\vec{A}$. The $\vec{B}$ vector will be represented on the computer screen by a brown arrow.

```
> BB:=arrow(A,B,0.2,1,0.2,color=brown):
```

The third vector $\vec{C}$ must then be placed with its tail at the vector sum of $\vec{A}$ and $\vec{B}$. A coral colored arrow is used to represent $\vec{C}$.

```
> CC:=arrow(A+B,C,0.2,1,0.25,color=coral):
```

The resultant vector $\vec{R}$ is a vector with tail at the origin and tip R located at the vector sum of $\vec{A}$, $\vec{B}$, and $\vec{C}$. It is represented by a black arrow.

```
> RR:=arrow([0,0,0],R,0.25,1,0.2,color=black):
```

The four plots are superimposed with the display command,

```
> display({AA,BB,CC,RR},axes=normal,orientation=[55,74],
  tickmarks=[2,2,2],labels=["x","y","z"]);
```

the resulting picture being shown in Figure 4.1.

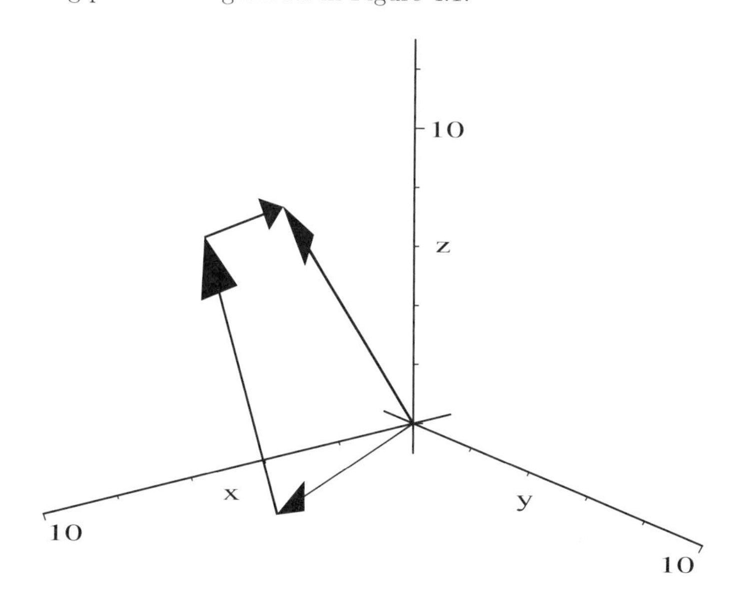

Figure 4.1: Graphical illustration of the vector sum of $\vec{A}$, $\vec{B}$, and $\vec{C}$.

By clicking on the plot with the mouse, the plot can be rotated by dragging on it to view the vector addition from different perspectives.

With these preliminaries under her belt, Vectoria is ready to tackle some simple physical applications of vectors. This will be done in the following story.

PROBLEMS:

Problem 4-1: Angle between vectors

Determine the angle in degrees between the vectors $\vec{A}$ and $\vec{B}$ in each of the following cases:

(a) $\vec{A} = 3\,\hat{e}_x + 4\,\hat{e}_y + \hat{e}_z, \quad \vec{B} = 0\,\hat{e}_x + 2\,\hat{e}_y - 5\,\hat{e}_z;$

(b) $\vec{A} = \hat{e}_x + 0\,\hat{e}_y + 3\,\hat{e}_z, \quad \vec{B} = 5\,\hat{e}_x + 2\,\hat{e}_y - 6\,\hat{e}_z.$

In each case, make a graph showing the vectors $\vec{A}$ and $\vec{B}$ and their sum. Also calculate the areas of the parallelograms with $\vec{A}$, $\vec{B}$ as edges.

Problem 4-2: Vector manipulations

Consider the three vectors

$$\vec{P} = 2\,\hat{e}_x + 0\,\hat{e}_y - \hat{e}_z, \quad \vec{Q} = 2\,\hat{e}_x - \hat{e}_y + 2\,\hat{e}_z, \quad \vec{R} = 2\,\hat{e}_x - 3\,\hat{e}_y + \hat{e}_z.$$

Make a graph showing $\vec{P}$, $\vec{Q}$, $\vec{R}$ and their sum. Then determine:

(a) $(\vec{P} + \vec{Q}) \times (\vec{P} - \vec{Q})$;

(b) $\vec{Q} \cdot (\vec{R} \times \vec{P})$;

(c) $\vec{P} \cdot (\vec{Q} \times \vec{R})$;

(d) angle between $\vec{Q}$ and $\vec{R}$;

(e) $\vec{P} \times (\vec{Q} \times \vec{R})$;

(f) the component of $\vec{P}$ along $\vec{Q}$.

Problem 4-3: Area of a triangle

Determine the area of the triangle with vertices $A(1, 4, 6)$, $B(-2, 5, -1)$, and $C(1, -1, 1)$.

Problem 4-4: Coplanar vectors

Show that

$$\vec{A} = \hat{e}_x + 4\,\hat{e}_y - 7\,\hat{e}_z, \quad \vec{B} = 2\,\hat{e}_x - \hat{e}_y + 4\,\hat{e}_z, \quad \vec{C} = -9\,\hat{e}_y + 18\,\hat{e}_z$$

are coplanar, i.e., the vectors lie in the same plane. *Hint*: Show that the volume of the parallepiped formed by $\vec{A}$, $\vec{B}$, and $\vec{C}$ is zero.

Problem 4-5: Unit vector

Determine the unit vector perpendicular to the plane that contains the vectors $\vec{A} = 2\,\hat{e}_x - 6\,\hat{e}_y - 3\,\hat{e}_z$ and $\vec{B} = 4\,\hat{e}_x + 3\,\hat{e}_y - \hat{e}_z$.

Problem 4-6: Torque

When a force $\vec{F}$ acts on a rigid body at a point given by a position vector $\vec{r}$ relative to a given origin of coordinates, the torque (or moment) $\vec{\tau}$ with respect to the origin is defined by the cross product $\vec{\tau} = \vec{r} \times \vec{F}$. The torque measures the tendency of the body to rotate about the origin. Calculate the torque if $\vec{r} = \hat{e}_x + 3\,\hat{e}_y + 2\,\hat{e}_z$ meters and $\vec{F} = -40\,\hat{e}_x - 20\,\hat{e}_y + 40\,\hat{e}_z$ newtons. What direction does the torque vector point in?

Problem 4-7: Trig identity
Consider the two vectors

$$\vec{P} = \cos\theta_1\,\hat{e}_x + \sin\theta_1\,\hat{e}_y, \quad \vec{Q} = \cos\theta_2\,\hat{e}_x + \sin\theta_2\,\hat{e}_y.$$

(a) Prove that $\vec{P}$, $\vec{Q}$ are unit vectors making angles θ_1 and θ_2 with the x-axis.

(b) Using the dot product, obtain a well-known trig identity for $\cos(\theta_2 - \theta_1)$.

Problem 4-8: Volume of a parallelepiped
As already noted, any three distinct vectors $\vec{A}$, $\vec{B}$, and $\vec{C}$ whose tails have a common origin or vertex may be thought of as defining a volume element having six faces, parallel in pairs: a parallelepiped. The volume V of such a parallepiped is given by $V = |\vec{A} \cdot (\vec{B} \times \vec{C})|$. A parallepiped with one vertex at the origin is described by three vectors whose tips are located at the vertices $(10, -5, 3)$, $(3, -4, 7)$, and $(-5, -6, 3)$, respectively, in rectangular coordinates. Distances are in centimeters. Calculate V for the corresponding parallelepiped. Make a graph showing the three vectors, each with its tail at the origin.

Problem 4-9: Angle
Determine the angle between the central diagonal of a cube and one of its edges. Express your answer in degrees and in radians.

Problem 4-10: Vector identity
For general three-dimensional vectors, show that

$$\vec{A} \times (\vec{B} \times \vec{C}) = \vec{B}(\vec{A} \cdot \vec{C}) - \vec{C}(\vec{A} \cdot B).$$

4.1.2 Vectoria and Fowles's Fly

The fly that does not want to be swatted is safest if it sits on the fly-swat.
G. C. Lichtenberg, German physicist, philosopher (1742–1799)

In an elementary mechanics text by Fowles and Cassiday [FC99], Vectoria runs across a simple kinematic application of vectors. It is stated that a fly moves along a path given by the time-dependent position vector

$$\vec{r}(t) = \hat{e}_x\, d\, \sin(\omega\, t) + \hat{e}_y\, d\, \cos(\omega\, t) + \hat{e}_z\, e\, t^2, \tag{4.5}$$

where d, e and ω are real parameters and $\hat{e}_x$, $\hat{e}_y$, and $\hat{e}_z$ are unit vectors along the x-, y-, and z-axes, respectively. She is asked to show that the magnitude of the acceleration is constant. In addition she decides to plot the path traced out by the fly, the distance it travels along the path, and its displacement from the starting point, for representative values of the parameters.

Loading the necessary library packages, with the warning level set to zero,

```
> restart: interface(warnlevel=0):
> with(plots): with(VectorCalculus): with(LinearAlgebra):
```

and entering the position vector,

```
> pos:=<d*sin(omega*t),d*cos(omega*t),e*t^2>;
```

$$pos := d\sin(\omega\,t)\,e_x + d\cos(\omega\,t)\,e_y + e\,t^2\,e_z$$

the velocity and acceleration vectors are calculated by taking the first and second time derivatives of the position.

```
> vel:=diff(pos,t); accel:=diff(pos,t,t);
```

$$vel := d\cos(\omega\,t)\,\omega\,e_x - d\sin(\omega\,t)\,\omega\,e_y + 2\,e\,t\,e_z$$

$$accel := -d\sin(\omega\,t)\,\omega^2\,e_x - d\cos(\omega\,t)\,\omega^2\,e_y + 2\,e\,e_z$$

The dot product of the acceleration vector with itself is carried out, assuming that $d > 0$, $e > 0$, $\omega > 0$, and $t > 0$.

```
> acceldotaccel:=DotProduct(accel,accel)
                    assuming d>0,e>0,omega>0,t>0;
```

$$acceldotaccel := d^2\sin(\omega\,t)^2\,\omega^4 + d^2\cos(\omega\,t)^2\,\omega^4 + 4\,e^2$$

The previous line is simplified using the trig option of the `combine` command, and the magnitude of the acceleration obtained by taking the square root.

```
> accelmag:=sqrt(combine(%,trig));
```

$$accelmag := \sqrt{d^2\,\omega^4 + 4\,e^2}$$

Vectoria notes that she could have calculated the magnitude of the acceleration in a different manner, using the `VectorNorm` command (contained in the LinearAlgebra package) and simplifying with the same assumptions.

```
> accelmag2:=VectorNorm(accel,2);
```

$$accelmag2 := \sqrt{|d\sin(\omega\,t)\,\omega^2|^2 + |d\cos(\omega\,t)\,\omega^2|^2 + 4\,|e|^2}$$

```
> accelmag2:=simplify(%) assuming d>0,e>0,omega>0,t>0;
```

$$accelmag2 := \sqrt{d^2\,\omega^4 + 4\,e^2}$$

In either case, since the output does not contain the time t, the acceleration magnitude is indeed constant.

To plot the path traced out by the fly, Vectoria chooses the nominal parameter values $d = 2$, $e = 1/20$, and $\omega = 3$, and creates an operator `F` to evaluate an arbitrary quantity V with these parameter values.

```
> params:={d=2,e=1/20,omega=3}: F:=V->eval(V,params):
```

Using `F`, the position, velocity, and acceleration vectors take the following forms.

```
> pos:=F(pos); vel:=F(vel); accel:=F(accel);
```

$$pos := 2\sin(3\,t)\,e_x + 2\cos(3\,t)\,e_y + \frac{t^2}{20}\,e_z$$

$$vel := 6\cos(3\,t)\,e_x - 6\sin(3\,t)\,e_y + \frac{t}{10}\,e_z$$

$$accel := -18\sin(3\,t)\,e_x - 18\cos(3\,t)\,e_y + \frac{1}{10}\,e_z$$

Extracting the three components of the position vector and forming them into a Maple list, the `spacecurve` command is used to plot the 3-dimensional path (spatial coordinates in meters) traveled by the fly over a time interval of 20 s.

```
> spacecurve([pos[1],pos[2],pos[3]],t=0..20,numpoints=200,
  axes=normal,tickmarks=[2,2,2],labels=["x","y","z"]);
```

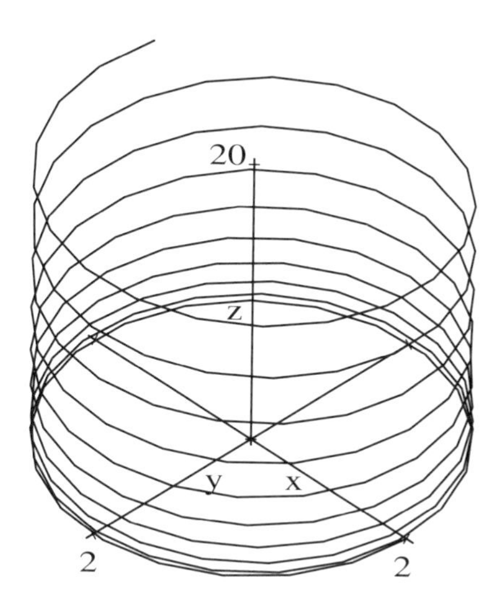

Figure 4.2: Helical path of the fly.

The resulting picture is shown in Figure 4.2, revealing that the fly moves along a helical path. Vectoria is curious as to how much distance the fly covers moving along this path. Because of the nature of the trajectory, this distance will be considerably larger than the magnitude of the displacement vector pointing from the starting position to the fly's position at $t = 20$ s. The distance along the helix must be equal to the time integral of the speed, i.e., $\int v\,dt$. To calculate the speed, Vectoria applies the `VectorNorm` and `simplify(symbolic)` commands.

```
> v:=VectorNorm(vel,2);
```

$$v := \frac{1}{10}\sqrt{3600\,|\cos(3\,t)|^2 + 3600\,|\sin(3\,t)|^2 + |t|^2}$$

```
> v:=simplify(v,symbolic);
```

$$v := \frac{\sqrt{t^2 + 3600}}{10}$$

The distance traveled in the time interval $t = 0$ to 20 seconds is calculated by performing the following time integration.

```
> distance:=Int(v,t=0..20)=int(v,t=0..20);
```

$$distance := \int_0^{20} \frac{\sqrt{t^2+3600}}{10}\,dt = -\frac{90\left(-\frac{2\sqrt{\pi}\sqrt{10}}{9} - 2\sqrt{\pi}\,\mathrm{arcsinh}\left(\frac{1}{3}\right)\right)}{\sqrt{\pi}}$$

Vectoria numerically evaluates the right-hand side of the previous output, to 4 digits accuracy.

```
> distance:=evalf(rhs(%),4);
```

$$distance := 122.2$$

The fly has covered about 122 meters along the helix in the 20 second time interval. Its displacement from its starting point at $t = 0$ is, of course, much less. The displacement vector is calculated by evaluating the position vector at $t = 20$ and $t = 0$ and subtracting.

```
> displ_vector:=evalf(eval(pos,t=20)-eval(pos,t=0));
```

$$displ\_vector := -0.6096212422\,\mathrm{e_x} - 3.904825961\,\mathrm{e_y} + 20.\,\mathrm{e_z}$$

Using the `VectorNorm` and `evalf` commands, the magnitude of the displacement is evaluated to 3 significant figures.

```
> displ_mag:=evalf(VectorNorm(%,2),3);
```

$$displ\_mag := 20.4$$

The fly is only about 20 meters from its starting position.

PROBLEMS:

Problem 4-11: Flight of the bumblebee

A bumblebee goes out from its hive along a path given in plane polar coordinates by

$$r = a\,t^2, \quad \theta = b\,t,$$

where a and b are positive constants. Using Cartesian coordinates,

(a) Plot the path traced out by the bumblebee for $a = b = 1$.

(b) Calculate the velocity and acceleration vectors.

(c) Show that the angle between the velocity and acceleration is constant.

Problem 4-12: Polar plots

Using Maple Help, look up the command structure to make a polar plot. Then make polar plots of the following for a few different positive a values:

(a) The flight of the bumblebee in the previous problem;

(b) $r = a\,\theta$; **(c)** $r = a\cos(2\,\theta)$; **(d)** $r = a^2\sin(2\,\theta)$; **(e)** $r = a\sin(3\,\theta)$.

Problem 4-13: A flight of fantasy

A mythical creature of mass 1 kilogram is flying in such a way that its position vector in meters is given at time t seconds by

$$\vec{r} = t\,\hat{\mathrm{e}}_\mathrm{x} + \left(t + \frac{1}{2}t^2\right)\hat{\mathrm{e}}_\mathrm{y} - \frac{4}{\pi^2}\sin\left(\frac{\pi t}{2}\right)\hat{\mathrm{e}}_\mathrm{z}.$$

(a) Plot the creature's position as a function of time for $t=0$ to 100.

(b) Calculate the creature's velocity, acceleration, and kinetic energy as a function of time. Evaluate these quantities at $t=1$ second.

(c) Calculate the creature's distance from the origin at $t=1$ second.

4.1.3 Ain't She Sweet

Ain't she nice? Just cast an eye in her direction.
Oh, me! Oh, my! Ain't that perfection?
1926 Hit song (Words by Jack Yellen, Music by Milton Ager)

On finishing with Fowles's fly, Vectoria looks out the window of the computer lab where she has been working and realizes what a beautiful sunny day it is outside. So, she decides to soak up some sunshine on a rustic bench located under a flowering cherry tree adjacent to an ivy covered brick wall near the math building. In the distance, she spots her classmate Mike jogging in an easterly direction along a straight sidewalk (see Figure 4.3), which passes through an arch in the wall.

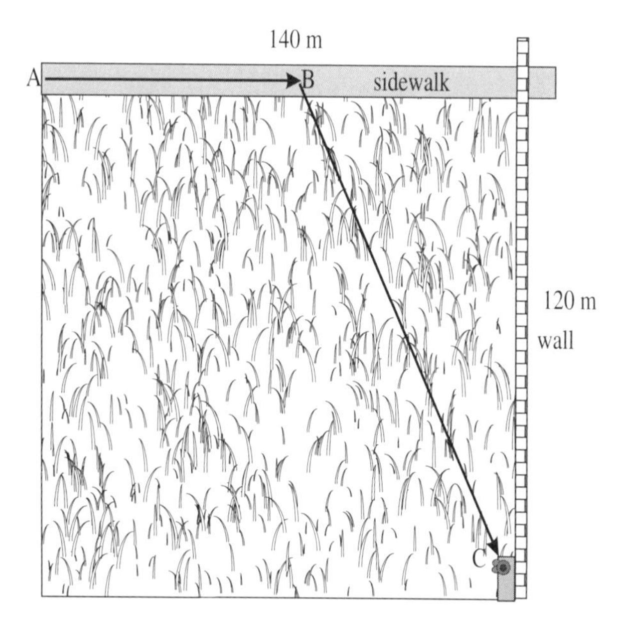

Figure 4.3: Schematic drawing of Mike's path.

She has heard from her friend Colleen that Mike is training for an upcoming triathlon. Vectoria has been interested in Mike for some time, but he has been too shy to approach her and, unlike many of her contemporaries, she doesn't want to appear too forward and ask him out.

At the point A in the figure, some 140 m from the wall, Mike has spotted Vectoria and finally worked up his courage to ask her out on a date. He wishes to join her posthaste before she possibly gets up and leaves. Mike is running at a speed of 5 m/s along the sidewalk, and at some point B must cut across the rough grassy field to get to the bench at C, located 120 m south of the sidewalk. However, his speed on the rough grassy field is only 3 m/s, so the question is what path ABC must Mike follow to reach Vectoria in minimum time, and what is the minimum time? Mike is too busy jogging and thinking of Vectoria, so let's answer these questions for him and, as a bonus, animate his motion along the path.

Let's first load the plots and VectorCalculus packages.

```
> restart: with(plots): with(VectorCalculus):
```

We will let $t1$ be the time it takes Mike to run from A to B. For times $t \leq t1$, Mike's displacement (with x corresponding to east) from his starting point is

```
> d1:=<5*t,0>;
```

$$d1 := 5\,t\,\mathrm{e_x}$$

Let $T - t1$ be the time it takes him to run from B to C, where T is the total time. The goal is to make T a minimum. If θ is the acute angle that the path segment BC makes with the sidewalk, then Mike's displacement (with y corresponding to south) from his starting point for the interval $t1 \leq t \leq T$ is

```
> d2:=<5*t1+3*cos(theta)*(t-t1),3*sin(theta)*(t-t1)>;
```

$$d2 := (5\,t1 + 3\cos(\theta)\,(t - t1))\,\mathrm{e_x} + 3\sin(\theta)\,(t - t1)\,\mathrm{e_y}$$

The relevant kinematic equations are obtained by evaluating the first (x) and second (y) components of $d2$ at time $t = T$ and equating the results to 140 and 120, respectively. The pair of equations are entered as a Maple set.

```
> eqs:={eval(d2[1],t=T)=140,eval(d2[2],t=T)=120};
```

$$eqs := \{5\,t1 + 3\cos(\theta)\,(T - t1) = 140,\ 3\sin(\theta)\,(T - t1) = 120\}$$

We analytically solve eqs for $t1$ and T in terms of the unknown angle θ.

```
> sol:=solve(eqs,{t1,T});
```

$$sol := \left\{ t1 = \frac{4\,(7\sin(\theta) - 6\cos(\theta))}{\sin(\theta)},\ T = \frac{4\,(7\sin(\theta) - 6\cos(\theta) + 10)}{\sin(\theta)} \right\}$$

The solution is assigned,

```
> assign(sol):
```

and the time T differentiated with respect to θ and set equal to zero.

```
> eq3:=diff(T,theta)=0;
```

$$eq3 := \frac{4\,(7\cos(\theta) + 6\sin(\theta))}{\sin(\theta)} - \frac{4\,(7\sin(\theta) - 6\cos(\theta) + 10)\cos(\theta)}{\sin(\theta)^2} = 0$$

The transcendental equation $eq3$ is numerically solved for the angle θ that minimizes the total time.

```
> theta:=fsolve(eq3,theta);
```

$$\theta := 0.9272952180$$

The angle is approximately 0.93 radians, or on converting θ to degrees and keeping 3 digits,

```
> Theta:=evalf(convert(theta,degrees),3);
```

$$\Theta := 53.1 \; degrees$$

about 53°. The angle could be obtained in a completely different manner by generalizing Fermat's principle to this example. This principle states that a light ray will travel along a path that minimizes the time. Using this idea, one can prove Snell's law for a light ray passing from one medium to another through a planar interface dividing them. If a light ray with speed v_1 in medium 1 is incident on the interface at an angle θ_1 to the interface, and is refracted at an angle θ_2 in medium 2 (where the speed is v_2), then Snell's law is

$$\frac{\cos\theta_1}{\cos\theta_2} = \frac{v_1}{v_2}. \tag{4.6}$$

For $\theta_2 = 0$ radians, the refracted ray will travel along the interface. Then $\cos(\theta_2) = \cos(0) = 1$, and the critical angle in medium 1 is given by $\theta_1 = \theta_{\rm cr} = \arccos(v_1/v_2)$. Mentally reversing Mike's path, and taking the grass to be medium 1 (where Mike's speed is $v_1 = 3$ m/s) and the sidewalk to be medium 2 (where his speed is $v_2 = 5$ m/s), the critical angle to minimize the time is $\theta_{\rm cr} = \arccos(3/5)$. This relation is now entered and numerically evaluated,

```
> angle_check:=evalf(arccos(3/5)); #Snell's law
```

$$angle\_check := 0.9272952180$$

yielding exactly the same value (in radians) for the angle as before.

The time *t1* from A to B, the total time T, and the distance from A to B are evaluated.

```
> t1:=t1; T:=T; distance:=5*t1;
```

$$t1 := 10.00000000 \qquad T := 60.00000000 \qquad distance := 50.00000000$$

So the minimum (total) time for Mike to reach Vectoria is 60 seconds. He must run for 10 seconds along the sidewalk, traveling a distance of 50 meters, before cutting across the field.

To create a plot of Mike's path, his horizontal and vertical coordinates for the first and second legs of the path ABC are rewritten as follows.

```
> d1b:=(d1[1],120-d1[2]): d2b:=(d2[1],120-d2[2]):
```

The entire path can then be expressed as the following piecewise function.[2]

```
> d:=piecewise(t<=t1,[d1b],t>=t1,[d2b]);
```

$$d := \begin{cases} [5\,t,\ 120] & t \le 10.000 \\ [32.000 + 1.800\,t,\ -2.400\,t + 144.000] & 10.000 \le t \end{cases}$$

The first portion, AB, of the path is plotted as a thick red line,

```
> plot||1:=plot([d1b,t=0..t1],color=red,thickness=2):
```

[2]To fit into the page width, superfluous zeros have been removed from the decimal output.

the second leg, BC, as a thick green line,

```
> plot||2:=plot([d2b,t=t1..T],color=green,thickness=2):
```

and the labels A, B, C included with the `textplot` command.

```
> plot||3:=textplot([[5,115,"A"],[45,115,"B"],[130,5,"C"]]):
```

The 3 plots are superimposed with constrained scaling to produce Mike's path.

```
> path:=display({seq(plot||i,i=1..3)},tickmarks=[3,3],
          axes=box,view=[0..140,0..120],scaling=constrained):
```

Mike's run along this path is now animated, Mike being represented by a size-20 blue circle. The first frame of the animation is shown in Figure 4.4, the circle located at (0,120). Click on the computer plot and the start arrow to initiate Mike's run. The time for each frame will be displayed as the animation runs.

```
> animate(pointplot,[[eval(d,t=tau)],symbol=circle,symbolsize
  =20,color=blue],tau=0..T,frames=100,background=path);
```

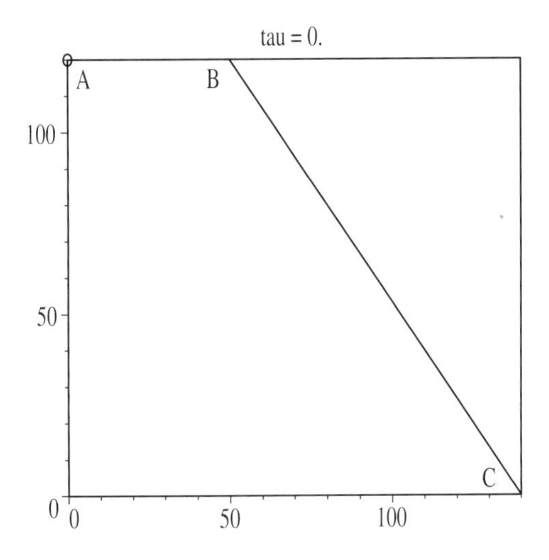

Figure 4.4: Opening frame of animation of Mike's run along the path.

Dripping with sweat, Mike reaches the bench, and persuades a not-too-reluctant Vectoria to go to a movie with him on Friday night.

PROBLEMS:

Problem 4-14: Relative velocity

A gravel truck belonging to the Boffo Trucking Company is traveling due north and descending a hill that has a 10% grade at a constant speed of 90 km/hr. At the bottom of the hill, the road is level and heads 30° east of north. A southbound police car, with a radar unit, is traveling at 80 km/hr along the level stretch at the base of the hill and is approaching the truck.

What is the relative velocity of the truck with respect to the police car? What is its relative speed? Make a 3-dimensional plot showing the relative velocity vector and the velocity vectors for the truck and the police car.

Problem 4-15: A brainteaser

Our sales manager friend Colleen was recently on a sea cruise with her ship traveling steadily east at 15 knots (knot is the abbreviation for nautical mile (6076.1 ft)) per hour. At some instant in time, she observed a naval ship 6 nautical miles due south of her, traveling on a steady course at a speed of 26 knots. Some time later the naval ship was observed to pass somewhere behind her ship, the distance of closest approach being 3 nautical miles. Her brainteaser for you is to answer the following set of questions using a vector approach, given only the above information:

(a) What was the course of the naval ship?

(b) What was the time elapsed between the first sighting and the time of minimum distance?

(c) Making an appropriate plot, labeled with the compass directions, deduce the location of each ship (relative to the cruise ship's initial position) at the time when the minimum separation occurs.

(d) Animate the motion of the ships.

Problem 4-16: Coulomb's force law

The electrical force $\vec{F}$ (in N) exerted on a point charge q (in C) in free space located at the position $\vec{r}$ (in m) due to n other charges q_i located at $\vec{r}_i$, $i = 1, 2, \ldots, n$, is given by Coulomb's law,

$$\vec{F} = \frac{q}{4\pi\epsilon_0} \sum_{i=1}^{n} \frac{q_i\,(\vec{r} - \vec{r}_i)}{|\vec{r} - \vec{r}_i|^3},$$

where $\epsilon_0 = 8.85 \times 10^{-12}\,\mathrm{C}^2/(\mathrm{N}\cdot\mathrm{m}^2)$ is the permittivity of free space. The electric field $\vec{E}$ at the point $\vec{r}$ due to the n charges is defined as $\vec{E} = \vec{F}/q$ (in V/m).

Point charges $q_1 = 1\,\mathrm{mC}$ and $q_2 = -2\,\mathrm{mC}$ are located at $\vec{r}_1 = 3\,\hat{e}_x + 2\,\hat{e}_y - \hat{e}_z$ and $\vec{r}_2 = -\hat{e}_x - \hat{e}_y + 4\,\hat{e}_z$, respectively. (Note: $1\,\mathrm{mC} = 10^{-3}\,\mathrm{C}$, $1\,\mathrm{nC} = 10^{-9}\,\mathrm{C}$.)

(a) What is the angle in radians and degrees between the vectors $\vec{r}_1$ and $\vec{r}_2$?

(b) Calculate $\vec{F}$ in mN on a 10-nC charge that is located at $\vec{r}_0 = 3\,\hat{e}_y + \hat{e}_z$.

(c) Determine $\vec{E}$ in kV/m at $\vec{r}_0$. Calculate the angle that $\vec{E}$ makes with the positive x-, y-, and z-axes at that point.

Problem 4-17: Another Coulomb's law problem

Point charges with charge 5 nC and −2 nC are located at $(2, 0, 4)$ and $(-3, 0, 5)$.

(a) Make a 2-dimensional plot showing the position vectors of these charges.

(b) What is the angle in radians and degrees between the two position vectors?

(c) Determine the force on a 4-nC point charge located at $(1, -3, 7)$.

(d) What is the electric field $\vec{E}$ at $(1, -3, 7)$? Calculate the angle that $\vec{E}$ makes with the positive x-, y-, z-axes at this point.

Problem 4-18: Electric force
A particle of mass 2 kg and charge 3 C starts at $t=0$ at the point $(1, -2, 0)$ with velocity $\vec{v}=4\,\hat{e}_x + 3\,\hat{e}_z$ m/s in an electric field $\vec{E}=12\,\hat{e}_x + 10\,\hat{e}_y$ V/m.

(a) Determine the particle acceleration, velocity, and position at arbitrary time $t > 0$.

(b) Make a three-dimensional plot showing the trajectory of the particle over the time interval $t=0$ to 1 second. Animate the motion of the particle over this interval, using the plot as the background.

4.1.4 Born Curl-Free

None who have always been free can understand the terrible fascinating power of the hope of freedom to those who are not free.
Pearl S. Buck, American novelist, *What America Means to Me (1943)*

Since our previous encounter, Vectoria Product has advanced in her physics degree program and is currently learning all about the vector operator triad, gradient, divergence, and curl, as well as how to calculate line integrals, in her vector calculus and intermediate electromagnetics courses. Recall that Vectoria's learning philosophy is to first solve a number of simple examples by hand in order to understand the underlying concepts and then explore what can be done with a CAS. Then, any tedious or difficult manipulations that she may have to do can be done accurately and quickly and the results suitably plotted. After her first date with Mike, Vectoria's friendship with him has blossomed into something more serious and now they can often be found working together in the computer lab. In response to her request, Mike has found or invented some interesting examples of vector fields[3] that they can hone their computer algebra skills on. We shall now eavesdrop on one of their work sessions.

"I have already done some preliminary calculations with these new vector operators," Mike remarks, "and I suggest that we load the plots and VectorCalculus library packages. In addition to the gradient, the VectorCalculus package contains the divergence, curl, and line integral operators, all of which we will be using in this vector field example.

```
> restart: with(plots): with(VectorCalculus):
```

To apply any of the vector operators, the coordinate system must generally be specified. We will start with a typical example, from elementary electromagnetic theory, involving a specified electric field vector $\vec{E}$ given in Cartesian coordinates. So I will set the coordinates to be Cartesian.

[3] A vector field is a vector quantity whose value varies from point to point in space.

```
> SetCoordinates('cartesian'[x,y,z]):
```

Using the VectorField command, let's enter the following polynomial expression for an electric field vector $\vec{E}$,

```
> E:=VectorField(<2*a*x*y*z-3*a*z,4*b*x^2*z+4*b*z,
    a*x^2*y-3*x+2*c*y>);
```

$$E := (2\,a\,x\,y\,z - 3\,a\,z)\,\bar{e}_x + (4\,b\,x^2\,z + 4\,b\,z)\,\bar{e}_y + (a\,x^2\,y - 3\,x + 2\,c\,y)\,\bar{e}_z$$

with the real parameters a, b, and c unknown. The overbars on the unit vectors in the output indicate that we are dealing with a vector field.

The question is, what values should a, b, and c have in order that the electric field be derivable from an electrostatic potential Φ, i.e., that we can write $\vec{E} = -\nabla\Phi$. From your lectures, you know that a potential will exist if the vector field is irrotational, i.e., curl $\vec{E} = 0$ everywhere. In the language of physics, we want to choose the parameters so as to make $\vec{E}$ a conservative field. Or to paraphrase the title of a movie that was popular in my parents' more youthful days, we want our vector field to be born curl-free. In Cartesian coordinates, the curl of a general electric field $\vec{E}$ takes the determinantal form

$$\text{curl}\,\vec{E} = \nabla \times \vec{E} = \begin{vmatrix} \hat{e}_x & \hat{e}_y & \hat{e}_z \\ \dfrac{\partial}{\partial x} & \dfrac{\partial}{\partial y} & \dfrac{\partial}{\partial z} \\ E_x & E_y & E_z \end{vmatrix}. \tag{4.7}$$

Although we could easily calculate the curl of our particular electric field by hand, let's let the computer determine curl $\vec{E}$.

```
> CurlE:=Curl(E);
```

$$CurlE := (a\,x^2 + 2\,c - 4\,b\,x^2 - 4\,b)\,\bar{e}_x + (-3\,a + 3)\,\bar{e}_y + (8\,b\,x\,z - 2\,a\,x\,z)\,\bar{e}_z$$

We can obtain the same result by using the Del operator to calculate $\nabla \times \vec{E}$.

```
> Del &x E;
```

$$(a\,x^2 + 2\,c - 4\,b\,x^2 - 4\,b)\,\bar{e}_x + (-3\,a + 3)\,\bar{e}_y + (8\,b\,x\,z - 2\,a\,x\,z)\,\bar{e}_z$$

For the curl to vanish, each curl component must be set equal to zero. Again, for our example, this is easily done by hand. But we may ultimately have to solve more complex examples, so let's continue with our computer algebra approach. We can impose the zero curl condition with the following command line and simultaneously solve for a, b, and c. Note that solve's default is to assume that the curl components are zero even though this is not specified.

```
> sol:=solve({CurlE[1],CurlE[2],CurlE[3]},{a,b,c});
```

$$sol := \left\{a = 1,\, b = \frac{1}{4},\, c = \frac{1}{2}\right\}$$

So, curl $\vec{E} = 0$ and an associated potential exists if we choose $a = 1$, $b = 1/4$, and $c = 1/2$. Evaluating $\vec{E}$ with *sol*, we see that the conservative electric field $\vec{E}$ is now completely determined."

```
> E:=eval(E,sol);
```

$$E := (2\,x\,y\,z - 3\,z)\,\bar{e}_x + (x^2\,z + z)\,\bar{e}_y + (x^2\,y - 3\,x + y)\,\bar{e}_z$$

"Mike," Vectoria interjects, "now that $\vec{E}$ is known, we can calculate the charge density ρ that produces this field as well as the potential itself. In MKS units, the field and charge density are related through the differential form of Gauss's law,

$$\operatorname{div} \vec{E} = \nabla \cdot \vec{E} = \rho/\epsilon_0, \tag{4.8}$$

where ϵ_0 is the permittivity of free space. So using the `Divergence` command,[4] we can calculate ρ from $\vec{E}$.

```
> rho:=epsilon[0]*Divergence(E);
```

$$\rho := 2\,\varepsilon_0\,y\,z$$

The charge density is proportional to the product $y\,z$ but is independent of x. Notice that the charge density is positive if y and z are both positive or both negative, but is negative if y and z are of opposite sign. We can get an even better feeling for the charge density by plotting it. The `plot3d` command can be used to graph the normalized charge density ρ/ϵ_0, with contours corresponding to different values of this quantity.

```
> plot3d(rho/epsilon[0],y=-5..5,z=-5..5,axes=framed,
  contours=[-25,-20,-15,-10,-1,1,10,20,30],tickmarks=[3,3,3],
  style=patchcontour,labels=["y","z","rho"],
  orientation=[-15,70]);
```

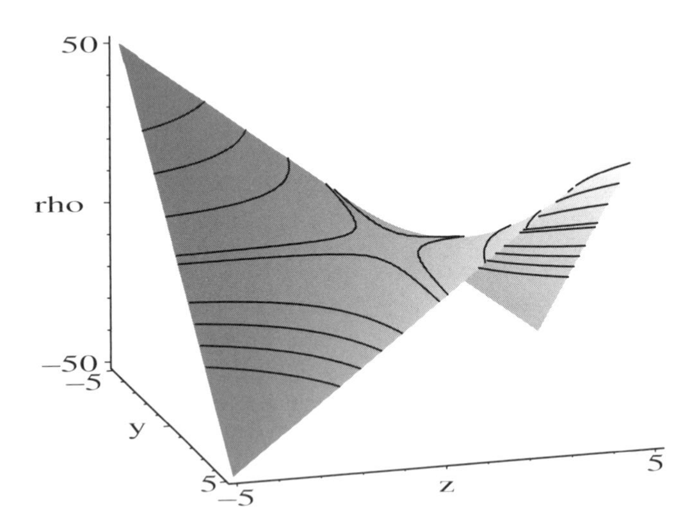

Figure 4.5: Plot of the normalized charge density.

The charge density distribution (which is reproduced in Figure 4.5) resembles a horse saddle. If we rotate the saddle and view it in the y–z, plane, the rectangular hyperbolas corresponding to holding the product $y\,z$ equal to a constant can be more clearly seen.

[4]Alternatively, one can use `Del . E`.

To determine the potential function, we have to carry out the line integral

$$\Phi = -\int \vec{E} \cdot d\vec{s}, \tag{4.9}$$

where $d\vec{s} = \hat{e}_x\, dx + \hat{e}_y\, dy + \hat{e}_z\, dz$ is the vector element of length. This line integral is readily carried out with the following `LineInt` command, the integration being along the straight line from the origin $(0, 0, 0)$ to an arbitrary point (x,y,z). Since the field is irrotational, Φ doesn't depend on the path chosen. The form of Φ can be simplified by applying `simplify` to the line integral result.

```
> Phi:=simplify(-LineInt(E,Line(<0,0,0>,<x,y,z>)));
```

$$\Phi := -x^2\, y\, z + 3\, x\, z - y\, z$$

So the electrostatic potential Φ is now known to within an arbitrary additive constant. If we start at some point (x, y, z) and integrate the electric field around a closed contour ending at the same point, the line integral will be zero and we will have come back to the same value of $\Phi(x, y, z)$. This zero "circulation" of the electric field is the signature of a curl-free situation. I don't know about you, Mike, but I have no feeling for what this potential or the associated electric field really look like. Do you think that we should plot these functions as well?"

"You're right. We are dealing with a three-dimensional vector field and I would feel more comfortable with a picture showing us the electric field distribution and a couple of representative equipotential surfaces. Let's consider two equipotential surfaces corresponding to $\Phi = \Phi_1 = 8$ V and $\Phi = \Phi_2 = -8$ V.

```
> Phi1:=Phi=8; Phi2:=Phi=-8;
```

$$\Phi1 := -x^2\, y\, z + 3\, x\, z - y\, z = 8 \qquad \Phi2 := -x^2\, y\, z + 3\, x\, z - y\, z = -8$$

The three-dimensional equipotential surface corresponding to a given value of Φ can be generated using the `implicitplot3d` command. Let's create a functional operator `ip` to carry out this task for an implicitly defined surface `v`. To create differently colored surfaces, the shading `s` must also be supplied.

```
> ip:=(v,s)->implicitplot3d(v,x=-5..5,y=-5..5,z=-3..3,
      grid=[25,25,25],style=patchcontour,axes=boxed,shading=s):
```

For the equipotentials Φ_1 and Φ_2, let's use the shadings `zgreyscale` and `zhue`, respectively, to color the surfaces. Although not too distinguishable in a textbook, the surfaces will look quite distinct on the computer screen.

```
> ip1:=ip(Phi1,zgreyscale): ip2:=ip(Phi2,zhue):
```

We can use the `fieldplot3d` command to graph the electric field vectors. The field at a point will be represented by a thick black arrow indicating the direction of the electric field at that point and whose length is proportional to the magnitude of the electric field.

```
> fp:=fieldplot3d(E,x=-5..5,y=-5..5,z=-3..3,axes=framed,
      arrows=THICK,orientation=[-121,43],color=black):
```

The two equipotential surfaces and the electric field arrows can be superimposed with the `display` command. Including the option `style=hidden` makes the arrows hollow rather than filled. (The eavesdropping reader is referred to Figure 4.6 for the resulting picture.)

```
> display({ip1,ip2,fp},orientation=[45,45],style=hidden);
```

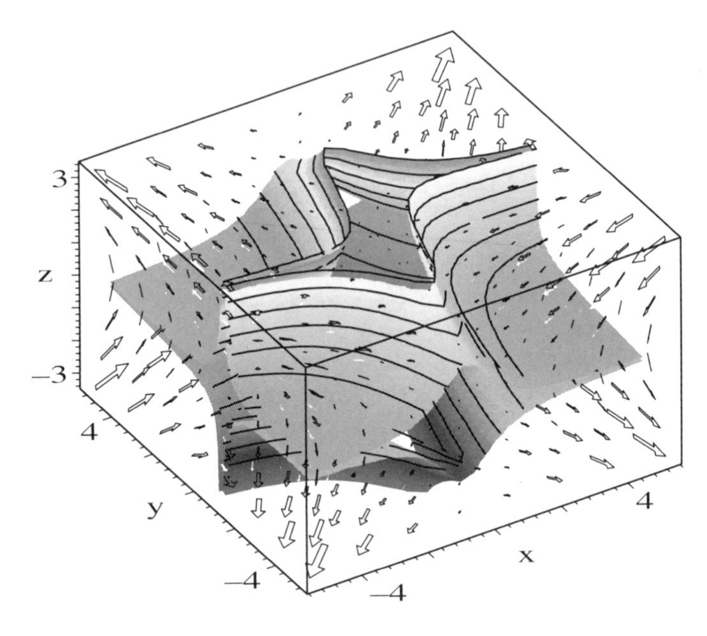

Figure 4.6: The arrows indicate $\vec{E}$, the surfaces the equipotentials.

Boy, am I glad that we created this picture. I had no idea that the electric field and equipotential surfaces were so complicated. For each value of the potential, there are two distinct surfaces, which are best viewed by rotating the viewing box on the computer screen. By looking from different angular perspectives, we can see that the electric field arrows are perpendicular to the equipotential surfaces as they should be. If we rotate the viewing box to see the y–z plane, I can see that the circulation of the electric field could be zero for some choices of contour path, but if we hadn't imposed the curl-free condition in the first place I am not confident that I would have been absolutely sure of the field's conservative nature."

"Mike, I am thirsty. Let's take a short break and go over to the student union cafeteria and get something to drink. Then you can show me some examples in non-Cartesian coordinate systems as well some applications of Stokes's theorem and Gauss's theorem."

PROBLEMS:

Problem 4-19: Conservative field

Consider the vector field $\vec{A}=(x+2\,y+a\,z)\,\hat{e}_x+(b\,x-3\,y-z)\,\hat{e}_y+(4\,x+c\,y+2\,z)\,\hat{e}_z$.

(a) Determine the values of a, b, and c that make $\vec{A}$ a conservative field.

(b) Determine the divergence of $\vec{A}$.

(c) Determine the potential function Φ such that $\vec{A} = \nabla\Phi$.

(d) Make a 3-dimensional plot of $\vec{A}$, showing the field arrows, axes, etc.

(e) Plot a few of the equipotential surfaces for values of your own choosing.

(f) Superimpose the two previous plots in the same figure and thus demonstrate that the field arrows are perpendicular to the equipotential surfaces.

Problem 4-20: A possible irrotational field
Consider the electric field with components $E_x = 6\,x\,y$, $E_y = 3\,x^2 - 3\,y^2$, $E_z = 0$.

(a) Calculate the curl of the electric field. Is the field irrotational? Explain.

(b) Calculate the associated charge density.

(c) Determine the electrostatic potential function.

(d) Make a three-dimensional plot of the electric field, clearly showing the field arrows.

(e) Plot a few of the equipotential surfaces for values of your own choosing.

(f) Superimpose the two previous plots in the same figure and thus demonstrate that the electric field is perpendicular to the equipotential surfaces.

Problem 4-21: Radial field
Consider the radial electric field $\vec{E} = \vec{r}/r^2$, where $\vec{r} = x\,\hat{e}_x + y\,\hat{e}_y + z\,\hat{e}_z$ is the position vector.

(a) Without doing any calculation, present an argument that shows that $\vec{E}$ is conservative.

(b) Confirm your argument by calculating $\text{curl}\,\vec{E}$ in Cartesian coordinates.

(c) Determine the potential function Φ such that $\vec{E} = -\nabla\Phi$ and $\Phi(a) = 0$, where $a > 0$. Express your answer in terms of r and a, simplifying Φ as much as possible.

Problem 4-22: Another conservative field

(a) Show that $\vec{A} = (6\,x\,y + z^3)\,\hat{e}_x + (3\,x^2 - z)\,\hat{e}_y + (3\,x\,z^2 - y)\,\hat{e}_z$ is conservative.

(b) Calculate the divergence of $\vec{A}$.

(c) Find the corresponding potential function Φ such that $\vec{A} = \nabla\Phi$.

(d) Create a suitably colored three-dimensional plot that contains representative field lines and equipotentials.

4.1.5 Of Coordinates and Circulation Too

The denunciation of the young is a necessary part of the hygiene of older people, and greatly assists the circulation of their blood.
Logan Pearsall Smith, American essayist, aphorist (1865–1946)

On returning to the computer lab with Vectoria, Mike begins a new worksheet by loading the following library packages. He first sets the interface warnlevel to zero to avoid displaying the multiple warnings that would otherwise occur.

```
> restart: interface(warnlevel=0):
> with(plots): with(VectorCalculus): with(plottools):
```

"In addition to the packages we used in the last worksheet, plottools is needed because it has commands for drawing contours involving circular arcs and arrows. What type of example would you like to see first?"

"Mike, vector operator computations are relatively easy to do in Cartesian coordinates, unless the functions are very complicated. For other coordinate systems, such as spherical polar, the calculations can be considerably harder. Suppose, for example, that we are given the potential function $U = 10\, r\, \sin^2\theta\, \cos\phi$ in terms of the spherical polar coordinates r, θ, and ϕ.

```
> U:=10*r*sin(theta)^2*cos(phi);
```

$$U := 10\, r \sin(\theta)^2 \cos(\phi)$$

I know that r, θ, and ϕ are related to the Cartesian coordinates by the relations

$$x = r \sin\theta \cos\phi, \quad y = r \sin\theta \sin\phi, \quad z = r \cos\theta, \tag{4.10}$$

where r is the radial distance from the origin, θ is the angle between the radius vector and the z-axis, and ϕ is the angle that the projection of the radius vector into the x–y plane makes with the x-axis.[5] I also know that the directions of the unit vectors in any other coordinate system than Cartesian depend on position. However, I can't remember the forms of the gradient, divergence, or curl operators in the spherical polar coordinate system. Sure, I could either derive the forms by hand or look them up in a math text, but this is a waste of my time. Can Maple do the calculations for me?"

"Sure, we can set the coordinates to be spherical as follows. Alternatively, we would have to include the option `'spherical'[r,theta,phi]` in each of the vector operator commands.

```
> SetCoordinates('spherical'[r,theta,phi]):
```

Now let's calculate the electric field $\vec{E} = -\nabla U$ and then take its curl.

```
> E:=Gradient(-U); CurlE:=Curl(E);
```

$$E := -10 \sin(\theta)^2 \cos(\phi)\, \bar{e}_r - 20 \sin(\theta) \cos(\phi) \cos(\theta)\, \bar{e}_\theta + 10 \sin(\theta) \sin(\phi)\, \bar{e}_\phi$$

$$CurlE := 0\, \bar{e}_r$$

The existence of a potential guarantees a conservative (curl-free) electric field.

[5] This is the physics convention. Mathematicians and the default Maple reverse θ and ϕ.

If you are interested in calculating the gradient, curl, etc., in other coordinate systems, go to Maple's Help, enter `coords` in the Topic Search, and hit OK to find out what coordinate systems Maple supports. Many of the coordinate systems listed there are rarely used in physical problems.

As in our first Cartesian coordinate example, we could calculate the normalized charge density from the electric field. But we haven't used the Laplacian operator, $\nabla^2(\ ) \equiv \text{div grad}(\ ) \equiv \nabla \cdot \nabla(\ )$, yet. The normalized charge density is related to the potential U by the relation $\rho/\epsilon_0 = -\nabla^2 U$ and is readily calculated as follows."

```
> norm_rho:=-Laplacian(U);
```

$$norm\_rho := -\frac{40\, r \sin(\theta) \cos(\phi) \cos(\theta)^2 - 10\, r \sin(\theta) \cos(\phi)}{r^2 \sin(\theta)}$$

```
> norm_rho:=expand(%);
```

$$norm\_rho := -\frac{40 \cos(\phi) \cos(\theta)^2}{r} + \frac{10 \cos(\phi)}{r}$$

"Mike, this example as well as the one you showed me in the previous worksheet are of an electrostatic nature and characterized by zero curls. How about some examples with nonzero curls and possibly from other areas of physics? In fluid mechanics, for example, we know that the line integral or circulation of the velocity field around the center of a circular whirlpool is not equal to zero. Further, we haven't looked at any situations involving Stokes's theorem."[6]

"OK, if it's a fluid mechanics example that you want, let's choose to work in cylindrical coordinates r, θ, z,

```
> SetCoordinates('cylindrical'[r,theta,z]):
```

and consider the following z-independent fluid velocity vector field $\vec{V}$.

```
> V:=VectorField(<r*cos(theta),sin(theta),0>);
```

$$V := r \cos(\theta)\, \overline{e}_r + \sin(\theta)\, \overline{e}_\theta$$

Would you like to see Stokes's theorem applied to the fluid velocity field? You remember the mathematical form of Stokes's theorem, don't you?"

"Yes, we just covered it in our vector calculus course. For a vector field $\vec{V}$, Stokes's theorem is given by

$$\oint_L \vec{V} \cdot d\vec{s} = \int_S (\nabla \times \vec{V}) \cdot d\vec{A}. \tag{4.11}$$

It states that the circulation of $\vec{V}$ around a closed path L is equal to the surface integral of the curl of $\vec{V}$ over the open surface S bounded by L."

"Good. First, I will make a plot of the velocity field in the x–y plane as well as the path L that will be used to confirm Stokes's theorem. To plot the velocity field, let's use the `MapToBasis` command to convert $\vec{V}$ into Cartesian coordinates x, y, z. This command makes use of the fact that the cylindrical and Cartesian coordinates are related by $x = r\cos\theta$, $y = r\sin\theta$, and $z = z$.

[6]Named after the Irish mathematical physicist George Stokes (1819–1903).

```
> Vb:=MapToBasis(V,'cartesian'[x,y,z]);
```

$$Vb := \left(\frac{x^2}{\sqrt{x^2+y^2}} - \frac{y^2}{x^2+y^2}\right)\bar{e}_x + \left(\frac{x\,y}{\sqrt{x^2+y^2}} + \frac{y\,x}{x^2+y^2}\right)\bar{e}_y$$

In the plot I am going to create a simple closed path. The first leg of the path is an arc of radius 2 centered at the origin and spanning the angular range $\pi/6$ radians (30°) to $\pi/3$ radians (60°). The arc command is used to plot this leg.

```
> a:=arc([0,0],2,Pi/6..Pi/3,thickness=3):
```

For the second leg, I will use the arrow command to indicate an integration along the line $\theta = \pi/6$ from $r = 2$ to $r = 5$.

```
> b:=arrow([2*cos(Pi/6),2*sin(Pi/6)],
    [5*cos(Pi/6),5*sin(Pi/6)],.05,.4,.1):
```

The third leg of the path is an arc of radius 5, centered at the origin and spanning the same angular range as the first arc.

```
> c:=arc([0,0],5,Pi/6..Pi/3,thickness=3):
```

The contour is closed with a second arrow command joining the outer arc to the inner one along the line $\theta = \pi/3$.

```
> d:=arrow([5*cos(Pi/3),5*sin(Pi/3)],
    [2*cos(Pi/3),2*sin(Pi/3)],.05,.4,.1):
```

The textplot command is used to label the various legs of the path L.

```
> tp:=textplot([[3.4,1.45,"theta=Pi/6"],[1.3,3.6,"theta=Pi/3"],
    [3.9,3.65,"r=5"],[1,1.4,"r=2"]],color=black):
```

The fieldplot command produces a picture of the vector field.

```
> fp:=fieldplot([Vb[1],Vb[2]],x=0.1..5,y=0.1..5,arrows=thick,
    grid=[10,10],color=blue):
```

All the graphs are now superimposed, the integration path being colored red.

```
> display({a,b,c,d,tp,fp},view=[0..5,0..5],color=red,
  scaling=constrained,labels=["x","y"],tickmarks=[3,3]);
```

The resulting picture is displayed in Figure 4.7. It would appear that the velocity field has a nonzero curl. This is easily verified.

```
> CurlV:=Curl(V);
```

$$CurlV := \frac{\sin(\theta) + r\sin(\theta)}{r}\,\bar{e}_z$$

Let's use this result to check Stokes's theorem for the contour of Figure 4.7. First we will evaluate the line integral around the contour, taking the inner arc as our first leg. Along this arc, the element of length is $ds = r\,d\theta$ with $r = 2$. The line integral is then $\int_{\pi/3}^{\pi/6} V_\theta\, r\, d\theta = \int_{\pi/3}^{\pi/6} 2\,\sin\theta\, d\theta$, which is easily evaluated.

```
> L1:=Int(subs(r=2,V[2]*r),theta=Pi/3..Pi/6)
  =int(subs(r=2,V[2]*r),theta=Pi/3..Pi/6);
```

$$L1 := \int_{\frac{\pi}{3}}^{\frac{\pi}{6}} 2\sin(\theta)\,d\theta = -\sqrt{3} + 1$$

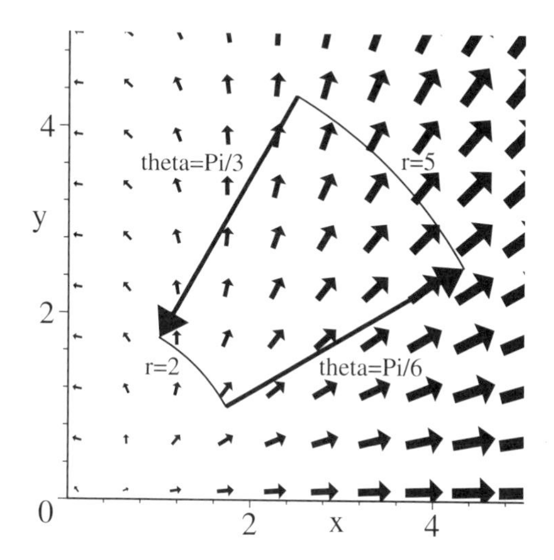

Figure 4.7: Velocity field vectors and integration path for line integral.

The second leg involves the integral $\int_2^5 V_r\, dr$ along the line $\theta = \pi/6$.

```
> L2:=Int(subs(theta=Pi/6,V[1]),r=2..5)
     =simplify(int(subs(theta=Pi/6,V[1]),r=2..5));
```

$$L2 := \int_2^5 r\cos\left(\frac{\pi}{6}\right)\, dr = \frac{21\sqrt{3}}{4}$$

The third integral is like the first, except now $r = 5$ and the angular integration is in the opposite sense.

```
> L3:=Int(subs(r=5,V[2]*r),theta=Pi/6..Pi/3)
     =int(subs(r=5,V[2]*r),theta=Pi/6..Pi/3);
```

$$L3 := \int_{\frac{\pi}{6}}^{\frac{\pi}{3}} 5\sin(\theta)\, d\theta = \frac{5\sqrt{3}}{2} - \frac{5}{2}$$

The fourth integral is like the second, except that it is along $\theta = \pi/3$ and the integration direction is reversed.

```
> L4:=Int(subs(theta=Pi/3,V[1]),r=5..2)
     =simplify(int(subs(theta=Pi/3,V[1]),r=5..2));
```

$$L4 := \int_5^2 r\cos\left(\frac{\pi}{3}\right)\, dr = \frac{-21}{4}$$

The total line integral is the sum of the four contributions.

```
> Line_int:=rhs(L1+L2+L3+L4);
```

$$Line\_int := \frac{27\sqrt{3}}{4} - \frac{27}{4}$$

According to Stokes's theorem, the same number should result if we carry out the surface integral

$$\int_{\pi/6}^{\pi/3} \int_2^5 (\nabla \times \vec{V})_z \, r \, dr \, d\theta.$$

Forming the surface integral,

```
> Surface_int:=Int(Int(r*CurlV[3],r=2..5),theta=Pi/6..Pi/3);
```

$$Surface\_int := \int_{\frac{\pi}{6}}^{\frac{\pi}{3}} \int_2^5 \sin(\theta) + r\sin(\theta) \, dr \, d\theta$$

and evaluating it, we obtain the same result as for the line integral.

```
> Surface_int:=value(%);
```

$$Surface\_int := \frac{27\sqrt{3}}{4} - \frac{27}{4}$$

Thus, we have confirmed Stokes's theorem for our fluid velocity field."

PROBLEMS:

Problem 4-23: Gradient of a potential

Calculate the electric field $\vec{E} = -\nabla V$ if the electric potential function is given in spherical polar coordinates by $V = \ln r \cos\theta \, \sin\phi + r^2 \, \phi$.

Problem 4-24: Divergence and curl of an electric field

Consider an electric field given in spherical polar coordinates by

$$\vec{E} = \frac{1}{r^2} \cos\theta \, \hat{e}_r + r \sin\theta \cos\phi \, \hat{e}_\theta + \cos\phi \, \hat{e}_\phi.$$

Calculate the divergence of the electric field and the charge density associated with the field. Calculate the curl of the electric field.

Problem 4-25: Divergence and curl

Determine the divergence and curl of the following vector fields expressed in Cartesian, cylindrical, and spherical polar coordinates, respectively. Also evaluate them at the specified points:

(a) $\vec{A} = y\,z\,\hat{e}_x + 4\,x\,y\,\hat{e}_y + y\,\hat{e}_z$, at $(1, -2, 3)$;

(b) $\vec{B} = \rho\,z\,\sin\phi\,\hat{e}_\rho + 3\,\rho z^2\,\cos\phi\,\hat{e}_\phi + 0\,\hat{e}_z$, at $(5, \pi/2, 1)$;

(c) $\vec{C} = 2\,r\,\cos\theta\,\cos\phi\,\hat{e}_r + \sqrt{r}\,\hat{e}_\phi$, at $(1, \pi/6, \pi/3)$.

Problem 4-26: Vector identity

Consider a general vector field $\vec{A} = A_1(x, y, z)\,\hat{e}_x + A_2(x, y, z)\,\hat{e}_y + A_3(x, y, z)\,\hat{e}_z$. Prove the vector identity $\nabla \cdot (\nabla \times \vec{A}) = 0$. Prove that the identity also holds in spherical and cylindrical coordinates.

4.1.6 All Is Flux

All is flux, nothing stays still.
Heraclitus, Greek philosopher (c. 540–c. 480 B.C.)

Continuing with their computer algebra session, Vectoria asks, "Mike, do you also have a good example of applying Gauss's theorem. For a vector field $\vec{W}$, this theorem states that for a closed volume V having a bounding surface S,

$$\oint_S \vec{W} \cdot d\vec{A} = \int_V \nabla \cdot \vec{W}\, dV. \tag{4.12}$$

That is to say, the total outward *flux* of a vector field $\vec{W}$ (the lhs of (4.12)) is equal to the volume integral of the divergence of $\vec{W}$ (the right-hand side)."

"Actually, I ran into an interesting math example in Stewart's *Calculus* text [Ste87], although it's in Cartesian coordinates. Taking the same library packages as in the last recipe and setting the coordinates to be Cartesian,

```
> restart: interface(warnlevel=0):
> with(plots): with(VectorCalculus): with(plottools):
> SetCoordinates('cartesian'[x,y,z]):
```

consider the following vector field $\vec{W}$.

```
> W:=VectorField(<x*y,y^2+exp(x*z^2),sin(x*y)>);
```

$$W := x\,y\,\bar{e}_x + (y^2 + e^{(x\,z^2)})\,\bar{e}_y + \sin(x\,y)\,\bar{e}_z$$

We can easily make an informative plot of W using the `fieldplot3d` command.

```
> fieldplot3d([W[1],W[2],W[3]],x=-1..1,y=0..2,z=0..1,
  axes=framed,arrows=THICK,orientation=[63,47],shading=zhue);
```

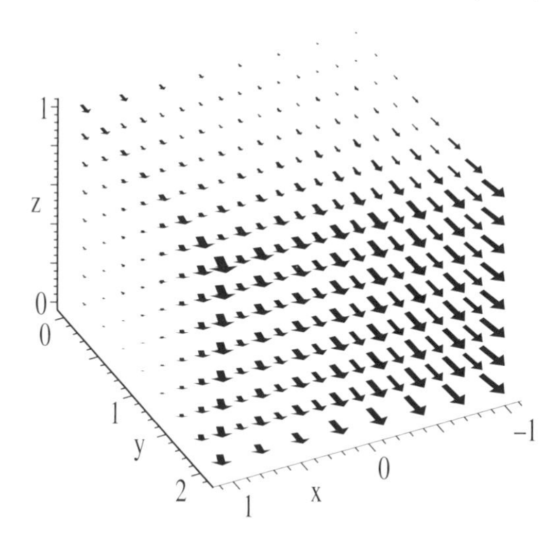

Figure 4.8: Arrows indicate the direction of the vector field.

From Figure 4.8, we can easily see that the vector lengths are changing in the y-direction, so we can anticipate that $\nabla \cdot \vec{W}$ will be nonzero. Let's check to see whether this is so.

```
> DivW:=Del . W;
```

$$DivW := 3\,y$$

The divergence is indeed nonzero, increasing linearly with y. Calculating $\nabla \times \vec{W}$, we find that the curl is also nonzero.

```
> CurlW:=Del &x W;
```

$$CurlW := (\cos(x\,y)\,x - 2\,x\,z\,e^{(x\,z^2)})\,\overline{e}_x - \cos(x\,y)\,y\,\overline{e}_y + (z^2\,e^{(x\,z^2)} - x)\,\overline{e}_z$$

The example given by Stewart is to consider a volume bounded by the parabolic cylinder $z = 1 - x^2$ and the planes $z = 0$, $y = 0$, and $y + z = 2$ and calculate the flux of the vector field out through the four bounding surfaces. Again, a plot is useful to visualize the volume that we are considering. The following command line creates a parabolic cylindrical surface $z = 1 - x^2$ in the ranges $x = -1$ to 1 and $y = 0$ to 2.

```
> pl||1:=plot3d(1-x^2,x=-1..1,y=0..2):
```

The plane $z = 2 - y$ is plotted for $x = -1$ to 1 and $y = 1$ to 2, and colored red.

```
> pl||2:=plot3d(2-y,x=-1..1,y=1..2,style=patchnogrid,
          color=red):
```

The PLOT3D(POLYGONS) command is used to create a planar segment in the $z = 0$ plane with vertices $(-1, 0, 0)$, $(1, 0, 0)$, $(1, 2, 0)$, $(-1, 2, 0)$.

```
> pl||3:=PLOT3D(POLYGONS([[-1,0,0],[1,0,0],[1,2,0],[-1,2,0]])):
```

Similarly, a plot is created for a segment in the $y = 0$ plane.

```
> pl||4:=PLOT3D(POLYGONS([[-1,0,0],[1,0,0],[1,0,1],[-1,0,1]])):
```

The textplot3d command is used to add text to the three-dimensional plot, labeling the four surfaces involved in the integrations.

```
> pl||5:=textplot3d([[0,1.8,1.1,"z=1-x^2"],[1,0,1.1,"y=0"],
          [0,0.45,1.25,"y=2-z"],[0.45,1.25,-0.15,"z=0"]],
          align=LEFT,color=black):
```

The five plots are superimposed with the display command,

```
> display([seq(pl||i,i=1..5)],axes=NONE,scaling=constrained,
  labels=["x","y","z"],orientation=[42,100]);
```

the resulting picture being shown in Figure 4.9. The volume of interest is the interior of the region bounded by the indicated surfaces. To directly calculate the net flux of the vector field out through the four surfaces would involve some difficult surface integrals.

A simpler procedure, suggested by Stewart, is to make use of the divergence, or Gauss's, theorem. The divergence of $\vec{W}$ turned out to be simpler than the vector field itself. Thus, the flux can be calculated by invoking the divergence theorem and performing the volume integral over the divergence of $\vec{W}$.

```
> FLUX:=Int(Int(Int(DivW,y=0..2-z),z=0..1-x^2),x=-1..1);
```

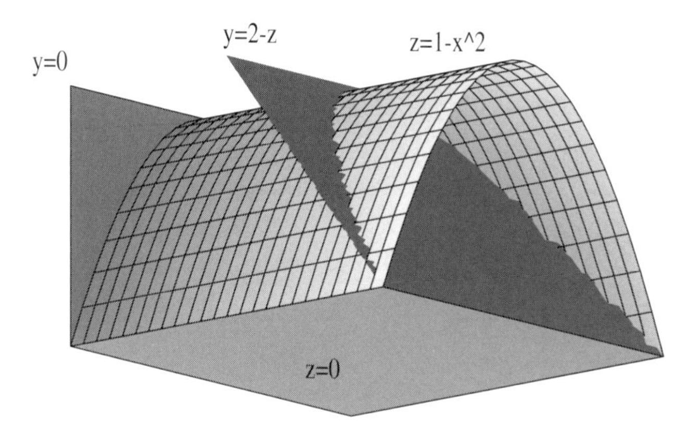

Figure 4.9: Surfaces through which the flux of the vector field passes.

$$FLUX := \int_{-1}^{1} \int_{0}^{1-x^2} \int_{0}^{2-z} 3\, y\, dy\, dz\, dx$$

The value of the flux is obtained by applying the `value` command to `Flux`.

```
> FLUX:=value(%);
```

$$FLUX := \frac{184}{35}$$

Well, Vectoria, we could look at lots of other aspects of vector calculus using Maple, but this should be enough for now."

PROBLEMS:

Problem 4-27: Gauss's law

If $\vec{E}$ is the electric field, Gauss's law states that the net charge enclosed by a closed surface S is $Q = \epsilon_0 \int_S \vec{E} \cdot d\vec{A}$. For $\vec{E} = x\, \hat{e}_x + y\, \hat{e}_y + 2\, z\, \hat{e}_z$ use Gauss's law and the divergence theorem to find the charge contained in the solid hemisphere $x^2 + y^2 + z^2 \leq a^2$, $z \geq 0$.

Problem 4-28: Electric flux

Consider the electric field $\vec{E} = y\, e^{z^2}\, \hat{e}_x + y^2\, \hat{e}_y + e^{x\, y}\, \hat{e}_z$. Use the divergence theorem to calculate the electric flux out of the volume bounded by the cylindrical surface $x^2 + y^2 = 9$ and the planes $z = 0$ and $z = y - 3$.

Problem 4-29: Tetrahedron

A solid tetrahedron is bounded by the four planes $x = 0$, $y = 0$, $z = 0$, and $x + y + z = 1$. Make a colored plot showing the four polygonal surfaces of the tetrahedron with labels attached.

4.2 Matrix Models

A general matrix A of order $m \times n$ is given by

$$A = \begin{bmatrix} a_{11} & a_{12} & a_{13} & \cdots & a_{1n} \\ a_{21} & a_{22} & a_{23} & \cdots & a_{2n} \\ \cdots & \cdots & \cdots & \cdots & \cdots \\ \cdots & \cdots & \cdots & a_{jk} & \cdots \\ \cdots & \cdots & \cdots & \cdots & \cdots \\ a_{m1} & a_{m2} & a_{m3} & \cdots & a_{mn} \end{bmatrix}$$

If $m = n$, the matrix is *square*. A *row matrix* or *row vector* has only one row, while a *column matrix* (*column vector*) has only one column. The *transpose* of $A \equiv [\, a_{jk} \,]$ is $A^T \equiv [\, a_{kj} \,]$, while its complex conjugate is $A^* = [\, a^*_{jk} \,]$. The *Hermitian conjugate* (or *adjoint*) matrix $A^\dagger$ is defined to be $(A^T)^*$. A square matrix is *Hermitian* if $A^\dagger = A$. Some basic matrix properties are as follows:

- If matrices A and B are of the same order, then $A \pm B = [\, a_{jk} \pm b_{jk} \,]$.
- If λ is a scalar, then $\lambda\, A = A\, \lambda = [\, \lambda\, a_{jk} \,]$. The product $A\,B$ (or $A \cdot B$) of an $m \times n$ matrix A and an $n \times p$ matrix B is an $m \times p$ matrix C, with matrix elements $c_{jk} = \sum_{i=1}^{n} a_{ji}\, b_{ik}$.
- If A is a *nonsingular* matrix (i.e., has a nonzero determinant), then the *inverse* matrix A^{-1} is given by $[\, A_{jk} \,]^T/\text{determinant}(A)$, where $[\, A_{jk} \,]$ is the matrix of cofactors A_{jk}. (The *cofactor* A_{jk} is equal to $(-1)^{j+k}$ times the resulting determinant of A obtained by removing all the elements of the jth row and kth column.) It follows that $A\, A^{-1} = A^{-1}\, A = I$, where I is the *unit* or *identity matrix* with each element along its principal diagonal equal to 1 and all off-diagonal elements 0.

4.2.1 Secret Message Revisited

Always do right – this will gratify some and astonish the rest.
Mark Twain, Message to the Young People's Society, New York City, 1901

While waiting for Vectoria to show up the following day for a Maple session on matrix manipulations, Mike recalls the data matrix that he received when he first arrived for his summer job at the archaeological site near Machu Pichu. His new colleagues played a joke on him, giving Mike a square array of data allegedly recording the location and number of artifacts in each squared-off area of the site. He was further informed that the rows and columns of the data were inadvertently interchanged and should be transposed. On plotting the transposed data matrix, a "secret message" was revealed.

Until Vectoria arrives, he decides to "play around" with the secret message data. He loads the `LinearAlgebra` package and sets `rtablesize` to infinity in the `interface` command so as to explicitly display the large (larger than 10×10) matrix A that will be entered.

```
> restart: with(plots): with(LinearAlgebra):
> interface(rtablesize=infinity):
```

The original square data matrix A is entered,

```
> A:=Matrix([[1,0,1,0,1,1,2,1,0,2,1],[1,2,2,2,1,1,2,0,3,1,0],
            [1,1,2,0,1,1,3,1,0,1,0],[0,1,9,0,10,1,1,7,7,8,1],
            [1,2,8,2,10,3,2,1,9,2,0],[1,1,7,10,9,2,1,0,9,1,0],
            [1,1,7,1,9,2,1,2,8,0,1],[0,2,9,2,10,3,1,9,9,8,2],
            [2,1,2,1,0,0,3,2,0,1,0],[1,2,3,0,1,1,2,1,1,0,0],
            [0,1,2,1,0,3,1,0,2,1,0]]):
```

and then transposed to produce the matrix B.

```
> B:=Transpose(A);
```

$$B := \begin{bmatrix} 1 & 1 & 1 & 0 & 1 & 1 & 1 & 0 & 2 & 1 & 0 \\ 0 & 2 & 1 & 1 & 2 & 1 & 1 & 2 & 1 & 2 & 1 \\ 1 & 2 & 2 & 9 & 8 & 7 & 7 & 9 & 2 & 3 & 2 \\ 0 & 2 & 0 & 0 & 2 & 10 & 1 & 2 & 1 & 0 & 1 \\ 1 & 1 & 1 & 10 & 10 & 9 & 9 & 10 & 0 & 1 & 0 \\ 1 & 1 & 1 & 1 & 3 & 2 & 2 & 3 & 0 & 1 & 3 \\ 2 & 2 & 3 & 1 & 2 & 1 & 1 & 1 & 3 & 2 & 1 \\ 1 & 0 & 1 & 7 & 1 & 0 & 2 & 9 & 2 & 1 & 0 \\ 0 & 3 & 0 & 7 & 9 & 9 & 8 & 9 & 0 & 1 & 2 \\ 2 & 1 & 1 & 8 & 2 & 1 & 0 & 8 & 1 & 0 & 1 \\ 1 & 0 & 0 & 1 & 0 & 0 & 1 & 2 & 0 & 0 & 0 \end{bmatrix}$$

It is confirmed that B has the dimension 11, 11, i.e., 11 rows and 11 columns.

```
> dimension:=Dimension(B);
```

$$dimension := 11, 11$$

In some applications, it is important to be able to extract a specific matrix element, row, or column. Mike now extracts from B the element corresponding to the fourth row, sixth column, then extracts the second row, and finally the fifth column, which he transposes into a row to save on space.

```
> element:=B[4,6]; row:=Row(B,2); col:=Transpose(Column(B,5));
```

$$element := 10 \qquad row := [0, 2, 1, 1, 2, 1, 1, 2, 1, 2, 1]$$

$$col := [1, 2, 8, 2, 10, 3, 2, 1, 9, 2, 0]$$

The trace of a matrix is equal to the sum of the elements along the central diagonal. Mike next determines the trace of B. He also calculates the determinant of B, a calculation which is trivial with Maple but very tedious to do by hand.

```
> tr:=Trace(B); det:=Determinant(B);
```

$$tr := 27 \qquad det := -100759$$

Another formidable task to do by hand is to calculate the inverse matrix B^{-1}. However, this is easily accomplished with the `MatrixInverse` command. (Most of the output is artificially suppressed here.)

```
> Binverse:=MatrixInverse(B);
```

$$Binverse := \left[\frac{3451}{5927}, \frac{16113}{100759}, \frac{-155}{100759}, \frac{903}{100759}, \ldots\ldots, \frac{-26548}{100759}, \frac{22278}{100759}, \frac{61571}{100759}\right]$$
$$\left[\frac{-3181}{5927}, \frac{6580}{100759}, \frac{981}{100759}, \frac{-4415}{100759}, \ldots\ldots, \frac{43212}{100759}, \frac{6565}{100759}, \frac{45204}{100759}\right]$$
$$\cdots\cdots\cdots\cdots\cdots\cdots\cdots\cdots$$
$$\cdots\cdots\cdots\cdots\cdots\cdots\cdots\cdots$$
$$\left[\frac{-729}{5927}, \frac{-41712}{100759}, \frac{30961}{100759}, \frac{-3557}{100759}, \ldots\ldots, \frac{16648}{100759}, \frac{953}{100759}, \frac{-6760}{100759}\right]$$

As a check, the matrix B is multiplied by the inverse matrix, the output (not displayed here) being the expected unit matrix.

```
> check:=Multiply(B,Binverse);
```

In the matrix B, the outer rows and columns can be deleted without altering the secret message. Mike first deletes the first, second, and eleventh rows and then the first, second, third, ninth, tenth, and eleventh columns.

```
> C:=DeleteColumn(DeleteRow(B,[1,2,11]),[1,2,3,9,10,11]);
```

$$C := \begin{bmatrix} 9 & 8 & 7 & 7 & 9 \\ 0 & 2 & 10 & 1 & 2 \\ 10 & 10 & 9 & 9 & 10 \\ 1 & 3 & 2 & 2 & 3 \\ 1 & 2 & 1 & 1 & 1 \\ 7 & 1 & 0 & 2 & 9 \\ 7 & 9 & 9 & 8 & 9 \\ 8 & 2 & 1 & 0 & 8 \end{bmatrix}$$

Seeing Vectoria enter the computer lab, Mike uses the `matrixplot` command to graphically display the new matrix C as 3-dimensional colored boxes, each box colored according to its height, i.e., to the magnitude of the corresponding matrix element.

```
> matrixplot(C,heights=histogram,style=patch,shading=zhue,
  orientation=[-72,10],tickmarks=[0,0,0]);
```

The simple message that Vectoria sees on the computer screen hides the deeper feeling that Mike has begun to experience whenever Vectoria is around.

PROBLEMS:

Problem 4-30: Matrix Operations

Given the 4×4 matrix

$$A = \begin{bmatrix} 2 & 1 & -1 & 4 \\ -2 & 3 & 2 & -5 \\ 1 & -2 & -3 & 2 \\ -4 & -3 & 2 & -2 \end{bmatrix},$$

(a) Confirm the dimensions of A.

(b) Calculate the determinant of A, first mimicking a hand calculation and then with the `Determinant` command.

(c) Find A^{-1}.

(d) Calculate $A^5 \equiv A\,A\,A\,A\,A$.

(e) Repeat the above steps for a new matrix obtained by deleting one row and one column. You are free to choose which row and which column, as long as the resulting matrix is not singular.

Problem 4-31: More Operations

Consider the two square matrices, $A = \begin{bmatrix} 1 & 2 & -1 \\ 3 & 0 & 2 \\ 4 & 5 & 0 \end{bmatrix}$, $B = \begin{bmatrix} 1 & 0 & 0 \\ 2 & 1 & 0 \\ 0 & 1 & 3 \end{bmatrix}$.

Find A^{-1} and B^{-1}. Verify that $(A\,B)^T = B^T\,A^T$ and $(A\,B)^{-1} = B^{-1}\,A^{-1}$.

Problem 4-32: Matrix Inverse

(a) Find the inverse of the matrix $A = \begin{bmatrix} 3 & -2 & 2 \\ 1 & 2 & -3 \\ 4 & 1 & 2 \end{bmatrix}$.

(b) Check the answer by showing that $A\,A^{-1} = I$.

(c) Repeat the above steps for a new matrix obtained by adding one new row and one new column. The choice of new matrix elements is up to you, as long as the new matrix is not singular.

4.2.2 A Fishy Tale

I'd rather have a bottle in front of me, than a frontal lobotomy.
anonymous, observed on a sign in a seaside pub

After joining him in the computer lab, Vectoria asks Mike to show her a simple illustration of solving simultaneous linear equations, using a matrix approach. So Mike considers the following fishy tale.

A single fish of species 1 consumes 10 grams of food 1, 5 grams of food 2, and 3 grams of food 3 per day. A fish of species 2 consumes 6 grams of food 1, 4 grams of food 2, and 2 grams of food 3 per day. Finally, a fish of species 3 consumes 7 grams of food 1, 3 grams of food 2, and 1 gram of food 3 per day. If 2.304, 1.161, and 0.561 kilograms of foods 1, 2, and 3, respectively, are available daily, what population sizes of the three fish species will consume exactly all of the available food?

To answer this question using a matrix approach, Mike begins by loading the LinearAlgebra library package.

```
> restart: with(LinearAlgebra):
```

Taking $x1$, $x2$, and $x3$ to be the unknown population sizes of species 1, 2, and 3, respectively, he forms a column vector X with the population numbers as the matrix elements. He uses a "shorthand" syntax to enter the vector. Since the consumption of food is expressed in grams, the available food is also expressed in grams, i.e., 2304, 1161, and 561 grams of foods 1, 2, and 3 are available daily. Mike expresses the available food as a column matrix A.

```
> X:=<<x1,x2,x3>>; A:=<<2304,1161,561>>;
```

$$X := \begin{bmatrix} x1 \\ x2 \\ x3 \end{bmatrix} \qquad A := \begin{bmatrix} 2304 \\ 1161 \\ 561 \end{bmatrix}$$

He next forms a 3×3 "consumption matrix" C, expressing the daily consumptions of the three types of food by a single fish of species 1, 2, and 3. The first column in C is the daily consumption of foods 1, 2, and 3 by a single fish of species 1, and so on. A shorthand syntax is again used for entering the matrix.

```
> C:=<<10,5,3>|<6,4,2>|<7,3,1>>; #consumption matrix
```

$$C := \begin{bmatrix} 10 & 6 & 7 \\ 5 & 4 & 3 \\ 3 & 2 & 1 \end{bmatrix}$$

The values of $x1$, $x2$, and $x3$ can be obtained by solving the matrix equation $C\,X = A$ for X. The matrix equation is now entered, Mike choosing to use the shorthand dot product notation to perform the matrix multiplication.

```
> eq:=C . X=A;
```

$$eq := \begin{bmatrix} 10\,x1 + 6\,x2 + 7\,x3 \\ 5\,x1 + 4\,x2 + 3\,x3 \\ 3\,x1 + 2\,x2 + x3 \end{bmatrix} = \begin{bmatrix} 2304 \\ 1161 \\ 561 \end{bmatrix}$$

Mentally equating the lhs and rhs of eq, row by row, one clearly has three simultaneous linear equations for the unknown population numbers $x1$, $x2$, and $x3$. Although one could extract the three equations and solve them using the `solve` command, Mike takes an easier approach. He directly solves the matrix equation $C\,X = A$ for X using the `LinearSolve` command. The matrix C is given as the first argument, the matrix A as the second argument.

```
> sol1:=X=LinearSolve(C,A);
```

$$sol1 := \begin{bmatrix} x1 \\ x2 \\ x3 \end{bmatrix} = \begin{bmatrix} 93 \\ 75 \\ 132 \end{bmatrix}$$

So there are 93 of species 1, 75 of species 2, and 132 of species 3. Mike can derive exactly the same answer by using the inverse matrix C^{-1}. To see this, mentally multiply the matrix equation from the left by C^{-1}, so that $C^{-1}\,C\,X = C^{-1}\,A$. But $C^{-1}\,C = I$, the identity matrix, and $I\,X = X$, so $X = C^{-1}\,A$. Mike calculates X using this last result.

```
> sol2:=X=MatrixInverse(C) . A;
```

$$sol2 := \begin{bmatrix} x1 \\ x2 \\ x3 \end{bmatrix} = \begin{bmatrix} 93 \\ 75 \\ 132 \end{bmatrix}$$

If desired, the population numbers can be removed from the matrix format. For example, Mike finishes the solution of the problem by extracting the population number $x2$ of species 2 from $sol2$.

```
> lhs(sol2)[2,1]=rhs(sol2)[2,1];
```

$$x2 = 75$$

PROBLEMS:

Problem 4-33: System of Linear Equations

Solve the following system of linear equations using the matrix approach:

$$\begin{aligned} 2\,x_1 + 3\,x_2 - 4\,x_3 &= -3 \\ 3\,x_1 - 2\,x_2 + 5\,x_3 &= 24 \\ x_1 + 4\,x_2 - 3\,x_3 &= -6 \end{aligned}$$

Problem 4-34: Electrical Network

The currents i_1, i_2, i_3, and i_4 in an electrical network satisfy the following system of equations. Determine all four currents using the matrix approach.

$$\begin{aligned} 3\,i_1 + 2\,i_2 - i_4 &= 65 \\ 2\,i_1 - i_2 + 4\,i_3 + 3\,i_4 &= 160 \\ -7\,i_1 - 4\,i_2 - 2\,i_4 &= 23 \\ 5\,i_1 - i_2 - 2\,i_3 + i_4 &= 3 \end{aligned}$$

Problem 4-35: Birds Munch Aphids [AL79]

Three species of birds eat aphids from different parts of trees. Species 1 feed half of the time on the top levels and half of the time on the middle levels of the trees. Species 2 feed half on the middle levels and half on the lower levels. Species 3 feed entirely on the lower levels. There are equal numbers of aphids available on the middle and lower levels, but only half this number available on the upper levels. What should be the relative sizes of the populations of the three species in order that the supply of aphids will be entirely consumed?

Problem 4-36: Kirchhoff Returns

Solve the dc network example in Section 3.1.2 using a matrix approach.

4.2.3 Population Waves

It's just a job. Grass grows, birds fly, waves pound the sand.
I beat people up.
Muhammad Ali, American boxer (1942–)

Bernadelli [Ber41] has introduced a simple model of "population waves" in a beetle species, the natural life span of the beetles being three years. The female beetle has a survival rate of 1/2 in the first year of life, a survival rate of 1/3 from the second to third years, and gives birth in the third year to an average of six new females before dying at the end of the third year. The contribution of an individual female beetle, in a probabilistic sense, to the female population number can be summarized in the following matrix A,

$$A = \begin{bmatrix} 0 & 0 & 6 \\ 1/2 & 0 & 0 \\ 0 & 1/3 & 0 \end{bmatrix}.$$

The matrix element a_{ij} in A denotes the contribution that a single female of age j will make to the next year's female population of age i.

(a) If there are initially 6000 female beetles in each of the three age groups (ages 1, 2, and 3), show that the model leads to a cyclic variation in population number in each age group.

(b) Determine the eigenvalues and associated eigenvectors of A. Are any of these real?

(c) If a sample of this species was needed for laboratory test purposes that could have a constant proportion in each age group from year to year, what criteria could be imposed on the initial female population to ensure that this would be satisfied?

To answer these questions, Mike loads the LinearAlgebra library package and enters the matrix A and the initial number N in the three age groups.

```
> restart: with(LinearAlgebra):
> A:=<<0|0|6>,<1/2|0|0>,<0|1/3|0>>; N:=6000*<<1,1,1>>;
```

$$A := \begin{bmatrix} 0 & 0 & 6 \\ 1/2 & 0 & 0 \\ 0 & 1/3 & 0 \end{bmatrix} \qquad N := \begin{bmatrix} 6000 \\ 6000 \\ 6000 \end{bmatrix}$$

The number of females in the three age groups n years hence is obtained by calculating $A^n\,N$. Mike forms a sequence S of these population numbers for n=0 to 12, and assigns the result.

```
> S:=seq(P||n=Multiply(A^n,N),n=0..12); assign(S):
```

$$S := P0 = \begin{bmatrix} 6000 \\ 6000 \\ 6000 \end{bmatrix},\ P1 = \begin{bmatrix} 36000 \\ 3000 \\ 2000 \end{bmatrix},\ P2 = \begin{bmatrix} 12000 \\ 18000 \\ 1000 \end{bmatrix},\ P3 = \begin{bmatrix} 6000 \\ 6000 \\ 6000 \end{bmatrix}, \cdots$$

Then, for example, *P1* indicates that one year later there are 36000 female beetles of age 1, 3000 of age 2, and 2000 of age 3. The cyclic variation in population number as n increases can be clearly seen.

To plot the data, Mike introduces a functional operator `F` to form a list of lists of plotting points for each of the three age groups. The row number r in each column vector Pn must be specified.

```
> F:=r->[seq([n,P||n[r,1]],n=0..12)]:
```

Using `F`, the population numbers for the three age groups are now plotted, different colors and line styles being chosen for each curve.

```
> plot([F(1),F(2),F(3)],color=[red,blue,green],
  linestyle=[1,2,3],thickness=2,tickmarks=[4,4]);
```

The resulting picture is shown in Figure 4.10, the population number of each age group being shown one, two, etc., years later.

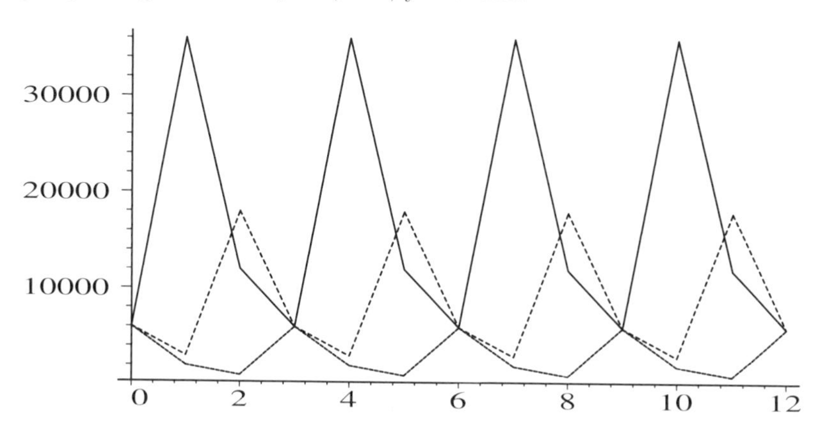

Figure 4.10: Cyclic variation of beetle population in the three age groups.

The solid curve is the number of age 1, the intermediate (dotted) curve the number of age 2, and the lower (dashed) curve the number of age 3.

Next, Mike determines the eigenvalues and eigenvectors of A. If X is an n-element column vector, the matrix equation $A\,X = \lambda\,X$ has nontrivial solutions for X if and only if the *characteristic matrix* $A - \lambda\,I$ has a zero determinant. Expanding $|A - \lambda\,I| = 0$ yields an nth-order *characteristic polynomial* equation for the *eigenvalues* λ. Corresponding to each of the n eigenvalues will be a nontrivial *eigenvector* X.

The characteristic matrix CM is easily generated for A,

```
> CM:=CharacteristicMatrix(A,lambda);
```

$$CM := \begin{bmatrix} \lambda & 0 & -6 \\ -1/2 & \lambda & 0 \\ 0 & -1/3 & \lambda \end{bmatrix}$$

as is the characteristic polynomial CP.

```
> CP:=CharacteristicPolynomial(A,lambda)=0;
```

$$CP := \lambda^3 - 1 = 0$$

Then CP is solved for λ, yielding only one real eigenvalue, viz., λ=1.

```
> eivs:=[solve(CP,lambda)];
```

$$eivs := \left[1, -\frac{1}{2} + \frac{1}{2} I\sqrt{3}, -\frac{1}{2} - \frac{1}{2} I\sqrt{3}\right]$$

The eigenvalues can alternatively be obtained by applying the `Eigenvalues` command directly to A. The option `output='list'` puts the eigenvalues in a list format. The default is a column style.

```
> eivs2:=Eigenvalues(A,output='list');
```

$$eivs2 := \left[1, -\frac{1}{2} + \frac{1}{2} I\sqrt{3}, -\frac{1}{2} - \frac{1}{2} I\sqrt{3}\right]$$

To obtain the eigenvectors mimicking a hand calculation, the three-element column eigenvector X and the zero column vector are entered.

```
> X:=<<x1,x2,x3>>: zero:=<<0,0,0>>:
```

The eigenvector corresponding to the real eigenvalue (first entry in *eivs*) is determined to within an arbitrary constant $\_t_{1,1}$.

```
> sol:=X=LinearSolve(eval(CM,lambda=eivs[1]),zero);
```

$$sol := \begin{bmatrix} x1 \\ x2 \\ x3 \end{bmatrix} = \begin{bmatrix} 6\,\_t_{1,1} \\ 3\,\_t_{1,1} \\ \_t_{1,1} \end{bmatrix}$$

The arbitrary constant is set equal to 1 on the rhs of *sol*, yielding the real eigenvector V corresponding to λ=1.

```
> V:=subs(_t[1,1]=1,rhs(sol));
```

$$V := \begin{bmatrix} 6 \\ 3 \\ 1 \end{bmatrix}$$

An easier way of simultaneously obtaining both the eigenvalues and eigenvectors is to apply the `Eigenvectors` command to A.

```
> sol2:=Eigenvectors(A);
```

$$sol2 := \begin{bmatrix} 1 \\ -\frac{1}{2} + \frac{1}{2} I\sqrt{3} \\ -\frac{1}{2} - \frac{1}{2} I\sqrt{3} \end{bmatrix}, \begin{bmatrix} 6 & \frac{6}{-\frac{1}{2} + \frac{1}{2} I\sqrt{3}} & \frac{6}{-\frac{1}{2} - \frac{1}{2} I\sqrt{3}} \\ 3 & -\frac{3\left(\frac{1}{2} + \frac{1}{2} I\sqrt{3}\right)}{-\frac{1}{2} + \frac{1}{2} I\sqrt{3}} & -\frac{3\left(\frac{1}{2} - \frac{1}{2} I\sqrt{3}\right)}{-\frac{1}{2} - \frac{1}{2} I\sqrt{3}} \\ 1 & 1 & 1 \end{bmatrix}$$

The column vector to the left of the comma on the rhs of *sol2* gives the eigenvalues, the 3×3 matrix on the right the corresponding eigenvectors. The first column in the latter gives the eigenvector corresponding to the first eigenvalue (top entry in the column vector), the second column the eigenvector corresponding to the second eigenvalue, etc. The order of the entries varies from one run to the next.

To extract the real eigenvector from *sol2*, the three individual columns are first obtained from the second entry in *sol2* and put into a list.

```
> vectors:=[seq(Column(sol2[2],n),n=1..3)]:
```

The `remove` command is used to remove the imaginary eigenvectors from *vectors*.

```
> V2:=remove(has,vectors,I)[];
```

$$V2 := \begin{bmatrix} 6 \\ 3 \\ 1 \end{bmatrix}$$

The answer to part **(c)** above is to choose the initial population for the three age groups to be a multiple of *V2*. Mike suggests that you confirm this by taking the initial population in the three age groups to be, e.g., 6000, 3000, and 1000. You will observe that because $\lambda = 1$, the number in each age group will remain the same year after year.

PROBLEMS:

Problem 4-37: Eigenvalues and Eigenvectors

Determine the eigenvalues and eigenvectors for each of the following matrices:

$$\textbf{(a)}\ A = \begin{bmatrix} 5 & 7 & -5 \\ 0 & 4 & -1 \\ 2 & 8 & -3 \end{bmatrix}, \qquad \textbf{(b)}\ B = \begin{bmatrix} -2 & -9 & 5 \\ -5 & -10 & 7 \\ -9 & -21 & 14 \end{bmatrix}.$$

First mimic a hand calculation and then use Maple's shortcut commands.

Problem 4-38: Diagonalization

For each matrix in the previous problem, determine a matrix M that *diagonalizes* the matrix (i.e., puts the eigenvalues on the central diagonal with all remaining matrix entries equal to 0). Note: The matrix M that transforms A into the diagonal form $A2$ satisfies the *similarity transformation* $M^{-1}\,A\,M = A2$, or, equivalently, $A\,M - M\,A2 = 0$.

Problem 4-39: More eigenvalues and eigenvectors

Determine the eigenvalues and eigenvectors of the following matrix:

$$A = \begin{bmatrix} 1 & 2 & 3 & 4 & 5 \\ 6 & 7 & 8 & 9 & 10 \\ 11 & 12 & 13 & 12 & 11 \\ 10 & 9 & 8 & 7 & 6 \\ 5 & 4 & 3 & 2 & 1 \end{bmatrix}.$$

Does A have an inverse? Explain your answer by calculating the determinant.

Problem 4-40: A different beetle species

Suppose that a beetle species has a life span of four years, a female beetle in the first year having a survival rate of 1/2, in the second year a survival rate of 1/3, and in the third year a survival rate of 1/4. On the average, a female beetle gives birth to four new females in the third year and to eight females in the fourth year, before expiring from the effort.

(a) Construct the relevant matrix A where the entry in the ith row and jth column denotes the probabilistic contribution that a female of age j makes to the next year's female population of age i.

(b) If there are initially 6000 female beetles in each of the four age groups (ages 1, 2, 3, and 4), determine the behavior of the population number in each age group as the number of years is made large. Plot the population numbers for the four age groups in the same figure.

(c) Determine the eigenvalues and associated eigenvectors of A. Are any of these real?

(d) If a sample of this species was needed for laboratory test purposes that could have a constant proportion in each age group from year to year, what criteria could be imposed on the initial female population to ensure that this would be satisfied?

Chapter 5

Linear ODE Models

"And if you take one from three hundred and sixty-five, what remains? Three hundred and sixty-four, of course." Humpty Dumpty looked doubtful. "I'd rather see that done on paper," he said.
Lewis Carroll, *Through the Looking Glass* (1872)

Dynamic models for which the independent variable, e.g., the time t, is continuous are governed by one or more ordinary differential equations (ODEs). Linear ODE models are described by ODEs that are *linear* in the dependent variable and its derivatives. An nth-order linear ODE for a dependent variable $x(t)$ has the general structure

$$\frac{d^n x}{dt^n} + a_{n-1}(t)\,\frac{d^{n-1}x}{dt^{n-1}} + \cdots + a_1(t)\,\frac{dx}{dt} + a_0(t)\,x = f(t), \tag{5.1}$$

where n takes on integer values. The value $n = 1$ produces a first-order ODE, $n = 2$ a second-order ODE, and so on. If $f(t) = 0$, the ODE is said to be *homogeneous*. Otherwise, it is *inhomogeneous*.

The first ODEs that science students usually encounter are those with constant coefficients (a_0, etc., independent of t), which can be solved by assuming a solution to the homogeneous equation of the form $x = e^{\lambda t}$. On substituting x into (5.1), one obtains $\lambda^n + a_{n-1}\,\lambda^{n-1} + \cdots + a_1\,\lambda + a_0 = 0$, which is solved for the n roots. Using standard methods (see, e.g., Stewart [Ste87]), a *particular* solution is then found to account for the $f(t)$ term.

Later in their academic careers, students are introduced to "special" ODEs with variable coefficients, which are solved [Boa83] using series methods, yielding so-called *special function* (Bessel, Legendre, Hermite, etc.) solutions.

The goal of this chapter is not to teach you about the various methods for solving ODEs, but instead to illustrate how Maple's `dsolve` command can be used to easily accomplish the same task. Before attempting to solve a second-order ODE (or two coupled first-order ODEs) with specified initial or boundary conditions, it is sometimes useful to obtain a graphical overview of the possible solutions by creating a *phase-plane portrait*, the subject of the first section.

5.1 Phase-Plane Portraits

A picture may instantly present what a book could set forth only in a hundred pages.
Ivan Turgeniev, Russian novelist (1818–1883)

Consider a system of two coupled first-order ODEs of the general structure

$$\dot{x} \equiv \frac{dx}{dt} = P(x, y, t), \quad \dot{y} \equiv \frac{dy}{dt} = Q(x, y, t), \tag{5.2}$$

where P and Q are known functions of the dependent variables x and y and, perhaps, the independent variable time t. For compactness, the dot notation is often used to denote a time derivative. In some model equations, the independent variable could be a spatial variable. If P and Q do not explicitly depend on the independent variable, the system (5.2) is said to be *autonomous*. Otherwise, the system is referred to as being *nonautonomous*.

Some models are naturally described in the *standard* form (5.2), while second-order ODE models such as those arising from Newton's second law of motion can be recast into that form. With x the displacement and F the force per unit mass, Newton's second law is of the general structure

$$\ddot{x} = F(x, \dot{x}, t). \tag{5.3}$$

Setting the velocity $\dot{x}$ equal to y, then (5.3) can be rewritten as the system

$$\dot{x} = y, \qquad \dot{y} = F(x, y, t), \tag{5.4}$$

so in this case $P = y$ and $Q = F(x, y, t)$.

Whether linear or nonlinear, a graphical approach can be used to view *all possible solutions* of those ODE systems that can be put into the standard form (5.2) and are autonomous. This graphical procedure has proved especially important in the investigation of nonlinear systems, where analytical solutions are usually impossible to obtain. (See the **Advanced Guide**.)

If Equations (5.2) do not depend explicitly on t, the independent variable can be eliminated by dividing one equation by the other to form the ratio

$$\frac{dy}{dx} = \frac{Q(x, y)}{P(x, y)}. \tag{5.5}$$

Except at a *stationary point* where $\dot{x} = 0 = P$ and $\dot{y} = 0 = Q$, this ratio represents the slope of the *trajectory* of the ODE system at an *ordinary* point (x, y) in the y versus x plane. At a stationary point, the ratio is $0/0$ and the slope is indeterminate. The x–y plane is referred to as the *phase plane* and the trajectory as a *phase-plane trajectory*. For ODEs governed by Newton's second law, the phase plane is a plot of velocity ($\dot{x}$) against displacement (x).

The time evolution of any possible motion of the ODE system may be pictured by systematically filling the phase plane with a grid of uniformly spaced arrows indicating the direction of increasing time and the slope at each grid point. For a given set of initial conditions, the subsequent temporal evolution of the system can be traced out by moving from one arrow to the next and drawing an appropriate line for the trajectory. Pictures created in this manner are called *phase-plane portraits*. Since an arrow at any ordinary point in the phase plane is tangent to the trajectory at that point, the grid of arrows is often referred to as the *tangent field*.

Analysis of the behavior of the trajectories at ordinary points near a stationary point reveals that only four types of stationary points are possible for linear[1] autonomous ODE systems. These are referred to as *focal* (or *spiral*), *nodal*, *vortex*, and *saddle* points. The behavior of the trajectories near these points is schematically indicated in Figure 5.1. As $t \to \infty$, the trajectories for

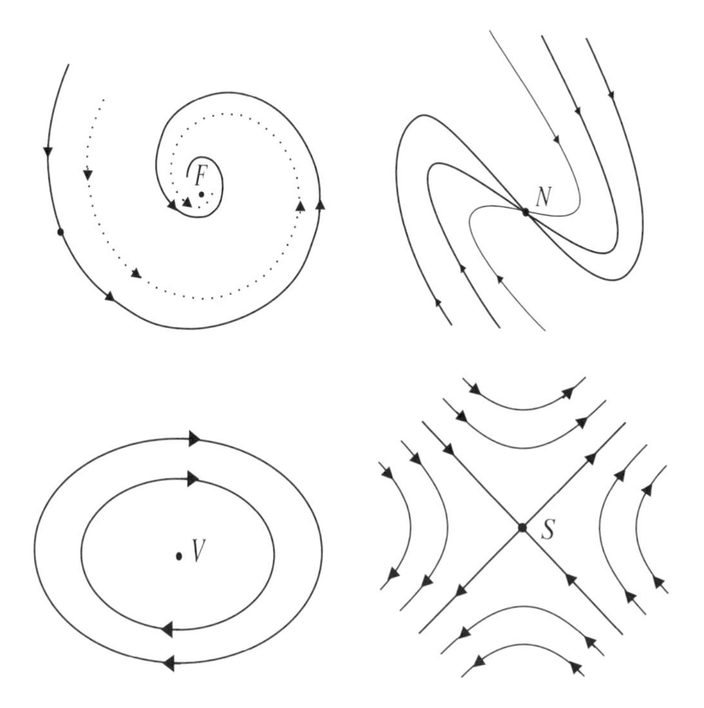

Figure 5.1: Curves near a focal (F), nodal (N), vortex (V), saddle (S) point.

the focal and nodal points in the picture approach those points. In this case, they are referred to as *stable* focal and nodal points. For unstable focal and nodal points, the sense of the arrows would be reversed.

If the location and nature of the stationary points is known, it is possible to sketch the phase-plane portrait by hand, even for nonlinear systems. The interested reader is referred, e.g., to [EM00], where the analysis of stationary points is carried out. In this section, we will be content to identify the stationary points after we have used Maple to draw the tangent field. In the following two linear ODE models, all four types of stationary points will be observed.

[1]For nonlinear ODE systems, additional types of stationary points are also possible.

5.1.1 Tenure Policy at Erehwon University

A great many people think they are thinking when they are merely rearranging their prejudices.
William James, American psychologist and philosopher (1842–1910)

At time t, the engineering faculty of Erehwon Institute of Technology is made up of $x(t)$ untenured and $y(t)$ tenured[2] professors. The engineering school has the following hiring and renewal policy. Each year a number of new untenured professors are hired equal to one-tenth of the entire engineering faculty. Also, one-tenth of the untenured professors are given tenure and one-tenth are turfed out (not given tenure) each year. Historically, for this engineering school, 5% or one-twentieth of the tenured professors retire or leave for other reasons each year. Our goal is to create a phase-plane picture that displays the temporal evolution of $x(t)$ and $y(t)$ over, say, a 30 year time span, making the undoubtedly dubious assumption that the above policy is maintained over this time interval.

To plot a tangent field and make a phase-plane portrait, the `DEtools` package, which contains the `dfieldplot` and `phaseportrait` commands, is loaded.

```
> restart: with(DEtools):
```

First, the ODE describing the rate of change of $x(t)$ (number of untenured professors at time t) with time is entered.[3] The first term $((x(t) + y(t))/10)$ on the right-hand side of the `eq1` input represents the rate of increase in x due to the hiring policy, the second term $(-x(t)/10)$ the rate of decrease in x due to promotion to the tenured rank, and the last term $(-x(t)/10)$ the rate of decrease in x due to the weeding out of unsuitable untenured professors. Applying the `simplify` command produces a simplified output, which could of course have been mentally done before entering the ODE.

```
> eq1:=diff(x(t),t)=simplify((x(t)+y(t))/10-x(t)/10-x(t)/10);
```

$$eq1 := \frac{d}{dt}\,x(t) = -\frac{1}{10}\,x(t) + \frac{1}{10}\,y(t)$$

The next command line states the rate of change of y with time, the first term on the right-hand side indicating the gain in y due to promotion from the untenured ranks, the second term reflecting the loss due to retirement and other causes.

```
> eq2:=diff(y(t),t)=x(t)/10-y(t)/20;
```

$$eq2 := \frac{d}{dt}\,y(t) = \frac{1}{10}\,x(t) - \frac{1}{20}\,y(t)$$

Either by inspection or by using Maple as follows,

---

[2]Tenure is a permanent employment status granted to professors after a probationary period and is based on satisfactory academic accomplishments and teaching performance. It is intended to protect professors from dismissal, except for serious misconduct or incompetence, even if their views are unpopular. It does not protect them, however, against budget cuts.

[3]Since discrete time intervals for the official hiring and renewal process usually prevail and professor numbers are integers, this problem would be better modeled by a difference equation. Difference equation models are the subject of Chapter 6.

```
> stat_point:=solve({rhs(eq1),rhs(eq2)},{x(t),y(t)});
```

$$stat\_point := \{x(t) = 0,\ y(t) = 0\}$$

we see that the coupled first-order ODE system, which is in standard form, has a single stationary point at the origin. The mathematical nature of this stationary point can be established by creating a tangent field of arrows over a region that includes the origin, e.g., $x = -2$ to 2, $y = -2$ to 2.

The tangent field for this region, which is shown in Figure 5.2, is produced using the `dfieldplot` command and entering the two equations and the two dependent variables as separate Maple lists. The time interval is taken to be 30 years. The `dirgrid` option specifies the number of horizontal and vertical mesh points to use for the arrows. The minimum is $[2, 2]$ and $[20, 20]$ is the default if the option is omitted. Here we have taken $[25, 25]$ for a total of $25 \times 25 = 625$ arrows in the picture. The option `arrows=MEDIUM` produces full arrowheads rather than the default half-arrowheads.

```
> dfieldplot([eq1,eq2],[x(t),y(t)],t=0..30,x=-2..2,y=-2..2,
  dirgrid=[25,25],arrows=MEDIUM);
```

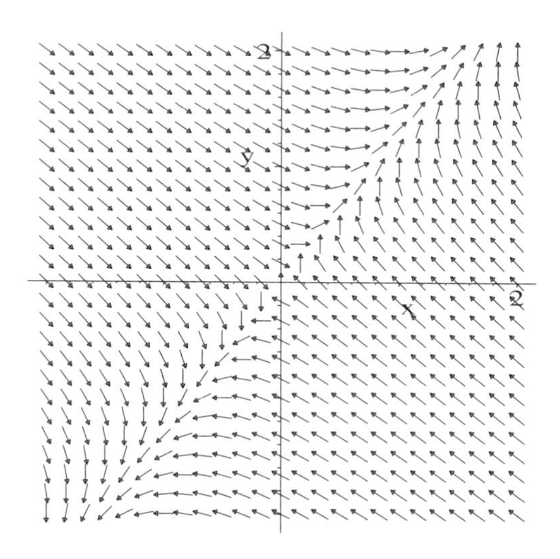

Figure 5.2: Tangent field for the tenure example.

Comparing the resulting tangent field in Figure 5.2 with the earlier singular point pictures in Figure 5.1, we see that the origin appears to be a saddle point. In the first ($x > 0$, $y > 0$) and third ($x < 0$, $y < 0$) quadrants the time arrows diverge from the origin, while in the second ($x < 0$, $y > 0$) and fourth ($x > 0$, $y < 0$) quadrants they approach the origin. From a physical viewpoint, negative values of x and y are not permitted for the tenure problem, so we need only concern ourselves with the first quadrant. As already noted, in this quadrant the time arrows tend to "flow" away from the origin. Thus, no matter what nonzero initial condition is chosen in the first quadrant, the corresponding trajectory will move away from the origin with increasing time.

We can choose a number of different initial conditions in the first quadrant and create a plot of the resulting phase-plane trajectories superimposed on the tangent field to create a complete phase-plane portrait. A functional operator `f` is formed to enter the values `A` and `B` of $x(0)$ and $y(0)$, respectively.

```
> f:=(A,B)->(x(0)=A,y(0)=B):
```

Using `f`, four different initial conditions, labeled `ic||1` to `ic||4`, will be considered. For example, in `ic||1` we consider 30 untenured and 10 tenured engineering professors at time $t = 0$.

```
> ic||1:=f(30,10): ic||2:=f(10,20): ic||3:=f(45,10):
  ic||4:=f(10,40):
```

The `phaseportrait` command is now used to create the desired phase-plane picture. As for the `dfieldplot` command, the system of two equations is entered as a Maple list as are the dependent variables. Next appears the time interval $t = 0$ to 30 years and the sequence of four initial conditions entered as a list of lists. Since Maple will numerically solve the system as a function of time, a small time step of 0.2 years is entered to obtain reasonable accuracy. Finally, we choose a blue line color for each of the four trajectories, the default being an unattractive (in our opinion) shade of yellow.

```
> phaseportrait([eq1,eq2],[x(t),y(t)],t=0..30,[seq([ic||j],
  j=1..4)],stepsize=0.2,dirgrid=[25,25],linecolor=blue);
```

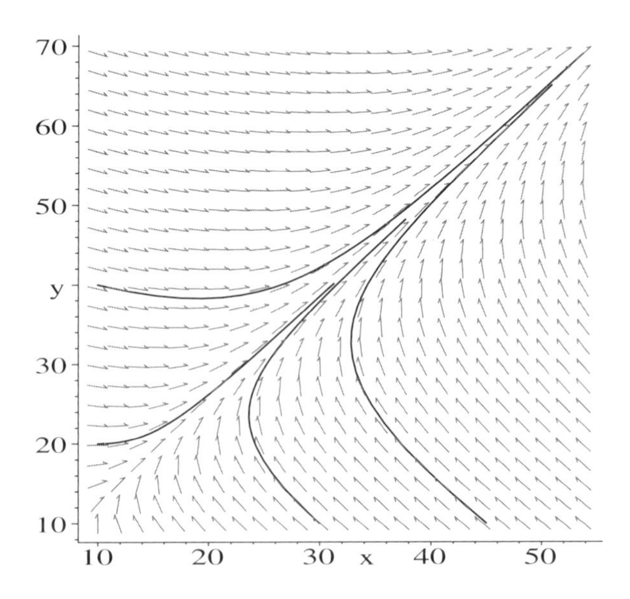

Figure 5.3: Phase-plane portrait for the engineering faculty.

The result of the `phaseportrait` command is shown in Figure 5.3. The arrows (with default half-arrowheads) indicate the temporal evolution of the ODE system. The solid lines depict the trajectories corresponding to the four different initial conditions. On Erehwon, there is a folk saying that, when translated, roughly parallels the Earth saying "All roads lead to Rome." No matter which of the four initial conditions is considered, after a sufficiently long time all trajectories in Figure 5.3 approach a single straight line in the phase plane.

Although one can deduce the approximate slope of the asymptotic straight line from the picture, we can use the fact that the ODE system is linear and ask Maple to seek an analytic solution using the `dsolve` command. If no initial conditions are specified, and an analytic solution exists, the general solution in terms of two arbitrary constants will be produced. The values of the two constants are fixed by the two initial conditions on $x(0)$ and $y(0)$. The initial conditions can also be fed directly into the `dsolve` command line. Since, in the problem at hand, all the other initial conditions yield trajectories that approach the same straight line, we select one representative set of initial condition, e.g., `ic||1`, and solve *eq1* and *eq2* for $x(t)$ and $y(t)$.

```
> sol:=dsolve({eq1,eq2,ic||1},{x(t),y(t)});
```

$$sol := \{x(t) = \left(15 + \frac{5\sqrt{17}}{17}\right) e^{\left(\left(\frac{-3+\sqrt{17}}{40}\right)t\right)} + \left(15 - \frac{5\sqrt{17}}{17}\right) e^{\left(-\frac{(3+\sqrt{17})\,t}{40}\right)}, \cdots\}$$

Only the analytic solution for $x(t)$ is displayed here in the text, the form of $y(t)$ being similar. If desired, the solution can be put into floating-point form by applying the `evalf` command to the last output.

```
> sol:=evalf(%);
```

$$sol := \{x(t) = 16.21267812\, e^{(0.02807764065\, t)} + 13.78732188\, e^{(-0.1780776406\, t)},$$
$$y(t) = 20.76481562\, e^{(0.02807764065\, t)} - 10.76481563\, e^{(-0.1780776406\, t)}\}$$

From the output, it can be seen that both $x(t)$ and $y(t)$ are made up of a linear combination of two exponentials, one increasing with time, the other decreasing with time. As $t \to \infty$ the exponentials with positive exponents will dominate and the ratio of $y(t)$ to $x(t)$ will approach a fixed number, which is equal to the line's slope. The slope can also be extracted by assigning the solution,

```
> assign(sol):
```

and taking the limit as $t \to \infty$ of the ratio $dy/dx = (dy/dt)/(dx/dt)$.

```
> limit(diff(y(t),t)/diff(x(t),t),t=infinity);
```

$$1.280776406$$

After a sufficiently long time has elapsed, the ratio of tenured to untenured faculty members in the Erehwon University engineering school will be 1.28, independent of the initial conditions chosen, provided that both $x(0)$ and $y(0)$ are not equal to zero.

The `dsolve` command comes with a number of options, e.g., `numeric` for numerical solutions, `laplace` for applying the Laplace transform method, etc.

PROBLEMS:

Problem 5-1: Recasting into standard form

For each of the following second-order ODEs, recast the ODE into the standard form (5.2) and indicate whether the system is linear or nonlinear. All coefficients are real and positive and primes denote derivatives with respect to x.

(a) simple pendulum equation: $\ddot{\theta} + 2\,\gamma\,\dot{\theta} + \omega^2\,\sin\theta = 0$;

(b) RLC circuit equation: $\ddot{V} + (R/L)\,\dot{V} + (1/LC)\,V = 0$;

(c) "hard" spring equation: $\ddot{x} + \omega_0^2\,(1 + a^2 x^2)\,x = 0$;

(d) Bessel equation: $x^2\,y'' + x\,y' + (x^2 - p^2)\,y = 0$;

(e) Legendre equation: $\left[(1 - x^2)\,y'\right]' + n\,(n+1)\,y = 0$.

(f) Van der Pol equation: $\ddot{x} - \epsilon\,(1 - x^2)\,\dot{x} + x = 0$;

Problem 5-2: Time to reach the asymptotic line

For each of the four initial conditions in the text, determine how long it takes for the corresponding trajectory to approach within 1% of the asymptotic straight line of slope 1.28.

Problem 5-3: Increasing the rejection rate

Explore the effect of increasing the rejection rate for untenured professors, i.e., the rate at which they are turfed out. You may want to change the time interval and/or the initial conditions. For example, what happens when the rejection rate is 30%?

Problem 5-4: Increased rate of retirement

Suppose that the term describing the rate of retirement (or departure) of the tenured professors is of the form $-y(t)^2/20$.

(a) Is the resulting system of equations linear or nonlinear? Explain.

(b) How many stationary points does the new system have and where are they located?

(c) Using the `dfieldplot` command, plot the tangent field for the new system over a suitable range in the first quadrant and identify the nature of the stationary points.

(d) Using the `phaseportrait` command and appropriate initial conditions, support the identification made in part (c). What is the long-time behavior of the system?

(e) Can the new system of equations be solved analytically with the `dsolve` command for the initial condition `ic||1`?

5.1.2 Vectoria Investigates the RLC Circuit

The desire of knowledge, like the thirst of riches, increases ever with the acquisition of it.
Laurence Sterne, *Tristram Shandy* (1760)

A "bread and butter" topic [Pur85] in any intermediate course in electromagnetic theory is investigating the dynamical behavior of an electric circuit consisting of an initially charged capacitor C placed in series with an inductor L and a resistor R as shown in Figure 5.4. At time $t = 0$ a switch (not shown)

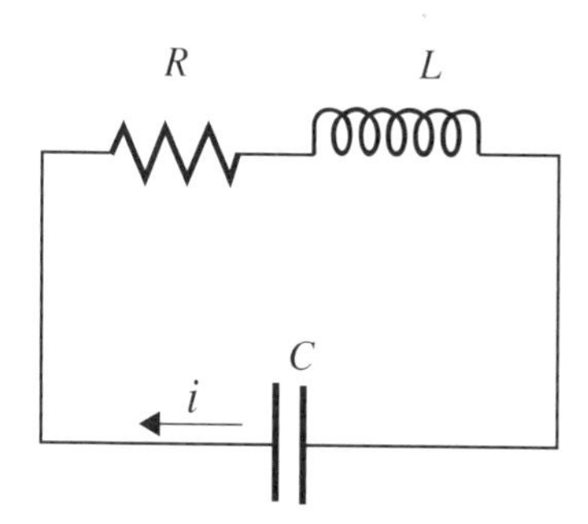

Figure 5.4: An RLC electrical circuit.

is closed, and charge and therefore a current i begin to flow around the circuit. The solution of the RLC circuit problem is one of the earliest physical examples involving a second-order linear ODE with constant coefficients that engineering and physics students encounter. Although she knows that this problem can be solved in a straightforward manner with pen and paper, Vectoria wants to continue to improve her computer algebra skills and is convinced that in applying her growing knowledge in this area to the RLC circuit problem she will gain a deeper feeling for the underlying physics.

Since she wishes to use the symbol γ for the damping (due to the resistor R) coefficient, Vectoria unprotects γ from its Maple representation as Euler's constant of mathematics.

```
> restart: with(plots): unprotect(gamma):
```

The DEtools and PDEtools library packages are called up in order that the `phaseportrait` and `dchange` commands, respectively, can be used. Vectoria intends to use the latter command to make a convenient change of variables in the resulting ODE.

```
> with(DEtools): with(PDEtools):
```

From the definition of capacitance, the charge q on the capacitor is given by

```
> q:=C*v(t);
```

$$q := C\,v(t)$$

where $v(t)$ is the voltage across C at time t. The instantaneous current in the circuit, due to the discharge of the capacitor, is $i = -dq/dt$. The minus sign is a reflection of the fact that the charge on the capacitor must initially decrease ($dq/dt < 0$) as charge begins to flow around the circuit. Inclusion of the minus sign ensures that i starts increasing for $t > 0$ in a positive sense.

```
> i:=-diff(q,t);
```

$$i := -C\left(\frac{d}{dt}v(t)\right)$$

Applying Kirchhoff's potential rule, the sum of the potentials around the circuit must add to zero at any instant t. With the instantaneous current direction as indicated, and starting at the inductor, the potential changes across the inductor, resistor, and capacitor are, respectively, $-L\,(di/dt)$, $-R\,i$, and $v(t)$.

```
> de:=-L*diff(i,t)-R*i+v(t)=0;
```

$$de := L\,C\left(\frac{d^2}{dt^2}v(t)\right) + R\,C\left(\frac{d}{dt}v(t)\right) + v(t) = 0$$

Noting that the expression for the current was automatically substituted into *de*, Vectoria observes that the output is a linear second-order ODE with constant coefficients. To be mathematically consistent, each term in the equation must have exactly the same dimension. Comparing the first and third terms, this implies that $L\,C$ has the dimension of time squared or inverse frequency squared. So Vectoria introduces the frequency $\omega_0 = 1/\sqrt{L\,C}$ by substituting $L = 1/(\omega_0^2\,C)$ into the differential equation *de*.

```
> de2:=subs(L=1/(omega[0]^2*C),de);
```

$$de2 := \frac{\frac{d^2}{dt^2}v(t)}{\omega_0{}^2} + R\,C\left(\frac{d}{dt}v(t)\right) + v(t) = 0$$

Vectoria wishes to transform from the original variables into a new set expressed in terms of the dimensionless time τ, defined through the relation $t = \tau/\omega_0$. Also setting $v(t) \equiv x(\tau)$, the variable transformation *tr* is as follows.

```
> tr:={t=tau/omega[0],v(t)=x(tau)};
```

$$tr := \left\{t = \frac{\tau}{\omega_0},\ v(t) = x(\tau)\right\}$$

Vectoria uses the transformation *tr* in the following `dchange` (change of variables) command to express *de2* in terms of the new variables.

```
> de3:=dchange(tr,de2,[tau,x(tau)]);
```

$$de3 := \left(\frac{d^2}{d\tau^2}x(\tau)\right) + R\,C\left(\frac{d}{d\tau}x(\tau)\right)\omega_0 + x(\tau) = 0$$

Now she notes that in *de3* there is effectively only one parameter in the problem, namely the product $R\,C\,\omega_0$ which is the coefficient of the second term. Labeling this product as $2\,\gamma$, Vectoria substitutes $R = 2\,\gamma/(\omega_0\,C)$ in *de3*,

```
> de4:=subs(R=2*gamma/(omega[0]*C),de3);
```

$$de4 := \left(\frac{d^2}{d\tau^2}x(\tau)\right) + 2\,\gamma\left(\frac{d}{d\tau}x(\tau)\right) + x(\tau) = 0$$

yielding a damped simple harmonic oscillator (SHO) equation, with γ identified as the damping coefficient. Vectoria decides to create a phase-plane portrait, by setting $dx(\tau)/d\tau = -y(\tau)$,

```
> de5:=diff(x(tau),tau)=-y(tau);
```

$$de5 := \frac{d}{d\tau}\, x(\tau) = -y(\tau)$$

and substituting *de5* into *de4* to produce a second first-order linear ODE, *de6*.

```
> de6:=expand(subs(de5,de4));
```

$$de6 := -\left(\frac{d}{d\tau}\, y(\tau)\right) - 2\,\gamma\, y(\tau) + x(\tau) = 0$$

Since the dependent variable y is proportional to i, a phase-plane portrait showing y versus x is actually a picture illustrating the behavior of the current versus the voltage. By inspecting *de5* and *de6*, Vectoria sees that there is a single stationary point at $y = 0$, $x = 0$.

Holding L and C fixed, she decides to investigate the effect of increasing the resistance R by assigning different values to the damping coefficient γ. In the next command line, four increasing γ values are entered, namely $\gamma 1 = 0$, $\gamma 2 = 0.1$, $\gamma 3 = 1$, and $\gamma 4 = 2$.

```
> gamma||1:=0: gamma||2:=0.1: gamma||3:=1: gamma||4:=2:
```

Vectoria creates a functional operator `eq` to substitute the nth γ value into *de6*.

```
> eq:=n->subs(gamma=gamma||n,de6):
```

She also forms a phase portrait operator `pp` to apply the `phaseportrait` command to *de5* and *eq*(n), the time range being $\tau = 0$ to 40. The initial condition is taken to be $y(0) = 0$ (zero initial current) and $x(0) = 1$. The latter condition corresponds to "normalizing" the voltage by dividing $v(t)$ by its initial value $v(0)$. The `scene=[A,B]` option is introduced with `A` and `B` to be specified. If `A` and `B` are chosen to be x and y, then y versus x will be displayed. The arguments in the scene option may also be taken to be τ and x in order to produce a plot of $x(\tau)$, or τ and y to plot $y(\tau)$. Since she intends to place all four phase-plane portraits in the same figure, Vectoria chooses to use a coarser grid (15 by 15) of black (rather than the default red) arrows than would be given by the default grid (20 by 20) in order that each tangent field can be clearly seen. The line color `C` of the phase-plane trajectory must also be specified.

```
> pp:=(n,A,B,C)->phaseportrait([de5,eq(n)],[x(tau),y(tau)],
     tau=0..40,[[x(0)=1,y(0)=0]],scene=[A,B],x=-1..1,y=-1..1,
     stepsize=0.03,dirgrid=[15,15],linecolor=C,
     color=black,arrows=MEDIUM,axes=normal):
```

Making use of the functional operator `pp`, a sequence of phase-plane portraits (y versus x) is created for all four γ values and assigned. The trajectory line color is taken to be blue.

```
> portraits:=seq(pl[n]=pp(n,x,y,blue),n=1..4):
  assign(portraits):
```

The `array` command is used to arrange the four phase-plane plots, `pl[1]` to pl[4], in a 2 by 2 viewing format,

```
> Graphs:=array(1..2,1..2,[[pl[1],pl[2]],[pl[3],pl[4]]]):
```

which is displayed in Figure 5.5.

```
> display(Graphs,tickmarks=[2,2]);
```

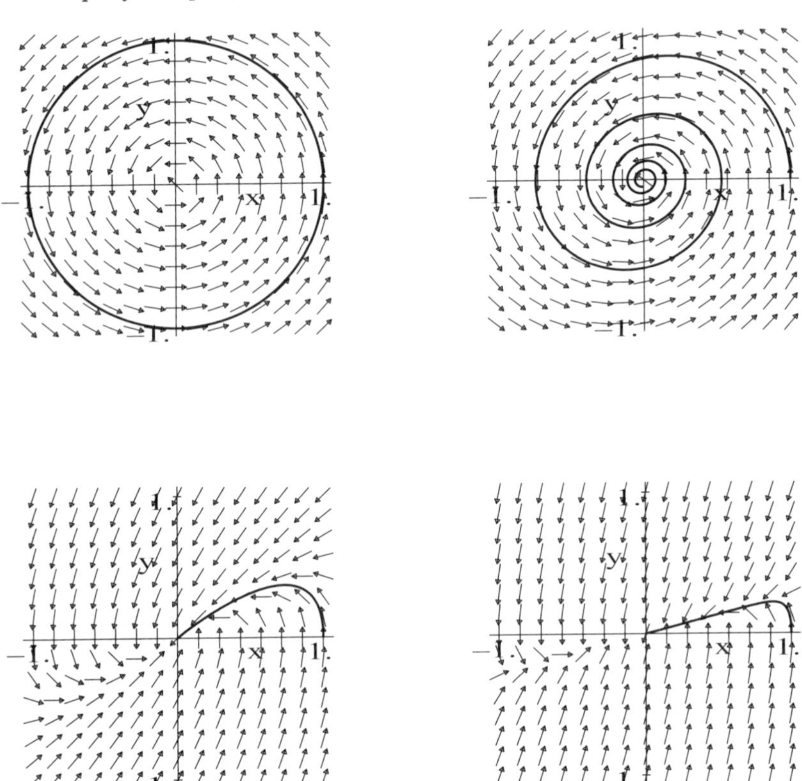

Figure 5.5: Phase-plane portraits for an RLC electrical circuit. Top left: $\gamma = 0$; top right: $\gamma = 0.1$; bottom left: $\gamma = 1$; bottom right: $\gamma = 2$.

The top left phase-plane portrait corresponds to $\gamma = 0$, i.e., zero resistance. In this case, the trajectory is a closed loop with a vortex stationary point at the origin. The amplitude of the oscillations neither grows nor decreases with time.

In the top right plot, $\gamma = 0.1$, corresponding to a small nonzero resistance. The trajectory now spirals into a stable focal point at the origin, x decreasing in an oscillatory manner with time. Vectoria realizes that this focal point situation must correspond to the "underdamped" SHO solution discussed in her electromagnetics class.

In the bottom left plot for $\gamma = 1$, the trajectory appears to shoot directly into a stable nodal point at the origin, x decreasing to zero without any oscillations. Vectoria will shortly show that this case corresponds to "critical damping."

Finally, in the bottom right plot of Figure 5.5, γ has been further increased so that there is definitely no overshoot of the origin before the trajectory proceeds to the stable nodal point. The SHO is in the "overdamped" regime.

Vectoria now focuses on one of the four cases, namely the underdamped situation for $\gamma = 0.1$. Setting $i = 2$ in the operator pp, Vectoria changes the scene option in pl[5] and pl[6] to tau,x and tau,y to produce plots of $x(\tau)$ and $y(\tau)$, respectively. She colors the curves blue and red, so that they can be easily distinguished on the computer screen.

```
> pl[5]:=pp(2,tau,x,blue): pl[6]:=pp(2,tau,y,red):
```

Since the plots pl[5] and pl[6] are to be superimposed, and may not be distinguishable if printed in black and white, Vectoria uses the textplot command to add the figure labels x and y.

```
> pl[7]:=textplot([[6.1,0.6,"x"],[8.6,0.57,"y"]]):
```

The graph that results on combining pl[5], pl[6], and pl[7] is displayed in Figure 5.6.

```
> display({pl[5],pl[6],pl[7]},tickmarks=[2,2],
  labels=["tau"," "]);
```

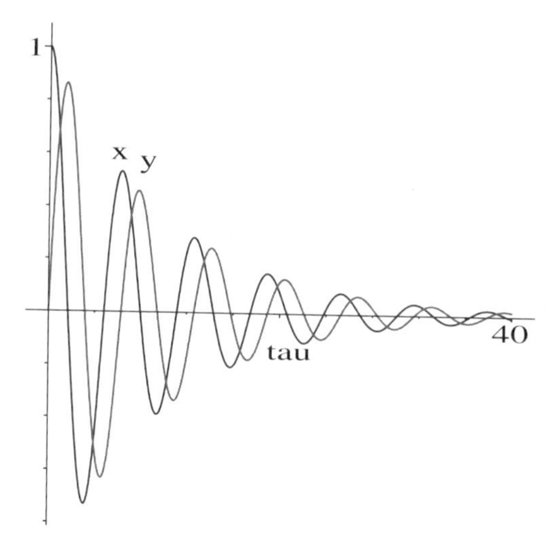

Figure 5.6: Damped oscillatory behavior of voltage (x) and current (y).

The current (proportional to y) reaches its maxima at later times than the voltage (x) and is said to "lag the voltage."

As in the tenure example, an analytic solution is obtainable for the voltage $x(\tau)$ and the current because the damped SHO equation is linear with constant coefficients. Vectoria applies the dsolve command to *de4*, subject to the initial conditions $x(0) = 1$ and the time derivative of the voltage (proportional to the current) equal to zero (entered as `D(x)(0)=0`, where `D` is the differential operator).[4]

```
> Voltage:=dsolve({de4,x(0)=1,D(x)(0)=0},x(tau));
```

$$Voltage := x(\tau) = \frac{\gamma\, e^{(-\gamma\,\tau)} \sin(\sqrt{-\gamma^2+1}\,\tau)}{\sqrt{-\gamma^2+1}} + e^{(-\gamma\,\tau)} \cos(\sqrt{-\gamma^2+1}\,\tau)$$

For $\gamma < 1$ (e.g., the $\gamma = 0.1$ case), the square root $\sqrt{1-\gamma^2}$ is real and the associated damped oscillatory motion is underdamped. For $\gamma > 1$ (e.g., $\gamma = 2$), the square root $\sqrt{1-\gamma^2}$ is purely imaginary and the nature of the solution changes. This is the overdamped case. For $\gamma = 1$, the solution collapses to one term, and a second independent solution must be obtained. Vectoria has left this critical damping case as a problem for you to solve.

With the analytic form of $x(\tau)$ known, the current is easily obtained by noting that

$$i = -C\,\frac{dv(t)}{dt} = -C\,\omega_0\,\frac{dx(\tau)}{d\tau}. \tag{5.6}$$

Vectoria enters this relation,

```
> Current:=-C*omega[0]*diff(rhs(Voltage),tau);
```

$$Current := -C\,\omega_0 \left(-\frac{\gamma^2\, e^{(-\gamma\,\tau)} \sin(\sqrt{-\gamma^2+1}\,\tau)}{\sqrt{-\gamma^2+1}} - e^{(-\gamma\,\tau)} \sin(\sqrt{-\gamma^2+1}\,\tau)\,\sqrt{-\gamma^2+1} \right)$$

and simplifies the output,

```
> Current:=simplify(Current);
```

$$Current := \frac{C\,\omega_0\, e^{(-\gamma\,\tau)} \sin(\sqrt{-\gamma^2+1}\,\tau)}{\sqrt{-\gamma^2+1}}$$

obtaining the desired formula for the current.

On completion of her Maple investigation of the RLC circuit, Vectoria feels pleased with her increased understanding of the underlying physics and her enhanced computer algebra skills. If more linear circuit elements and loops are added to the simple RLC circuit, she knows that she can easily extend the above analysis with little extra effort to handle the more complex situation.

PROBLEMS:

Problem 5-5: Critical damping

Determine the complete analytic solution for the voltage and current for $\gamma = 1$ in the RLC circuit.

Problem 5-6: Energy source

A 10-V battery is added to the RLC circuit. Investigate the behavior of the circuit for $\gamma = 1$ and discuss how adding the battery changes the nature of the solution.

[4] The differential operator `D` is more general than `diff`. It can represent derivatives evaluated at a point and can differentiate procedures.

Problem 5-7: A math problem
Consider the linear ODE system

$$\dot{x} = -4\,x - 3\,y + 5, \quad \dot{y} = 5\,x - 6\,y - 3, \quad \text{with } x(0) = 5,\ y(0) = -5.$$

(a) Create a phase-plane portrait showing the tangent field and the trajectory corresponding to the initial condition.

(b) Locate and identify the stationary point.

(c) Using appropriate scene options, plot $x(t)$ and $y(t)$ in the same graph.

(d) Analytically determine $x(t)$ and $y(t)$.

Problem 5-8: Erehwonian aardwolves
Two species of Erehwonian aardwolves, genetically altered relatives of those found on Earth,[5] are in competition for the same food supply. Initially there are $x = 2000$ gray aardwolves and $y = 1600$ red aardwolves, and the population equations are

$$\dot{x} = 3\,x - 2\,y, \quad \dot{y} = -2\,x + 3\,y,$$

with time measured in units of 50 years.

(a) Produce a tangent field plot and locate and identify the nature of the stationary point.

(b) Produce a phase-plane portrait for the time interval $t = 0$ to 0.6 for the given initial condition.

(c) Produce a plot of $y(t)$ for the same time interval using a suitable scene option.

(d) At what time do the red aardwolves become extinct? Use the previous plot to obtain an approximate value.

(e) Use the `dsolve` command to analytically determine $x(t)$ and $y(t)$.

(f) Use the `fsolve` command to precisely determine the time at which the red aardwolves go extinct.

(g) How many gray aardwolves exist at the time of extinction?

(h) What was the maximum number of red aardwolves and at what time did this occur?

Problem 5-9: Drug exchange
Consider the exchange of a particular prescription drug between the blood and the tissue of a human body. Let x be the concentration of the drug in the bloodstream and let y be the concentration in the body tissue. Including the extraction of the drug from the bloodstream by the kidneys, the relevant system of ODEs is

$$\dot{x} = k_1\,(y - x) - p\,x, \quad \dot{y} = k_2\,(x - y),$$

with the rate constants k_1 and k_2, as well as p, positive.

[5] An aardwolf ("earth wolf" in Afrikaans) is a South African flesh-eating mammal somewhat like the hyena and the civet.

(a) Explain the structure of the various terms appearing in the coupled ODE system.

(b) Taking the nominal values $k_1 = k_2 = 1$, $p = 1/2$, $x(0) = 1$, and $y(0) = 0$, create a phase-plane portrait showing the tangent field and the trajectory in the phase plane.

(c) Locate and identify the nature of the stationary point.

(d) Confirm your identification of the stationary point by using the scene option to form plots of $x(t)$ and $y(t)$ and superimpose the two plots in the same picture.

(e) Analytically solve for $x(t)$ and $y(t)$ for the specified initial condition.

(f) Plot the analytical form of $y(t)$ and compare it with the graphical solution by superimposing the two results in the same picture.

Problem 5-10: Romeo and Juliet

Steven Strogatz [Str88] [Str94] has suggested a simple linear dynamic model to create different scenarios for the love affair between Romeo and Juliet. In his model, $R(t)$ and $J(t)$ represent Romeo's love/hate for Juliet and Juliet's love/hate for Romeo, respectively, at time t. Positive values of R and J indicate love, while negative values indicate hate. The love affair equations take the form

$$\dot{R} = a\,R + b\,J, \quad \dot{J} = c\,R + d\,J,$$

with a, b, c, d real coefficients. For each of the following cases:

- produce a tangent field plot using the `dfieldplot` command and identify each stationary point;
- produce a phase-plane portrait with the specified initial condition;
- use the `phaseportrait` command and scene options to plot $J(t)$ and $R(t)$;
- use the `dsolve` command to derive the analytic solution;
- discuss how the love affair evolves with time.

(a) $\dot{R} = J, \quad \dot{J} = -R + J$, with $R(0) = 1, J(0) = 0$.

(b) $\dot{R} = 2\,R + J, \quad \dot{J} = R + 2\,J$, with $R(0) = 5, J(0) = -2$.

(c) $\dot{R} = J, \quad \dot{J} = R$ with $R(0) = 5, J(0) = -2$.

(d) $\dot{R} = J, \quad \dot{J} = -R$, with $R(0) = 5, J(0) = -2$.

(e) $\dot{R} = 2\,R + J, \quad \dot{J} = -R - 2\,J$, with $R(0) = -2, J(0) = 10$.

5.2 First-Order ODE Models

We have already seen Maple's ODE solver, `dsolve`, in action. In this and the following section, various interesting first- and second-order ODE models are analytically solved using this powerful command.

5.2.1 There Goes Louie's Alibi

Elementary my dear Watson, elementary.
Attributed to Sherlock Holmes, the fictional detective created by the English novelist and physician, Sir Arthur Conan Doyle (1859–1930). However, the quotation is not found in this form in any of Doyle's books.

A prominent elderly citizen of Metropolis was murdered in his luxurious climate-controlled penthouse and his body was found on Monday at 6:00 p.m. The homicide detectives assigned to the case suspect that the hit man was Louie the Louse and that the murder was committed early on Monday, or perhaps Sunday evening. When the body was discovered the temperature of the penthouse was 20 °C (68 °F) and the temperature of the body was 23.5 °C. The detectives ask Pat, the police department's forensic scientist, to establish the time of death. Pat knows that the normal body temperature of a living person is 37 °C. To ascertain the approximate time of death, Pat will make use of *Newton's law of cooling*, which he enters on his laptop computer.

```
> restart: de:=diff(T(t),t)=-k*(T(t)-Ts);
```

$$de := \frac{d}{dt}\,T(t) = -k\,(T(t) - Ts)$$

The rate of cooling at time t is proportional (proportionality constant k) to the difference in the instantaneous temperature $T(t)$ of the body and the ambient temperature Ts of the surroundings. Pat sets Ts to 20 °C,

```
> Ts:=20;
```

$$Ts := 20$$

and solves Newton's law of cooling with $T(0) = 23.5\,°\mathrm{C}$ in order to determine the theoretical curve that the victim's body temperature should obey as a function of time as it continues to cool. This assumes that the ambient temperature is maintained.

```
> sol:=dsolve({de,T(0)=23.5}, T(t));
```

$$sol := T(t) = 20 + \frac{7}{2}\,e^{(-k\,t)}$$

While the detectives are searching for clues and waiting for the fingerprint experts and police photographers to show up, Pat decides to determine the value of k by measuring the victim's body temperature every half-hour starting at 6:30 p.m. He manages to obtain 10 temperature measurements before the victim's body is removed just after 11:00 p.m., which he enters as a list.

```
> temp:=[23.3,23.1,23.0,22.8,22.7,22.6,22.5,22.3,22.2,22.1]:
```

After unprotecting `time` from its Maple meaning, he creates a list of the times in hours after 6:00 p.m. at which the temperature measurements were made.

```
> unprotect(time): time:=[seq(0.5*i,i=1..10)];
```

$$time := [0.5, 1.0, 1.5, 2.0, 2.5, 3.0, 3.5, 4.0, 4.5, 5.0]$$

In order to calculate k using the above data, Pat decides to use a least squares fitting routine. He rearranges the theoretical equation *sol*,

$$T(t) - 20 = \frac{7}{2} e^{-k\,t}, \tag{5.7}$$

and takes the natural logarithm of both sides, yielding

$$\ln(T(t) - 20) = \ln(7/2) - k\,t. \tag{5.8}$$

If he forms $\ln(T - Ts)$ with his temperature data and plots it against the time data, a straight line should result with a slope $-k$. Pat now computes $(T - Ts)$ for each of his ten temperature data points,

```
> temp2:=[seq(temp[i]-Ts,i=1..10)];
```

$$temp2 := [3.3, 3.1, 3.0, 2.8, 2.7, 2.6, 2.5, 2.3, 2.2, 2.1]$$

and uses the `map` command to apply the log to each operand of the *temp2* list. This command generate a new list, *temp3*, of the $\ln(T - Ts)$ values.

```
> temp3:=map(log,temp2);
```

$$temp3 := [1.193922468, 1.131402111, 1.098612289, 1.029619417, 0.9932517730, 0.9555114450, 0.9162907319, 0.8329091229, 0.7884573604, 0.7419373447]$$

The two lists *time* and *temp3* are zipped into a list of lists called *points*.

```
> pair:=(time,temp3)->[time,temp3];
```

$$pair := (time,\ temp3) \rightarrow [time,\ temp3]$$

```
> points:=zip(pair,time,temp3);
```

$$points := [[0.5, 1.193922468], [1.0, 1.131402111], [1.5, 1.098612289], [2.0, 1.029619417], [2.5, 0.9932517730], [3.0, 0.9555114450], [3.5, 0.9162907319], [4.0, 0.8329091229], [4.5, 0.7884573604], [5.0, 0.7419373447]]$$

Pat loads the Statistics and plots packages,

```
> with(Statistics): with(plots):
```

and then uses the `Fit` command to find the best-fitting (to 10 digits) straight line to the *points* data.

```
> eq:=evalf(Fit(a*t+b,time,temp3,t),10);
```

$$eq := -0.09908753428\,t + 1.240682126$$

To see how well the best-fitting straight line actually fits the data, Pat plots *eq* over the 5 hour measurement span.

```
> Gr:=plot(eq,t=0..5):
```

He also creates a plot of *points*,

```
> p0:=plot(points,style=point,symbol=circle,symbolsize=12,
      color=blue):
```

and superimposes p0 and Gr with the display command to produce Figure 5.7.

```
> display({p0,Gr},labels=["t","log(T-Ts)"]);
```

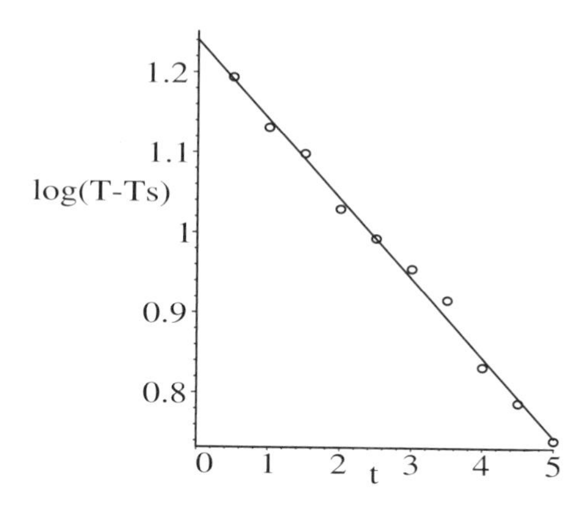

Figure 5.7: Best-fitting straight line to Pat's observational data.

The best-fitting straight line fits Pat's observational data quite well, so he feels confident that he can obtain a good estimate of the cooling coefficient k. Its value is found by taking minus the coefficient of t in *eq*.

```
> k:=-coeff(eq,t);
```

$$k := 0.09908753428$$

So, $k \approx 0.1\ \text{hr}^{-1}$ for the cooling coefficient. With this parameter determined, Pat solves the differential equation *de* using an initial temperature of 37 °C.

```
> eq:=dsolve({de,T(0)=37}, T(t));
```

$$eq := T(t) = \frac{49543767150}{2477188357} + \frac{42112202059}{2477188357}\, e^{\left(-\frac{2477188357\, t}{25000000000}\right)}$$

Pat notes how Maple has converted all floating-point numbers to rational exact numbers in order to speed up the analytic ODE-solving routine. He doesn't particularly like this structure of the solution, however, so he converts it back to a 10-digit floating-point form,[6]

```
> eq:=evalf(eq);
```

$$eq := T(t) = 20.00000000 + 17.00000000\, e^{(-0.09908753428\, t)}$$

and makes a plot of the exponential decrease of the body temperature for 20 hours after death.

```
> plot(rhs(eq),t=0..20,labels=["time","temp"],tickmarks=[3,5]);
```

[6] Alternatively, a solution in terms of floating-point numbers can be obtained by including the option convert_to_exact=false in the dsolve command.

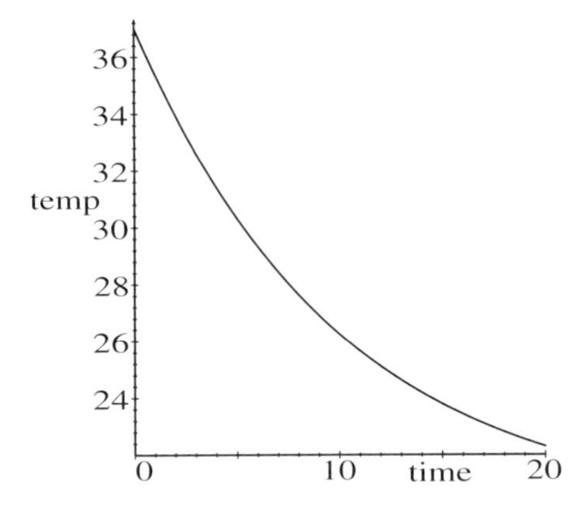

Figure 5.8: Exponential decrease in body temperature after death.

Although Pat could click on the Maple plot to find the approximate number of hours elapsed after death for the body to cool to 23.5°, an analytic solution is more accurate. This temperature is inputted into *eq*,

```
> T:=23.5: eq;
```

$$23.5 = 20.00000000 + 17.00000000\, e^{(-0.09908753428\, t)}$$

and, using the 24 hour time system with 18:00 corresponding to 6:00 p.m., the time of death would have been at $(18.00 - t)$ hr, where t is the solution of *eq*.

```
> Death_occurred:=18.00-solve(eq,t), AM_Monday_morning;
```

$$Death\_occurred := 2.04995757,\ AM\_Monday\_morning$$

So the estimated time of death was shortly after 2:00 a.m., Monday morning.

The detectives are not happy to hear this, since Louie the Louse appears to have an airtight alibi for this time period. He was observed returning to his apartment by neighbors at around 1:00 a.m. this Monday morning and apparently didn't leave after that.

Having done all that he could do, Pat calls it a night and heads home for a much needed good night's sleep. However, so convinced are the detectives of Louie's guilt that they go back to the penthouse and seek out the manager of the apartment complex. Here they learn that the temperature of the penthouse was computer controlled and that the thermostat had been previously set to 23 °C because it was winter and the elderly gentleman liked to be comfortably warm. However, this was the day that he had been about to leave for Palm Springs for the winter season. Thus, the thermostat had been programmed to linearly decrease the temperature of the penthouse from 23° to 20° between 8:30 a.m. and 9:00 a.m. Monday morning and then hold the temperature thereafter at 20°. Excited by this new information, they use their cell phone to contact Pat, who has just managed to fall asleep, at home.

Feeling that he isn't destined to get much sleep that night anyway, Pat enters the new ambient temperature data. The problem is broken into three parts, first to find the temperature of the body at 9:00 a.m. Monday morning, then the body temperature at 8:30 a.m., and finally the time it took for the body to cool from 37° to the 8:30 a.m. body temperature.

Between 9:00 a.m. Monday morning and 6:00 p.m. Monday afternoon, the penthouse temperature was 20 °C. Pat unassigns the value of T from the value 23.5 it had earlier and reenters Newton's law of cooling, de.

```
> unassign('T'): de;
```

$$\frac{d}{dt} T(t) = -0.09908753428\, T(t) + 1.981750686$$

The initial body temperature at 9.00 a.m. is labeled $T9$ and de is solved.

```
> dsol:=evalf(dsolve({de,T(0)=T9}, T(t)));
```

$$dsol := T(t) = 20.00000000 + e^{(-0.09908753428\, t)}\, (T9 - 20.00000000)$$

Substituting the body temperature of 23.5° at 6.00 p.m. Monday and noting that this is 9 hours later,

```
> eq:=subs({T(t)=23.5,t=9},dsol);
```

$$eq := 23.5 = 20.00000000 + e^{(-0.8917878085)}\, (T9 - 20.00000000)$$

eq is solved for the body temperature $T9$.

```
> T9:=solve(eq,T9);
```

$$T9 := 28.53820483$$

The body temperature at 9:00 a.m. Monday morning was about 28.5 °C. This temperature is then used as the final body temperature for the next time interval 8:30 a.m. to 9:00 a.m. Monday. So that the same symbols can be used again, t, Ts, and T are unassigned.

```
> unassign('t','Ts','T'):
```

Over a half-hour interval, the penthouse temperature decreased linearly from 23° to 20°, so in this interval the ambient temperature is given by $Ts = 23 - 6\,t$.

```
> Ts:=23-6*t;
```

$$Ts := 23 - 6\,t$$

Newton's cooling law is entered,

```
> de1:=diff(T(t),t)=-k*(T(t)-Ts);
```

$$de1 := \frac{d}{dt} T(t) = -0.09908753428\, T(t) + 2.279013288 - 0.5945252057\, t$$

and is solved with the initial body temperature now labeled as $T830$.

```
> dsol1:=evalf(dsolve({de1,T(0)=T830},T(t)));
```

$$dsol1 := T(t) = 83.55252099 - 6.000000000\, t + e^{(-0.09908753428\, t)}\, (T830 - 83.55252099)$$

The final body temperature at the end of the half-hour interval, $T9$, is entered,

```
> eq1:=subs({T(t)=T9,t=0.5},dsol1);
```

$$eq1 := 28.53820483 = 80.55252099 + e^{(-0.04954376714)}\, (T830 - 83.55252099)$$

and the body temperature *T830* determined.

```
> T830:=solve(eq1,T830);
```

$$T830 := 28.89631546$$

The body temperature at 8:30 a.m. Monday morning was slightly under 29 °C. Again unassigning the following variables,

```
> unassign('t','Ts','T'):
```

Newton's law of cooling is solved for an ambient temperature of 23° and an initial body temperature at the time of death of 37°.

```
> Ts:=23:
> de2:=diff(T(t),t)=-k*(T(t)-Ts);
```

$$de2 := \frac{d}{dt}\,T(t) = -0.09908753428\,T(t) + 2.279013288$$

```
> dsol2:=evalf(dsolve({de2,T(0)=37}, T(t)));
```

$$dsol2 := T(t) = 23.00000000 + 14.00000000\,e^{(-0.09908753428\,t)}$$

Substituting the new final body temperature *T830*,

```
> eq2:=subs({T(t)=T830},dsol2);
```

$$eq2 := 28.89631546 = 23.00000000 + 14.00000000\,e^{(-0.09908753428\,t)}$$

eq2 is solved for the number of hours before 8:30 a.m. that death occurred.

```
> Hours_before830am:=solve(eq2,t);
```

$$Hours\_before830am := 8.726926938$$

The time of death was approximately 8.73 hr before 8:30 a.m. Monday.

```
> Death_occurred:=12-(%-8.5), PM_Sunday_night;
```

$$Death\_occurred := 11.77307306,\ PM\_Sunday\_night$$

So, the victim was murdered just before midnight on Sunday. Pat phones the homicide detectives and tells them that Louie's alibi is no longer a good one and they might want to bring him in for further questioning.

PROBLEMS:

Problem 5-11: Newton's law of cooling

An object is originally 100 °C hotter than its surroundings. After 15 minutes, the temperature difference has fallen to 60 °C. How many minutes will it take the object to reach a temperature 10 °C above its surroundings? What will the temperature of the object be after 25 minutes?

Problem 5-12: Infectious disease

An infectious disease spreads slowly through a large population. The fraction f of the population that has been exposed to the disease within t years of its introduction is given by

$$\dot{f}(t) = 0.2\,(1 - f(t)).$$

If $f(0)=0$, determine the fraction of the population infected after t years. How long does it take for 75% of the population to be infected?

Problem 5-13: Population growth with immigration

The growth of large populations can be modeled over short time periods by assuming that the population number $N(t)$ grows continously with time at a rate proportional to $N(t)$. If the birth rate is b and there is also immigration into the population at a constant rate r:

(a) Write out the relevant ODE. Discuss the model.

(b) Analytically solve the ODE for $N(t)$.

(c) If initially $N(0) = 1$ million people are present, and 435 thousand individuals immigrate into the community in the first year, and 1.564 million people are present at the end of the first year, determine the birth rate b.

(d) Determine the population number 2 years later; 10 years later.

(e) Plot the population number over the time interval $t=0$ to 10 years.

(f) How would you include the death rate and emigration in the ODE?

Problem 5-14: Fick's law

The diffusion of a solute across a cell membrane is given by *Fick's law*, which takes the form

$$\dot{C}(t) = \kappa\,(C_S - C(t)).$$

Here $C(t)$ is the concentration of solute in the cell at time t, κ a constant which depends on the size of the cell and on the membrane properties, and C_S the concentration of solute outside the cell.

(a) For analysis purposes, one can set $\kappa=1$ without loss of generality. Explain why this can be done.

(b) Taking $\kappa = 1$ and an initial concentration $C(0) = C_S/100$, analytically solve the ODE for $C(t)$.

(c) Plot the ratio $C(t)/C_S$ over the time interval $t=0$ to 10.

(d) Making use of the `solve` command, determine how long it takes the concentration to attain the value $C_S/10$.

Problem 5-15: Population variation with emigration

The initial population of a city is 1.1 million people. The city is growing at 5% per year due to births, but losing population due to emigration at the rate of $50000\,(1 + \cos(\pi t/5))$, where t is in years.

(a) Write out the ODE describing the city's population number $N(t)$.

(b) Derive the analytic form of $N(t)$ for $t > 0$.

(c) Plot the solution over the range $t=0$ to 10 years.

(d) In the plotted range, at what time is the city population a minimum? a maximum?

(e) What is the city's population at $t=5$ years?

Problem 5-16: Radioactive chains

The nucleus of uranium 238 is unstable, decaying via α-emission into thorium 234, which in turn β-decays into palladium 234 and so on until the stable isotope lead 206 is created. Such a sequence of disintegrations is referred to as a radioactive chain. Let $N_1(t)$, $N_2(t)$, $N_3(t)$, etc., be the number of species-1, species-2, etc., atoms in a radioactive chain at time t. The decay rates of the various species can be described by the following coupled equations:

$$\dot{N}_1 = -\lambda_1 N_1, \quad \dot{N}_2 = \lambda_1 N_1 - \lambda_2 N_2, \quad \dot{N}_3 = \lambda_2 N_2 - \lambda_3 N_3, \quad \text{etc.}$$

(a) Explain the structure of these equations.

(b) If initially $N_1 = N$, $N_2 = 0$, $N_3 = 0$, simultaneously solve the system of equations to find an analytic expression for $N_3(t)$, i.e., the number of species-3 atoms at time t.

(c) Taking the nominal values $N=1000$, $\lambda_1=1$, $\lambda_2=2$, and $\lambda_3=3$, plot $N_3(t)$ for $t=0$ to 3.

(d) At what time is N_3 a maximum? How many nuclei of type 3 are there at this time? Determine your answer by clicking on the plot to find the coordinates of the maximum. Check your answer analytically.

Problem 5-17: Lead poisoning

Batschelet, Brand, and Steiner [BBS79] have formulated a model for the ingestion of lead, a semitoxic chemical, by the human body. For modeling purposes, the body is "divided" into the three "compartments," blood, tissue, and bones, as in the following figure. The semitoxic chemical enters the bloodstream at

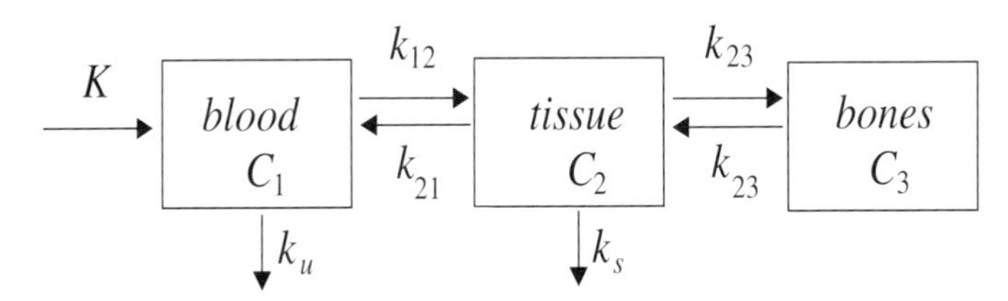

the rate K. It is distributed within the body, passing from blood to tissue to bones with the rate constants shown. It is lost from compartment 1 (blood) in the form of urine at the rate k_u and from compartment 2 (tissue) in the form of sweat at the rate k_s. Let C_1, C_2, and C_3 be the concentrations of lead in the blood, tissue, and bones. All the exchanges between compartments are assumed to be linear. The chemical rate equation for C_1 is

$$\dot{C}_1 = K - k_u\, C_1 - k_{12}\, C_1 + k_{21}\, C_2.$$

(a) Write out the rate equations for C_2 and C_3.

(b) Taking the initial concentrations to be all zero and letting $K = 2.0$ and $k_u = k_s = k_{12} = k_{21} = k_{23} = k_{32} = 1.0$, analytically solve for the three concentrations. Plot $C_3(t)$ for $t=0$ to 20 and discuss the result.

Problem 5-18: Price and supply

An important problem in economics is the interaction between the price P and supply S of a given commodity as a function of time t. A simple model for this interaction is described by the two coupled equations

$$\dot{P} = I(t) - k_1\,(S - S_0), \quad \dot{S} = k_2\,(P - P_0).$$

Here $I(t)$ is the positive inflation factor, k_1 and k_2 are positive proportionality coefficients, and P_0 and S_0 are the equilibrium price and supply, respectively. In the first equation, if $S > S_0$ the supply is too large and the price tends to decrease, while if $S < S_0$ the converse holds. In the second equation, if $P > P_0$ the right-hand side of the equation is positive, corresponding to the manufacturer increasing his supply to take advantage of the higher price (think, for example, about an oil producer). If $P < P_0$, the right-hand side is negative and the supply is decreased because the price is too low.

(a) Assuming that the inflation factor $I(t)$ is equal to a constant α and initially $P(0) = P_0$, $S(0) = S_0$, analytically solve the system of equations for $P(t)$, $S(t)$, using the Laplace transform option, `method=laplace`, in the `dsolve` command.

(b) Taking the nominal values $\alpha = 0.05$, $k_1 = k_2 = S_0 = P_0 = 1$, plot the price and supply curves in the same graph for the range $t = 0$ to 25. Does the supply curve lead or lag the price curve?

Problem 5-19: Specified initial conditions

Solve the following ODEs for the specified initial conditions:

(a) $\dot{x} + 5\,x = \cos(t) + e^{-t}, \quad x(0) = 1;$

(b) $\dot{x} = -4\,x + y + 3, \quad \dot{y} = -4\,x - 4\,y + 5, \quad x(0) = 1,\ y(0) = -1;$

Plot each solution over a suitable time range that includes the steady-state regime. In each case, estimate the time interval over which the transient lasts.

Problem 5-20: Predator–prey interaction

A predator (population number x)–prey (population number y) interaction between two species is modeled by the time-dependent ODE system

$$\dot{x} = x + y, \quad \dot{y} = y - 9\,x.$$

If the initial population numbers are $x(0) = 100$ and $y(0) = 1000$, find the population numbers at arbitrary time $t > 0$. Show that the prey become extinct after a certain time. Determine the time at which extinction takes place. Make a plot of the population numbers over this time interval.

Problem 5-21: Yeast growth

Yeast is growing in a sugar solution. If the rate of increase in the weight of yeast is equal to one-third of the weight already formed (when t is given in hours), write down the ODE describing the change in weight. If the weight is initially 14 grams, what is the weight after 2 hours? Plot the temporal evolution of the weight over this time.

5.2.2 The Water Skier

Just when you thought it was safe to go back in the water.
Publicity for *Jaws 2* (1978 film)

Russell, the engineer who was transferred to Phoenix, is also an avid water skier and often spends the blazingly hot Arizona summer weekends up at Teddy Roosevelt Lake engaging in this sport. His ever-active mind speculates on what path a hypothetical water skier would trace out if the situation depicted in Figure 5.9 prevailed and on the feasibility of actually carrying out the stunt.

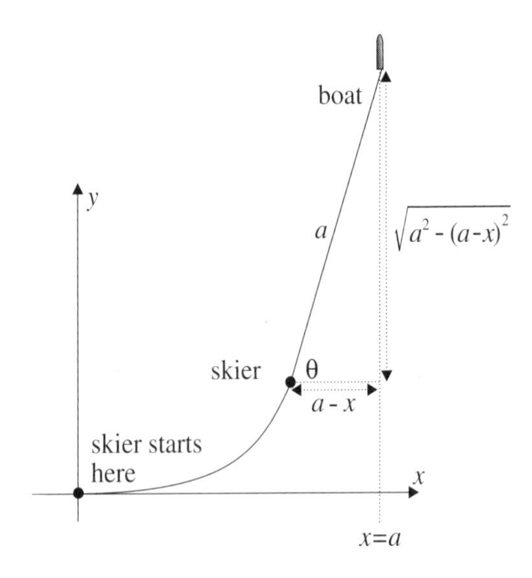

Figure 5.9: The water skier geometry.

Initially, the water skier is at the origin and holding onto a taut, inextensible, rope of length a attached to a motorboat that starts at $y = 0$ and moves at constant speed in the positive y-direction along the line $x = a$. The water skier is assumed to always point his skis toward the boat.

To carry out the calculation, Russell goes to his computer and starts typing in the formulation of the problem. He first wants to see what the skier's trajectory looks like and then obtain an analytic solution. To carry out the former, a call is made to both the plots and DEtools packages.

```
> restart: with(plots): with(DEtools):
```

From Figure 5.9, the rope has a slope given by $dy/dx = \sqrt{a^2 - (a-x)^2}/(a-x)$, which is entered.

```
> de:=diff(y(x),x)=sqrt(a^2-(a-x)^2)/(a-x);
```

$$de := \frac{d}{dx}\, y(x) = \frac{\sqrt{2\,a\,x - x^2}}{a - x}$$

Russell notes how Maple has automatically simplified the input expression in its output. To plot the path, he assumes a rope length of 10 m. The x-coordinate of the trajectory must lie between $x = 0$ and $x = 10$. At the latter value, the slope will become infinite. To avoid any possible mathematical difficulty, Russell uses the `assume` command to limit the x-range between 0 and 10. Although he is confident that his assumptions are as entered, he checks them by asking Maple about the status of x.

```
> a:=10: assume(0<x,x<a): about(x);
```

Originally x, renamed x~: is assumed to be: RealRange(Open(0),Open(10))

The output confirms the assumed range of x. The slope equation, which is a first-order linear ODE, now takes the form

```
> de;
```

$$\frac{d}{dx}\,y(x) = \frac{\sqrt{20\,x - x^2}}{10 - x}$$

Before seeking an analytic solution, Russell obtains a preliminary idea of the nature of the path by using the `DEplot` command to create a plot `pl` of the trajectory and the tangent field,

```
> pl:=DEplot(de,{y(x)},x=0..a-0.1,y=0..40,[[y(0)=0]],stepsize
      =0.1,dirgrid=[25,25],arrows=MEDIUM,linecolor=blue):
```

and displaying it in Figure 5.10. The skier's trajectory is the solid curve shown in the picture.

```
> display(pl,labels=["x","y"],tickmarks=[4,4]);
```

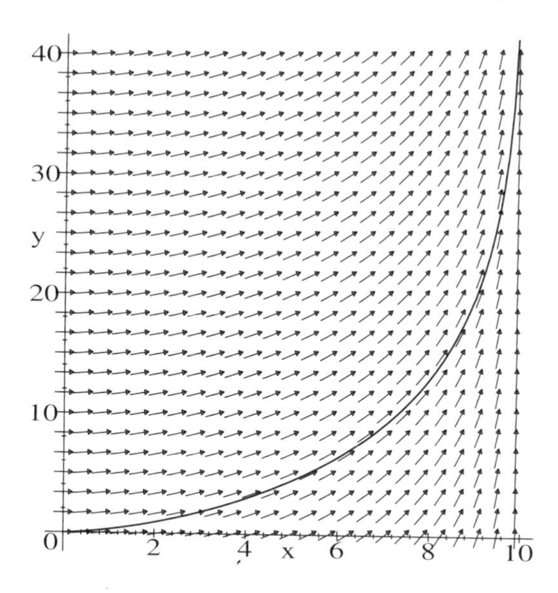

Figure 5.10: The solid line is the path of the water skier.

Looking at the structure of de, Russell realizes that he could obtain the analytic form $y(x)$ of the skier's trajectory by simply integrating the right-hand side of the equation from 0 to some value $x < a$. However, he decides instead to use the dsolve command to solve the first-order ODE de.

```
> sol:=dsolve({de,y(0)=0},y(x));
```

$$sol := \mathrm{y}(x) = -\sqrt{x\,(20-x)} + 10\,\mathrm{arctanh}\left(\frac{10}{\sqrt{x\,(20-x)}}\right) + 5\,I\,\pi$$

Since he has already graphically solved the problem and knows therefore that the analytic solution must be real, Russell is surprised by the appearance of the imaginary term, $5\,I\,\pi$, at the end of the output line. In an attempt to determine what is going on, he applies the symbolic complex evaluation command, evalc, to split the right-hand side of *sol* into real and imaginary parts.

```
> Y:=evalc(rhs(sol));
```

$$Y := -\sqrt{20\,x - x^2} + \frac{5}{2}\ln\left(\frac{\left(\frac{10}{\sqrt{20\,x-x^2}}+1\right)^2}{\left(\frac{10}{\sqrt{20\,x-x^2}}-1\right)^2}\right) + \left(-\frac{5}{2}\left(1-\mathrm{signum}\left(1-\frac{10}{\sqrt{20\,x-x^2}}\right)\right)\pi + 5\,\pi\right)I$$

The solution has been converted from an inverse trig form into a log form, but with the appearance of the signum function[7] in the imaginary term. The signum function appears because the sign of the argument inside the signum function is not known. If the sign were such that the coefficient of I became $-\frac{5}{2}(1+1)\pi + 5\pi = 0$, the solution would be completely real. This must be the case here.

To check this out, Russell uses the select command to extract the imaginary term containing I from Y.

```
> term:=select(has,Y,I);
```

$$term := \left(-\frac{5}{2}\left(1-\mathrm{signum}\left(1-\frac{10}{\sqrt{20\,x-x^2}}\right)\right)\pi + 5\,\pi\right)I$$

He then evaluates *term* for a representative value, e.g., $x = 4$, from the allowed range $0 \le x < 10$.

```
> Term:=eval(term,x=4);
```

$$Term := 0$$

The imaginary term is zero and may be removed from Y. This can be accomplished by substituting $term = Term$ into Y.

```
> Y:=subs(term=Term,Y);
```

[7]The signum function of x is defined as signum(x)= $x/|x|$ for $x \neq 0$.

$$Y := -\sqrt{20\,x - x^2} + \frac{5}{2}\ln\left(\frac{\left(\frac{10}{\sqrt{20\,x - x^2}} + 1\right)^2}{\left(\frac{10}{\sqrt{20\,x - x^2}} - 1\right)^2}\right)$$

The radical expression Y can be further simplified by applying the radical simplification (radsimp) command, yielding the final analytic formula for the skier's path.

```
> skier_path:=radsimp(Y);
```

$$skier\_path := -\sqrt{x\,(20 - x)} + \frac{5}{2}\ln\left(\frac{\left(10 + \sqrt{x\,(20 - x)}\right)^2}{\left(-10 + \sqrt{x\,(20 - x)}\right)^2}\right)$$

The skier's path is plotted as a thick (using the option thickness=2) blue line over the same x range as was used in producing Figure 5.10.

```
> skier_plot:=plot(skier_path,x=0..a-0.1,color=blue,
                  thickness=2):
```

The path of the boat along the straight line $x = a$ between $y = 0$ and 40 is plotted as a thick red line,

```
> boat_plot:=plot([[a,0],[a,40]],style=line,color=red,
                 thickness=2):
```

and displayed along with the skier's trajectory in Figure 5.11. The water skier's trajectory is, of course, the same as obtained graphically with DEplot.

```
> display({boat_plot,skier_plot},view=[0..a,0..40],
  labels=["x","y"],tickmarks=[4,4]);
```

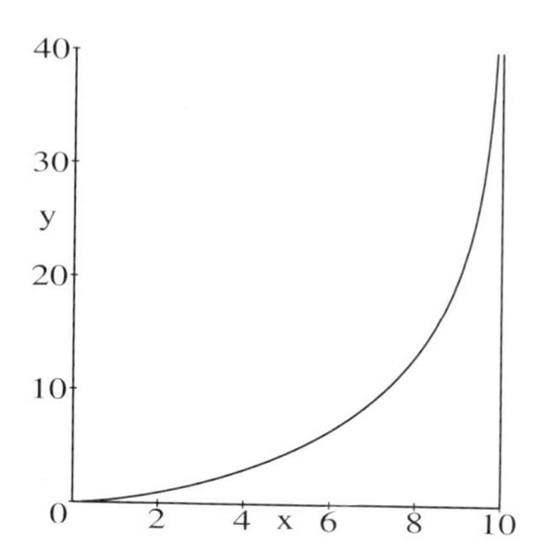

Figure 5.11: Plot of the skier's and boat's paths.

Having satisfied his mathematical curiosity, Russell is eager to get out of the Phoenix heat and onto the cool water of Roosevelt Lake. Maybe he can even attempt to approximately duplicate the situation depicted in the calculation.

PROBLEMS:
Problem 5-22: Solving by direct integration
By directly integrating the right-hand side of the relevant ODE, derive an analytic real expression for the skier's trajectory and plot it. At what value of x does the trajectory make a 45° angle with the x-axis?

5.2.3 Ready to Charge

Lawyers are like rhinoceroses: thick skinned, short-sighted, and always ready to charge.
David Mellor, British Conservative politician (1949–)

The differential equation obeyed by the charge $q(t)$ on a capacitor C connected in series with a resistor R and a voltage source V is

$$R\,\dot{q} + \frac{q}{C} = V. \tag{5.9}$$

Suppose that $V = A\,(t/\tau)\,e^{-t/\tau}$, where A is a positive constant and τ is a characteristic time. Analytically, determine $q(t)$ if the initial charge on the capacitor is zero. Taking $R = 5$ ohms, $C = 2$ farads, $A = 3$ volts, and $\tau = 1$ second, plot the growth of the charge on the capacitor for $t = 0$ to 70 seconds. At what time is the charge on the capacitor a maximum?

The analytic form of the voltage source V is entered.

```
> restart:
> V:=A*(t/tau)*exp(-t/tau);
```

$$V := \frac{A\,t\,e^{\left(-\frac{t}{\tau}\right)}}{\tau}$$

On entering the ODE in Equation (5.9), V is automatically substituted.

```
> ode:=R*diff(q(t),t)+q(t)/C=V;
```

$$ode := R\left(\frac{d}{dt}\,q(t)\right) + \frac{q(t)}{C} = \frac{A\,t\,e^{\left(-\frac{t}{\tau}\right)}}{\tau}$$

To solve *ode*, subject to the initial condition $q(0)=0$, let's first mimic the main steps that one would carry out in a hand calculation. Loading the DEtools library package, the *integrating factor* F for the first-order ODE is obtained.

```
> with(DEtools): F:=intfactor(ode);
```

$$F := e^{\left(\frac{t}{C\,R}\right)}$$

With the integrating factor known, the hand calculation proceeds by multiplying the ODE by the integrating factor and integrating the result. This step is now carried out by multiplying *ode* by F and applying the first integral (`firint`) command.

```
> firint(F*ode);
```

$$e^{\left(\frac{t}{C\,R}\right)}\,q(t)+\frac{e^{\left(-\frac{t}{\tau}+\frac{t}{C\,R}\right)}\,A\,(R\,C\,\tau+t\,C\,R-t\,\tau)\,C}{(-\tau+C\,R)^2}+\_C1=0$$

Using the `isolate` command allows us to isolate $q(t)$ to the left-hand side of the equation, thus giving us a solution with one undetermined constant $\_C1$.

```
> sol:=isolate(%,q(t));
```

$$sol:=q(t)=\frac{-\frac{e^{\left(-\frac{t}{\tau}+\frac{t}{C\,R}\right)}\,A\,(R\,C\,\tau+t\,C\,R-t\,\tau)\,C}{(-\tau+C\,R)^2}-\_C1}{e^{\left(\frac{t}{C\,R}\right)}}$$

The initial condition is applied by evaluating the rhs of sol at $t=0$ and equating it to zero. Then one analytically solves for the constant $\_C1$.

```
> _C1:=solve(eval(rhs(sol),t=0)=0,_C1);
```

$$\_C1:=-\frac{A\,R\,C^2\,\tau}{(-\tau+C\,R)^2}$$

Noting that the constant is automatically substituted, the solution is obtained. To obtain a nice form, the result is simplified with the exponential option.

```
> simplify(sol,exp);
```

$$q(t)=\left(-\frac{e^{\left(-\frac{t\,(-\tau+C\,R)}{R\,C\,\tau}\right)}\,A\,(R\,C\,\tau+t\,C\,R-t\,\tau)\,C}{(-\tau+C\,R)^2}+\frac{A\,R\,C^2\,\tau}{(-\tau+C\,R)^2}\right)\,e^{\left(-\frac{t}{C\,R}\right)}$$

We will now bypass all the steps above and derive the same result using the `dsolve` command. It is sometimes useful to ask Maple what steps it carried out in arriving at the solution. Even if the ODE is such that no analytic solution exists, as is often the case with nonlinear ODEs, it is instructive to know what methods Maple tried. The relevant information will be produced in the output of the `dsolve` command if the following `infolevel` command is entered. The number on the right of the colon can be an integer between 1 and 5. The larger the number, the more information is usually provided. Here, we have taken the integer to be 5.

```
> infolevel[dsolve]:=5:
```

Using `dsolve`, *ode* is analytically solved for $q(t)$, subject to the initial condition.

```
> q(t):=rhs(dsolve({ode,q(0)=0},q(t)));
```

Methods for first order ODEs:
— Trying classification methods —
trying a quadrature
trying 1st order linear
<– 1st order linear successful

$$q(t):=\left(-\frac{e^{\left(-\frac{t\,(-\tau+C\,R)}{R\,C\,\tau}\right)}\,A\,(R\,C\,\tau+t\,C\,R-t\,\tau)\,C}{(-\tau+C\,R)^2}+\frac{A\,R\,C^2\,\tau}{(-\tau+C\,R)^2}\right)\,e^{\left(-\frac{t}{C\,R}\right)}$$

The solution is identical to that obtained earlier. Maple has recognized the ODE as being first-order and linear and has presumably solved it by standard techniques, such as carried out in our "hand" calculation. Now $q(t)$ is evaluated with the given set of parameter values,

```
> q(t):=eval(q(t),{R=5,C=2,A=3,tau=1});
```

$$q(t) := \left(-\frac{2}{27}\,e^{\left(-\frac{9\,t}{10}\right)}\,(10+9\,t)+\frac{20}{27}\right)\,e^{\left(-\frac{t}{10}\right)}$$

and plotted over the range $t=0$ to 70.

```
> plot(q(t),t=0..70,labels=["t","q"]);
```

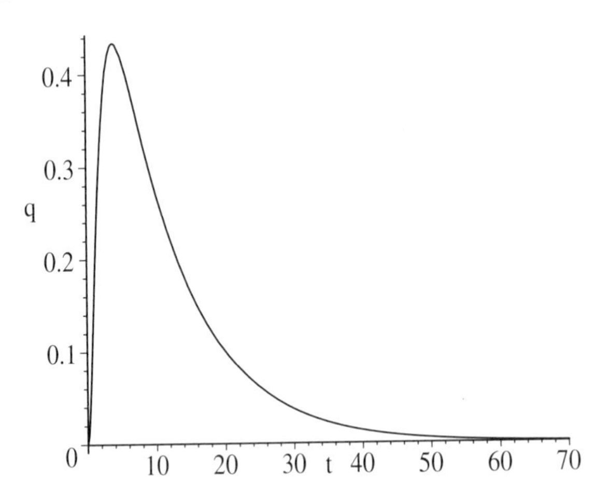

Figure 5.12: Charge on the capacitor as a function of time.

The resulting picture is shown in Figure 5.12. The charge on the capacitor initially builds up for about 4 seconds and then ultimately decreases to zero because of the presence of the resistor in the circuit. The time at which the maximum charge on the capacitor occurs can be more precisely obtained by setting $dq/dt=0$ and solving for the time using the floating-point solve command. The option telling Maple to avoid $t=0$ is included. Otherwise, the answer 0 is obtained for T, which corresponds to a minimum, rather than a maximum of the curve.

```
> T:=fsolve(diff(q(t),t)=0,t,avoid={t=0});
```

$$T := 4.016611586$$

The maximum charge on the capacitor occurs at 4.02 seconds.

PROBLEMS:

Problem 5-23: Charging a capacitor

Solve the text example again, but with a voltage source given by $V=A\,(t/\tau)^2\,e^{-t/\tau}$.

Problem 5-24: Purely Math

Obtain the general solutions of the following ODEs by **(i)** finding the integrating factor and the first integral, **(ii)** using the `dsolve` command. The primes denote spatial derivatives.

(a) $x\,y' + (1 - x)\,y = x\,e^x$;

(b) $x\,y' - k\,y = x^2$ (consider $k = 2$ separately);

(c) $y' - y\tan(x) = x$.

5.3 Second-Order ODE Models

5.3.1 Shrinking the Safety Envelope

It is always safe to learn, even from our enemies;
seldom safe to venture to instruct, even our friends.
C. C. Colton, English author, clergyman (1780–1832)

In Chapter 3, the envelope of safety around an erupting volcanic mountain in the Cascade Range of the Pacific Northwest was calculated. Recall that Sheelo, a part-time *National Geographic* photographer, wanted to approach as close as possible to the eruption in a rented plane. However, the effect of air resistance on the ejected rocks was completely neglected. Qualitatively, the inclusion of air drag should shrink the envelope and allow the plane to approach more closely. In this section, we shall estimate how much this shrinkage would be.

The simplest model of air drag is *Stokes's law of resistance*, for which the drag force is given by $\vec{F}_d = -k\,m\,\vec{v}$, where k is a positive constant, m the mass of the projectile, and $\vec{v}$ the velocity vector. Note that in this model the force law is linear in the velocity and directed in the opposite direction to the motion. In reality, the drag force is more complicated, and quadratic terms in the velocity can become important, depending on the speed. However, Stokes's law is a reasonable approximation [MT95] at very low speeds and also at speeds much greater than the speed of sound (about 340 m/s at 20 °C and 1 atmosphere). Further, it leads to a linear ODE that is easily solved, whereas inclusion of quadratic terms results in a nonlinear ODE that is more formidable to deal with. So, since the emphasis in this chapter is on analytically solving linear ODEs, let's assume that Stokes's law of air drag prevails.

```
> restart: with(plots):
```

Then, on mentally canceling the mass from both sides, Newton's equation of motion in the horizontal (x) direction gives

```
> xeq:=diff(x(t),t,t)=-k*diff(x(t),t);
```

$$xeq := \frac{d^2}{dt^2}\,x(t) = -k\left(\frac{d}{dt}\,x(t)\right)$$

A given rock starts out at the origin traveling at an angle θ with respect to the horizontal and with an initial speed Vo. Thus the initial x-component of velocity is $Vo\cos\theta$ and the vertical (y) component is $Vo\sin\theta$. The dsolve command is used to solve xeq for $x(t)$, subject to $x(0)=0$, $\dot{x}(0)=Vo\cos\theta$.

```
> xsol:=dsolve({xeq,x(0)=0,D(x)(0)=Vo*cos(theta)},x(t));
```

$$xsol := x(t) = \frac{Vo\cos(\theta)}{k} - \frac{Vo\cos(\theta)\,e^{(-k\,t)}}{k}$$

The collect command is employed to successively (note the use of a Maple list) collect the coefficients of $\cos(\theta)$, Vo, and $1/k$ in the above output.

```
> xsol:=collect(%,[cos(theta),Vo,1/k]);
```

$$xsol := x(t) = \frac{(1 - e^{(-k\,t)})\,Vo\cos(\theta)}{k}$$

Including drag and gravity, the equation of motion in the vertical direction is

```
> yeq:=diff(y(t),t,t)=-k*diff(y(t),t)-g;
```

$$yeq := \frac{d^2}{dt^2}\,y(t) = -k\left(\frac{d}{dt}\,y(t)\right) - g$$

which is solved, subject to $y(0)=0$, $\dot{y}(0)=Vo\sin\theta$, for $y(t)$.

```
> ysol:=dsolve({yeq,y(0)=0,D(y)(0)=Vo*sin(theta)},y(t));
```

$$ysol := y(t) = -\frac{(g + Vo\sin(\theta)\,k)\,e^{(-k\,t)}}{k^2} - \frac{g\,t}{k} + \frac{g + Vo\sin(\theta)\,k}{k^2}$$

The terms in the solution are regrouped with the following collect command.

```
> ysol:=collect(%,[sin(theta),Vo,1/k,g]);
```

$$ysol := y(t) = \frac{(1 - e^{(-k\,t)})\,Vo\sin(\theta)}{k} - \frac{g\,t}{k} + \frac{(1 - e^{(-k\,t)})\,g}{k^2}$$

Recall that in our earlier discussion in Chapter 3 the initial speed was taken to be $Vo = 700$ m/s and $g \approx 10$ m/s^2, where Vo was the maximum speed of the rocks. Rocks of lower speed will lie inside the envelope of safety calculated with the maximum speed. As a representative value, the coefficient of friction is taken to be $k = 0.01$ s^{-1} and the total time elapsed for generating the plot taken to be $T = 120$ seconds. (More realistically, k would decrease with altitude.)

```
> Vo:=700: g:=10: k:=0.01: T:=120:
```

A uniform range of angles between 0.01 and 3.01 radians is considered for plotting the different possible trajectories. The graphing procedure makes use of the spacecurve command, and the 3-dimensional viewing box is oriented so as to show the vertical coordinate, $y(t)$, versus the horizontal coordinate, $x(t)$.

```
> for i from -20 to 20 do
> theta:=1.51+(i*.075):
> pl[i]:=spacecurve([rhs(xsol),rhs(ysol),t],t=0..T,axes=normal,
          labels=["x","y","t"],orientation=[-90,0],color=red);
> end do:
```

The family of parabolic trajectories is displayed in Figure 5.13.

```
> display(seq(pl[i],i=-20..20),tickmarks=[4,4,2]
  view=[-30000..30000,0..20000,0..T]);
```

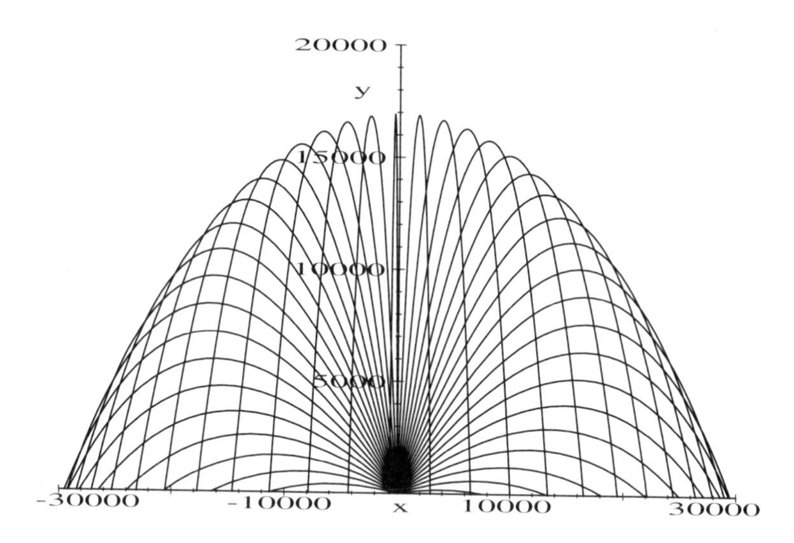

Figure 5.13: Family of parabolic trajectories when air resistance is considered.

If this figure is compared with Figure 3.16 in Chapter 3, the envelope of safety has shrunk considerably, from a horizontal radius of 50 km to about 30 km and a maximum altitude of 17 km compared to 24 km without air resistance. Thus, Sheelo can fly closer to the eruption and perhaps obtain a more spectacular picture to grace the cover of the *National Geographic* magazine. If the damping coefficient k is larger (smaller) than the representative value used here, the envelope is correspondingly smaller (larger).

The time of flight of a rock ejected at a given angle can be calculated as well as the speed with which the rock strikes the ground. For simplicity, the height of the volcano above the surrounding level land is neglected.

The angle θ is unassigned and then given the value $\theta = 45°$, or $\pi/4$ radians.

```
> unassign('theta'): theta:=evalf(Pi/4):
```

When the rock hits the ground again (remember that we are neglecting the height of the volcano), the right-hand side of `ysol` must be zero.

```
> eq:=rhs(ysol)=0;
```

$$eq := 149497.4747 - 149497.4747\, e^{(-0.01\, t)} - 1000.\, t = 0$$

Clearly one root of the above transcendental equation is $t = 0$, the time at which the rock is ejected from the volcano. Inserting the option `avoid={t=0}` in the `fsolve` command, the time of flight for the rock is found to be

```
> T:=fsolve(eq,t,avoid={t=0},0..120);
```

$$T := 86.63836788$$

about 87 seconds. The x-and y-components of velocity are calculated at arbitrary time t by differentiating the right-hand side of $xsol$ and $ysol$, respectively, with respect to time.

```
> V[x]:=diff(rhs(xsol),t); V[y]:=diff(rhs(ysol),t);
```

$$V_x := 494.9747468\, e^{(-0.01\, t)}$$

$$V_y := 1494.974747\, e^{(-0.01\, t)} - 1000.$$

Forming the speed, $\sqrt{V_x^2 + V_y^2}$, and substituting the time of flight, T,

```
> Speed:=evalf(subs(t=T,sqrt(V[x]^2+V[y]^2)))*meters/second;
```

$$Speed := \frac{425.7455131\, meters}{second}$$

yields a speed of about 426 m/s when the ejected rock strikes the ground.

PROBLEMS:

Problem 5-25: Varying damping coefficient

With all other parameters the same as in the text recipe, determine the maximum height and range for k values varying in 0.01 increments from $k=0$ to $k=0.1$. Make a plot summarizing each set of results.

Problem 5-26: Time of flight

For each angle used in generating the envelope of safety, determine the time of flight and make a plot summarizing your results. Hold all other parameters as in the text recipe.

Problem 5-27: The Paris gun

In World War I, the Germans used a long-range gun named the "Paris gun" to shell the city of Paris. The muzzle velocity was 1450 m/s. If the Paris gun was fired at an angle of 55° to the horizontal and the drag coefficient was $k=0.005$ s^{-1}, modify the text code (take $g=9.8$ m/s^2) to answer the following questions:

(a) What was the horizontal range in km of a projectile fired by this gun?

(b) To what maximum height in km did the projectile rise?

(c) Assuming that the launch site and Paris are at the same elevation, how long was the projectile in the air before striking its target?

Problem 5-28: How deep is that well?

A stone of mass m is dropped down a deep well. The stone is heard striking the water 10.0 s after it is released from rest. Assume that sound travels at a speed of 340 m/s, and that on its way down the stone experiences a drag force $F=-b\, m\, v$, where v is the speed and $b=0.1$ s^{-1}. Take $g=9.8$ m/s^2.

(a) How deep is the well?

(b) How long does the stone take to hit the water?

(c) Plot the stone's position and velocity over this time interval.

(d) What is the stone's speed when it hits the water?

Problem 5-29: Formulas for time of flight and range

For the text recipe, the time T that the projectile is in the air (the time of flight) is the solution of the transcendental equation

$$T = \frac{k\,V_0 \sin\theta + g}{g\,k}(1 - e^{-k\,T}).$$

(a) Assuming that k is sufficiently small, use an approximation procedure to derive the following approximate formula for the time of flight:

$$T \approx \frac{2\,V_0 \sin\theta}{g}\left(1 - \frac{k\,V_0 \sin\theta}{3\,g}\right).$$

(b) Making use of part **(a)**, show that the range is given approximately by

$$R \approx \frac{V_0^2 \sin(2\,\theta)}{g}\left(1 - \frac{4\,k\,V_0 \sin\theta}{3\,g}\right).$$

Problem 5-30: Greg's falling quarter-pounder

A careless young boy, named Greg, accidentally drops his still-wrapped quarter-pound[8] hamburger (mass $m = 1/4$ lb, or 113.5 g) from a restaurant balcony located atop one of Metropolis's taller buildings. If the viscous drag (coefficient k) is proportional to the velocity and the acceleration due to gravity is g:

(a) Write out the analytical form of the ODE describing the quarter-pounder's height $y(t)$ above the street after t seconds.

(b) If the quarter-pounder falls from rest from an initial height $y(0) = h$ meters above the street below, determine the height $y(t)$ of the falling burger above the street level after t seconds.

(c) If $h = 100$ m, the acceleration due to gravity is $g = 9.8$ m/s^2, and the viscous drag coefficient is $k = 0.1$ s^{-1}, determine the time it takes Greg's quarter-pounder to hit the street.

(d) Determine the speed of the burger when it hits the street.

(e) Plot the height of the burger above the street as a function of time.

Problem 5-31: Sarah the diver

Sarah, a young girl of mass $m = 30$ kg, dives into an ocean lagoon from a low cliff, entering the water perpendicular to the smooth surface at a speed of 10 m/s. She makes no swimming motion to ascend but instead lets the buoyant force of the water bring her back to the surface in 20 s.

(a) Assuming that the drag force due to the water is given by $F_{\text{drag}} = -k\,v$, where v is the velocity, and that Sarah's specific gravity is 0.95, determine the drag coefficient k. Take $g = 9.8$ m/s^2. *Note*: The specific gravity of a body is the ratio of the density of that object to that of water. According to Archimedes' principle, the water will exert an upward buoyant force equal to Sarah's weight divided by the specific gravity.

[8]Technically, this usually refers to the actual meat patty before cooking. However, you can assume that this is the mass of the entire cooked burger with all of its fixings.

(b) Plot Sarah's vertical distance $y(t)$ relative to the water surface over the 20-second interval that she is below the surface.

(c) Determine the time for Sarah to reach maximum depth. How long does it take her then to regain the surface if she makes no swimming motion?

(d) What is the maximum depth?

Problem 5-32: General solutions

Use `dsolve` to determine the general solutions of the following second-order ODEs, identifying any functions in the solution that are not "elementary":

(a) $\frac{d^2y}{dx^2} + 3\,\frac{dy}{dx} + 2\,y = e^x;$

(b) $x^2\,\frac{d^2y}{dx^2} - 6\,y = x^3\,\ln\,x;$

(c) $\frac{d^2y}{dx^2} - 2\,x\,\frac{dy}{dx} + (m-1)\,y = 0;$

(d) $x\,\frac{d^2y}{dx^2} - \frac{dy}{dx} + 4\,x^3\,y = 0;$

(e) $\frac{d^2y}{dx^2} + y = \tan x;$

(f) $x^2\frac{d^2y}{dx^2} + x\,\frac{dy}{dx} - (x^2 + \frac{1}{4})\,y = 0.$

Problem 5-33: Specified initial conditions

Use the `dsolve` command to find the solutions of the following time-dependent ODEs for the specified initial conditions:

(a) $\ddot{y} - 5\,\dot{y} + 6\,y = 0, \quad y(0) = 2,\ \dot{y}(0) = 5;$

(b) $9\,\ddot{y} + 6\,\dot{y} + y = 5, \quad y(0) = 6,\ \dot{y}(0) = 1;$

(c) $5\,\ddot{y} + 2\,\dot{y} + y = \sin^2(t), \quad y(0) = -3,\ \dot{y}(0) = 1;$

(d) $\ddot{y} + \dot{y} = 2\cos^4(t)\,e^{-t}, \quad y(0) = 0,\ \dot{y}(0) = -2;$

(e) $\ddot{y} + 2\,\dot{y} = t\cos(t)\,e^{-t} + t^2\cos(2\,t)\,e^{-2t}, \quad y(0) = 0,\ \dot{y}(0) = 0.$

Plot each solution over a suitable time range that includes the steady-state regime. In each case, estimate the time interval over which the transient lasts.

Problem 5-34: Third-order equations

Determine the general solutions of the following third-order ODEs:

(a) $\frac{d^3y}{dt^3} + \frac{d^2y}{dt^2} + 2\,\frac{dy}{dt} = t^2 + 3\,t + 1;$

(b) $\frac{d^3y}{dt^3} + 3\,\frac{d^2y}{dt^2} + 4\,\frac{dy}{dt} + 2\,y = 20\cos t.$

Identify the steady-state solutions, simplifying if necessary.

5.3.2 Frank N. Stein Is Not Heartless

I beheld the wretch–the miserable monster whom I created.
Frankenstein, Chapter 5, by Mary Shelley (1797–1851)

Cardiology involves the study of the heart, its functions and its diseases. The mathematical modeling of the heart as it pumps blood through the circulatory system has a long history, and good models can be very complex. In this section, we shall look at a very simple model built on mechanical principles.

Hugo, a modern eccentric scientist, has created a ghoulish replica of a human that he has named Frank N. Stein, in memory of the title character in the novel written in 1818 by Mary Shelley. Frank is lying flat on a table that has a spring arrangement that allows the table to move horizontally, but not vertically. To get Frank's heart going, Hugo gives it a kick start with an artificially created lightning bolt. When the heart begins to beat, the table will undergo very small horizontal vibrations as a consequence of the pumping of the heart. Hugo realizes that these vibrations could be monitored, thus acting as a mechanical analogue of an electrocardiograph. By studying the vibrations of the table, Hugo hopes to gain information about the vibrations of Frank's heart. Now, before the heartless reader pokes holes in this unorthodox approach, let him or her be reminded that Hugo is eccentric, and anyway, it's the authors' story.

To model the heart's pumping action, Hugo assumes that m kg of blood is pumped out of the heart on each vibration and $y(t)$ is the instantaneous center of mass position of this mass of blood. Hugo has a mathematical background and knows that y can be quite generally expressed as a Fourier series, i.e., a superposition of harmonic waves of different frequencies and amplitudes. Taking the frequency of the heart to be Ω,

```
> restart: with(plots):
```

Hugo writes out a Fourier sine series for $y(t)$, keeping only two terms.

```
> y(t):=add(a[i]*sin(i*Omega*t),i=1..2);
```

$$\mathrm{y}(t) := a_1 \sin(\Omega\, t) + a_2 \sin(2\,\Omega\, t)$$

The amplitude coefficients a_1, a_2,... will be such that the first harmonic or fundamental contribution (subscript 1) is dominant, with higher contributions being progressively less important. The forcing function due to heart vibrations then will be of the form $F(t) = m\,\ddot{y}$. Letting the mass of Frank and that portion of the table free to vibrate be M kg, the damping coefficient $2\,\beta$, and the natural frequency ω, the equation of motion of the table (plus Frank) is given by

```
> de:=M*diff(x(t),t,t)+2*beta*diff(x(t),t)+omega^2*x(t)
    =m*diff(y(t),t,t);
```

$$de := M\left(\frac{d^2}{dt^2}x(t)\right) + 2\,\beta\left(\frac{d}{dt}x(t)\right) + \omega^2\,x(t)$$
$$= m\,(-a_1 \sin(\Omega\, t)\,\Omega^2 - 4\,a_2 \sin(2\,\Omega\, t)\,\Omega^2)$$

Hugo assumes that the displacement and velocity of Frank + table are zero at $t = 0$, the instant at which Frank's heart begins to beat due to the lightning bolt strike. With this initial condition, an analytic solution for $x(t)$ is obtained by using the `dsolve` command and collecting sin and cos terms that appear in the quite formidable answer (whose exact form varies from run to run).

```
> sol:=dsolve({de,x(0)=0,D(x)(0)=0},x(t)):
> sol:=collect(sol,{sin,cos});
```

$$sol := x(t) = \frac{2\,m\,\Omega^3\,\beta\,a_1\cos(\Omega\,t)}{\Omega^4\,M^2 + (4\,\beta^2 - 2\,\omega^2\,M)\,\Omega^2 + \omega^4} + \frac{16\,m\,\Omega^3\,a_2\,\beta\cos(2\,\Omega\,t)}{16\,\Omega^4\,M^2 + (-8\,\omega^2\,M + 16\,\beta^2)\,\Omega^2 + \omega^4} - \frac{m\,\Omega^2\,(\omega^2 - \Omega^2\,M)\,a_1\sin(\Omega\,t)}{\Omega^4\,M^2 + (4\,\beta^2 - 2\,\omega^2\,M)\,\Omega^2 + \omega^4} - \frac{4\,m\,\Omega^2\,a_2\,(\omega^2 - 4\,\Omega^2\,M)\sin(2\,\Omega\,t)}{16\,\Omega^4\,M^2 + (-8\,\omega^2\,M + 16\,\beta^2)\,\Omega^2 + \omega^4} - \frac{1}{2}e^{\left(\frac{(-\beta+\sqrt{\beta^2-\omega^2\,M})\,t}{M}\right)}\,m\,\Omega^3(-72\,\Omega^4\,M^3\,a_2\,\omega^2 + 2\,a_1\,\beta^2\,\omega^4 + \cdots) + \frac{1}{2}e^{\left(-\frac{(\beta+\sqrt{\beta^2-\omega^2\,M})\,t}{M}\right)}(32\,\Omega^6\,M^4\,a_2 + 16\,\Omega^6\,M^4\,a_1 + \cdots)$$

The $\cos(\Omega\,t)$, $\cos(2\,\Omega\,t)$, $\sin(\Omega\,t)$, and $\sin(2\,\Omega\,t)$ terms make up the steady-state part of the solution, the transient exponential terms vanishing in the limit $t \to \infty$. The dots in the above output indicate that not all the transient terms are shown here in the text.

Then, Hugo inputs the following parameters: (Frank + table) mass $M = 100$ kg, blood mass $m = 100$ g $= 0.1$ kg, heartbeat frequency $\Omega = 6.3$ radians per second (corresponding to about 60 heartbeats per minute), damping parameter $\beta = 250$ kg/s, (Frank + table) natural frequency $\omega = \sqrt{3400} \approx 58.3$ radians/s, amplitude coefficient $a_1 = 0.05$ m, and $a_2 = 0.02$ m.

```
> M:=100.0: m:=0.1: Omega:=6.3: beta:=250.0:
  omega:=sqrt(3400.0); a[1]:=0.05: a[2]:=0.02:
```

$$\omega := 58.30951895$$

The complete analytic solution is then given by the output of the following command line. The complex evaluation command helps to simplify the result, but again it should be noted that occasionally the exact form may differ from that shown here in the text. However, the resulting picture is always the same.

```
> sol:=evalc(sol);
```

$$sol := x(t) = 0.00006100932818\cos(6.3\,t) + 0.00001024045306\cos(12.6\,t) + 0.00001102041515\sin(6.3\,t) + 0.00002027934800\sin(12.6\,t) - 0.00007124978124\,e^{(-2.500000000\,t)}\cos(5.267826876\,t) - 0.00009549912424\,e^{(-2.500000000\,t)}\sin(5.267826876\,t)$$

Hugo has the (Frank + table) system connected to a recording apparatus, which traces out a graph of the motion. This allows comparison of the model solution with what is actually observed. The model solution is now plotted in Figure 5.14.

```
> pl:=plot(rhs(sol),t=0..5,numpoints=500,thickness=2):
  display(pl,labels=["t","x"],tickmarks=[4,3],
  view=[0..5,-0.0001..0.0001]);
```

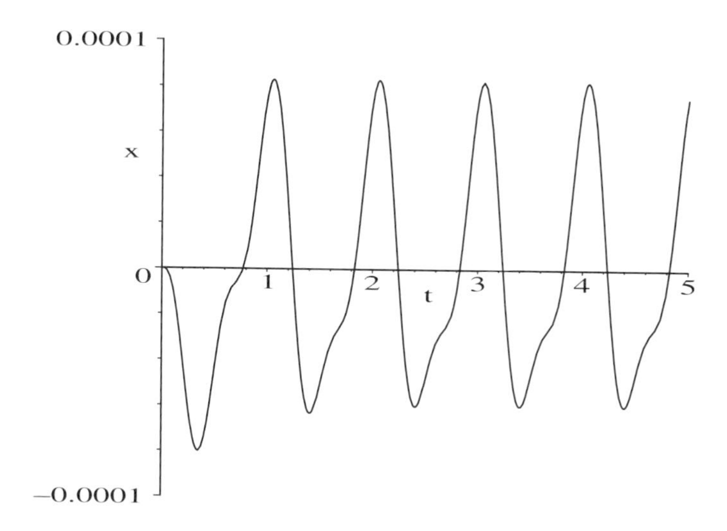

Figure 5.14: Vibrations of the table due to Frank N. Stein's beating heart.

A steady state is quickly achieved, and the dominant frequency in $x(t)$ is just the frequency of Frank's heart. What features in the graph of $x(t)$ are due to the second term in the Fourier series driving term? You can check your conclusion by running the code with $a_2=0$.

PROBLEMS:

Problem 5-35: Frank N. Stein's heartbeat

Investigate the influence of keeping higher-order terms in the Fourier series, taking numerical values of the amplitude coefficients of your own choosing. Remember that the amplitudes should be substantially less than the amplitude of the fundamental contribution.

Problem 5-36: It's a bumpy road ahead, Rob

Rob, a unicyclist of mass m traveling at a constant horizontal speed v along a smooth road, enters a bumpy region as shown in Figure 5.15 at time $t=0$. The spring supporting the unicycle seat has a spring constant k and the shock absorber introduces damping of any vertical motion of the seat. The damping force is proportional to the vertical velocity of the seat, the damping coefficient

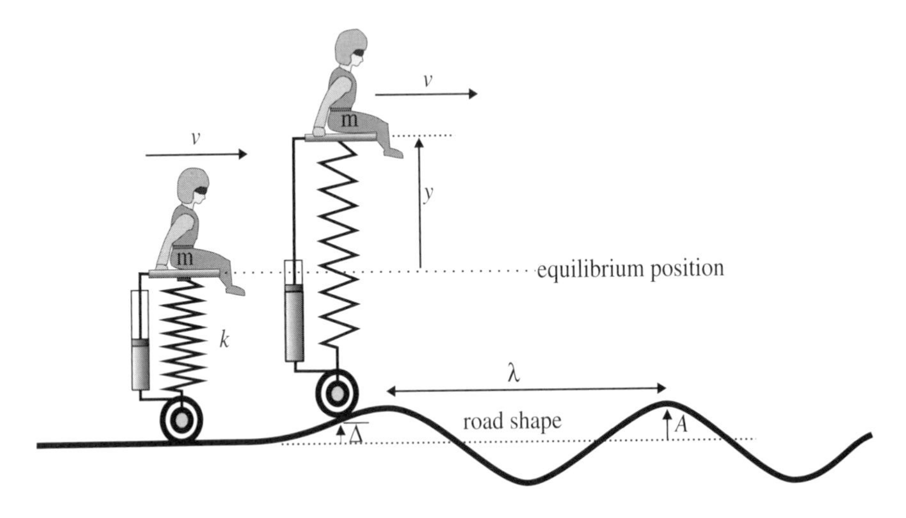

Figure 5.15: Rob, the unicyclist, entering a bumpy road region.

being c. The bumpy region of the road is described by a sinusoidal function with amplitude A and wavelength λ.

(a) Assuming that the horizontal speed is maintained in the bumpy region, show that the equation for the vertical displacement $y(t)$ of the seat from equilibrium in this region is given by

$$\ddot{y} + 2\,\gamma\,\dot{y} + \omega^2 y = 2\,\gamma\,\Omega\,A\cos(\Omega\,t) + \omega^2 A\sin(\Omega\,t),$$

with $\omega = \sqrt{k/m}$, $\Omega = 2\,\pi\,v/\lambda$, and $\gamma = c/(2\,m)$.

(b) Analytically solve the ODE for $y(t)$, given that $y(0)=0$ and $\dot{y}(0)=0$, and identify the steady-state and transient parts of the solution.

(c) Given $v=7$ m/s, $\lambda=5$ m, $\omega=10$ rads/s, $\gamma=1/5\ \mathrm{s}^{-1}$, and $A=0.05$ m, plot $y(t)$ for a sufficiently long time that steady state is essentially achieved.

(d) What is the amplitude of the vertical displacement $y(t)$ in steady state?

(e) How long does it take for $y(t)$ to get within 1% of steady state?

(f) At what speed v does the amplitude of $y(t)$ attain its maximum value?

(g) Plot $y(t)$ for this latter case and determine the maximum amplitude.

(h) How long does it take to achieve steady state in this latter case?

5.3.3 Halley's Comet

Mankind is not a circle with a single center but an ellipse with two focal points of which facts are one and ideas the other.
Victor Hugo, French poet, novelist, *Les Misrables* (1862)

One of Isaac Newton's many great accomplishments was in applying the law of universal gravitation to the motion of the planets of our solar system and to comets. Comets are such small objects that they are not easily observed until they reach the inner regions of the solar system. Newton's interest in comets began as an undergraduate in 1664, subsided for several years, and was rekindled in the early 1680s by the appearance of a spectacular comet moving away from the sun toward the outer limits of the solar system. This comet was also being followed by John Flamsteed, the first Astronomer Royal, at the Royal Observatory at Greenwich. Flamsteed believed it to be the same less brilliant comet that he had observed moving toward the sun a month earlier. Evidently, at first Newton resisted Flamsteed's idea because he found the apparent complete reversal of direction of the comet difficult to understand. However, he finally came to the conclusion that the orbit of a comet is either hyperbolic or a very elongated ellipse. In the former case, the comet would make only one close encounter with the sun, while in the latter it would return periodically. Using Newton's ideas, Edmund Halley, who had also observed the comet in 1682, predicted that it would return in 1758, 76 years after its previous appearance. The successful reappearance of the comet ensured Halley's fame and the naming of the comet in his honor. Earth's last rendevouz with Halley's comet was in 1986.

Vectoria is in Jennifer's classical mechanics class and has asked her to derive the equation for the trajectory traced out by Halley's comet using Maple's computer algebra system. They have already studied the "hand derivation" carried out in their mechanics text, Marion and Thornton [MT95]. Agreeing to this request, Jennifer's lecture on this topic involves executing the computer algebra steps on her laptop and projecting the results onto a large screen for the students to view. Let us eavesdrop and listen in on Jennifer's lecture.

"We begin by loading the plots and VectorCalculus packages, the latter being needed for deriving the form of the acceleration vector in polar coordinates.

```
> restart: with(plots): with(VectorCalculus):
```

To derive the acceleration of a celestial object as it moves about the sun, we let $r(t)$ be its radial distance from the sun at time t and $\theta(t)$ the angle that the radius vector makes with the X-axis. The Cartesian coordinates (X, Y) and polar coordinates are related as follows.

```
> X:=r(t)*cos(theta(t)); Y:=r(t)*sin(theta(t));
```

$$X := \mathrm{r}(t)\cos(\theta(t)) \qquad Y := \mathrm{r}(t)\sin(\theta(t))$$

The acceleration vector is calculated by differentiating X and Y twice with respect to t and applying the `VectorField` command in Cartesian coordinates.

```
> accel:=VectorField(<diff(X,t,t),diff(Y,t,t)>,'cartesian'[x,y]):
```

Because the Cartesian unit vectors are fixed in magnitude and direction they were not differentiated. The MapToBasis command is next used to transform or "map" the acceleration into polar unit (basis) vectors.

```
> accel:=MapToBasis(accel,'polar'[r,theta]):
```

The result of the above transformation is to introduce $\cos\theta$ and $\sin\theta$ terms in the output with no time dependence. We can rectify this situation by making the following substitution and simplifying the result.

```
> accel:=simplify(subs({cos(theta)=cos(theta(t)),
        sin(theta)=sin(theta(t))},accel));
```

$$accel := \left(\frac{d^2}{dt^2}r(t) - r(t)\left(\frac{d}{dt}\theta(t)\right)^2\right)\overline{e}_r + \left(2\,\frac{d}{dt}r(t)\,\frac{d}{dt}\theta(t) + r(t)\,\frac{d^2}{dt^2}\theta(t)\right)\overline{e}_\theta$$

The acceleration vector is the standard result for polar coordinates. Now, suppose that the force f per unit mass exerted on the celestial object by the sun is entirely radial, i.e., $f = f(r)$. Using Newton's second law, its radial and tangential accelerations are given by *eq1* and *eq2*, respectively.

```
> eq1:=accel[1]=f(r); eq2:=accel[2]=0;
```

$$eq1 := \frac{d^2}{dt^2}r(t) - r(t)\left(\frac{d}{dt}\theta(t)\right)^2 = f(r)$$

$$eq2 := 2\left(\frac{d}{dt}r(t)\right)\left(\frac{d}{dt}\theta(t)\right) + r(t)\left(\frac{d^2}{dt^2}\theta(t)\right) = 0$$

Yes, Vectoria, I see that you have your hand up. Do you have a comment on the derivation or a question?"

"Why didn't you immediately specialize to the inverse square law by taking $f(r) = -K/r^2$ in the gravitational force formula, where $K = G\,M$ with G the gravitational constant and M the mass of the sun?"

"That's a good question. By leaving the form of $f(r)$ general for the moment, we can use this recipe to study other radial force laws. For example, to account for the observed slow precession of Mercury's elliptical orbit, Einstein formulated an $f(r)$ that has a correction term to the inverse square law.

Any more questions? If not, let me continue with the derivation. I will now obtain an integral expression for $\theta(t)$ by applying the dsolve command to *eq2* subject to the initial conditions $\theta(0) = 0$ and $\dot{\theta}(0) = V/A$. Here $r = A$ is the initial distance of the object from the sun and V is its initial tangential velocity.

```
> eq2b:=dsolve({eq2,theta(0)=0,D(theta)(0)=V/A},theta(t));
```

$$eq2b := \theta(t) = \frac{V\,r(0)^2}{A}\int_0^t \frac{1}{r(\_z1)^2}\,d\_z1$$

We need to eliminate $d\theta/dt$ from *eq1*, so let's substitute $r(0) = A$ into *eq2b*, perform the time differentiation, and then substitute *eq2c* into *eq1*.

```
> eq2c:=diff(subs(r(0)=A,eq2b),t);
```

$$eq2c := \frac{d}{dt}\theta(t) = \frac{A\,V}{r(t)^2}$$

```
> eq1b:=subs(eq2c,eq1);
```

$$eq1b := \frac{d^2}{dt^2}r(t) - \frac{A^2\,V^2}{r(t)^3} = f(r)$$

We now assume that an inverse square force law prevails, i.e., $f(r) = -K/r(t)^2$, which is automatically substituted into *eq1b*.

```
> f(r):=-K/r(t)^2; eq1c:=eq1b;
```

$$f(r) := -\frac{K}{r(t)^2} \qquad eq1c := \frac{d^2}{dt^2}r(t) - \frac{A^2\,V^2}{r(t)^3} = -\frac{K}{r(t)^2}$$

eq1c is a nonlinear ODE that cannot be analytically solved for $r(t)$. However, a simple closed form can be obtained for $r(\theta)$ by carrying out the following procedure. First a change of variables is made by setting $p = dr/dt$ and rewriting

$$\frac{d^2r}{dt^2} = \frac{dp}{dt} = \left(\frac{dp}{dr}\right)\left(\frac{dr}{dt}\right) = p\left(\frac{dp}{dr}\right). \tag{5.10}$$

This variable change is substituted into *eq1c*, and we also set $r(t) = r$.

```
> eq1d:=subs({r(t)=r,diff(r(t),t,t)=p(r)*diff(p(r),r)},eq1c);
```

$$eq1d := p(r)\,\frac{d}{dr}p(r) - \frac{A^2\,V^2}{r^3} = -\frac{K}{r^2}$$

At the initial radial distance $r = A$, the radial velocity $dr/dt = p$ is taken to be zero, i.e., $p(r = A) = 0$. That is, the celestial mass is initially placed at one of its turning points. Then *eq1d* is solved for $p(r)$ subject to this initial condition.

```
> sol:=dsolve({eq1d,p(A)=0},p(r));
```

$$sol := p(r) = \frac{\sqrt{-A\,(A^3\,V^2 - 2\,K\,r\,A - V^2\,r^2\,A + 2\,K\,r^2)}}{A\,r},$$
$$p(r) = -\frac{\sqrt{-A\,(A^3\,V^2 - 2\,K\,r\,A - V^2\,r^2\,A + 2\,K\,r^2)}}{A\,r}$$

Except for a possible rotation of 180° in the final picture, which of the two $p(r)$ solutions is chosen is immaterial. I will select the positive square root. Since $p = dr/dt$, the ratio $p/(d\theta/dt)$ is equal to $dr/d\theta$. This ratio is calculated in the following command line. From the structure of the terms in $p(r)$, it is clear that the parameter K has the same dimension as $A\,V^2$, so for later convenience I will set $K = A\,V^2/(1+\epsilon)$, where ϵ is a dimensionless constant. Also in the command line, all r dependence is expressed as $r(\theta)$.

```
> eq3:=diff(r(theta),theta)=subs({K=A*V^2/(1+epsilon),
        r(t)=r(theta),r=r(theta)},rhs(sol[1])/rhs(eq2c));
```

$$eq3 := \frac{d}{d\theta}r(\theta) = \frac{\sqrt{-A\left(A^3V^2 - \frac{2\,V^2A^2\;r(\theta)}{\epsilon+1} - V^2r(\theta)^2A + \frac{2\,V^2A\,r(\theta)^2}{\epsilon+1}\right)}\;r(\theta)}{A^2V}$$

Then *eq3* is analytically solved for $r(\theta)$, assuming that $A > 0$, $\epsilon > 0$, and $V > 0$.

```
> sol2:=dsolve(eq3,r(theta)) assuming A>0,epsilon>0,V>0;
```

$$sol2 :=$$
$$\theta - \arctan\left(\frac{-A^2 - A^2\epsilon + A\,r(\theta)}{A\sqrt{(-A^2\epsilon - A^2 + 2\,A\,r(\theta) + r(\theta)^2\,\epsilon - r(\theta)^2)\,(\epsilon+1)}}\right) + \_C1 = 0$$

Since an implicit solution has been generated, the `isolate` command is applied to *sol2* to obtain $r(\theta)$. The messy output is suppressed.

```
> R:=isolate(sol2,r(theta)):
```

The right-hand side of R is converted to sines and cosines.

```
> R:=convert(rhs(R),sincos):
```

By choosing the constant $\_C1$ in R to be $\pi/2$, so that $r(0)\!=\!A$,

```
> R:=simplify(subs(_C1=Pi/2,R),symbolic);
```

$$R := \frac{A\,(\epsilon+1)}{\epsilon\cos(\theta)+1}$$

the answer reduces to the standard mathematical form for an ellipse with eccentricity ϵ and perihelion distance A. The perihelion is the point of closest approach of the orbiting mass to the sun. The distance from the sun to the perihelion is related to the semimajor axis a of the ellipse by the relation $A\!=\!a\,(1-\epsilon)$, which is now substituted into R.

```
> R:=subs(A=a*(1-epsilon),R);
```

$$R := \frac{a\,(-\epsilon+1)\,(\epsilon+1)}{\epsilon\cos(\theta)+1}$$

For $\epsilon = 0$, we have $R = a$, the equation for a circular orbit of radius a. A functional operator `s` is formed for evaluating R for different values of the semimajor axis and eccentricity.

```
> s:=(a0,e0)->eval(R,{a=a0,epsilon=e0}):
```

The semimajor axis of Earth's slightly elliptical ($\epsilon = 0.0167$) orbit about the Sun is 1.495×10^8 km, or approximately 93 million miles. If we choose to measure distances in units of Earth's semimajor axis, then $a\!=\!1$ for Earth and $a = 17.9$ for Halley's comet. The semimajor axis and eccentricity values are now entered in order for Halley's comet, Jupiter, Saturn, Uranus, Neptune, and Pluto. On the scale of these outer planets, the orbits of Earth and the inner planets would be quite small in the subsequent plot, so are not included.

```
> a||0:=17.9: a||1:=5.20: a||2:=9.54: a||3:=19.19:
  a||4:=30.06: a||5:=39.53:
> e||0:=0.967: e||1:=0.0483: e||2:=0.0560: e||3:=0.0461:
  e||4:=0.0100: e||5:=0.248:
```

The `polarplot` command is used to generate the orbits of the above-mentioned celestial bodies, the result being shown in Figure 5.16.

```
> polarplot([seq(s(a||i,e||i),i=0..5)],theta=0..2*Pi,
  scaling=constrained,thickness=2);
```

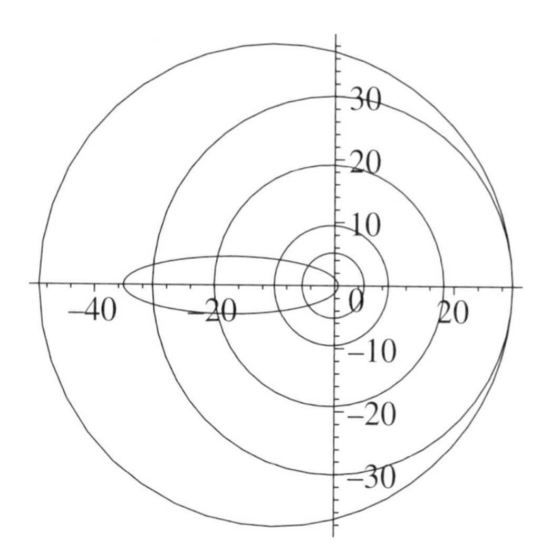

Figure 5.16: Orbits of Halley's comet and the planets Jupiter out to Pluto.

It should be noted that although the orbits are planar, they in fact do not all lie in the same plane. The elongated elliptical orbit that passes close to the sun at the origin and stretches out to between Neptune (second-largest orbit) and Pluto (largest orbit) is that of Halley's comet.

I notice that our lecture time is up, so if you have any questions bring them up at the beginning of the next lecture. In the meantime, you should start working on the next assignment (given below) based on this recipe."

PROBLEMS:

Problem 5-37: Asteroid Eros

The asteroid Eros has a semimajor axis $a=1.46$ and eccentricity $\epsilon=0.22$. Use the text file to create a plot of the orbits of Eros, Mercury ($a=0.39$, $\epsilon=0.21$), Venus ($a=0.72$, $\epsilon=0.0068$), Earth ($a=1.0$, $\epsilon=0.017$), and Mars ($a=1.52$, $\epsilon=0.093$) in the same graph.

Problem 5-38: Inverse cube law

Consider an object of mass m starting at $(x=R, y=0)$ with velocity V in the positive y-direction and moving under the influence of an attractive central force whose magnitude is given by $F=k/r^3$. Take $k=\alpha\, m\, R^2\, V^2$.

(a) Derive an analytic form for the orbit.

(b) What shape is the orbit for $\alpha=1$?

(c) Plot the orbit for $\alpha=1.01$, $R=1$, and $V=1$, and time $t=0$ to 9.999. What shape is the orbit?

Problem 5-39: Orbital Precession

According to Fowles and Cassiday [FC99], orbital precession occurs for the motion of a mass moving in the gravitational field of an oblate spheroid of mass M, the equation of motion being given by

$$\frac{d^2\vec{r}}{dt^2} = -\frac{G\,M}{r^2}\left(1 - \frac{\alpha\,G\,M}{c^2\,r}\right)\hat{r},$$

where c is the speed of light and α a small dimensionless constant.

(a) Taking $G\,M = 4\,\pi^2$, $\alpha/c^2 = 10^{-3}$, an initial radius $R = 1$, an initial tangential velocity $V = 1/\sqrt{10}$, and the integration constant equal to zero in the $r(\theta)$ integration, determine the analytic form of $r(\theta)$ and plot the orbit. Your plot should look like an n-leaf clover with n to be determined.

(b) How many orbital "leaves" occur for $V = 1/\sqrt{20}$? for $V = 1/\sqrt{30}$?

5.3.4 Wheel of misFortune

Praise without end the go-ahead zeal of whoever it was invented the wheel; but never a word for the poor soul's sake that thought ahead, and invented the brake.

Howard Nemerov, American poet and novelist (1920–1991)

Having arrived in Los Alamos on the weekend before an engineering conference begins, Russell decides to spend the day mountain biking down various trails in the vicinity. After driving up from Phoenix, he needs the exercise and, more importantly, the practice, since he intends to enter the local Jemez Mountain Bike Challenge in late May and another bike competition later in the summer in Telluride, Colorado. While zooming down a winding, dusty trail he fails to successfully fly over a small obstacle lying across his path and wipes out. The good news is that he survives the fall without incurring any broken bones and with only minor scrapes. The bad news is that he has put a sizeable bow in the rim of his front wheel and has to take it into the bicycle shop for straightening. Although he has had the misfortune to damage his wheel, he notes that at least he didn't lose the magnetic sensor attached to the spokes that measures his speed, distance, etc. While waiting in his motel room for his bike to be repaired, he flicks on the TV, but only game shows and comedy reruns are on. So, Russell decides to while away his time on his laptop computer by simulating the motion of a small, loose sensor sliding along a rotating bike spoke under different assumptions. Based on his own bike wheel, he takes the length L of a spoke to be 25 cm and the initial distance between the sensor (mass m) and the axis of rotation to be $r = r0 = 10$ cm. For simplicity, he considers the wheel to be rotating (with θ the angle of rotation) in the horizontal plane so the effect of gravity on the sensor can be ignored.

A call is first made to the plots package and the radial (a_r) and transverse (a_{tr}) acceleration components are entered in plane polar coordinates.

```
> restart: with(plots):
> a[r]:=diff(r(t),t,t)-r(t)*diff(theta(t),t)^2;
```

$$a_r := \frac{d^2}{dt^2} r(t) - r(t) \left(\frac{d}{dt} \theta(t) \right)^2$$

```
> a[tr]:=r(t)*diff(theta(t),t,t)+2*diff(r(t),t)*diff(theta(t),t);
```

$$a_{tr} := r(t) \left(\frac{d^2}{dt^2} \theta(t) \right) + 2 \left(\frac{d}{dt} r(t) \right) \left(\frac{d}{dt} \theta(t) \right)$$

Russell next creates a functional operator to calculate the angular coordinate ($\theta = \omega\, t + \frac{1}{2} \alpha\, t^2$) of the spoke on which the sensor is located for a given time t. Here ω is the initial angular velocity and α the angular acceleration.

```
> theta:=t->omega*t+(1/2)*alpha*t^2:
```

Applying Newton's second law in the radial direction to the sensor yields

$$m\, a_r = F_{\text{friction}} = -\mu\, N, \tag{5.11}$$

with the frictional force F_{friction} proportional to the normal force N, μ being the coefficient of friction. But for the rotating spoke, N is related to the transverse acceleration a_{tr} by $N = m\, a_{tr}$, so that $a_r = -\mu\, a_{tr}$, which Russell enters.

```
> ode:=a[r]=-mu*a[tr];
```

$$ode := \frac{d^2}{dt^2} r(t) - r(t)\, (\omega + \alpha\, t)^2 = -\mu \left(2 \left(\frac{d}{dt} r(t) \right) (\omega + \alpha\, t) + r(t)\, \alpha \right)$$

If $\mu = 0$ then the radial acceleration is zero. But since a_r is made up of two terms (see the lhs of *ode*), the radial coordinate, r, of the sensor can still change with time if ω and/or α is nonzero. Russell recalls solving the constant angular velocity case in an elementary mechanics course. To generate the relevant solution, he sets $\alpha = 0$ and $\mu = 0$, thus reducing *ode* to the form *ode1*.

```
> alpha:=0: mu:=0: ode1:=ode;
```

$$ode1 := \frac{d^2}{dt^2} r(t) - r(t)\, \omega^2 = 0$$

Assuming that the initial condition is that the sensor is at the radial position $r0$ and has no radial velocity at $t = 0$,

```
> ic1:=r(0)=r0,D(r)(0)=0:
```

ode1 is analytically solved with the Laplace transform option. This yields a trig (hyperbolic cosine) solution, rather than the default exponential form.

```
> rsol:=dsolve({ode1,ic1},r(t),method=laplace);
```

$$rsol := r(t) = r0 \cosh(\omega\, t)$$

Russell enters the spoke length $L = 25$ cm and the sensor's initial distance $r0 = 10$ cm from the axis of rotation. He takes the nominal angular velocity $\omega = 0.20\ \text{s}^{-1}$, and displays the radial solution.

```
> L:=25: r0:=10: omega:=0.20: rsol:=rhs(rsol);
```

$$rsol := 10 \cosh(0.20\, t)$$

The time T it takes the sensor to move from its initial position to the rim end of the spoke is numerically determined,

```
> T:=fsolve(rsol=L,t,0..10);
```

$$T := 7.833996185$$

and found to be about 7.8 seconds. Using the `polarplot` command, Russell plots the position of the sensor at 40 successive time steps between $t=0$ and T. Including the option `adaptive=false` switches off the default adaptive plotting scheme so the time interval between steps (circles) remains the same.

```
> pp:=polarplot([rsol,theta(t),t=0..T],style=point,symbol=
      circle,adaptive=false,coords=polar,numpoints=40):
```

The command `polarplot(L,color=blue)` plots the rim of the wheel as a blue circle, the plot being superimposed with `pp` to produce Figure 5.17. The scaling is constrained so that the rim is plotted as a circle, the original shape of the rim before Russell wiped out.

```
> display({pp,polarplot(L,color=blue)},scaling=constrained);
```

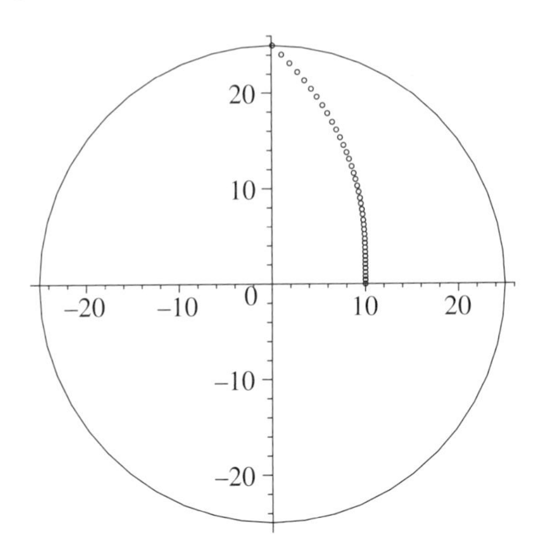

Figure 5.17: Successive sensor positions for the case $\mu=0$, $\omega=0.2$, and $\alpha=0$.

The successive positions of the sensor are clearly seen as it spirals out from its initial position to the circular rim. The spokes are omitted from the plot.

Looking at his watch, Russell wonders whether his bike rim has been straightened. On phoning the bike shop, he finds out that it will be another hour before the job is completed, so he returns to his calculation and decides to tackle the harder problem of a nonzero friction coefficient μ and a nonzero acceleration α. He unassigns these quantities along with ω and *r0*. Assuming that $\omega=0$,

```
> unassign('alpha','mu','omega','r0'): omega:=0:
```

the ODE then takes the form shown in *ode2*. For the zero-friction case, the sensor began to immediately slip at $t=0$. With friction present there will be a time delay before the sensor begins to slide outward. Until slipping begins, the radial position of the sensor will remain equal to a constant value, say k. Then *ode2* is evaluated with this value in *eq*.

```
> ode2:=ode; eq:=eval(ode2,r(t)=k);
```

$$ode2 := \left(\frac{d^2}{dt^2}\, r(t)\right) - r(t)\,\alpha^2\,t^2 = -\mu\,\left(2\,\left(\frac{d}{dt}\, r(t)\right)\,\alpha\,t + r(t)\,\alpha\right)$$

$$eq := -k\,\alpha^2\,t^2 = -\mu\,k\,\alpha$$

Then *eq* is solved to find the initial time $t = Ti$ at which the bead begins to slip outward along the spoke. Positive and negative square root solutions are generated. Since the time should be positive, the argument [1] is used to pick out the positive square root. At the time Ti that slipping begins, the radial position of the sensor is $r0$ and the radial velocity zero. These initial conditions are entered,

```
> Ti:=solve(eq,t)[1]; ic2:=r(Ti)=r0,D(r)(Ti)=0:
```

$$Ti := \frac{\sqrt{\alpha\,\mu}}{\alpha}$$

and the equation of motion *ode2* solved for $r(t)$ assuming that $\alpha > 0$, $\mu > 0$.

```
> rsol2:=simplify(dsolve({ode2,ic2},r(t)))
        assuming alpha>0,mu>0;
```

$$rsol2 := r(t) = r0\, e^{\left(-\frac{\mu\,(\alpha\,t^2 - \mu)}{2}\right)}\sqrt{t}\,\left(\mathrm{BesselI}\left(\frac{1}{4}, \%2\right)\mu\,\mathrm{BesselK}\left(\frac{1}{4}, \%1\right)\right.$$
$$+\mathrm{BesselI}\left(\frac{1}{4}, \%2\right)\mathrm{BesselK}\left(\frac{3}{4}, \%1\right)\sqrt{1+\mu^2} - \mathrm{BesselK}\left(\frac{1}{4}, \%2\right)\mu\,\mathrm{BesselI}\left(\frac{1}{4}, \%1\right)$$
$$\left. + \mathrm{BesselK}\left(\frac{1}{4}, \%2\right)\mathrm{BesselI}\left(\frac{-3}{4}, \%1\right)\sqrt{1+\mu^2}\right)\alpha^{(1/4)} \Bigg/ \left(\sqrt{1+\mu^2}\right.$$
$$\left.\left(\mathrm{BesselK}\left(\frac{1}{4}, \%1\right)\mathrm{BesselI}\left(\frac{-3}{4}, \%1\right) + \mathrm{BesselI}\left(\frac{1}{4}, \%1\right)\mathrm{BesselK}\left(\frac{3}{4}, \%1\right)\right)\mu^{(1/4)}\right)$$

$$\%1 := \frac{\sqrt{1+\mu^2}\,\mu}{2} \qquad \%2 := \frac{\alpha\,\sqrt{1+\mu^2}\,t^2}{2}$$

The answer is expressed in terms of two "special" functions, namely *modified Bessel functions* of the first kind (BesselI) and of the second kind (BesselK). If these functions are unfamiliar to you, you can learn more about them by highlighting, e.g., BesselK in the computer output, then clicking on Help, and finally on Help on BesselK. In "standard" mathematical notation, the modified Bessel functions of the first and second kinds appearing in the above solution would be written as $I_{1/4}(z)$ and $K_{1/4}(z)$, where z is one of the two subexpressions, %1 or %2. The subscript is referred to as the *order* of the Bessel function.

For plotting purposes, Russell takes $\alpha = 0.05$ s^{-2} and $\mu = 1.0$. The latter coefficient value is approximately that for steel sliding on steel. Setting $r0 = 10$ cm once again, the radial solution then is as follows.

```
> alpha:=0.05: mu:=1.0: r0:=10: rsol2:=rhs(rsol2);
```

$$rsol2 := 2.364354022\, e^{(-0.02500000000\, t^2 + 0.5000000000)} \sqrt{t}\ (1.873897406 \\ \text{BesselI}\left(1/4, 0.03535533905\, t^2\right) + 0.3595650184\, \text{BesselK}\left(1/4, 0.03535533905\, t^2\right))$$

The initial time *Ti* at which slipping begins is numerically determined, as well as the final time *Tf* at which the sensor reaches the rim.

```
> Ti:=Ti; Tf:=fsolve(rsol2=L,t,Ti..15);
```

$$Ti := 4.472135954 \qquad Tf := 12.97023282$$

The sensor starts to slide at about 4.5 seconds after rotational motion of the wheel begins, reaching the rim at 13 seconds.

To animate the motion of the sensor along a spoke, the time interval $Tf - Ti$ is divided into $N = 50$ equally spaced time steps and the time step size $\Delta = (Tf - Ti)/N$ calculated.

```
> N:=50: Delta:=(Tf-Ti)/N;
```

$$\Delta := 0.1699619373$$

The step size is 0.17 seconds. Functional operators `R` and `Theta` are formed to evaluate the radial and angular position of the sensor at time $t = Ti + n\,\Delta$, where the number n of the time step must be specified.

```
> R:=n->eval(rsol2,t=Ti+n*Delta):
> Theta:=n->eval(theta(t),t=Ti+n*Delta):
```

Russell calculates the angles, expressed in degrees, through which the spoke on which the sensor is located has rotated at the initial and final times.

```
> start_angle:=eval(theta(t),t=Ti)*180/evalf(Pi)*degrees;
```

$$start\_angle := 28.64788974\ degrees$$

```
> end_angle:=eval(theta(t),t=Tf)*180/evalf(Pi)*degrees;
```

$$end\_angle := 240.9673406\ degrees$$

The sensor begins to slip when the spoke has rotated through 29° and reaches the rim when the spoke has rotated through 241°.

To animate the motion of the sensor from the initial time of slipping until it reaches the rim, the solution is converted back into rectangular coordinates. A functional operator `F` is created to plot the location of the sensor as a size 16 black circle on the nth time step.

```
> F:=n->plot([[R(n)*cos(Theta(n)),R(n)*sin(Theta(n))]],
      style=point,symbol=circle,symbolsize=16,color=black):
```

A second operator `G` is formed to plot the spoke on which the sensor is located as a thick blue line on the nth time step.

```
> G:=n->plot([[0,0],[L*cos(Theta(n)),L*sin(Theta(n))]],
      color=blue,thickness=2):
```

The operator `P` superimposes the sensor on the spoke on the nth time step. The rim is plotted as well and the scaling is constrained.

```
> P:=n->display({F(n),G(n),polarplot(L)},scaling=constrained):
```

Making use of the insequence=true option, the following command line produces the animation. Click on the computer plot and on the start arrow in the tool bar to see the sensor move along the rotating spoke.

```
> display(seq(P(n),n=0..N),insequence=true);
```

Russell is pleased with how easy it was to solve this second, considerably harder, case and animate the results. However, he is even more pleased to learn that his bike is finally ready and he can zoom down the local trails once again.

PROBLEMS:

Problem 5-40: A more general case

Determine the analytic solution for the motion of the sensor when ω, α, and μ are all not equal to zero. Plot your answer for $r0 = 10$ cm, $L = 25$ cm, $\omega = 0.25$ s^{-1}, $\alpha = 0.05$ s^{-2}, and $\mu = 0.5$. What is the radial velocity of the sensor when it reaches the rim? and what is the total velocity?

Problem 5-41: Modified Bessel functions

Plot $I_n(z)$, $K_n(z)$ for a few real n values over suitable ranges of z. Calculate $dI_n(z)/dz$ and $\int z^{n+1} I_n(z)\, dz$ for arbitrary real n.

Problem 5-42: Bessel function solutions

Find the general solution of the following ODEs, each of which has a solution involving Bessel functions: **(a)** $x\,y'' - 3\,y' + x\,y = 0$; **(b)** $x\,y'' + (2x+1)\,(y'+y) = 0$; **(c)** $x\,y'' - y' - x\,y = 0$. Identify the type and order of each Bessel function.

Problem 5-43: Rayleigh's criterion and the Mafia boss

An escaped convict, intent on gaining revenge on the Mafia boss who framed him for murder, drives into an enclosed valley in Rainbow County and is blown up in a booby trap. One of the inhabitants of this valley, who lives 9.60 km from the scene of the explosion, claims that while sitting on his verandah he saw the twin headlights of the vehicle come over the ridge at the entrance to the valley just before the explosion took place. Using the information that the distance between the headlights was 1.52 m, that the diameter of the pupil of the eye is 3 mm, and that the mean wavelength of the light emitted by the headlights was 5.20×10^{-7} m, Pat, the forensic scientist, has deduced that the witness is lying and suspects him of being the Mafia boss. Your task is to confirm Pat's suspicions by answering the following questions:

(a) Light of wavelength λ and incident intensity I_0 on passing through a circular aperture of diameter a is diffracted with an intensity distribution I given by $I = I_0 (J_1(x)/x)^2$ with $x = \pi\, a \sin\theta/\lambda$. Here $J_1(x)$ is the first-order Bessel function of the first kind (entered as BesselJ(1,x)) and the angle θ is in rads. Confirm that most of the diffracted light is contained in the central maximum by plotting I/I_0 as a function of θ for $a = 4\,\lambda$.

(b) At what angle does the first minimum in the diffraction pattern occur? Show that this angle satisfies $a\,\sin\theta = 1.22\,\lambda$. This is a general result.

(c) Show that light arriving from two distant small sources (the two headlights) and passing through a small circular aperture (the pupil) cannot be distinguished if the angular separation is less than $\Delta\theta = 1.22\,\lambda/a$. Use this so-called *Rayleigh's criterion* to show that the witness was lying.

5.3.5 The Weedeater

Ignorance is an evil weed, which dictators may cultivate among their dupes, but which no democracy can afford among its citizens.
William Beveridge, British economist (1879–1963)

When Jennifer purchased her first house near the MIT campus, she discovered the joys of gardening, as well as more prosaic chores such as cutting and trimming the lawn and edging the flower beds. To carry out the latter function, she purchased a "weedeater," which consists of two nylon filaments that whirl rapidly in a circle and lop off the grass blades. The rotational motion of a single filament with its accompanying small transverse vibrations reminded Jennifer of the following problem [Mor48], whose solution she will now present.

A very light filament of length L and uniform mass density (mass per unit length) ρ is whirling in a horizontal circle about a pivot point ($x = 0$) at one end of the filament. Air drag is completely neglected. If the filament is slightly perturbed it can execute small vibrations transverse to the plane of rotation. What are the allowed transverse normal modes of vibration of the filament?

Because the filament is moving in a circle, each length element dx of the filament experiences a centripetal force pointing toward the center of the circle. This force is equal to the mass ($\rho\,dx$) of the element times the radial distance (x) from the pivot point times the square of the angular frequency (ν) of rotation. The tension T at arbitrary x along the filament is given by the sum of the forces on all the elements of the filament from x out to L and is now calculated.

```
> restart: with(plots):
> T:=Int(rho*nu^2*x,x=x..L)=int(rho*nu^2*x,x=x..L);
```

$$T := \int_x^L \rho\,\nu^2\,x\,dx = \frac{\rho\,\nu^2\,(L^2 - x^2)}{2}$$

Thus, the tension varies from zero at $x = L$ to its maximum value at $x = 0$. With the form of T determined, Jennifer considers the transverse vibrations of the filament, these vibrations being superimposed on its horizontal rotational motion. Equating the transverse force, due to T, to mass times acceleration, the transverse displacement $\psi(x,t)$ is given by the *partial differential equation*

$$\frac{\partial}{\partial x}\left(T\,\frac{\partial\psi}{\partial x}\right) = \rho\,\frac{\partial^2\psi}{\partial t^2}. \tag{5.12}$$

Jennifer now enters (5.12), the tension T being automatically substituted.

```
> pde:=diff(rhs(T)*diff(psi(x,t),x),x)=rho*diff(psi(x,t),t,t);
```

$$pde := -\rho\,\nu^2 x\left(\frac{\partial}{\partial x}\,\psi(x,t)\right) + \frac{1}{2}\,\rho\,\nu^2(L^2 - x^2)\left(\frac{\partial^2}{\partial x^2}\,\psi(x,t)\right) = \rho\left(\frac{\partial^2}{\partial t^2}\,\psi(x,t)\right)$$

To obtain a normal mode of frequency ω, $\psi(x,t)=X(x)\cos(\omega t)$ is assumed.

```
> psi(x,t):=X(x)*cos(omega*t);
```

$$\psi(x,\,t) := X(x)\cos(\omega\,t)$$

Then *pde* reduces to an ODE, which is divided by $\rho\,\nu^2\cos(\omega\,t)$ and simplified.

```
> ode:=simplify(pde/(rho*nu^2*cos(omega*t)));
```

$$ode := -x\left(\frac{d}{dx}X(x)\right) + \frac{1}{2}\left(\frac{d^2}{dx^2}X(x)\right)L^2 - \frac{1}{2}\left(\frac{d^2}{dx^2}X(x)\right)x^2 = -\frac{X(x)\,\omega^2}{\nu^2}$$

After the information level on the dsolve command is set to 2,

```
> infolevel[dsolve]:=2:
```

a general analytic solution is sought to *ode*.

```
> X:=rhs(dsolve(ode,X(x)));
```

Methods for second order ODEs:
— *Trying classification methods* —
....
–> *Trying a solution in terms of special functions:*
....
<– *Legendre successful*

$$X := \_C1\ \text{LegendreP}\left(\frac{\sqrt{\nu^2+8\,\omega^2}-\nu}{2\,\nu},\,\frac{x}{L}\right)+\_C2\ \text{LegendreQ}\left(\frac{\sqrt{\nu^2+8\,\omega^2}-\nu}{2\,\nu},\,\frac{x}{L}\right)$$

A solution is obtained in terms of special functions, namely a linear combination of *Legendre functions* of the first (LegendreP) and second (LegendreQ) kinds. The latter diverge to ∞ at $x=L$, which is unphysical, so they are removed.

```
> X:=remove(has,X,LegendreQ);
```

$$X := \_C1\ \text{LegendreP}\left(\frac{\sqrt{\nu^2+8\,\omega^2}-\nu}{2\,\nu},\,\frac{x}{L}\right)$$

For the filament, Jennifer takes $L = 10$ cm, or $1/10$ m. Since X depends on the ratio ω/ν, by choosing $\nu=1$ then ω can be interpreted as the "normalized" frequency. Since she is not interested in a general solution, Jennifer uses the operand command to remove the coefficient *_C1* from X.

```
> L:=1/10: nu:=1: X:=X/op(1,X);
```

$$X := \text{LegendreP}\left(\frac{\sqrt{1+8\,\omega^2}}{2}-\frac{1}{2},\,10\,x\right)$$

At $x=0$, $X=0$, which is entered as a boundary condition.

```
> bc:=subs(x=0,X)=0;
```

$$bc := \text{LegendreP}\left(\frac{\sqrt{1+8\,\omega^2}}{2}-\frac{1}{2},\,0\right) = 0$$

The allowed frequencies of the normal modes are obtained by solving the boundary condition for ω. To guide her in the numerical search, Jennifer plots the lhs of *bc* over the range $\omega=0$ to 10, the resulting picture being shown in Figure 5.18.

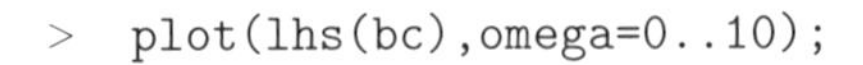

```
> plot(lhs(bc),omega=0..10);
```

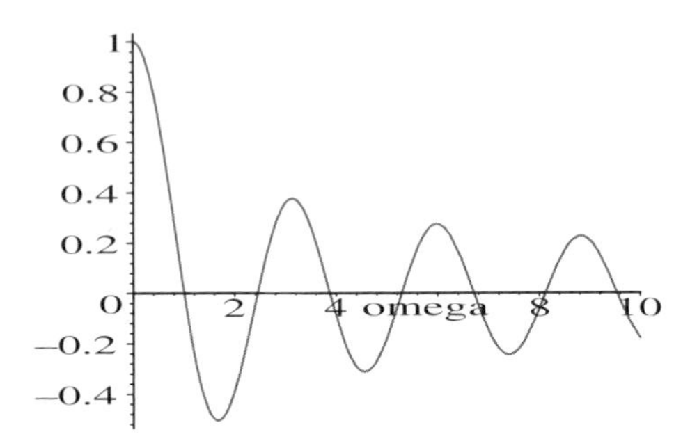

Figure 5.18: The zeros determine the allowed frequencies.

The eigenfrequencies are determined by the zeros. The lowest frequency occurs at $\omega = 1$, which yields $(1/2)\sqrt{1+8} - 1/2 = 1$ for the first argument in LegendreP. So, the lowest spatial mode is LegendreP$(1, 10\,x)$, or in standard mathematical notation, $P_1(10\,x)$. This suggests that the n values on the other allowed P_n will also be positive integers. P_2 is not possible, however, because its implies a zero at $\omega \approx 1.73$, which doesn't occur in Figure 5.18. The next zero is at $\omega \approx 2.45$, corresponding to P_3. This is confirmed in the following do loop where the ω_{2n+1} are found numerically and substituted into X to yield P_{2n+1} for $n = 0, 1, 2, 3$.

```
> for n from 0 to 3 do
> w[2*n+1]:=fsolve(op(1,X)=2*n+1,omega);
> P[2*n+1]:=subs(omega=w[2*n+1],X);
> end do;
```

$$w_1 := 1.000000000 \qquad P_1 := \mathrm{LegendreP}(1.000000000,\ 10\,x)$$

$$w_3 := 2.449489743 \qquad P_3 := \mathrm{LegendreP}(3.000000000,\ 10\,x)$$

$$w_5 := 3.872983346 \qquad P_5 := \mathrm{LegendreP}(5.000000000,\ 10\,x)$$

$$w_7 := 5.291502622 \qquad P_7 := \mathrm{LegendreP}(7.000000000,\ 10\,x)$$

These frequencies agree with the observed zeros in Figure 5.18. Any one of the normal modes may be animated, e.g., $P_7 \cos(w_7\,t)$ by choosing $n=3$.

```
> n:=3:  #select mode
> animate(P[2*n+1]*cos(w[2*n+1]*t),x=0..L,t=0..10,frames=100,
  numpoints=300,thickness=2,tickmarks=[2,0]);
```

PROBLEMS:

Problem 5-44: Planar vibrations

Derive the equation of motion for small vibrations in the plane of rotation. Determine the eigenfrequencies and normal modes. Plot the five lowest normal modes at $t = 0$ and animate one of them.

5.3.6 Can an Unstable Spring Find Stability?

The law must be stable, but it must not stand still.
Roscoe Pound, *Introduction to the Philosophy of Law* (1999)

A unit mass is fastened to the origin by a linear spring whose spring constant is $k(t) = \sin(r\,t)/(2+\cos(r\,t))$, r being a positive parameter. The mass is constrained to move along the x-axis, subject to the initial conditions $x(0)=1$ and $\dot{x}(0)=0$. Our goal is to show that a critical value $r_{\rm cr}$ exists such that for $r < r_{\rm cr}$, the motion is unstable (unbounded growth occurs), while for $r > r_{\rm cr}$ the motion is stable. After $r_{\rm cr}$ is found, a phase-plane trajectory will be plotted for r just below and just above $r_{\rm cr}$. Even though the relevant ODE (viz., $\ddot{x}(t)+k(t)\,x(t)=0$) is linear in x, it does not have a closed-form analytic solution, so a numerical answer must be obtained for $x(t)$.

After the initial condition, `ic`, is entered an operator `ode` is formed to generate the ODE (with $k(t)$ substituted) for a specified r.

```
> restart: ic:=x(0)=1,D(x)(0)=0:
> ode:=r->diff(x(t),t,t)+(sin(r*t)/(2+cos(r*t)))*x(t)=0;
```

$$ode := r \to \frac{d^2}{dt^2}x(t) + \frac{\sin(r\,t)\,x(t)}{2+\cos(r\,t)} = 0$$

An operator `sol` is introduced to numerically solve the ODE for a given r. The option `output=listprocedure` allows us to evaluate the displacement $x(t)$ and the velocity $v = dx(t)/dt$ at an arbitrary time t.

```
> sol:=r->dsolve({ode(r),ic},x(t),numeric,output=listprocedure):
```

The operator `X` evaluates $x(t)$ at $t=T$ for a specified r. For example, `X(1,300)`

```
> X:=(r,T)->eval(x(t),sol(r))(T): X(1,300);
```

$$0.461690284540523871\;10^{25}$$

yields $x(T=300) \approx 0.46 \times 10^{25}$ for $r=1$. The large number is indicative of an unstable solution. Using `X`, the log of the absolute value of $x(T=300)$ is plotted

```
> plot([seq([i/200,ln(abs(X(i/200,300)))],i=1..300)]);
```

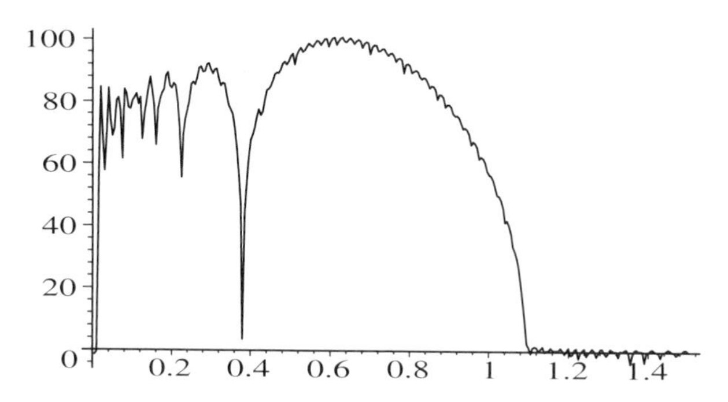

Figure 5.19: Vertical scale: ln(abs(x(300))). Horizontal scale: r.

as a function of r up to $r=1.500$ in steps of size $1/200=0.005$. The resulting picture is shown in Figure 5.19. From the figure, the transition from instability to stability is seen to occur at $r_{\rm cr} \approx 1.10$. To confirm this interpretation, let's plot the trajectory of the unit mass in the phase ($x(t)$ vs. $v(t)$) plane. The operator V evaluates the velocity $v(t)=dx(t)/dt$ at $t=T$ for a specified r.

```
> V:=(r,T)->eval(diff(x(t),t),sol(r))(T):
```

Then, a graphing operator **gr** is formed to plot x vs. v over the time range $T=0$ to 300 for a specified r. To obtain a smooth trajectory the number of plotting points is taken to be 1000. Graphs are then produced for $r=1.09$ and $r=1.10$, the resulting pictures being shown in Figure 5.20.

```
> gr:=r->plot([X(r,T),V(r,T),T=0..300],numpoints=1000,
          labels=[x,v]):
> gr(1.09); gr(1.10);
```

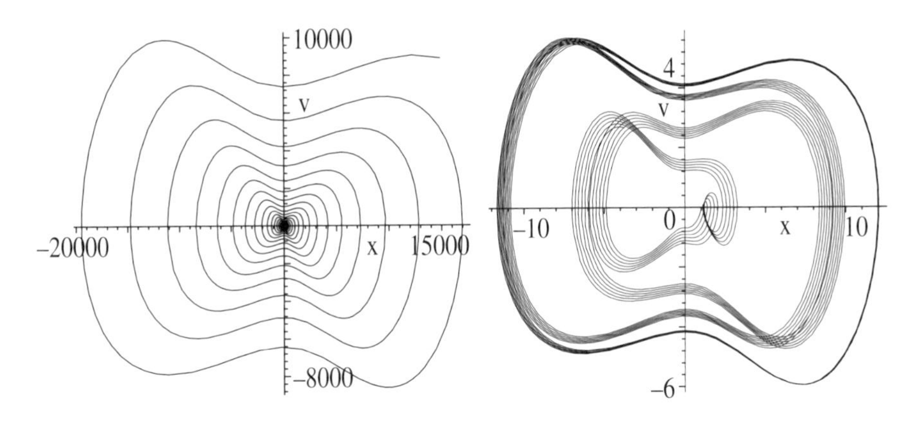

Figure 5.20: Left: Unstable solution for $r=1.09$. Right: Stable for $r=1.10$.

For $r=1.09$, the trajectory displays unbounded growth, unwinding off the origin as time increases. For $r=1.10$, a bounded oscillatory solution results.

PROBLEMS:

Problem 5-45: Other parameter values

Explore the stability and nature of the solution in the text recipe for other values of the parameter r, in particular in the neighborhood of the downward spike at $r=0.38$. Discuss your results.

Problem 5-46: Other time-dependence

Explore other time-dependence of $k(t)$ and discuss your results.

Chapter 6

Difference Equation Models

Everybody is ignorant, only on different subjects.
Will Rogers, American humorist (1879–1935)

If the independent variable, e.g., the time t, in a dynamic model is not continuous but characterized by finite time intervals or steps, the model will be described by one or more *difference equations*. Difference equations can be found in many areas of science.

For example, in a biological context, oceangoing salmon return to their original freshwater streams every 4 years to lay their eggs and die. The fish biologist might record the salmon population at each of these 4-year intervals. If $t=0$ corresponds to the present year, $t=1$ to one time interval (4 years here) later, and so on, the fish population number N_{t+1} at time $t+1$ will be related to the number N_t at time t by a difference equation of the structure

$$N_{t+1} = f(N_t), \tag{6.1}$$

where f is a mathematical function that is either created phenomenologically to account for the observed fish numbers or produced from first principles. For other animal population counts, the time interval will generally be different and f could also depend on what happened two or more intervals ago.

Similarly, in the world of physics, the experimentalist might record measurements at regular time intervals. In the world of finance, the Dow Jones industrial average and the price of stocks are recorded at the end of each trading day and reported in the financial pages of the daily newspapers. All of these situations could be modeled by difference equations.

As with ODEs, there exist both linear and nonlinear difference equation models. Linear models are those in which the equation is linear in the dependent variable(s). Linear models can be analytically solved with Maple, using the recurrence equation solve (`rsolve`) command. We shall demonstrate this for a number of first- and second-order difference equations.

Nonlinear models, on the other hand, cannot generally be solved analytically. In this case, we can proceed by iterating the nonlinear difference equation using a do loop construction. Several interesting nonlinear models will be presented.

6.1 Linear Models

A general pth-order linear difference, or recurrence, equation in the dependent variable x typically is of the structure

$$a_0\,x_{n+p} + a_1\,x_{n+p-1} + \cdots + a_p\,x_n = h_n, \tag{6.2}$$

relating the value of x in a given generation $n+p$ to its value p generations earlier. If not in this "standard" form, any difference equation can be transformed into it by a suitable change of the subscripts. Most commonly, science students encounter first- and second-order difference equations, i.e., equations with $p=1$ and $p=2$, viz.,

$$\begin{aligned} a_0\,x_{n+1} + a_1\,x_n &= h_n, \\ a_0\,x_{n+2} + a_1\,x_{n+1} + a_2\,x_n &= h_n. \end{aligned} \tag{6.3}$$

If $h_n = 0$, the difference equation is said to be *homogeneous*. Otherwise it is *inhomogeneous*. If all of the coefficients $a_0, a_1, \ldots$ are constant, the standard approach is to assume a solution of the form $x_n = \lambda^n$. For example, on substituting the assumed form into a second-order homogeneous equation and simplifying, this procedure yields the quadratic equation

$$a_0\,\lambda^2 + a_1\,\lambda + a_2 = 0 \tag{6.4}$$

in λ, which has in general two roots λ_1 and λ_2. The general solution then is the linear combination $x_n = A\,(\lambda_1)^n + B\,(\lambda_2)^n$ with two arbitrary constants A and B. The two constants are determined by specifying two of the x values, for example, x_0 and x_1. For an inhomogeneous equation, the solution will be made up of a *particular solution* to account for the extra term plus the solution to the homogeneous equation. Maple has a linear recurrence equation solver that can take care of solving linear difference equations, whether they be homogeneous or inhomogeneous.

6.1.1 Those Dratted Gnats

Float like a butterfly, sting like a bee.
Muhammad Ali, American boxer (1942–)

On the planet Erehwon, there exists a colony of ferocious gnats, feared because of their beelike sting, which live in a particularly swampy region, referred to as the Big Bad Bog. The normalized number x_n of gnats in the nth generation is governed over a limited time interval by a difference equation of the structure

$$x_{n+1} = 2\,x_n + n^2. \tag{6.5}$$

The first term on the right-hand side represents the increase in population number from generation n to generation $n+1$ due to the natural birth rate. In this case the population number would double in each generation if no predators were present. The n^2 term represents the rapid influx (immigration) of gnats into the Big Bad Bog region from other areas. The difference equation in this case is a first-order, linear (x occurs only to the first power), inhomogeneous (because of the n^2 term) equation.

Assuming that the model is valid for $N+1$ generations (with $N=5$) and that the normalized number of gnats initially is $x_0 = 1$,

```
> restart: x[0]:=1: N:=5:
```

how many gnats are there in generations one, two, ..., six?

A general approach, one that we will be forced to use for almost all nonlinear difference equation models, is to iterate the recursive relation. A do loop is formed, the difference equation is iterated from $n=0$ to $N=5$, and the output gnat population numbers displayed. (To save on text space, they are placed on the same line here, instead of vertically.)

```
> for n from 0 to N do
> x[n+1]:=2*x[n]+n^2;
> end do;
```

$$x_1 := 2 \quad x_2 := 5 \quad x_3 := 14 \quad x_4 := 37 \quad x_5 := 90 \quad x_6 := 205$$

Thus in the sixth generation there are 205 times as many gnats as there were initially. By studying the structure of the numbers, one could try to come up with a formula telling us what the population number x_n should be as a function of n. A better approach is to assume that the solution will be of the structure $x_n = (\lambda)^n + ps$, where the particular solution, ps, and λ remain to be determined. An even better approach is to let Maple do the analytical work for us. First, we unassign n and x,

```
> unassign('n','x'):
```

and enter the difference equation. Note how parentheses (round brackets) are used in the following command line to enclose the generation numbers.

```
> eq:=x(n+1)=2*x(n)+n^2;
```

$$eq := x(n+1) = 2\,x(n) + n^2$$

The difference equation eq is solved for x in the nth generation, using the recurrence equation solve (`rsolve`) command, given the initial condition $x_0 = 1$.

```
> x:=rsolve({eq,x(0)=1},x);
```

$$x := 4\ 2^n - 1 - 2\,(n+1)\left(\frac{n}{2}+1\right) + n$$

In this case, we can identify $\lambda = 2$ in the first term of x. If there was no immigration, so that the n^2 term was not present in eq, the solution to the homogeneous equation will be $x_n = 2^n$. The factor of 4 (multiplying 2^n) in the first term of x and the remaining terms represent the effect of immigration.

The solution x can be simplified by applying the `expand` command.

```
> x:=expand(x);
```

$$x := 4\ 2^n - 3 - n^2 - 2\,n$$

As one can see, the structure of x is a nontrivial function of n. As a check, let's use this analytic expression to calculate the gnat number x_6.

```
> n:=6: number:=x;
```

$$number := 205$$

The formula correctly produces the same number for x_6. Now, the gnat number x_n could be plotted as a function of n using the already known plotting procedures. But let's introduce a new command structure for graphing the solution of linear difference equations. Again unassigning n and x,

```
> unassign('n','x'):
```

the linear recurrence equation (LRE) tools library package is loaded.

```
> with(LREtools):
```

The recurrence equation plot (REplot) command is used to generate Figure 6.1.

```
> REplot(eq,x(n),{x(0)=1},1..6,labels=["n","x"],tickmarks=
  [3,2],style=point,symbol=circle,symbolsize=12,color=blue);
```

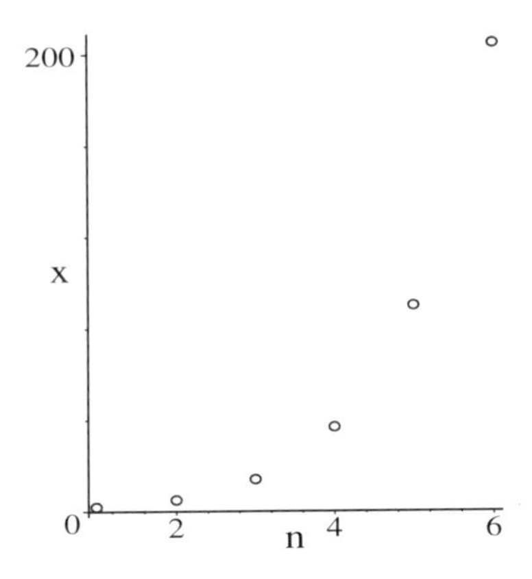

Figure 6.1: Number of gnats in the nth generation.

The explosive growth of the unchecked gnat population is clearly evident. Those Erehwonians living on the edge of the Big Bad Bog are in for a particularly nasty summer season unless something is done about those dratted gnats.

PROBLEMS:

Problem 6-1: Classification

For each of the following difference equations, state the order of the equation, identify whether the equation is linear or nonlinear, and whether it is homogeneous or otherwise:

(a) $x_{n+1} + 4\,x_{n-1} = 1/n$; **(b)** $x_{n+3} + (2/x_{n+2}) = n\,x_{n+2}$;
(c) $x_{n+4} + 3\,x_{n-1} = (n-2)^2$; **(d)** $x_{n+2} + 4\,x_n^2 = 0$;
(e) $x_{n-1} + \sin x_n = 1$; **(f)** $n\,x_{n-2} + n^2\,x_{n-3} = 2$.

Problem 6-2: Larger immigration
Replace the influx term in the text recipe with n^3 and execute the worksheet. Compare your results with the text results.

Problem 6-3: A nonlinear gnat population model
In the text recipe, replace the $2\,x_n$ term with $2\,x_n^2$. Using the do loop approach, calculate the population numbers in generations 1 to 6, given $x_0 = 1$. Attempt to obtain a general analytic solution using the `rsolve` command. Comment on the results.

Problem 6-4: Some difference equations
Solve the following first-order difference equations, identifying whether or not they are homogeneous:

(a) $x_{n+1} - n^2\,x_n = 0, \quad x_1 = 1;$

(b) $(n+1)\,x_{n+1} - n\,x_n = n^2, \quad x_1 = 1.$

Plot the solutions in each case over a suitable range of n.

Problem 6-5: Puffin explosion
A population of puffins (a type of seabird) on a northern island increases by 20% per year by natural growth and by 20 birds per year due to immigration.

(a) Write down the difference equation for the population number N_t after t years.

(b) Use the `rsolve` command to find the general analytic solution.

(c) Use the `REplot` command to plot the puffin population for the first 12 years if initially there are 100 puffins.

(d) What is the number of puffins on the island in the twelfth year?

Problem 6-6: Erehwon swamp fever
Each year, 1000 new cases of Erehwon swamp fever, a debilitating illness, occur and half of the existing cases are cured. At the end of the year 1990, there were 1200 cases of the disease.

(a) Write down the relevant difference equation.

(b) Use the `rsolve` command to find the general analytic solution at the end of year n.

(c) How many cases of swamp fever were there at the end of 2000?

(d) Use the `REplot` command to plot the number of cases over the period 1990 to 2000.

(e) What is the equilibrium number of cases as $n \to \infty$?

(f) What if there had been 3000 cases in 1990. What is the equilibrium number in this case?

6.1.2 Gone Fishing

Angling may be said to be so like the mathematics that it can never be fully learnt.
Isaac Walton, *The Compleat Angler* (1593–1683)

As a second example of a first-order difference equation, let's return to Earth and look at a problem of interest to fish biologists. The fish population in one of the Great Lakes initially consists of $N = 1$ million fish.

```
> restart: with(LREtools): N:=10^6:
```

The natural growth rate is such that in the absence of fishing, the fish population would increase by one-third each year. In units of 1 million fish, the linear difference equation describing the change in fish number x from year t to $t+1$ would be of the structure

$$x_{t+1} - x_t = \frac{1}{3} x_t \quad \text{or} \quad x_{t+1} = \frac{4}{3} x_t. \tag{6.6}$$

If the fish were harvested at exactly the same rate as their natural growth rate, the fish population would remain constant. However, suppose that current fishing regulations permit the harvesting of 350 thousand fish per year. In units of 1 million fish, the difference equation for the fish number would then be

$$x_{t+1} = R\, x_t - h \tag{6.7}$$

with $R = 4/3$ and $h = 0.35$ million fish.

```
> R:=4/3: h:=0.35:
> eq:=x(t+1)=R*x(t)-h;
```

$$eq := x(t+1) = \frac{4}{3}\, x(t) - 0.35$$

As a summer student working for the Fisheries Department, Heather is asked by her supervisor to answer the following questions related to the above model:

(a) What is the analytic formula for the normalized fish number x_t?

(b) How many years would it take before the fish population is depleted to less than half a million fish?

(c) How many years would it take before the fish population in the lake is wiped out?

(d) How many fish would remain in the year just before extinction?

In addition, her supervisor asks Heather to create a point plot showing the fish numbers up to the year of extinction if the harvesting policy were maintained.

Heather uses the `rsolve` command with $x(0) = 1$ million fish to solve the difference equation eq for $x(t)$.

```
> x:=rsolve({eq,x(0)=1},x(t));
```

$$x := -\frac{\left(\frac{4}{3}\right)^t}{20} + \frac{21}{20}$$

By setting $x = 0.5$ in the fsolve command, the time t for the population to decrease to less than 0.5 million fish,

```
> half_life:=fsolve(x=0.5,t);
```

$$half\_life := 8.335226635$$

is found to be slightly more than 8 years. Using the same command, but now solving for the time at which $x = 0$,

```
> extinction_time:=fsolve(x,t);
```

$$extinction\_time := 10.58294113$$

the extinction time is between 10 and 11 years. The number remaining after 10 years is obtained by taking $t = 10$, using the analytic formula for x and multiplying the resulting number by $N = 10^6$. The numerical value is rounded off to the nearest integer (nearest whole fish).

```
> t:=10: number_after_10_years:=round(evalf(x)*N);
```

$$number\_after\_10\_years := 162114$$

So 162,114 fish would remain after 10 years. To create a plot over the time interval 0 to 10 years, Heather unassigns x and t,

```
> unassign('x','t'):
```

and uses the REplot command to produce Figure 6.2.

```
> REplot(eq,x(t),{x(0)=1},0..10,style=POINT,symbol=circle,
  symbolsize=16,color=blue,labels=["year","number"],
  view=[0..10,0..1],tickmarks=[3,3]);
```

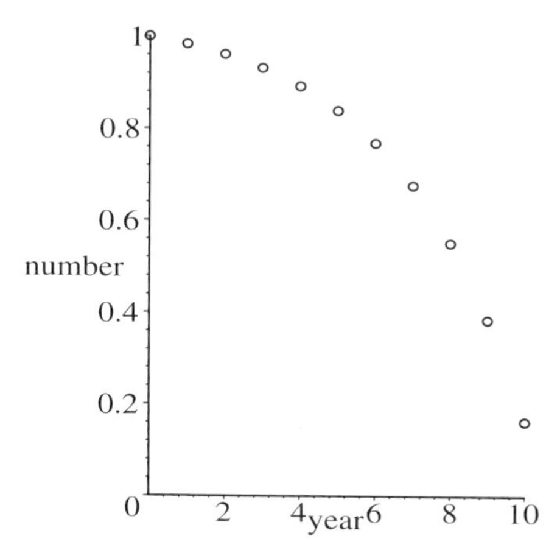

Figure 6.2: Number of fish (in millions) in the lake after t years.

On successfully completing her task, Heather is rewarded with a fishing weekend on the departmental boat, which is unofficially called the *Petty Bureaucrat*.

PROBLEMS:

Problem 6-7: Harvesting whales

For a certain species of whale, the initial population is 1000 whales and in the absence of harvesting increases by 20% per year. In the first year, 100 whales are harvested, but it is proposed to increased this number by 25 per year (125 harvested in the second year, 150 in the third year, and so on).

(a) Write down the relevant difference equation.

(b) Solve the difference equation analytically.

(c) Use the `fsolve` command to determine when the whale population is 500.

(d) Make a plot of the analytic solution using the `REplot` command over a time interval that allows you to graphically answer the following two questions.

(e) When does the maximum whale population occur?

(f) If the proposed harvesting policy were maintained, in what year would the whale population become extinct?

Problem 6-8: The latest news on those gnus

A herd of rare Erehwonian gnus, initially numbering 500 animals, increases by 10% each year due to the normal birth and death rate. If 20 gnus join the herd each year from other areas of Erehwon but $5\,t^2$ leave in year t:

(a) Write down the difference equation and solve it analytically.

(b) Make a plot of the analytic solution using the `REplot` command.

(c) In what year is the gnu population a maximum?

(d) How many gnus are there in this year?

(e) In what year do the gnus become extinct?

Problem 6-9: Hunting Erehwonian bandicoots

A population of pernicious Erehwonian bandicoots (a ratlike animal) initially numbers 1000 and naturally grows by 50% per year. If the population is reduced by hunting at the rate of 400 per year the first year, and the rate is increased by 12% each successive year (448 "removed" in the second year, etc.):

(a) What is the general difference equation?

(b) Use the `rsolve` command to determine the analytic solution.

(c) Use the `REplot` command to plot the bandicoot population over the time range that the bandicoot population exists.

(d) When does the bandicoot population reach a maximum number?

(e) When do the bandicoots become extinct?

6.1.3 Fibonacci's Adam and Eve Rabbit

Population, when unchecked, increases in a geometric ratio. Subsistence increases only in an arithmetical ratio.
Thomas Malthus, English clergyman and political economist (1766–1834)

In the year 1202, the Italian mathematician Leonardo of Pisa, more commonly known as Fibonacci,[1] began the study of population growth by proposing and solving a problem posed in terms of rabbit numbers that bears his name to this day. Let R_n be the number of pairs of rabbits in generation n. In generation 0, no rabbits exist ($R_0 = 0$), but in generation 1, Adam and Eve Rabbit, the ultimate father and mother of all rabbits, are spontaneously created. So, $R_1 = 1$. Fibonacci postulated that rabbits reach breeding maturity after 1 month and in each subsequent month each pair produces precisely one more pair (male and female). How many rabbit pairs are there in the nth generation, assuming that none die?

In the second month, Adam and Eve have reached maturity and produce one pair in the third month. Thus, $R_2 = 1$ and $R_3 = 2$. The new pair cannot breed for another month, so $R_4 = 3$ due solely to the efforts of Adam and Eve. In the fifth month, two offspring are produced so that $R_5 = 5$, and so on. The numbers in the sequence $0, 1, 1, 2, 3, 5, \ldots$ are called the *Fibonacci numbers.* As the reader may verify, they satisfy the second-order (homogeneous) linear recurrence, or difference equation,

$$R_{n+2} = R_{n+1} + R_n \tag{6.8}$$

for $n = 0, 1, 2, 3, \ldots$ and $R_0 = 0$, $R_1 = 1$. Although one could keep on iterating the difference equation, we shall seek an analytic solution. To be somewhat more general, let's assume that there are N isolated rabbit colonies, each with their own Adam and Eve, so that $R_1 = N$.

The LREtools library package is accessed so that `REplot` can be used.

```
> restart:with(LREtools):
```

The Fibonacci difference equation is now entered,

```
> eq:=R(n+2)=R(n+1)+R(n);
```

$$eq := \mathrm{R}(n+2) = \mathrm{R}(n+1) + \mathrm{R}(n)$$

and solved for $R(n)$ subject to the two initial conditions $R(0) = 0$ and $R(1) = N$.

```
> sol:=rsolve({eq,R(0)=0,R(1)=N},R(n));
```

$$sol := \frac{\sqrt{5}\,N\left(\frac{1}{2}+\frac{\sqrt{5}}{2}\right)^n}{5} - \frac{N\sqrt{5}\left(\frac{1}{2}-\frac{\sqrt{5}}{2}\right)^n}{5}$$

The reader should be able to identify the roots λ_1 and λ_2 as well as the coefficients A and B in the solution of the homogeneous equation. The solution could be left in the above form, or if desired it can be factored.

[1] "Son of good nature."

```
> number:=factor(sol);
```

$$number := \frac{\sqrt{5}\,N\left(\left(\frac{1}{2}+\frac{\sqrt{5}}{2}\right)^n-\left(\frac{1}{2}-\frac{\sqrt{5}}{2}\right)^n\right)}{5}$$

Suppose that there is only one ($N=1$) colony of rabbits and we want to know how many rabbit pairs there would be in the 24th generation (2 years later).

```
> N:=1; n:=24;
```

$$N := 1 \qquad n := 24$$

With the above input values, the number of rabbit pairs is given by

```
> Rabbit_pairs:=radnormal(number);
```

$$Rabbit\_pairs := 46368$$

Note the use of the `radnormal` command, which performs normalization of expressions containing algebraic numbers in radical notation. Leave it off and see what the answer would look like. So, in the 24th generation, there would be about 46,000 rabbit pairs! Of course, in the real world Fibonacci's oversimplified model must be modified to include such factors as the natural and unnatural death rates, the effects of overcrowding and depletion of food supplies, etc.

To plot the number of rabbit pairs as a function of generation n, we must `unassign` n so that it does not retain the value of 24 used above.

```
> unassign('n'):
```

Using the `REplot` command, and taking the default line style, Figure 6.3 is generated, showing R_n over the range $n=1$ to 24.

```
> REplot(eq,R(n),{R(0)=0,R(1)=1},1..24,
  labels=["n","R"],tickmarks=[3,3]);
```

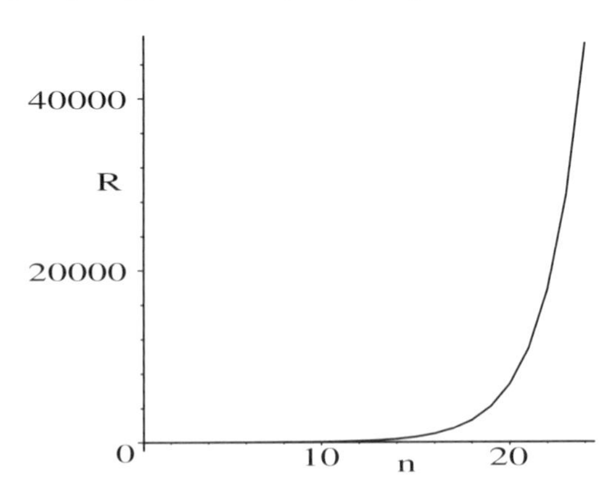

Figure 6.3: Number of rabbit pairs generated in the nth generation.

Using a line style in some cases can prove a bit deceptive, so if you prefer a point style, add the necessary options to the `REplot` command.

Because the Fibonacci sequence appears in many different science problems, it should be noted that Maple also has a built-in Fibonacci number generator, which can be invoked by loading the combinatorial function library package with the fibonacci subpackage.

```
> with(combinat,fibonacci):
```

For $n=24$, the `fibonacci` command,

```
> Fibonacci_number:=fibonacci(24);
```

$$Fibonacci\_number := 46368$$

yields exactly the same number, 46,368, of rabbit pairs as before.

PROBLEMS:

Problem 6-10: Malthus's quotation

Discuss the quotation by Thomas Malthus given at the start of this subsection, particularly the part about subsistence.

Problem 6-11: General solutions

Determine the general analytic solution of the following second-order equations:

(a) $x_{n+1} + 2\,x_{n-1} = 1/n$;

(b) $x_{n+1} - 2\,x_n \cosh(\phi) + x_{n-1} = 0$.

Problem 6-12: Specified initial conditions

Analytically solve the following second-order difference equations, subject to $x_0=1$ and $x_1=2$, and plot each solution over the range $n=0$ to 20.

(a) $3\,x_{n+2} - 6\,x_{n+1} + 4\,x_n = 0$;

(b) $4\,x_{n+2} + 4\,x_{n+1} + x_n = 0$.

Describe the behavior of each solution as n is increased.

Problem 6-13: Inhomogeneous equations

Analytically solve the following inhomogeneous equations, subject to $x_0=1$ and $x_1=2$, and plot each solution over the range $n = 0$ to 20 using a line style.

(a) $3\,x_{n+2} - 6\,x_{n+1} + 4\,x_n = \cos(n)$;

(b) $4\,x_{n+2} + 4\,x_{n+1} + x_n = n$.

Problem 6-14: Chebyshev polynomials

The first two Chebyshev polynomials are $T_0(x)=1$ and $T_1(x)=x$. The remaining polynomials for $n \geq 2$ can be found by solving the difference equation

$$T_{n+1}(x) - 2\,x\,T_n(x) + T_{n-1}(x) = 0.$$

(a) Solve the difference equation for the nth-order Chebyshev polynomial.

(b) Explicitly determine the Chebyshev polynomials corresponding to $n =$ 2, 3, 4, 5. You may have to do some algebraic manipulation to obtain the simplest forms of the polynomials.

(c) Comment on using Maple here instead of doing the problem by hand.

Problem 6-15: Population pressure

A wild goat population is decreasing because of the limited resources available. If the decrease in population number between the year $(n-1)$ and the year n is one-quarter the number in the year $(n-2)$:

(a) Write down the relevant difference equation for the wild goats.

(b) Assuming that the goat number is 20,000 in year 0 and 19,500 in year 1, use the `rsolve` command to find the number in year n.

(c) Determine the number of wild goats in year 8.

(d) Use the `REplot` command to plot the number between year 0 and year 8.

Problem 6-16: Third-order equation

Consider the third-order linear difference equation

$$x_{n+3} - 6\,x_{n+2} + 11\,x_{n+1} - 6\,x_n = 0.$$

(a) Determine the general solution of this difference equation.

(b) Taking $x_0=1$, $x_1=-2$, and $x_3=3$, plot the difference equation over the range $n=0$ to 4.

Problem 6-17: Flu epidemic

It is noted by an epidemiologist that during the spread of a flu epidemic, the number N of new cases occurring during the nth week is equal to twice the number of cases that existed at the end of week $(n-2)$.

(a) Write down the relevant difference equation for the flu epidemic.

(b) Determine the general solution of the difference equation.

(c) Determine the analytic solution when $N_0=N_1=1$.

(d) Plot the analytic solution over an appropriate time interval.

(e) Now assume that during the nth week half the cases that existed at the end of the previous week are cured. You may still assume that the number of new cases during that week is $2\,N_{n-2}$. Write down the new difference equation and obtain an analytic solution given that $N_0=1$ and $N_1=4$.

Problem 6-18: Legendre polynomials

The Legendre polynomials $P_n(x)$ satisfy the linear difference equation

$$(n+1)\,P_{n+1}(x) - (2\,n+1)\,x\,P_n(x) + n\,P_{n-1}(x) = 0,$$

with $P_0(x)=1$ and $P_1(x)=x$.

(a) Can the difference equation be solved using the `rsolve` command? The answer yes or no is insufficient. You must prove it.

(b) Using the `solve` command and a do loop construction, determine the Legendre polynomials for $n=2, 3, \ldots, 10$.

6.1.4 How Red Is Your Blood?

I have nothing to offer but blood, toil, tears and sweat.
Winston Churchill, wartime speech in the British House of Commons (1940)

In *Mathematical Models in Biology*, the mathematical biologist Leah Edelstein-Keshet [EK88] presents a simple difference equation model describing the number of red blood cells (RBCs) circulating in the blood. Since the RBCs carry oxygen throughout the body, their number must remain more or less constant.

```
> restart: with(plots):
```

Let R_n be the number of RBCs present in the blood on day n. The number R_{n+1} present on day $n+1$ will depend on two factors, the rate at which they are removed by the spleen and the rate at which they are produced by the bone marrow. If f is the fraction of RBCs removed by the spleen and M_n the number produced by the marrow on day n, then R_{n+1} is given by *eq1*.

```
> eq1:=R(n+1)=(1-f)*R(n)+M(n);
```

$$eq1 := R(n+1) = (1-f)\,R(n) + M(n)$$

If γ, which we now unprotect, is the number of RBCs produced per number lost,

```
> unprotect(gamma):
```

then the number of RBCs produced by the marrow on day $n+1$ is

```
> eq2:=M(n+1)=gamma*f*R(n);
```

$$eq2 := M(n+1) = \gamma\, f\, R(n)$$

Here, we have a system of two first-order linear difference equations. By letting $n \to n+1$ in *eq1* and substituting *eq2*, the system could be replaced by the second-order difference equation,

$$R_{n+2} = (1-f)\,R_{n+1} + \gamma\, f\, R_n. \tag{6.9}$$

Although, based on our experience with the Fibonacci equation, Equation (6.9) is readily solvable with Maple, let's work with the first-order system instead. The two coupled first-order equations are entered as a Maple set and solved for the set of two unknowns R_n and M_n. To avoid a lengthy page-consuming display, the output has been suppressed.

```
> sol:=rsolve({eq1,eq2},{R(n),M(n)}):
```

We shall only display the RBC number or count, R_n, on day n. To do this, the solution (`sol`) is assigned so that R_n is given in terms of $R(0)$ and $M(0)$, the initial conditions on day zero.

```
> assign(sol): RBC:=R(n);
```

$$RBC := -2\,\gamma\, f(-\%3\,\mathrm{R}(0) + \%3\, f\,\mathrm{R}(0) + \%3\,\mathrm{R}(0)\,\sqrt{\%1} - 2\,\%3\,\mathrm{M}(0) + \%2\,\mathrm{R}(0) - \%2\, f\,\mathrm{R}(0) + \%2\,\mathrm{R}(0)\,\sqrt{\%1} + 2\,\%2\,\mathrm{M}(0))\,/(\sqrt{\%1}\,(-1+f-\sqrt{\%1})(-1+f+\sqrt{\%1}))$$

$$\%1 := 1 - 2\,f + f^2 + 4\,\gamma\, f \quad \%2 := \left(\frac{2\,\gamma\, f}{-1+f+\sqrt{\%1}}\right)^n \quad \%3 := \left(\frac{2\,\gamma\, f}{-1+f-\sqrt{\%1}}\right)^n$$

To see what the complicated appearing solution gives, let's take the following nominal values.

```
> R(0):=1: M(0):=1: f:=1/2: gamma:=1: N:=15:
```

$R_0 = 1$ and $M_0 = 1$ could represent the normalized values of those quantities. We have assumed that $f = 1/2$, so that one-half of the RBCs are removed by the spleen. To attain a fixed level of RBCs, γ was set equal to one, so that the number of RBCs produced equals the number lost. The RBC number will be plotted up to day 15. The formula for the RBC count can then be simplified to yield the following compact result:

```
> RBC:=simplify(RBC);
```

$$RBC := \frac{(-1)^{(n+1)}\, 2^{(-n)}}{3} + \frac{4}{3}$$

For plotting purposes, the sequence command is used to make a list of lists for the $N+1$ plotting points.

```
> plotting_points:=[seq([n,RBC],n=0..N)]:
```

The graph is formed, but not displayed. To guide the reader's eye, the plotting points are joined by straight blue lines.

```
> Graph:=plot(plotting_points,style=line,color=blue):
```

The variation in the RBC count is now displayed in Figure 6.4.

```
> display(Graph,view=[0..N,0..2],labels=["n","RBC"],
  tickmarks=[3,3]);
```

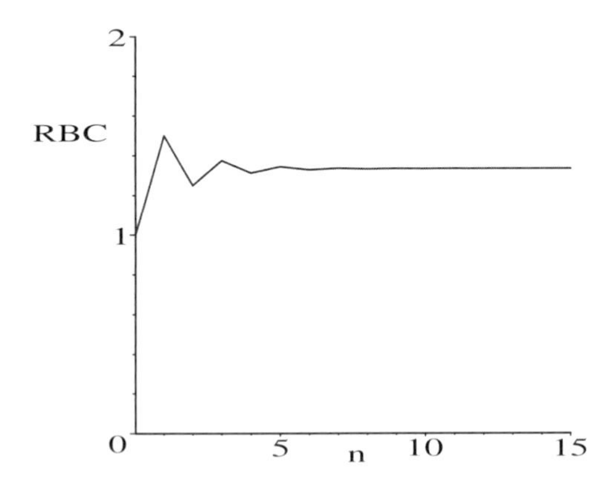

Figure 6.4: Red blood cell (RBC) count on day n.

After a transient period, the RBC count settles down to the constant value 1.33. If γ is not equal to one, the RBC count will either increase for $\gamma > 1$ or decrease for $\gamma < 1$. The reader can experiment with different parameter values.

According to Edelstein-Keshet, the description of RBC production can be made more accurate by treating the time as continuous, i.e., introducing an ODE model, and including a time delay.

PROBLEMS:

Problem 6-19: Solving the second-order RBC equation

Taking $f = 1/2$ and all other parameters as in the text, analytically solve the equivalent second-order RBC equation for the RBC number and plot the result.

Problem 6-20: Blood CO_2

According to Edelstein-Keshet, there is a steady production of CO_2 in the blood that results from the basal metabolic rate. On the other hand, CO_2 is lost by way of the lungs at a ventilation rate controlled by CO_2-sensitive chemoreceptors in the brain stem. In suitably normalized units, an oversimplified model yields the following equation for the blood CO_2 concentration C_n at time n:

$$C_{n+1} - C_n + a\, C_{n-1} = m,$$

where m is the constant production rate of CO_2 and a is a positive parameter.

(a) Determine the general analytic solution of the difference equation. Identify the particular solution.

(b) Determine the behavior of the solution for $4\,a < 1$ and $4\,a > 1$, taking initial conditions and parameter values of your own choice. Plot representative solutions for both cases. If a is large enough, show that the oscillations may increase in magnitude.

Problem 6-21: System of equations

Find the analytic solution of the following coupled difference equations,

$$x_{n+1} = -4\, y_n + 2\, x_n, \quad y_{n+1} = 2\, y_n + x_n,$$

with $x_0 = -3$ and $y_0 = 0$. Plot the solutions over a suitable range of n.

6.1.5 Fermi–Pasta–Ulam Is Not a Spaghetti Western

I wouldn't say when you've see one Western you've seen the lot; but when you've seen the lot you get the feeling you've seen one.
Katherine Whitehorn, English journalist (1928–)

Heather, who is enrolled in an undergraduate science program with the intention of going on to medical school, has been reading a popularized account of some historical developments in nonlinear dynamics that involve concepts beyond her current mathematical background. Her older mathematician sister, Jennifer, does not have any lectures to give this afternoon, so Heather drops into her sister's office and asks her to explain one of the topics that she has been reading about, namely the Fermi–Pasta–Ulam problem.

"Well," says Jennifer, "the mathematical details of this topic are quite involved, but as I recall, it basically involves the exchange of energy between the normal modes of a one-dimensional atomic lattice due to the inclusion of nonlinearities in the force law between the atoms. It was thought that, consistent with the zeroth law of thermodynamics, the energy exchange would ultimately lead to an equipartition of energy between the modes in a time-average sense. Enrico Fermi, the Nobel physics laureate, and his collaborators used the MANIAC I computer at Los Alamos to attempt to verify this conjecture numerically. Much to their surprise, the atomic system that they were simulating did not approach equilibrium, and their goal was to understand what was going on. I can see by the frown on your face that something about my explanation is bothering you."

"I am afraid that you lost me right at the beginning. I am not even sure what a normal mode is. Even though I have taken a number of physics and math courses, my premed program has concentrated more on biochemistry and biology," replies Heather.

"OK, I will tell you something about the normal modes of vibration in a one-dimensional atomic lattice governed by a linear force law. I could carry out the relevant derivation of the normal modes by hand, but I would also like to show you how a computer algebra system can save you a lot of work in this problem. But first a little bit of history. According to the classical mechanics text by Marion and Thornton [MT95], a mathematically similar problem was first studied as early as 1687 by that intellectual giant of his time (indeed of any time), Isaac Newton. Newton considered the small vibrations of an elastic string loaded with regularly spaced identical particles. The problem was again pursued by John Bernoulli and his son Daniel, beginning around 1727. In fact, out of this investigation came the formulation of the principle of linear superposition by Daniel in 1753. So you can see that the vibrational problem that Fermi, Pasta, and Ulam studied in the 1950s has a long and illustrious history behind it. Before you can begin to really understand the Fermi–Pasta–Ulam problem, you first have to know what happens in an atomic lattice modeled by linear interactions of the atoms with their nearest neighbors.

Consider the following oversimplified model of a one-dimensional solid made up of atoms all of the same type. The atoms, each of mass m, are regularly spaced an equilibrium distance d apart as shown in Figure 6.5. The complicated

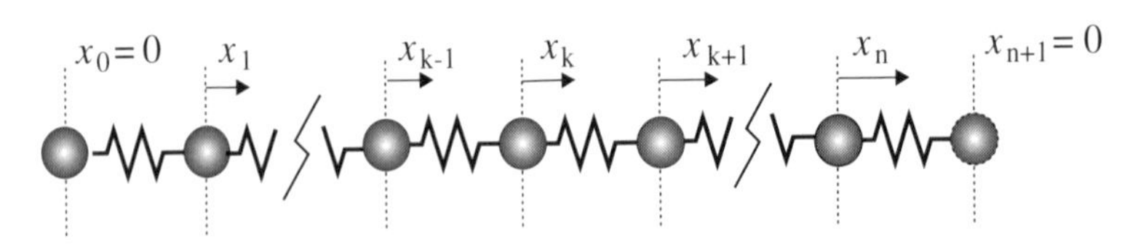

Figure 6.5: The one-dimensional atomic lattice.

electrical forces holding the solid together are approximated by identical linear springs between each of the atoms. The electrical forces drop rapidly with distance, so it is a good approximation to assume that a given atom interacts only with its neighbors. The restoring force, when an atom is displaced by a small amount x from equilibrium, is assumed to be given by Hooke's law, $F = -K\,x$, where K is the spring constant. The linear array of atoms shown in the figure is numbered from $k = 0$ to $k = n + 1$. The end atoms, $k = 0$ and $n + 1$, are pinned so that they cannot move, thus leaving n atoms, i.e., those from $k = 1$ to n, to vibrate. The general vibrational motion will be a linear superposition of longitudinal and transverse oscillations. For the former, the atoms vibrate back and forth along the direction of the chain, while for the latter they oscillate perpendicularly to the chain direction. We shall only explicitly consider longitudinal vibrations in our development, but the results for transverse oscillations are of a similar mathematical structure.

Consider the kth atom ($k = 1, 2, \ldots, n$), which has two nearest neighbors, $k-1$ and $k+1$. Let x_k, x_{k-1}, and x_{k+1} be their displacements from equilibrium. The net force on atom k will depend on the displacement of atom $k+1$ relative to k and on the displacement of atom k relative to $k-1$. If $x_{k+1} > x_k$, there will be a force contribution to the right in the figure. On the other hand, if $x_k > x_{k-1}$, there will be a force contribution to the left. Thus, Newton's second law yields the following equation of motion for the kth atom,

$$K\,(x_{k+1} - x_k) - K\,(x_k - x_{k-1}) = m\,\ddot{x}_k, \tag{6.10}$$

or, on setting $\kappa \equiv K/m$ and rearranging,

$$\kappa\,(x_{k-1} - 2\,x_k + x_{k+1}) = \ddot{x}_k, \quad k = 1, 2, \ldots, n. \tag{6.11}$$

To determine the displacement of the kth atom from equilibrium, we assume a solution to this *differential–difference equation* of the form $x_k = a_k \cos(\omega\,t)$, where a_k is the amplitude and ω the frequency. Since k can be any one of n atoms, this solution corresponds to all n atoms vibrating with the same frequency ω. This special solution is referred to as a *normal mode* of oscillation. Actually, Heather, we shall find that for n vibrating atoms, n different normal mode solutions are possible each with a different characteristic frequency, ω_i. These frequencies, with $i = 1, 2, \ldots, n$, are referred to as the *eigenfrequencies*. The normal modes are important because any motion of the chain will simply be a linear superposition of normal modes, the exact mixture of modes depending on how the n atoms of the chain are excited at time $t = 0$. With our assumed solution, Equation (6.11) reduces to the second-order linear difference equation

$$-\kappa\,a_{k-1} + 2\,\kappa\,a_k - \kappa\,a_{k+1} = \omega^2\,a_k \equiv \lambda\,a_k, \quad k = 1, 2, \ldots, n, \tag{6.12}$$

connecting the n amplitudes. Starting with $k = 1$, and remembering that $a_0 = a_{n+1} = 0$, since the end atoms are not allowed to move, the recurrence relation yields the following set of n equations:

$$\begin{aligned}
2\,\kappa\,a_1 - \kappa\,a_2 &= \lambda\,a_1, \\
-\kappa\,a_1 + 2\,\kappa\,a_2 - \kappa\,a_3 &= \lambda\,a_2, \\
\vdots \qquad\qquad & \quad \vdots \\
-\kappa\,a_{n-1} + 2\,\kappa\,a_n &= \lambda\,a_n.
\end{aligned} \tag{6.13}$$

Although we could solve this system of equations directly, it is more convenient (particularly for large n) and probably more instructive for me to express this coupled set in the following matrix form,

$$A\,V = \lambda\,I\,V, \tag{6.14}$$

where A is an n by n matrix, V a column vector with the a_n as the elements, I the identity matrix, and λ is referred to as the *eigenvalue*. You have covered matrices, haven't you?"

"Yes, I have taken a linear algebra course, so even though I am somewhat rusty I am following your explanation."

"OK, let me continue. After constructing the forms of A and V, we shall determine the eigenfrequencies and normal modes of vibration. A call is made to the LinearAlgebra package. For the sake of definiteness, let's take $n = 6$, i.e., 6 atoms will be allowed to oscillate. I will animate the third mode ($M = 3$).

```
> restart: with(plots): with(LinearAlgebra): n:=6: M:=3:
```

The column vector V is formed, but not displayed.

```
> V:=<seq(a[k],k=1..n)>:
```

The *tridiagonal* matrix A is created using the BandMatrix command. Each entry of the central diagonal is equal to $2\,\kappa$, while the entries on the first subdiagonals adjacent to the central diagonal are $-\kappa$. All other entries are 0. The 1 in the command indicates that there is one subdiagonal and n is the matrix size (6×6 here).

```
> A:=BandMatrix([-kappa,2*kappa,-kappa],1,n);
```

$$A := \begin{bmatrix} 2\,\kappa & -\kappa & 0 & 0 & 0 & 0 \\ -\kappa & 2\,\kappa & -\kappa & 0 & 0 & 0 \\ 0 & -\kappa & 2\,\kappa & -\kappa & 0 & 0 \\ 0 & 0 & -\kappa & 2\,\kappa & -\kappa & 0 \\ 0 & 0 & 0 & -\kappa & 2\,\kappa & -\kappa \\ 0 & 0 & 0 & 0 & -\kappa & 2\,\kappa \end{bmatrix}$$

By calculating the matrix equation $A\,V = \lambda\,I\,V$, using the shorthand dot notation, we regain Equation (6.13) for $n=6$, confirming our matrix formulation.

```
> eq:=A . V=(lambda*IdentityMatrix(n)) . V;
```

$$eq := \begin{bmatrix} 2\,\kappa\,a_1 - \kappa\,a_2 \\ -\kappa\,a_1 + 2\,\kappa\,a_2 - \kappa\,a_3 \\ -\kappa\,a_2 + 2\,\kappa\,a_3 - \kappa\,a_4 \\ -\kappa\,a_3 + 2\,\kappa\,a_4 - \kappa\,a_5 \\ -\kappa\,a_4 + 2\,\kappa\,a_5 - \kappa\,a_6 \\ -\kappa\,a_5 + 2\,\kappa\,a_6 \end{bmatrix} = \begin{bmatrix} \lambda\,a_1 \\ \lambda\,a_2 \\ \lambda\,a_3 \\ \lambda\,a_4 \\ \lambda\,a_5 \\ \lambda\,a_6 \end{bmatrix}$$

To obtain numerical values for the eigenfrequencies, we shall choose a nominal value for κ, namely $\kappa = 1.0$. The decimal zero is included to express the eigenfrequencies in floating-point form.

```
> kappa:=1.0:
```

Then, the eigenvalues of A can be determined using the `Eigenvalues` command. The output is expressed as a list, and the list entries are ordered from the smallest to the highest by applying the `sort` command.

```
> lambda:=sort(Eigenvalues(A,output='list'));
```

$$\lambda := [0.1980622642, 0.7530203963, 1.554958132, 2.445041868, 3.246979604, 3.801937736]$$

The eigenfrequencies are calculated using the following `map` command, which applies the square root to each operand of λ.

```
> omega:=map(sqrt,lambda);
```

$$\omega := [0.4450418679, 0.8677674782, 1.246979604, 1.563662965, 1.801937736, 1.949855824]$$

As I told you earlier, for $n = 6$ vibrating atoms there are six different eigenfrequencies. A normal mode solution will be associated with each frequency. All six normal modes are now obtained using the following do loop, which runs from $i=1$ to n.

```
> for i from 1 to n do
```

Equation (6.12) is entered for the ith eigenvalue.

```
> eq[i]:=-kappa*a(k-1)+2*kappa*a(k)-kappa*a(k+1)=lambda[i]*a(k):
```

Setting the amplitudes at the end of the chain to be zero, the recurrence relation can be solved for the kth amplitude using the `rsolve` command,

```
> sol[i]:=rsolve({eq[i],a(0)=0,a(n+1)=0},a(k));
```

and the do loop ended.

```
> end do:
```

The following command line will produce all the normal modes. Specifically, using a "nested" pair of sequence commands, the displacement of the atoms $k = 1, 2, \ldots, n$ are generated for each normal mode $i = 1, 2, \ldots, n$ at time t. The `radnormal` and `evalf` commands help to simplify the results (not displayed here because of the length). Since it turns out that the displacements are all expressed as a multiple of the displacement a_n, we can divide this arbitrary scale factor out.

```
> nms:=evalf(seq([seq(radnormal(sol[i]/a(n)),k=1..n)]
        *cos(omega[i]*t),i=1..n));
```

I will select only a particular normal mode, say the third one ($M=3$). We can change the M value at the beginning of the recipe if we want to examine some other normal mode.

```
> mode:=nms[M];
```

$$mode := [1.000000000, 0.4450418677, -0.8019377359, -0.8019377357, 0.4450418680, 1.] \cos(1.246979604\, t)$$

The above list gives the amplitude for atoms 1 to $n=6$ in the linear chain for the third normal mode. The list is multiplied by a cosine function with the appropriate eigenfrequency present. To animate the third normal mode, I will separate the atoms by an equilibrium spacing d. The spacing will be taken to be 4 times the maximum amplitude in the list. To accomplish this the `remove` command is used to first remove the cosine term from *mode*. Inclusion of the term [] also removes the list brackets. The maximum amplitude is obtained by applying the `max` command, and the number is then multiplied by 4.

```
> d:=4*max(remove(has,mode,cos)[]);
```

$$d := 4.000000000$$

The equilibrium horizontal coordinates of the atoms are then obtained.

```
> equilpos:=[seq(k*d,k=1..n)]:
```

The `ScatterPlot` command will be used to plot the atomic positions, so the Statistics library package is loaded.

```
> with(Statistics):
```

An operator `gr` is formed to plot the atoms as size-20 circles. The lists `h` and `v` of horizontal and vertical coordinates, and the color `c`, must be specified.

```
> gr:=(h,v,c)->display(ScatterPlot(h,v,symbol=circle,
                 symbolsize=20),color=c):
```

Using `gr`, the stationary end atoms are plotted as red circles.

```
> gr1:=gr([0,(n+1)*d],[0,0],red):  #end atoms
```

Let's animate the normal mode for $T=50$ time units, and create $N=200$ time frames. The time step size then is equal to $T/N=1/4$ time units.

```
> T:=50: N:=200: step:=T/N;
```

$$step := \frac{1}{4}$$

An operator `gr2` is created to plot the internal mobile atoms as blue circles on the ith time step.

```
> gr2:=i->gr(equilpos+eval(mode,t=step*i),[seq(0,k=1..n)],
         blue):
```

Still another operator, `pl`, will be used to superimpose graphs of the stationary end atoms and the internal mobile atoms on the ith time step.

```
> pl:=i->display({gr1,gr2(i)}):
```

Using the `insequence=true` option, the `display` command will produce an animation of the normal modes. I will also remove the coordinate axes.

```
> display(seq(pl(i),i=0..N),insequence=true,axes=none);
```

Finally let's execute the above command line, click on the computer plot and on the start arrow, and see what the motion of the atoms looks like for the third normal mode. What do you think of my animation?"

"That's pretty impressive, Jennifer."

"If you want, it might be a good idea to experiment with different numbers of atoms and different normal modes to get a good feeling for what is going on. However, before you do so, let me return to your original question about the

Fermi–Pasta–Ulam problem. If the displacements x of the atoms from equilibrium are not small, but still symmetric about $x = 0$, the force law can be taken of the form $F = -K\,x - \beta\,x^3$, with $\beta > 0$. Why are nonlinear terms required, you might ask? If the force law is linear ($\beta = 0$), then we have seen that normal modes occur. If a particular normal mode is excited, all of the energy remains in that normal mode. To exchange energy between normal modes some sort of coupling between the modes must be present. Nonlinear contributions to the force law can provide this coupling. Unfortunately, this leads to coupled nonlinear differential–difference equations, so the problem can no longer be handled analytically. We would have to resort to numerical simulation. Do you still want me to pursue this problem for you?"

"Thanks a lot for your help, Jennifer, but I think I will pass on that. It would be beyond my mathematical background, and further, all this thinking has tired me out and made me hungry! Let's go to the Pizza Palace and get something to eat. I'm treating today."

PROBLEMS:

Problem 6-22: First variations on the text example

Animate all the other normal modes for the text recipe and discuss the observed behavior of the mobile atoms.

Problem 6-23: Second variation on the text example

Using the text recipe, investigate the normal modes for atomic chains of different lengths and discuss your results.

Problem 6-24: Transverse vibrations

Modify the text recipe to handle transverse vibrations of the mobile atoms.

Problem 6-25: Vibrations of the CO_2 molecule

The carbon dioxide (CO_2) molecule is a linear symmetric array of three atoms with the carbon (C) atom located between the two oxygen (O) atoms. Let the spring constant between the C atom and either of its O neighbors be K, the mass of an oxygen atom be m, and the mass of the C atom be M. All three atoms are free to vibrate away from equilibrium along the atomic chain direction.

(a) Modify the text program and show that the eigenfrequencies are $\omega_1 = 0$, $\omega_2 = \sqrt{\kappa}$, and $\omega_3 = \sqrt{\kappa + 2\,\kappa'}$, where $\kappa = K/m$ and $\kappa' = K/M$.

(b) Calculate the frequency ratio ω_3/ω_2 for CO_2, for which $m/M = 16/12$.

(c) Solve the relevant equations for the amplitudes and plot each normal mode for a set of representative times and then animate the normal modes.

6.2 Nonlinear Models

... a study of very simple nonlinear difference equations ... should be part of high school or elementary college mathematics courses. They would enrich the intuition of students who are currently nurtured on a diet of almost exclusively linear problems.
R. M. May and G. F. Oster, mathematical biologists (1976)

With the above quotation of May and Oster in mind, we think that we would be remiss if we didn't provide you with at least a small sampling of first- and second-order nonlinear difference equations. For almost all nonlinear difference equations of physical interest, the `rsolve` and `REplot` commands are of no help. Analytic solutions generally are not obtainable and we must resort to a do loop procedure to systematically generate numerical solutions that can be plotted. In this section, the examples will range from population growth, to the mechanics of a bouncing ball, to a discussion of chaos and the outbreak of war.

6.2.1 Competition for Available Resources

We will now discuss in a little more detail the Struggle for Existence.
Charles Darwin, English naturalist (1809–1882)

Ecology [BHT90] is the scientific study of the distribution and abundance of biological organisms due to their interactions with each other and their environment. *Interspecific competition* is the competition for existence between different species, while *intraspecific competition* refers to the competition between members of a single species due to a finite supply of available resources (e.g., food). Difference equations can be used to model both interspecific and intraspecific competition for species characterized by discrete breeding seasons. In this and the following subsection, we shall look at two models of intraspecific competition that have been much studied in the biology and mathematics literature. The first model is due to Maynard-Smith and Slatkin [MSS73], [MS74] and will be referred to from now on as the MSS model.

To understand the origin of the MSS model, let us first consider the situation in which the net reproductive rate R of a single species is constant with respect to time. This rate coefficient takes into account the births of new individuals minus the deaths of existing ones. If factors such as immigration and harvesting can be ignored, the number density N_t at time t for the species satisfies the linear difference equation

$$N_{t+1} = R\,N_t, \tag{6.15}$$

with $t = 0, 1, 2, \ldots$ time units. This first-order equation is easily solved by iterating. If N_0 is the population at $t = 0$, then at $t = 1$, $N_1 = R\,N_0$, at $t=2$, $N_2 = R\,N_1 = R^2\,N_0$, and so on. Thus, the solution at arbitrary time t is $N_t = R^t\,N_0$. The same analytic answer follows on applying Maple.

Loading the requisite library packages and setting the initial condition to be $N(0)=N_0$, the difference equation *eq1* is solved with the `rsolve` command.

```
> restart: with(LREtools): with(plots): ic:=N(0)=N[0]:
> eq1:=N(t+1)=R*N(t);
```

$$eq1 := N(t+1) = R\,N(t)$$

```
> N_t:=rsolve({eq1,ic},N(t));
```

$$N\_t := N_0\,R^t$$

If $R > 1$, the population will grow for $t > 0$ because births exceed deaths, whereas if $R < 1$ it will decrease. The plotting operator `pl` uses the `REplot` command to create plots of the growth curves over the time range $t = 0$ to 5 for $R = 0.5\,i$, given $N(0) = 10$. If we take $i = 1, 2, 3$, the growth curves will correspond to $R = 0.5$, 1, and 1.5. Each curve will have a different line style, $i=1$ generating a solid curve, $i=2$ a dotted curve, and $i=3$ a dashed line.

```
> pl:=i->REplot(subs(R=0.5*i,eq1),N(t),{N(0)=10},0..5,
        color=blue,thickness=2,linestyle=i):
```

Using the `textplot` command, the values $R=0.5$, 1, 1.5 will be added to the resulting picture.

```
> pl4:=textplot([[2.7,40,"R=1.5"],[2.7,15,"R=1"],
      [2.7,5,"R=0.5"]]):
```

Using the sequence command, graphs are generated for $i = 1, 2, 3$, which are superimposed on `pl4` with the display command. The resulting picture is shown in Figure 6.6.

```
> display({seq(pl(i),i=1..3),pl4},labels=["t","N"]);
```

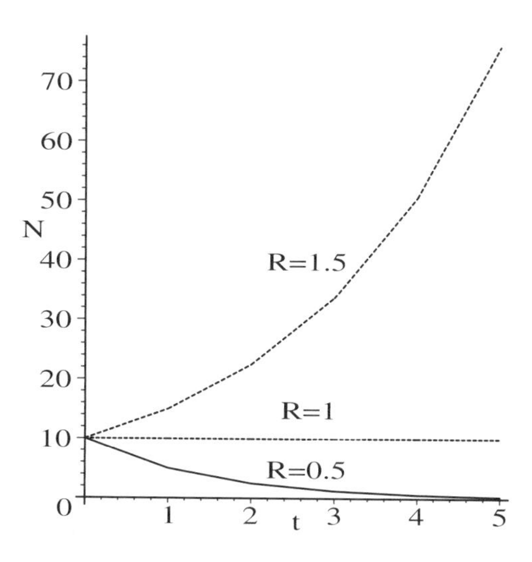

Figure 6.6: Population growth curves for three different R values.

For $R = 1.5$, the population number grows rapidly, whereas for $R = 0.5$ the population curve decays quickly to zero, since deaths exceed births. The value $R = 1$ corresponds to zero net growth.

This linear population model does not take into account any competition between its members for available resources, so the growth increases indefinitely for any $R > 1$. However, as the growing population puts more and more pressure on the usually finite resources, the rate coefficient may begin to decrease with time due to intraspecific competition. The constant R, which applies when no competition is present, must be replaced with a number-density-dependent coefficient $R(N_t)$, so that the governing equation is of the form

$$N_{t+1} = R(N_t)\, N_t. \tag{6.16}$$

Different choices of mathematical structure for $R(N_t)$ lead to different models. Nonlinear models occur when $R(N_t)$ depends explicitly on N_t. One such model is the MSS model, which has been successfully applied to intraspecific competition in beetle populations and certain other biological species. In the MSS model, $R(N_t)$ is taken to be of the form

$$R(N_t) = \frac{R}{1 + (a\, N_t)^b}, \tag{6.17}$$

where R is the reproductive rate with no competition, and a and b are positive parameters which can be determined by experimental observation.

To understand the origin of this form of $R(N_t)$, assume $R > 1$ and look at the ratio $N_t/N_{t+1} = 1/R(N_t)$. In Figure 6.7, we have taken N_t/N_{t+1} as the vertical axis and N_t as the horizontal axis. A horizontal line is drawn in the figure at a height corresponding to $N_{t+1} = N_t$, i.e., at a height 1.

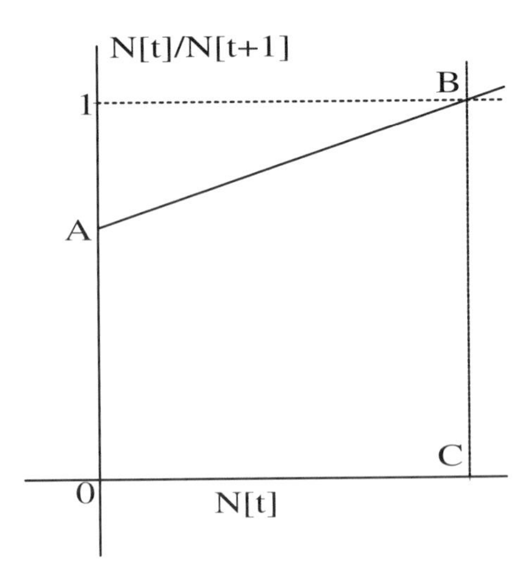

Figure 6.7: First step in "building" the MSS model.

When the population size is very small, i.e., N_t virtually zero, the competition is negligible, so $R(N_t)$ will be equal to R. Thus, the intercept point A on the vertical axis has a value equal to $1/R$. As the population number N_t increases, the effect of competition is to make $R(N_t)$ decrease or its reciprocal increase. The simplest phenomenological model for $1/R(N_t)$ is a straight-line model, i.e., write $1/R(N_t) = (1/R)(1 + a\,N_t)$, with $1/R$ the intercept value and a/R the slope. Then $N_t/N_{t+1} = 1/R(N_t)$ will intersect the horizontal line at some point B, the corresponding value of N_t being labeled C. The value of N_t for which the population number N_t stays fixed ($N_{t+1} = N_t$) is called the *carrying capacity* of that population. From Figure 6.7, the slope of the straight line is $a/R = (1 - 1/R)/C$, so that the straight-line model yields

$$N_{t+1} = \frac{RN_t}{1 + aN_t}, \quad \text{with } a = (R - 1)/C. \tag{6.18}$$

But using a straight line turns out to be quite limiting in the description of experimentally observed intraspecific competition. Maynard-Smith and Slatkin's model results on replacing the straight-line model $1/R(N_t) = (1/R)(1 + aN_t)$ with the curve $1/R(N_t) = (1/R)(1 + (a\,N_t)^b)$, with $b \geq 0$ and $a = (R - 1)/C$. Figure 6.8 shows three representative curves generated for the MSS model for

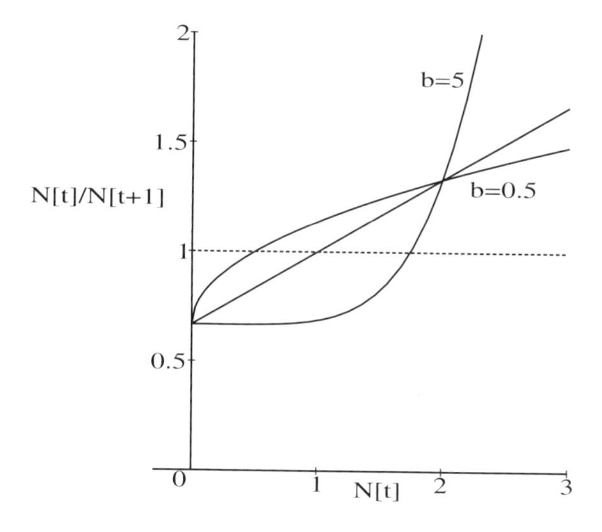

Figure 6.8: Determining the carrying capacity for $b = 0.5$, 1, and 5.

$R = 1.5$, $C = 1$, and therefore $a = 0.5$. The curves correspond to choosing $b = 5$, 1, and 0.5. The horizontal line $N_t/N_{t+1} = 1$ is also drawn. The value of N_t where each curve intersects the horizontal line is the carrying capacity for that b value. For $b = 1$, which is the straight-line model, the carrying capacity is $N_t = 1$, i.e., the input value of C. For $b = 5$, the carrying capacity is larger than C, having the value $N_t = 1.74$, while for $b = 0.5$, the carrying capacity is less than C.

Taking the same initial condition as before, we can try to solve the MSS difference equation, $N_{t+1}=R\,N_t/(1+(a\,N_t)^b)$, using the `rsolve` command.

```
> eq2:=N(t+1)=R*N(t)/(1+(a*N(t))^b);
```

$$eq2 := N(t+1) = \frac{R\,N(t)}{1+(a\,N(t))^b}$$

```
> rsolve({eq2,ic},N(t));
```

$$\text{rsolve}\left(\left\{\text{N}(t+1) = \frac{R\,\text{N}(t)}{1+(a\,\text{N}(t))^b},\ \text{N}(0) = N_0\right\},\ \text{N}(t)\right)$$

Maple cannot solve the nonlinear difference equation *eq2*. Of course, analytic forms can be generated at different times using an iterative approach. For example, let's take the total time to be $T=5$ and $N(0)=N_0$ still.

```
> T:=5: N(0):=N[0]:
```

The recurrence relation is iterated.

```
> for t from 0 to T do
> N(t+1):=R*N(t)/(1+(a*N(t))^b);
> end do;
```

$$\text{N}(1) := \frac{R\,N_0}{1+(a\,N_0)^b}$$

$$\text{N}(2) := \frac{R^2\,N_0}{(1+(a\,N_0)^b)\left(1+(\frac{a\,R\,N_0}{1+(a\,N_0)^b})^b\right)}$$

In the output, only the expressions for N_1 and N_2 are shown here in the text, those for N_3 to N_6 becoming increasingly lengthier. To go to even longer times is not very useful since the formulas are not compact or very revealing. It is better to pick some specific parameter values and plot the output. For comparison purposes with the linear model $N_{t+1}=R\,N_t$, we shall take $N_0=10$ and $R=1.5$. Recall that for this R value, the population number in the linear model grew indefinitely. For the MSS model, we also take $C=1$, $b=5$, and a total of 30 time steps. The value of a is calculated, using the relation $a=(R-1)/C$.

```
> N[0]:=10: R:=1.5: C:=1: a:=(R-1)/C; b:=5: Total:=30:
```

$$a := 0.5$$

The calculated value of a in the nonlinear model is 0.5. In the following do loop, the MSS equation is iterated and the triplet of numbers `[t+1,N[t],N[t+1]]` formed into a plotting point `pt` at each time step t.

```
> for t from 0 to Total do
> N[t+1]:=R*N[t]/(1+(a*N[t])^b);
> pt[t+1]:=[t+1,N[t],N[t+1]];
> end do:
```

For viewing convenience, the plotting points are joined by straight lines using the `spacecurve` command to create a three-dimensional plot that can be rotated. Choosing `orientation=[-90,0]` allows us to view N_t versus time.

```
> spacecurve([seq(pt[j+1],j=0..Total)],axes=boxed,
  tickmarks=[3,3,3],labels=["time","N[t]","N[t+1]"],
  color=red,thickness=2,orientation=[-90,0]);
```

The result is shown on the left of Figure 6.9. In this example, the population

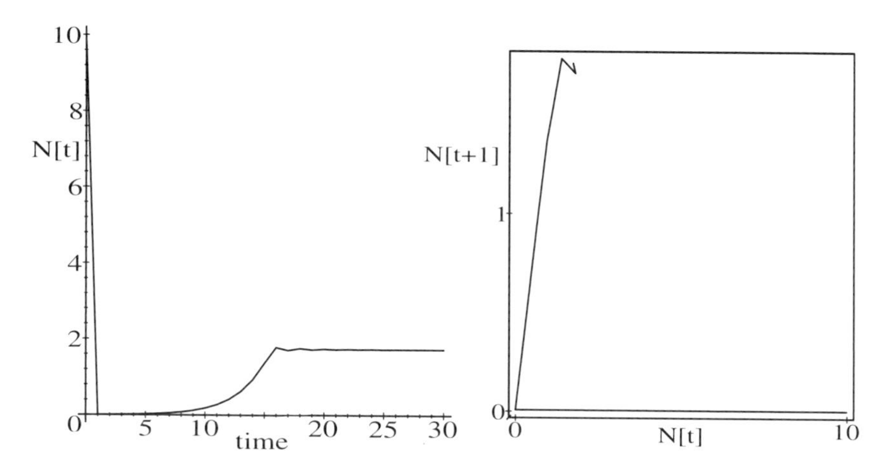

Figure 6.9: Solution of the MSS competition model for $R{=}1.5$, $a{=}0.5$, $b{=}5$.

nearly dies away before recovering and settling down to a constant value of N_t. The behavior is in sharp contrast to that in the linear model for the same parameter values. If the three-dimensional viewing box is rotated so that `orientation=[0,90]`, then a plot of N_{t+1} versus N_t results as shown on the right of Figure 6.9. In the language of mathematics, this plot represents a *mapping* of N_t into N_{t+1} as t advances. The first-order difference equation relating N_{t+1} to N_t is called a one-dimensional *nonlinear map*. For this map, we see a *trajectory* start at the point $N_0 = 10$, $N_1 = 0.004798464491$ and evolve with time toward the *fixed*, or *stationary, point* (located in the upper left-hand corner of Figure 6.9) corresponding to $N_{t+1} = N_t \equiv N^*$. The numerical value of the nonzero fixed point N^* is readily found with the `fsolve` command. The option `avoid` is used so that we do not obtain the trivial solution $X = 0$.

```
> fixed_pt:=fsolve(X=R*X/(1+(a*X)^b),X,avoid={X=0});
```

$$fixed\_pt := 1.741101127$$

From our earlier discussion, the fixed point $N_t = N^* \approx 1.74$ is identified as the carrying capacity of the population for the chosen parameter values. The population evolves toward a steady-state situation at the fixed point where births are balanced by deaths. According to Begon, Harper, and Townsend [BHT90], the MSS model has proven quite successful in accounting for the observed intraspecific competition between beetles such as *Stegobium panaceum*, *Tribolium confusum*, *Triboleum castaneum*, and the winter moth *Operophtera brumata*. In the laboratory experiments the parameters R, C, and b were chosen using the least squares procedure discussed in Chapter 2 to give best fits to the data.

PROBLEMS:

Problem 6-26: Carrying capacity

For the MSS model, with $R = 1.5$ and $C = 1$, what is the carrying capacity for $b = 0.5$? Run the Maple file for this situation and discuss the results.

Problem 6-27: Nonlinear growth

Attempt to find an analytic solution to the following nonlinear growth equation using the `rsolve` command:

$$N_{t+1} = N_t/(1 + \sqrt{N_t})^2.$$

Determine the form of the analytic solution by iterating the growth equation.

Problem 6-28: Population growth 1

A population grows according to the nonlinear difference equation

$$X_{t+1} = (R - a\,X_t - b\,X_t^2)\,X_t.$$

(a) Given that $R = 1.5$, $a = 0.003$, $b = 0.00002$, and $X_0 = 10$, plot X_t versus t up to $T = 15$ years. Determine the fixed point (steady state) by:
(i) clicking on the last point in the plot;
(ii) examining the plotting points output;
(iii) setting $X_{t+1} = X_t \equiv X^*$ in the equation and solving for X^*.

(b) Make a three-dimensional plot using the `spacecurve` option.

Problem 6-29: Logistic model

For a population of size P, the birth rate during the next year (i.e., the number of births as a fraction of the population) is equal to $(0.7 - 0.00005\,P)$ and the death rate per year is $(0.2 + 0.00015\,P)$. The change in population number will be equal to the number of births minus the number of deaths.

(a) Write down the difference equation for the growth of this population. This difference equation is an example of a logistic model.

(b) Plot the population number P as a function of year for $P(0) = 1000$ and determine the numerical value of the fixed point.

(c) At what minimum time will the population number be within 1% of the steady-state value?

(d) Repeat parts (b) and (c) for $P(0) = 4000$ and $P(0) = 6000$.

Problem 6-30: Logistic model 2

For a population of size P, the birth and death rates during a 1 year period are equal to $(0.5 - 0.0005\,P)$ and $(0.2 + 0.0005\,P)$, respectively.

(a) Write down the difference equation for the growth of this population.

(b) Plot the population number P as a function of year for a positive $P(0)$ of your choice and determine the numerical value of the fixed point.

Problem 6-31: Analytically solvable nonlinear difference equation
Making use of the `rsolve` command, show that the nonlinear growth equation

$$N_{t+1}\,(1+N_t) = N_t$$

has an analytic solution. Plot the solution for $N_0 = 10$. Relate the growth equation to the MSS model discussed in the text.

Problem 6-32: Adding text to Maple plots
As demonstrated a number of times in the text, the `textplot` command can be used to add text to a Maple plot. Guided by these examples, generate Figures 6.7 and 6.8.

6.2.2 The Logistic Map and Cobweb Diagrams

The map appears to us more real than the land.
T. E. Lawrence, English soldier and writer (1888–1935)

As a simple model of intraspecific competition in mathematical biology, which displays extraordinarily complex and interesting behavior, Robert May [May76] championed the introduction of the now famous *logistic model* of population growth into elementary mathematics courses. If N_n is the number in generation n, the number N_{n+1} in the next generation is given in this model by

$$N_{n+1} = \left(1 + r - \frac{r}{k}\,N_n\right)\,N_n, \tag{6.19}$$

where r is the real, positive growth coefficient and k is also a positive constant. If $r = 0$, then $N_{n+1} = N_n$ and no change in population number takes place from one generation to the next. The nonlinear term involving k is included to reduce the rate of growth due to overcrowding, limited resources, etc. In the limit that $k \to \infty$, the nonlinear term vanishes and the familiar linear difference equation $N_{n+1} = R\,N_n$ results, with $R \equiv 1 + r$.

As part of an assignment for a nonlinear dynamics course that he is taking from Jennifer, Mike is developing a Maple worksheet that investigates the behavior of the logistic model. He loads some necessary library packages,

```
> restart: with(plots): with(Statistics):
```

and decides to put the logistic equation (6.19) into a simpler form. For finite values of k, Mike realizes that the parameter k can be scaled out of the equation. To accomplish this, he introduces a new dependent variable x_n through the following operator `N`.

```
> N:=n->(1+r)*k*x[n]/r:
```

The logistic equation (6.19) is entered, the dependent variable transformation being automatically implemented.

```
> eq:=N(n+1)=(1+r-(r/k)*N(n))*N(n);
```

$$eq := \frac{(1+r)\,k\,x_{n+1}}{r} = \frac{(1+r-(1+r)\,x_n)\,(1+r)\,k\,x_n}{r}$$

Then, multiplying *eq* by the factor $r/(k\,(1+r))$,

```
> eq2:=r*eq/(k*(1+r));
```

$$eq2 := x_{n+1} = (1 + r - (1+r)\,x_n)\,x_n$$

substituting $r = a - 1$ in *eq2*,

```
> eq3:=subs(r=a-1,eq2);
```

$$eq3 := x_{n+1} = (a - a\,x_n)\,x_n$$

and collecting coefficients of a in *eq3*,

```
> logistic_map:=collect(eq3,a);
```

$$logistic\_map := x_{n+1} = (1 - x_n)\,x_n\,a$$

yields the "standard" mathematical form of the logistic difference equation. Only a single parameter, a, remains. For analysis purposes, Mike has read that mathematicians usually restrict the range of a to be $0 \leq a \leq 4$ and the input value x_0 to be between 0 and 1. This ensures that the variable x will always remain between 0 and 1, no matter how many iterations N are performed. With these restrictions, the nonlinear difference equation model is then referred to as the *logistic map.*

Mike begins by choosing some typical numbers in the "allowed" ranges, taking $a = 3.2$ and $x_0 = 0.1$. For later typing convenience, he decides to label the input value x_0 as b. The logistic map, being nonlinear, does not have an analytic solution, so must be solved iteratively. He considers $N = 119$ iterations,

```
> a:=3.2: N:=119: x[0]:=0.1: b:=x[0]:
```

and uses the arrow operator to create the right-hand side of the logistic equation. The name `F` is assigned to the operation,

```
> F:=x->a*x*(1-x):
```

and the logistic map $x_{n+1} = F(x_n)$ iterated with a do loop.

```
> for n from 0 to N do
> x[n+1]:=F(x[n]);
> end do:
```

Initially Mike had a semicolon on the `end do` command to see what the output numbers look like, but he has since replaced the semicolon with a colon to suppress the output. Knowing that Jennifer isn't going to want to look at a list of numbers, he decides to make a nice plot instead. To accomplish this, he places the x_n values and the n values into separate lists, and employs the `ScatterPlot` command, with the numerical points represented by blue circles.

```
> xpoints:=[seq(x[n],n=0..N)]:
> npoints:=[seq(m,m=0..N)]:
> display(Scatterplot(npoints,xpoints,symbol=circle,
  labels=["n","x[n]"],tickmarks=[3,3]),color=blue);
```

At the instant that the plot (reproduced in Figure 6.10) appears on the screen, Vectoria wanders into the computer lab where he is working.

"That's a nice picture, Mike. What does it represent?"

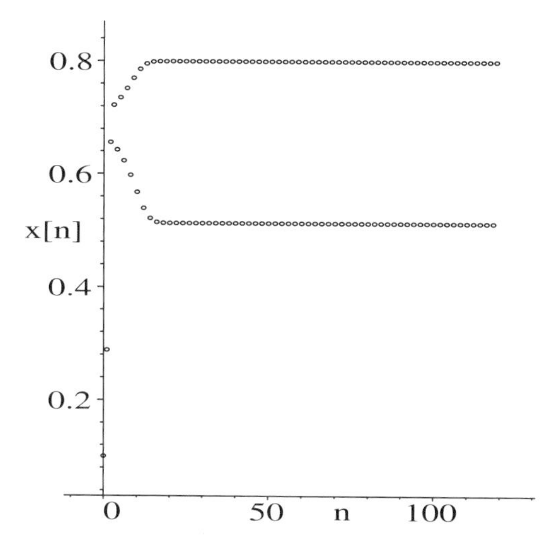

Figure 6.10: A period-two solution to the logistic map.

"I'm looking at the logistic map for a particular parameter value and this is the output. After a brief transient period, the solution settles down into a steady-state regime, oscillating between the two horizontal branches of points shown in the figure. The variable x takes on the values $x \approx 0.799$ and $x \approx 0.513$, returning to the same value every time n increases by two. Since the repeat period is two generations, this is an example of a *period-two* solution."

"You have chosen $a=3.2$. What happens if you change the a value?"

"As we can easily verify by executing the do loop again, taking $a < 3.0$ produces a graph of x_n versus n in which the steady-state regime consists of a single horizontal line of points, i.e., the same value of x occurs in every generation. This is referred to as a *period-one* solution.

Referring to the paper by May, period two prevails up to $a = 3.449490\ldots$, then period four (four horizontal lines will be plotted) up to $a = 3.544090\ldots$, period eight up to $a = 3.564407\ldots$, and so on. Irregular or chaotic behavior occurs when a passes through the value $a = 3.569946\ldots$ Notice how the ranges, or *windows*, of a values grow smaller and smaller as the periodicity increases toward the chaotic limit. This makes the discovery of higher-order periodicity a numerical challenge. At still larger a values, below $a = 4$, the steady-state response is characterized by windows of periodicity and chaos."

"Mike, I have seen what are called *cobweb* diagrams for nonlinear difference equations. Can you create a cobweb diagram for the logistic map?"

"Yes I can. A cobweb diagram is not only visually pleasing but it can geometrically reveal how the change in periodicity takes place as a is increased. I will have to tell you a bit about the stability of fixed or stationary points of the logistic map as I enter the relevant code.

Let's first form $f=F(x)$.

```
> f:=F(x);
```

$$f := 3.2\, x\, (1 - x)$$

The possible period-one stationary points correspond to the solutions of $f = x$, which are found on applying the `solve` command.

```
> f_fixed_points:=solve(f=x);
```

$$f\_fixed\_points := 0., 0.6875000000$$

There are two potential period-one fixed points, at $x = 0$ and $x = 0.6875$. But for $a = 3.2$, we have observed that period two occurs, not period one. So these fixed points must be "unstable," since the system is not "attracted" to either of these x values for large N.

Since period two corresponds to iterating twice before returning the same value, the fixed points for period two are obtained by solving $F(F(x)) \equiv F^{(2)}(x) = x$. We can create $F^{(2)}(x)$ using the function composition operator `@`.

```
> g:=(F@F)(x);
```

$$g := 10.24\, x\, (1 - x)\, (1 - 3.2\, x\, (1 - x))$$

Notice how x in f has been replaced with $3.2\, x\, (1 - x)$, thus producing g. Since two applications of F are applied in producing g, it is referred to as the *second iterate map*. In this language, f is the *first iterate map*, while three applications of F produce the *third iterate map*, and so on. The fixed points of the second iterate map are found by solving $g = x$ for x,

```
> g_fixed_points:=solve(g=x,x);
```

$$g\_fixed\_points := 0., 0.5130445095, 0.6875000000, 0.7994554905$$

yielding four possible period-two fixed points. Two of these points have values identical with those obtained for $f = x$. Since we observed that the logistic map evolved toward the remaining two ($x = 0.513$ and $x = 0.799$) fixed points, the latter must be "stable" fixed points for period two. Similarly, one could find the possible stationary points for higher-order periodicity by repeatedly applying the function composition operator to `F`, setting the result equal to x, and solving. Maple has no difficulty in analytically generating a function of a function of a function, as witnessed in the following command line, where $F^{(3)}(x)$ is produced.

```
> h:=(F@F@F)(x);
```

$$h := 32.768 x\, (1-x)\, (1-3.2\, x\, (1-x))\, (1-10.24\, x\, (1-x)\, (1-3.2\, x\, (1-x)))$$

Since only period two is observed for $a = 3.2$, again the fixed points corresponding to higher-order periodicity, such as those generated for the third iterate map on setting $h = x$, must be unstable."

"Mike, what is the mathematical criterion for stability here?"

"Oh, that's fairly easy to explain. Consider the general first-order map

$$x_{n+1} = F(x_n) \tag{6.20}$$

and label a possible fixed point as x^*. If one starts with some initial value, $x_0 = x^* + \epsilon$, close to x^* (i.e., ϵ small), the system will evolve toward x^* as n

increases if the fixed point is stable and away from it if it is unstable. Since ϵ is small, a single iteration of the x equation gives us, on Taylor expanding and keeping first order in ϵ:

$$x_1 = F(x^* + \epsilon) = F(x^*) + \epsilon \left(\frac{dF}{dx}\right)_{x^*} = x^* + (x_0 - x^*)\left(\frac{dF}{dx}\right)_{x^*}, \tag{6.21}$$

so that

$$|x_1 - x^*| = |(x_0 - x^*)| \left| \left(\frac{dF}{dx}\right)_{x^*} \right|. \tag{6.22}$$

If $|(dF/dx)_{x^*}| < 1$, i.e., the magnitude of the slope of $F(x)$ at the fixed point is less than one, then the iteration reduces the "distance" from the fixed point. Repeated iterations in this case eventually reduce the distance to zero, and the fixed point is stable. Conversely, the distance from the fixed point grows when the magnitude of the slope exceeds one, and the fixed point is unstable. This argument can be generalized to higher-order iterates, $F^{(2)}(x)$, $F^{(3)}(x)$, etc."

"OK, Mike, that was easy to understand. Can you use the above criterion in the context of a cobweb diagram?"

"Sure. Let's continue our Maple recipe, still with $a = 3.2$. I will produce a cobweb diagram for you by superimposing a number of plots. In the following plot command line pl||1, the first iterate map f is graphed as a thick red line.

```
> pl||1:=plot(f,x=0..1,y=0..1,style=line,thickness=2,
        color=red):
```

The second plot graphs the second iterate map g as a green dashed line.

```
> pl||2:=plot(g,x=0..1.6,y=0..1.6,style=line,linestyle=3,
        thickness=2,color=green):
```

Next we draw a thick 45° line, $y = x$, coloring it black.

```
> pl||3:=plot(x,x=0..1,y=0..1,style=line,thickness=2,
        color=black):
```

A vertical blue line at $x = b \equiv x_0$, spanning the range $y = 0$ to $y = F(b)$, is produced.

```
> pl||4:=plot([[b,0],[b,F(b)]],style=line,color=blue):
```

Using the repeated function composition operator @@, the following command line calculates repeated applications of the function F, given an initial value $b = 0.1$. The truncation operator trunc is used to stop the iterations at the integer value 20. We can change this value if we wish.

```
> pts:=[seq((F@@(trunc((i+2)/4)))(b),i=1..80)]:
```

The points pts are connected by straight lines with the pointplot command.

```
> pl||5:=pointplot(pts,style=line,view=[0..1,0..1],
        color=blue):
```

Finally, the sixth plot adds labels in suitable locations to the various curves.

```
> pl||6:=textplot([[.2,.84,"y=g"],[.45,.37,"y=x"],
        [.85,.3,"y=f"],[.65,.85,"cobweb"]]):
```

All six graphs are superimposed, producing Figure 6.11."

```
> display({seq(pl||i,i=1..6)},view=[0..1,0..1],
  tickmarks=[3,3],labels=["x","y"]);
```

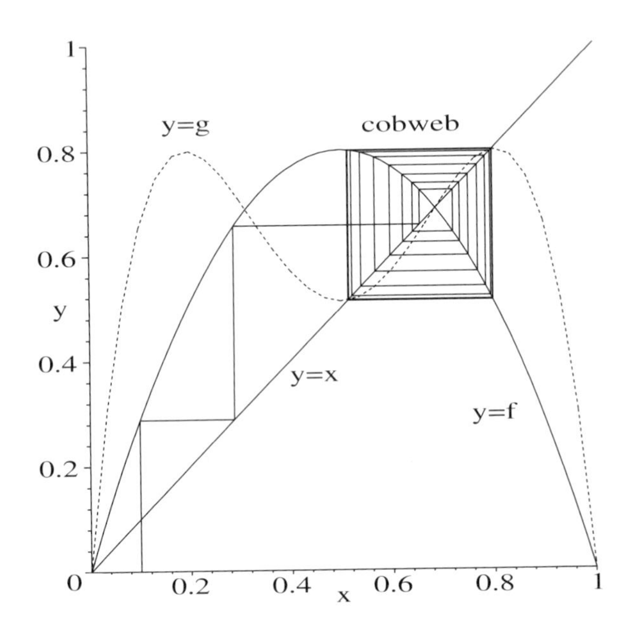

Figure 6.11: Cobweb diagram for a period-two plot.

"I can certainly see why it's called a cobweb diagram, but how do we interpret this picture?"

"In the iteration process, we have started at $x = b = 0.1$. In the first iteration, $F(x = b)$ is calculated, which geometrically corresponds to moving upward along the vertical blue line at $x = b$ to the parabola $y = f$. The resultant x value will be the new input for the second iteration. The new input value is produced by moving horizontally along the attached blue line to the black $y = x$ line. The second iteration then carries the system vertically to the $y = f$ line. Repeating this process, the trajectory winds in a cobweb fashion onto the heavy rectangle. This rectangle intersects the second iterate map $y = g$ at two locations at which the slope of g is less than $45°$, i.e., below the $y = x$ line. These two locations are the two stable stationary points that the period-two solution cycles between in steady state. The curve $y = g$ intersects $y = x$ at four locations, including the origin. At the other two intersection points, the slope of g is higher than the $y = x$ line, so these two points are unstable. Similarly, the two intersection points of the first iterate map f with $y = x$ have slopes whose magnitudes are greater than one, so the two possible stationary points for period one are unstable. Period one does not occur here."

"Mike, I can certainly see that the cobweb diagram is a useful visual tool for determining the stability of the stationary points and geometrically understanding how the logistic map evolves for different values of the parameter a. However, it's late on a Friday afternoon and you promised that we would meet the gang down at the Waddling Duck Pub. So why don't you save your file and let's be off."

PROBLEMS:

Problem 6-33: Periods 4 and 8

Taking $x_0 = 0.1$, confirm that period four occurs for $a = 3.5$ and period eight for $a = 3.56$ in the logistic map. First do this by making a plot of x_n versus n. Second, try the cobweb diagram approach. Discuss your results.

Problem 6-34: Function composition operator

Directly evaluate $(\tan @ \sin @ \sin @ \cos)(0.6)$. Confirm your answer by evaluating $\tan(\sin(\sin(\cos(0.6))))$.

Problem 6-35: Period?

Taking $x_0 = 0.1$ and making a suitable plot, determine the periodicity of the logistic map solution for $a = 3.83$. Up to what value of n does the transient part of the solution persist?

Problem 6-36: a = 3.8

Taking $x_0 = 0.2$ and $a = 3.8$ in the logistic map, create a plot of x_n versus n as well as a cobweb diagram. What is the probable nature of the solution? Explain.

Problem 6-37: Eighth iterate map

Use the function composition operator to generate the eighth iterate map of the logistic function $f = 3.5\,x\,(1 - x)$. Expand the result and determine the order of the highest-order polynomial in the expansion. Comment on doing this calculation by hand.

Problem 6-38: Feigenbaum constant

For the logistic map, suppose that period-2^k ($k = 1, 2, \ldots$) solutions are "born" at $a = a_k$. Feigenbaum [Fei78] was able to show that as k approaches infinity, the ratio $\delta_k = (a_k - a_{k-1})/(a_{k+1} - a_k)$ approaches the constant value $\delta = 4.6692\ldots$ This limiting value is called the *Feigenbaum constant*. Given $a_1 = 3.000000$, $a_2 = 3.449490$, $a_3 = 3.544090$, $a_4 = 3.564407$, $a_5 = 3.568759$, $a_6 = 3.569692$, $a_7 = 3.569891$, and $a_8 = 3.569934$, calculate the sequence of δ_k for $k = 2, 3, \ldots, 7$. Discuss your results in terms of loss of digits accuracy. Obtaining more digits is difficult for higher k, because the transient time gets longer and longer, so the number of iterations has to be greatly increased. The Feigenbaum constant is usually evaluated by alternative means. The Feigenbaum constant turns out to be a "universal constant," being obtained for any map characterized by a function f that is smooth, concave downward, and having a single maximum.

Problem 6-39: Onset of epileptic seizures

According to Glass (1995), a simple nonlinear difference equation that models the onset of epileptic seizures is

$$x_{t+1} = 4\,C\,x_t^3 - 6\,C\,x_t^2 + (1 + 2\,C)\,x_t,$$

where x_t is the fraction of neurons of a large neural network that fire at time t and C is a positive constant.

(a) Determine the fixed points x^* of this difference equation.

(b) Determine the stability of these fixed points and the C value at which they all become unstable.

(c) Determine the fixed points x^{**} corresponding to $x_{t+2} = x_t = x^{**}$ and the C value at which they all lose their stability.

(d) Take $x_0 = 0.45$ and the following values $C = 1.5$, 2.1, 2.5, 3.0, 3.3, 4.0. In each case solve the model equation for t running from 0 to $N = 500$ and create a three-dimensional plot of t versus N_t versus N_{t+1}. Determine the periodicity in each case and relate the results to those in parts (a) to (c). Relate the results to the idea that increasing the value for C leads to the onset of uncontrolled neuron firings characteristic of an epileptic seizure.

6.2.3 The Bouncing Ball Art Gallery

It's pretty, but is it Art?
Rudyard Kipling, Nobel laureate in literature (1907)

On their way to a concert at the MIT Music Academy, Colleen and Sheelo bump into Vectoria, who is shopping for a present for Mike's birthday. Sheelo, who is into computer-generated art, suggests that Vectoria contact Jennifer to see whether she can come up with an artistic piece that is not only art but arises

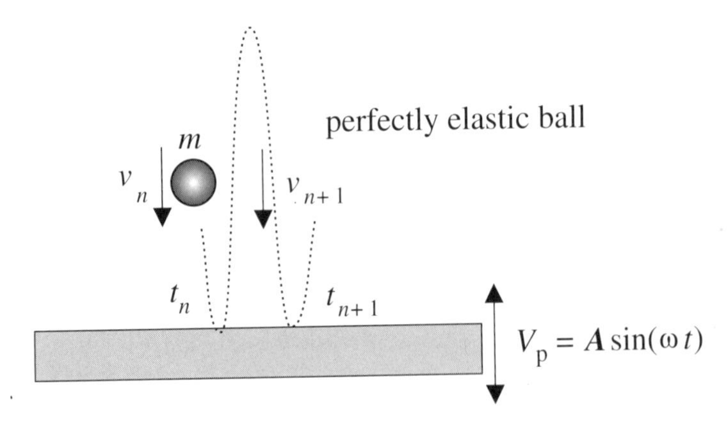

Figure 6.12: A schematic of the bouncing ball geometry.

from either physics or mathematics or both. After meeting with Vectoria, Jennifer came up with an artistic "masterpiece" from the "bouncing ball art gallery." With Jennifer's permission, we shall now provide the interested reader with the recipe for producing this work of physics-generated modern art.

A perfectly elastic ball bounces vertically on a horizontal vibrating plate (see Figure 6.12) whose vertical velocity is given by $V_p = A\ \sin(\omega\, t)$. Jennifer lets v_n be the velocity of the ball prior to the nth bounce at time t_n, v_{n+1} the velocity prior to bounce $(n+1)$ at t_{n+1}, and so on. For simplicity, she neglects the vertical displacement of the plate relative to the path traveled by the ball as well as air resistance. The velocity of the plate at time t_n is entered.

```
> restart:
> Vp:=A*sin(omega*t[n]);
```

$$Vp := A \sin(\omega\, t_n)$$

When the elastic ball strikes the plate, it experiences a change of velocity of amount $2\, V_p$. If v_n is the velocity of the ball just before the nth bounce, its velocity just before bounce $(n+1)$ is given by the difference equation

```
> vel_eq:=v[n+1]=v[n]+2*Vp;
```

$$vel\_eq := v_{n+1} = v_n + 2\, A \sin(\omega\, t_n)$$

For convenience, Jennifer sets $\omega\, t_n = \theta_n$ in the velocity equation.

```
> vel_eq:=subs(omega*t[n]=theta[n],%);
```

$$vel\_eq := v_{n+1} = v_n + 2\, A \sin(\theta_n)$$

This equation relates the change in velocity on successive bounces to the *phase* θ. Since two variables are involved, a second equation is needed relating the velocity and the phase. The change in phase on successive bounces is given by

```
> phase_eq:=theta[n+1]=theta[n]+omega*T[n+1];
```

$$phase\_eq := \theta_{n+1} = \theta_n + \omega\, T_{n+1}$$

where $T_{n+1} = t_{n+1} - t_n$ is the time interval between bounces n and $n+1$. To calculate T_{n+1} in terms of v_{n+1}, Jennifer uses one of Newton's kinematic equations, $d = u_0\, t + (1/2)\, a\, t^2$, where d is the displacement after time t when the initial velocity is u_0 and the acceleration is a. Setting $d = 0$, $u_0 = v_{n+1}$, $a = -g$, where g is the acceleration due to gravity, and $t = T_{n+1}$, this kinematic relation yields $T_{n+1} = 2\, v_{n+1}/g$, which is automatically substituted into *phase_eq*.

```
> T[n+1]:=2*v[n+1]/g; phase_eq;
```

$$T_{n+1} := 2\,\frac{v_{n+1}}{g} \qquad \theta_{n+1} = \theta_n + \frac{2\,\omega\, v_{n+1}}{g}$$

To simplify the phase equation, Jennifer substitutes $2\,\omega\, v_{n+1}/g \equiv V_{n+1}$.

```
> phase_eq:=subs(2*omega*v[n+1]/g=V[n+1],%);
```

$$phase\_eq := \theta_{n+1} = \theta_n + V_{n+1}$$

Of course, if v is rescaled to V in one equation, it must also be rescaled in the other. So, the velocity equation is rescaled in the next few lines.

```
> vel_eq:=expand(2*omega*vel_eq/g);
```

$$vel\_eq := 2\,\frac{\omega\, v_{n+1}}{g} = 2\,\frac{\omega\, v_n}{g} + \frac{4\,\omega\, A \sin(\theta_n)}{g}$$

```
> vel_eq:=subs({2*omega*v[n+1]/g=V[n+1],2*omega*v[n]/g=V[n]},%);
```

$$vel\_eq := V_{n+1} = V_n + \frac{4\,\omega\, A \sin(\theta_n)}{g}$$

Replacing $4\,\omega\, A/g$ with K, yields the final form of the velocity equation.

```
> vel_eq:=subs(omega=K*g/(4*A),vel_eq);
```

$$vel\_eq := V_{n+1} = V_n + K \sin(\theta_n)$$

For the reader's benefit, Jennifer has informed us that coupled equations identical in structure to those for V and θ arise in many other physical contexts, e.g., for the relativistic motion of an electron in a microton accelerator as well as in stellerator setups used in plasma fusion experiments [Jac90].

The two coupled first-order nonlinear difference equations that Jennifer has derived represent an example of a two-dimensional nonlinear map. This map cannot be solved analytically, but must be attacked using a do loop procedure. The range of the dependent variables depends on the physical context being considered, the bouncing ball, the microton accelerator, or whatever. Mathematicians like Jennifer define the *standard map* as a map of the above structure with both variables limited to a finite range, usually chosen to be 0 to 1. For the bouncing ball, the phase can be kept in the range 0 to $2\,\pi$ or, by setting $\theta_n = 2\,\pi\, y_n$, in the y range 0 to 1. This can be accomplished by the mathematical operation embodied in the Maple command `Y[n+1]:=frac(y[n+1]):`, which subtracts the integer part from y_{n+1}. That is to say, if y has the value 3.1, the `frac` command removes the integer 3, thus giving the value 0.1 to y. For convenience, Jennifer also sets $V_n = 2\,\pi\, x_n$ but does not limit the range of x to be between 0 and 1, since this is unphysical for the bouncing ball.

Keeping in mind that the derivation made an approximation in neglecting the vertical displacement of the plate relative to the flight of the ball, so regions of slightly unphysical behavior can occur, Jennifer has provided the following code to explore the physics of the bouncing ball. If the reader wants to obtain the mathematician's standard map, simply add the `frac` command at the point indicated by the appropriate comment.

```
> restart: with(plots): Digits:=15:
> x[0]:=0.6: y[0]:=0.1: K:=0.5: N:=1000: p:=evalf(Pi):
> for n from 0 to N do
> x[n+1]:=x[n]+K/(2*p)*sin(2*p*y[n]);
> y[n+1]:=y[n]+x[n]+K/(2*p)*sin(2*p*y[n]);
> t[n+1]:=n+1; #keep track of the bounce time
> X[n+1]:=x[n+1]; #add frac if desired
> Y[n+1]:=frac(y[n+1]);
> pt[n]:=[X[n+1],Y[n+1],t[n+1]];
> end do:
```

The plotting points are now created and plotted using the seq and pointplot3d commands and joined together with lines.

```
> plotpoints:=[seq(pt[j],j=N-500..N)]:
> pointplot3d(plotpoints,style=line,axes=boxed,tickmarks=
  [2,2,2],orientation=[-90,0],labels=["X","Y","T"]);
```

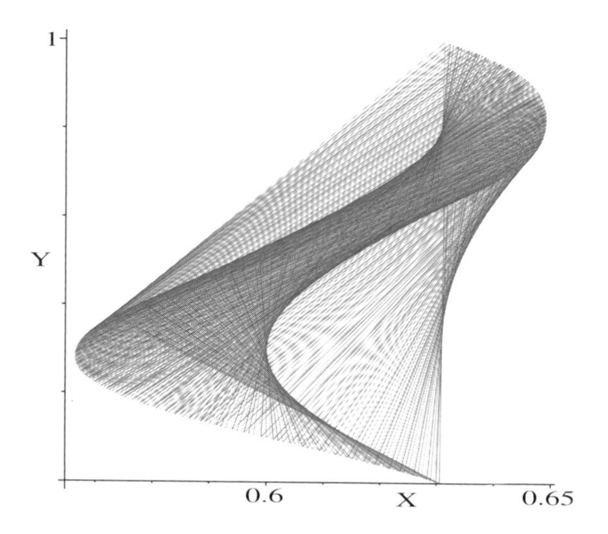

Figure 6.13: *Velocity vs. phase*, a masterpiece from the bouncing ball art gallery.

Two-dimensional maps that are properly oriented can display esthetically pleasing pictures traced out by the system's trajectory. One possible masterpiece that Mike might like is illustrated in Figure 6.13, where the X–Y plane has been selected. The black and white reproduction shown here in the text does not do justice to the spectrum of bluish colors observed in the actual computer plot when the zoom magnification is set at 200%. This particular picture is very delicate, however. If the viewing box is rotated even slightly the picture turns "ugly." Of course, the terms "beautiful" and "ugly" are subjective and depend on the prejudices of the viewer. See whether you can find other parameters and orientations that generate what you feel to be interesting, or even beautiful, masterpieces of computer-generated art. Start your own collection of masterpieces for the bouncing ball art gallery.

PROBLEMS:

Problem 6-40: Other masterpieces

Use the bouncing ball code with parameters of your own choosing to obtain five visually distinct masterpieces to add to the bouncing ball art gallery. You may wish to choose your own color options instead of accepting the default color as was done in the text recipe.

Problem 6-41: Other functional forms
In the text recipe, replace the `sine` function with the `cosine` function and see what the resulting picture looks like at a zoom magnification of 200%. Try different color options until you find the picture that most appeals to you.

Problem 6-42: Frac command
Taking `Digits:=20:`, evaluate each of the following numbers and then apply the `frac`, `trunc`, and `round` operators to each 20-digit number:

(a) the tangent of 120°;

(b) 123.456 raised to the fifth power.

6.2.4 Onset of Chaos: A Model for the Outbreak of War

"Let's fight till six, and then have dinner," said Tweedledum.
Lewis Carroll, *Alice's Adventures in Wonderland* (1865)

In a thought-provoking series of articles appearing in the journal *Nature* [Sap84] and a research text entitled *Chaos Theory in the Social Sciences* [KE96], Alvin Saperstein has created a number of interesting phenomenological nonlinear models involving the arms race between nations or groups of nations. He views the threshold of war as being signaled by the breakdown of predictability and control and the onset of unpredictability and chaos. Although nonlinear dynamical models are deterministic, they can display regions of parameter space where small perturbations or changes in the initial conditions produce large changes in the outcome. Since the initial conditions are never precisely known, predictability is lost and the solution to the underlying equations is chaotic. This is in contrast to the nonchaotic regime, in which a reasonable estimate of present conditions allows one to confidently predict the future. As a historical example of the extreme sensitivity in the chaotic regime to small perturbations, Saperstein cites the single assassination of the Archduke Francis Ferdinand at Sarajevo in 1917, which led to the great slaughter of World War I.

To understand the essence of his ideas, let's first look at an early Saperstein model of the arms race. The key dependent variable in the model is the *devotion* of a nation to arms spending. Devotion is defined as the ratio of arms expenditures to the gross national product (GNP) of that nation in a given budget cycle. Clearly the devotion can take on only values between 0 and 1. As a concrete example, Table 6.1 shows the devotion of a number of European countries in the mid-1930s, countries that were soon to be engaged in World War II. It should be noted that obtaining data in hindsight is much easier than trying to get accurate data at a time of impending conflict, a time when nations are not about to divulge how much they are spending on arms.

Saperstein's first model was a bilateral one, involving the arms competition between two nations. This is clearly an oversimplification for the situation in the 1930s, when several major countries were involved and alliances were being

Table 6.1: Fraction of arms expenditures to GNP prior to World War II.

Year	**France**	**Germany**	**Italy**	**United Kingdom**	**U.S.S.R.**
1934	0.0276	0.0104	0.0443	0.0202	0.0501
1935	0.0293	0.0125	0.0461	0.0240	0.0552
1936	0.0194		0.0298	0.0296	0.0781
1937	0.0248		0.0359	0.0454	0.0947

formed. However, the model is a good starting point for understanding Saperstein's ideas. So consider two nations X and Y that are engaged in an arms race and take x_n and y_n to be their devotions in budget cycle n. Since larger n corresponds to a later budget cycle, n plays the role of the time variable. It is not unreasonable to assume that the devotion of nation X in budget cycle n+1 is proportional to Y's devotion in the previous budget cycle n, and vice versa, i.e.,

$$x_{n+1} \propto y_n, \quad \text{and} \quad y_{n+1} \propto x_n. \tag{6.23}$$

As a further refinement, it is assumed that if X's rival Y has such a high devotion that Y's resources are stretched to the breaking point in the previous year and no more resources are available, then X's devotion may be decreased (and vice versa). Saperstein's arms race model then consists of the following coupled nonlinear difference equations:

$$x_{n+1} = 4\,a\,y_n\,(1-y_n) \equiv F_a(y_n), \qquad y_{n+1} = 4\,b\,x_n\,(1-x_n) \equiv F_b(y_n), \tag{6.24}$$

where the parameters a and b are to be determined from existing data, e.g., from Table 6.1. Consider, for example, France (X) and Germany (Y) and take 1934 to correspond to $n=0$ and 1935 to $n=1$. Then, from Table 6.1, $x_0 = 0.0276$, $y_0 = 0.0104$, $x_1 = 0.0293$, and $y_1 = 0.0125$, so that $a = x_1/(4\,y_0\,(1-y_0)) = 0.712$ and $b = y_1/(4\,x_0\,(1-x_0)) = 0.116$. Table 6.2 shows the various a and b values calculated in a similar manner from the data of Table 6.1. With a and b known, Equations (6.24) can be solved iteratively for larger n using a do loop.

Table 6.2: Bilateral arms race model parameters a and b.

Nations	**Input Years**	a	b
France–Germany	1934–35	0.712	0.116
France–Italy	1936–37	0.214	0.472
UK–Germany	1934–35	0.582	0.158
UK–Italy	1934–35	0.142	0.582
USSR–Germany	1934–35	1.34	0.0657
USSR–Italy	1936–37	0.819	0.125

Before we discuss the specific application of Saperstein's bilateral model to the data of Tables 6.1 and 6.2, some preliminary general analysis is in order. First, let's eliminate one of the devotion variables from Equations (6.24). By changing n to $n+1$ in the first equation and then using the second, we have

$$x_{n+2} = 4\,a\,y_{n+1}\,(1 - y_{n+1}) = 16\,a\,b\,x_n\,(1 - x_n)\,(1 - 4\,b\,x_n\,(1 - x_n)), \quad (6.25)$$

or, in more compact functional notation,

$$x_{n+2} = F_a(y_{n+1}) = F_a(F_b(x_n)) \equiv F_{ab}(x_n). \quad (6.26)$$

To create the operator $F_{ab}(x)$ with Maple, the arrow notation is used to define the functions F_a and F_b.

```
> restart: with(plots):
> Fa:=x->4*a*x*(1-x); Fb:=x->4*b*x*(1-x);
```

$$Fa := x \to 4\,a\,x\,(1 - x) \qquad Fb := x \to 4\,b\,x\,(1 - x)$$

Using the function composition operator, then $F_{ab}(x)$ is formed.

```
> Fab:=Fa@Fb: Fab(x);
```

$$16\,a\,b\,x\,(1 - x)\,(1 - 4\,b\,x\,(1 - x))$$

The output is of the same structure as the right-hand side of Equation (6.25). Having formed $F_{ab}(x)$, we can use it like any other function. For example, the analytic derivative of $F_{ab}(x)$ is needed shortly in order to determine the stability of the fixed points. By forming an operator `d` to calculate the derivative of an arbitrary function $F(x)$, the derivative of $F_{ab}(x)$ is obtained in *d1*.

```
> d:=F->diff(F(x),x): d1:=d(Fab);
```

$$d1 := 16\,a\,b\,(1 - x)\,(1 - 4\,b\,x\,(1 - x)) - 16\,a\,b\,x\,(1 - 4\,b\,x\,(1 - x))$$
$$+ 16\,a\,b\,x\,(1 - x)\,(-4\,b\,(1 - x) + 4\,b\,x)$$

Rather than obtain an equation for x alone, we could just as easily have derived the corresponding equation for y. It takes the form

$$y_{n+2} = F_b(F_a(y_n)) \equiv F_{ba}(y_n). \quad (6.27)$$

Since for each x_n there is a corresponding y_n, we will concentrate on the x_n equation. Equation (6.26) connects x values that are two time steps apart, i.e., the values x_n are mapped into x_{n+2}. Similarly x_n can be mapped into x_{n+4}, i.e., x values four time steps apart are connected, by applying F_{ab} twice:

$$x_{n+4} = F_{ab}(x_{n+2}) = F_{ab}(F_{ab}(x_n)) \equiv F2_{ab}(x_n). \quad (6.28)$$

Again, Maple can be used to analytically calculate $F2_{ab}(x)$ and its derivative *d2*. The very lengthy output of the latter has been suppressed here in the text.

```
> F2ab:=Fab@Fab: F2ab(x); d2:=d(F2ab);
```

$$256\,a^2\,b^2\,x\,(1 - x)\,(1 - 4\,b\,x\,(1 - x))\,(1 - 16\,a\,b\,x\,(1 - x)\,(1 - 4\,b\,x\,(1 - x)))$$
$$(1 - 64\,b^2\,a\,x\,(1 - x)\,(1 - 4\,b\,x\,(1 - x))\,(1 - 16\,a\,b\,x\,(1 - x)\,(1 - 4\,b\,x\,(1 - x))))$$

By forming $F2_{ab}@F2_{ab}(x)$, x_{n+8} is connected to x_n, and so on.

Let's first look at the mapping $x_{n+2} = F_{ab}(x_n)$. The fixed points x^* of this mapping relation correspond to $x_{n+2} = x_n \equiv x^*$. So x^* is the solution of

$$x^* = F_{ab}(x^*) = 16\, a\, b\, x^*\, (1 - x^*)\, (1 - 4\, b\, x^*\, (1 - x^*)), \tag{6.29}$$

which may be numerically solved for specific values of a and b. For example, let's first take $a = 0.8$ and $b = 0.4$, with 15-digit accuracy. We shall hold a fixed in the subsequent analysis and change the value of b (thus, the comment).

```
> Digits:=15: a:=0.8: b:=0.4: #adjust b
```

We form an operator `sol1` to solve the general fixed-point equation $F(x) = x$.

```
> sol1:=F->[solve(F(x)=x,x)]:
```

The `remove` command is used to remove any unphysical complex answers, containing I, that may occur in `sol1`. F must be specified.

```
> sol2:=F->remove(has,sol1(F),I)[]:
```

The `slope` operator substitutes the ith answer in `sol2(F)` into the derivative or slope `s`. Note that F, i, and s must be given.

```
> slope:=(F,i,s)->subs(x=sol2(F)[i],s):
```

Taking $F = F_{ab}$ in `sol2`,

```
> Sol:=[sol2(Fab)]; #adjust F
```

$$Sol := [0., 0.708211494730374]$$

we find that there are two fixed points of $x_{n+2} = x_n$, viz., $x^* = 0$ and 0.708. To obtain the fixed points of $x_{n+4} = x_n$, F_{ab} must be replaced with $F2_{ab}$.

The corresponding derivatives of $F_{ab}(x)$ (the slopes) at the above fixed points are calculated. (To obtain the derivatives of $F2_{ab}(x)$ at the fixed points of $x_{n+4} = x_n$, one must replace F_{ab} with $F2_{ab}$ and *d1* with *d2*.)

```
> Slopes:=seq(slope(Fab,i,d1),i=1..nops(Sol)); #adjust F,d
```

$$Slopes := 5.12, -0.722193896757209$$

The first fixed point at $x^* = 0$ has a slope of 5.12, which is greater than 1, so this fixed point is unstable. The second fixed point at $x^* = 0.708$ has a slope whose magnitude is less than one and is therefore stable. On iteration of the model equations, the system will evolve toward the single stable fixed point $x^* = .708$.

To confirm this conclusion, let's take, e.g., $x_0 = 0.01$, $y_0 = 0.05$, and iterate the model equations $N = 150$ times.

```
> x[0]:=0.01: y[0]:=0.05: N:=150: pt[0]:=[0,y[0],x[0]]:
> for n from 0 to N do
> x[n+1]:=Fa(y[n]); y[n+1]:=Fb(x[n]);
> pt[n]:=[n+1,y[n+1],x[n+1]];
> end do:
```

An operator `pp`, making use of `pointplot3d`, is formed to plot the points generated in the do loop. A line style is chosen and an orientation to view x_n versus t. Rotate the plot on the computer screen to view y_n versus t.

```
> pp:=v->pointplot3d(v,axes=boxed,style=line,color=red,
```

```
            tickmarks=[3,4,4],labels=["t","y","x"],
            view =[0..N, 0..1, 0..1],orientation=[-90,90]):
```

Forming a sequence of pt[n] from $n=0$ to N and applying the operator pp

```
> pp([seq(pt[n],n=0..N)]);
```

generates the picture shown on the left of Figure 6.14. After a short transient time, x evolves toward the constant solution $x=0.708$ (confirmed by looking at the output of the do loop) as predicted by the stability analysis.

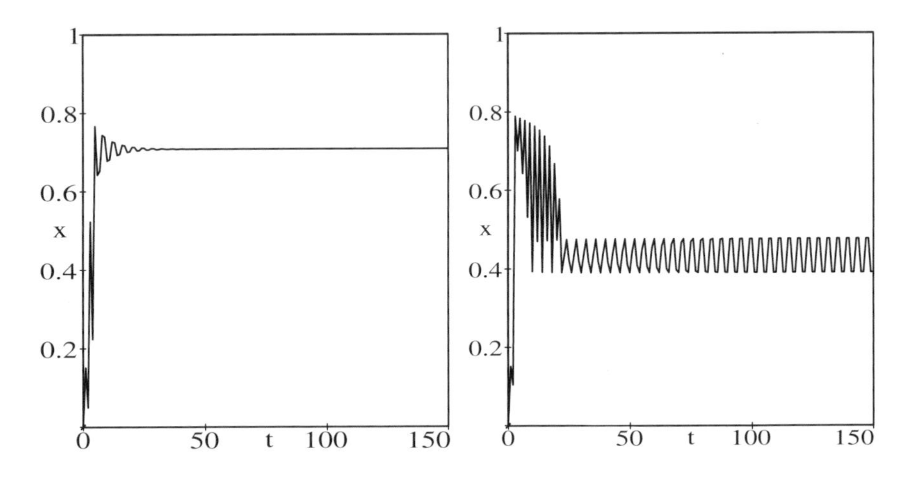

Figure 6.14: Devotion of nation X versus time t. Left: $b=0.4$. Right: $b=0.86$

With a held fixed, let's increase b. The location of the nonzero fixed point changes as does the slope of F_{ab} at that fixed point. When b is increased above about 0.85, the magnitude of the slope of F_{ab} increases to greater than one and this fixed point also becomes unstable. To confirm this, change b to 0.86 in the recipe. The fixed points of $x_{n+2}=x_n$ then are at $x^*=0$ and $x^* \approx 0.427$ with corresponding slopes of about 11 and -1.1. Since both slopes have magnitudes larger than one, both fixed points are unstable. What happens then?

One must examine the fixed points of $x_{n+4}=x_n$. Changing F_{ab} to $F2_{ab}$ and *d1* to *d2* at the appropriate places indicated by the comments in the recipe yields the fixed points 0, 0.390, 0.427, and 0.476 with slopes 121, 0.59, 1.21, and 0.59, respectively. So, for $b = 0.86$, $x_{n+4}=x_n$ has four fixed points, only two of which are stable, namely $x^*=0.390$ and 0.476. The two unstable fixed points are the same as for $F_{ab}(x)$ because $F2_{ab}$ obviously contains F_{ab}. For large enough times (budget cycles), the system will oscillate between the two stable fixed points. Again this can be confirmed by iterating the equations in the recipe and plotting the resulting curve, which is shown on the right of Figure 6.14. For large enough t (or n), the system oscillates between the two stable fixed points, a "period doubling" having occurred.

The period-doubling behavior displayed in the figure is consistent with the stability analysis. Again the future evolution of the system is quite predictable, since it keeps on repeating the periodic pattern.

As the reader may verify, when b is increased to 0.885 all the fixed points of $F2_{ab}$ become unstable and it turns out that a higher-period solution emerges. Between $b=0.885$ and 0.90, a sequence of further period doubling occurs until at $b=0.90$ one obtains a picture (not shown here, so you may want to execute the recipe) in which there is no apparent periodic pattern and the solution is chaotic. Using the stability approach just outlined, Saperstein was able to numerically find a curve relating the parameters a and b at which predictable periodic behavior is lost and chaos sets in. For example, from the above discussion, a point on the threshold curve is $a=0.80$ and $b \approx 0.90$.

For these latter parameter values and the input devotions $x_0 = 0.01$ and $y_0 = 0.05$ that we have been using, the difference equations yield $x_1 = 0.1520$ and $y_1 = 0.03564$. In other words, to get such a large a and b, nation X has noted that nation Y spent five times as much as it did on arms in the previous budget cycle and has increased its own arms spending fifteenfold in the next budget. Nation Y has actually cut back slightly, but too late to prevent instability.

What about the real data for the European countries in the 1930s? Using a and b values generated in the same manner as in Table 6.2, Saperstein concluded that the USSR–Germany arms race was already in the chaotic regime and the France–Germany and Italian–Soviet races were close to the threshold curve. World War II broke out shortly thereafter. Of course this calculation is done far after the historical time, and it is much easier to account for events in hindsight than to accurately predict the future.

A weakness of the bilateral model is clearly that when more than two countries are involved, as was the case in the 1930s, the model should be generalized to include more countries. Saperstein created a three-nation model that is a straightforward generalization of the bilateral one. Taking the devotion of the third country to be z, Saperstein's model equations are

$$\begin{aligned} x_{n+1} &= 4\,a\,y_n\,(1-y_n) + 4\,\epsilon\,z_n\,(1-z_n), \\ y_{n+1} &= 4\,b\,x_n\,(1-x_n) + 4\,\epsilon\,c\,z_n\,(1-z_n), \\ z_{n+1} &= 4\,\epsilon\,x_n\,(1-x_n) + 4\,\epsilon\,c\,y_n\,(1-y_n), \end{aligned} \tag{6.30}$$

with two additional parameters ϵ and c. The stability analysis for three coupled nonlinear difference equations is much more involved, and no attempt will be made to carry it out here. However, the generalization of the do loop for the bilateral model to the new tripolar model is trivial. Forming the function G,

```
> G:=x->4*epsilon*x*(1-x);
```

$$G := x \to 4\,\varepsilon\,x\,(1-x)$$

and keeping all parameters and initial conditions as in the bilateral model with $b=0.86$, we take $\epsilon=0.2$, $c=0.2$, and $z_0=0.02$.

```
> z[0]:=0.02: epsilon:=0.2: c:=0.2:
```

Executing the following do loop for the tripolar model equations,

```
> for n from 0 to N do
```

```
> x[n+1]:=Fa(y[n])+G(z[n]);
> y[n+1]:=Fb(x[n])+c*G(z[n]);
> z[n+1]:=G(x[n])+c*G(y[n]);
> pt2[n]:=[n,y[n],x[n]];
> end do:
```

and plotting the points,

```
> pp([seq(pt2[n],n=0..N)]);
```

will yield a chaotic solution for $b = 0.86$, rather than the periodic solution that occurred in the two-nations case. Including the third nation lowers the threshold curve, and chaos results instead. Inclusion of more countries leads to a greater chance of instability. Saperstein applied the comparison of the two nation and three-nation models to modern times and concluded that a tripolar world is more dangerous than a bipolar one. The latter was the situation when the world arms race was dominated by the United States and the Soviet Union.

PROBLEMS:

Problem 6-43: Bipolar model

Find the approximate threshold curve in the a–b plane for the onset of chaos.

Problem 6-44: Tripolar model

With all other parameter values the same, what is the approximate c value at which there is a change from periodicity to chaos? Explore the tripolar model for other values of ϵ and c. If nonchaotic, determine the periodicity.

Problem 6-45: Periodic cycles of disease

Anderson and May [AM82] have developed a difference equation model for the spread of disease that illustrates how periodic cycles of infection may arise in a given population. Let the basic unit of time t be the average time interval for infection and let C_t and S_t be the number of disease cases and number of susceptible people at time t. The Anderson–May model assumes:

- the number of new cases C_{t+1} at time $t+1$ is some fraction f of $C_t \times S_t$;
- a case lasts for only one time unit;
- the susceptible number S_t is increased at each time interval by a fixed number of births $B \neq 0$ and decreased by the number of new cases;
- individuals who recover from the disease are immune.

(a) Derive the difference equations corresponding to these assumptions. Determine the fixed point(s) of the model.

(b) Anderson and May state that in a third-world country typically $B = 36$ (births per 1000 people) and $f = 3 \times 10^{-5}$. Evaluate the fixed point(s).

(c) By solving the model equations, show that a small deviation away from the fixed point(s) results in a periodic cycle of disease incidence. Take the parameter values of part (b) and the initial values $S_0 = 33300$ and $C_0 = 20$.

Part III

THE DESSERTS

It's food too fine for angels; yet come, take
And eat thy fill! It's Heaven's sugar cake.
Edward Taylor, English poet (1664–1729)

Part of the secret of success in life is to eat
what you like and let the food fight it out inside.
Mark Twain, American humorist (1835–1910)

You won't need to follow recipes
when we've taught you how to cook.
Richard and George, your CAS chefs

Chapter 7

Monte Carlo Methods

Man can believe the impossible, but can never believe the improbable.
Oscar Wilde, Anglo-Irish writer (1854–1900)

The real strength of computer algebra systems compared to programming languages such as Fortran and C is in the ability to carry out symbolic computation and, where desired, easily plot out or even animate the resulting solution for specified parameter values. We would be remiss, however, if we didn't show that CASs can also prove quite useful in carrying out numerical simulations.

In this chapter we shall deal with a wide variety of Monte Carlo simulations, such as the random walk of a perfume molecule and the random-number evaluation of multidimensional integrals, that make use of a random-number generator. A random-number generator produces random numbers, integer or otherwise, over a specified range. Using the "best available" random-number generator is a concern of the serious scientific researcher, but Maple's built-in random number "procedure"[1] will suffice for our purposes.

The methods that rely on random-number generators are referred to as Monte Carlo methods, Monte Carlo being the gambling resort in Monaco where roulette and other games of chance are the featured attraction. The name Monte Carlo was introduced [KW86] by scientists working on the development of the atomic bomb at Los Alamos in the 1940s. The diffusion of fission-inducing neutrons can be simulated with a random-walk approach. Random-walk examples are the featured attraction in the next section. In presenting the numerical simulation recipes, we have tried to write programs whose structures are reasonably efficient in terms of time of execution, yet transparent to science and engineering readers who are not experts in computer programming. Writing the most efficient program is something of an art form, and if you feel that you can improve some of the recipes please feel free to do so.

Every Monte Carlo recipe in this chapter will start with the command `randomize()`. This will set the random number "seed" to a value based on the computer system clock.

[1] A pseudorandom sequence is produced, rather than truly random numbers.

```
> restart: randomize();
```

$$1120586799$$

The above output is the seed number. Note that since randomize( ) depends on the system clock, the same seed can be obtained if not enough time elapses between two consecutive calls to randomize( ). Since the specific seed number is usually of no interest to us, in our recipes we will suppress the output of `randomize()`.

The command `rand()` yields a random 12-digit nonnegative integer.

```
> r:=rand();
```

$$r := 575717670493$$

A fractional random number between 0 and 1 is obtained by dividing r by 10^{12},

```
> r2:=r/10^12;
```

$$r2 := \frac{575717670493}{1000000000000}$$

which can be put into (10-digit) floating-point form.

```
> r3:=evalf(%);
```

$$r3 := 0.5757176705$$

If it is desired to produce random integer numbers over a specified range, for example from 1 to 6 to simulate the outcomes produced by the rolling of an unbiased die, the command `rand(1..6)` is entered. This generates the relevent random-integer procedure.

```
> r4:=rand(1..6):
```

Then, using `r4()` in the following sequence command, 24 random numbers (assigned the name rn) between 1 and 6 are produced, simulating the rolling of a die. The numbers are put into a list format for statistical manipulation.

```
> rn:=[seq(r4(),i=1..24)];
```

$$rn := [5, 5, 1, 3, 3, 5, 5, 3, 4, 4, 5, 4, 4, 1, 3, 3, 6, 5, 6, 2, 3, 6, 5, 2]$$

After the Statistics library package is loaded, the `Tally` command is used to tally the number of times each random number rn turned up.

```
> with(Statistics): Tally(rn);
```

$$[1 = 2, 2 = 2, 3 = 6, 5 = 7, 4 = 4, 6 = 3]$$

In this particular run, the numbers 1 and 2 each occurred twice, the number 3 occurred 6 times, the number 5 occurred 7 times, and so on. The number of occurrences will vary from one run to the next.

In the recipes that follow, we will use variations on the above command structures, as well as introduce other statistical commands, to suit the specific aims of our files.

PROBLEMS:

Problem 7-1: Another range

Create a command structure that produces 10,000 random integers between 1 and 9. Tally the number of times each integer occurs.

Problem 7-2: Random negative decimal numbers
Create a command structure that produces random decimal numbers between -0.9 and -0.1. Then generate a sequence of 50 random numbers in this range.

7.1 Random Walks

In 1827, Robert Brown [Bro28a], [Bro28b], an English botanist, observed that the little grains of plant pollen he was studying under a microscope were motile in their suspending liquid. At first he thought these strange erratic dancing movements were self-induced, i.e., the particles were alive. However, after noticing that the same phenomena occurred in boiled water and in the water trapped inside crystals that were millions of years old, he ruled out biological causes. Brown was never able to explain this stochastic motion, now known as *Brownian motion*. The first person to present the explanation that the movements were due to the bombardment of the suspending liquid's molecules was Delsaux (1877), but the first precise measurements were made by Gouy [Gou88].

Early in the twentieth century a number of scientists conducted important research on Brownian motion, the foremost being Albert Einstein. Einstein published five papers (May, December 1905, 1906, 1907, 1908), giving mathematical explanations for Brownian movement in terms of the concept of a "random walk." A three-dimensional random walk of a particle is shown in Figure 7.1. We shall see how such a picture is produced shortly.

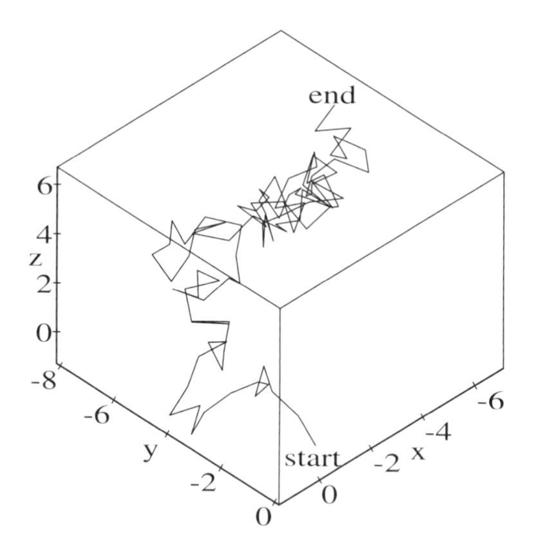

Figure 7.1: A three-dimensional random walk.

Einstein's papers are collected and published in a wonderful book,[2] which is highly recommended to the reader. In this collection of papers, the famous Einstein equation for the root mean square displacement d_x of bombarded particles of radius r suspended in a fluid of viscosity η in the direction of the x-axis in a time interval t, viz.,

$$d_x \equiv \sqrt{\langle x^2 \rangle} = \sqrt{\frac{t}{3\,\pi\,\eta\,r}\frac{RT}{N_a}} \equiv \sqrt{\frac{t}{3\,\pi\,\eta\,r}kT}, \tag{7.1}$$

appears for the first time. Here, enclosing x^2 in "angle brackets," i.e., $\langle x^2 \rangle$, indicates a statistical average over x^2, T is the absolute temperature, R is the ideal gas constant (8.31 J/(mol·K)), N_a is Avogadro's number (the number of molecules in 1 mole), and k is Boltzmann's constant (1.38×10^{-23} J/K).

The historical importance of Brownian motion cannot be overstated. The mathematical explanation of Brownian motion provided the culminating proof of the existence of atoms. Up to this time there were still scientists who doubted the existence of atoms. Einstein's publications explained the phenomenon of Brownian motion and suggested methods that could be used to make the elusive and invisible Cheshire cat of chemistry—the molecule—visible to all that wished to look. Even if Einstein had not produced his other two 1905 papers dealing with the photoelectric effect and the theory of special relativity, this work on Brownian motion would have established his reputation as a physicist of the first rank.

The French chemist Jean Perrin was able to duplicate Brownian motion in colloidal suspensions. Perrin was able to show that the jiggling particles obeyed the equipartition of energy theorem. He used Einstein's equation to make the first reasonable estimate of Avogadro's number ($N_a = 6.85 \times 10^{23}$ atoms per mole) [Sea58] and one of the first determinations of Boltzmann's constant.

Random walks at the molecular level are the basis for the physical process of diffusion. A localized concentration of a fragrant perfume will spread or diffuse in air due to the random walk of each perfume molecule as it bounces off air molecules. Even if the speeds of the diffusing molecules are high between collisions, the average distance of the molecules from the starting point after many collisions is considerably less than would be anticipated on the basis of speed alone because of the convoluted paths that the molecules travel. If a diffusing molecule had a constant speed and a fixed time interval between collisions, it can be shown that after n collisions the average distance traveled would be proportional to $\sqrt{n}$. This is considerably less than the distance the molecule would travel along a straight line without suffering collisions. In this latter case, the distance would be proportional to the elapsed time, which scales linearly with n.

[2]If you wish to explore the mathematics behind Brownian motion, you can do no better than start with this easily readable collection, entitled *Investigation on the Theory of the Brownian Movement* [Ein56]. As a bonus, there is an appendix that provides a short history and a list of the players involved in this drama. You are also directed to volume I of the famous Feynman trilogy, *Lectures on Physics* [FLS64], for a discussion of random walks.

To simulate a random walk problem with a computer, a random-number generator is used to produce random changes in direction of the diffusing particle. How this works will be illustrated with several simple examples, ranging from one- to three-dimensional motion.

PROBLEMS:

Problem 7-3: Estimating Avogadro's number

This problem illustrates the underlying method used by Perrin to estimate Avogadro's number. Figure 7.2 shows the successive positions at 30 second

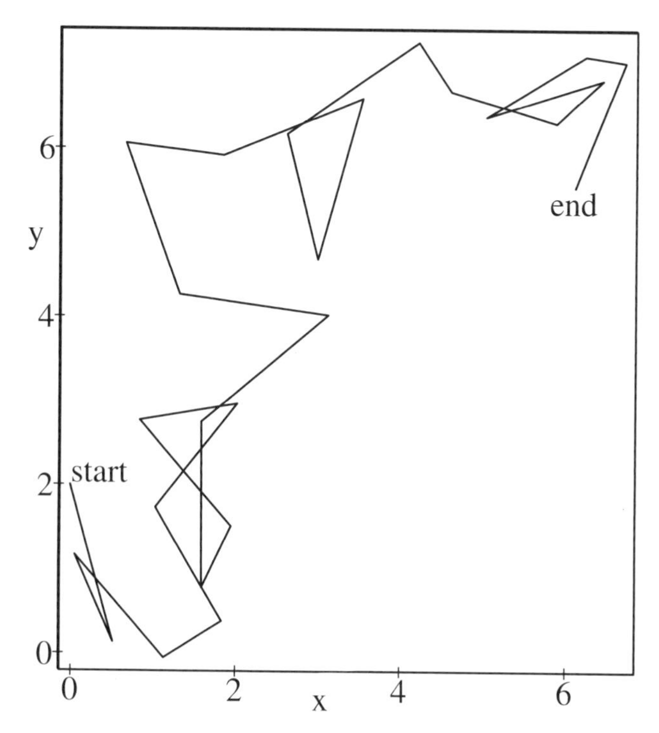

Figure 7.2: A random walk in two dimensions.

intervals of a small particle of radius $r = 0.1$ μm suspended in a water solution (viscosity $\eta = 1.00 \times 10^{-3}$ N·s/m^2) held at a temperature of 20 °C. In the figure, a distance of 1.0 cm corresponds to 1.0 μm.

(a) Measure the distance in cm of each of the first 10 steps. Using the scale factor, convert the measured distances into μm.

(b) Calculate the mean-square average distance $\langle d^2 \rangle$ by squaring each distance, summing the squares, and dividing by the 10 measurements.

(c) Calculate d_x using $\langle x^2 \rangle = \langle d^2 \rangle / 2$. Justify this relationship.

(d) Use Equation (7.1) to estimate Avogadro's number N_a.

(e) Check your answer by using the steps from 5 to 15, 10 to 20, or any other combination that you wish. Are all the values consistent? Average over the estimates and compare with the known value of N_a.

Problem 7-4: Einstein's estimate
In his May 1905 paper, Einstein made an estimate of what d_x should be in 1 second and in 1 minute for particles of diameter 1 μm suspended in water held at 17 °C. At this temperature, the viscosity of water is $\eta = 1.35 \times 10^{-3}$ N·s/m^2. Use Einstein's formula (7.1) to estimate d_x for $t = 1$ s; for $t = 1$ min.

7.1.1 The Soccer Fan's Drunken Walk

Some people think football is a matter of life and death ...
I can assure them it is much more serious than that.
Bill Shankly, former British football club manager (1914–1981)

When coauthor Richard was a postdoctoral fellow many eons ago, he spent a year carrying out theoretical calculations on the migration of electrons in a solid at the University of Liverpool in England. On Saturday afternoons, he would migrate to the local soccer (football, to the rest of the world outside North America) pitch to watch Liverpool, who were one of the premier soccer teams in the world at that time, play Manchester United and other fabled English and European soccer teams. At that time the stadium, which held some 50 to 60 thousand fans, had seats for only a small fraction of the crowd. Being a student, Richard would pay the minimum entry fee and stand with thousands of others in the ramped and cramped end zone, referred to as the Kop. The fans were jammed in like sardines and occasionally someone would faint and his or her body would be passed over the heads of the crowd down to the waiting St. John's ambulance people, who were ever present on the sidelines of the field. Leaving the game was equally hazardous, since one had to squeeze through narrow exits and down steep stairs, trying to avoid falling and therefore being trampled by the surging crowd. Once out on the street, the constabulary mounted on horses tried to direct the crowd in an orderly fashion away from the stadium. Often, it was mayhem! The boisterous fans then would retire to their favorite local pub and relive the exciting moments of the soccer match, the pitch of the arguments increasing as the beer mugs were drained. The migration of some of the inebriated fans to their nearby homes was almost that of a random walk or, should we perhaps say, a random stagger.

The simplest random-walk problem is the one-dimensional drunkard's walk, which is idealized as follows. Starting at the origin (the pub door), $X_0 = 0$, the drunken soccer fan is allowed to make either a step of length L to the right or to the left along the narrow street with equal probability. That is to say, the probability of a step to the left is $p = \frac{1}{2}$ and, of course, a probability of $\frac{1}{2}$ to the right. Thus, for example, after the first step, the fan's position would be either

$X_1 = X_0 + L = L$ or $X_1 = X_0 - L = -L$. Similarly, after the nth step, the fan's position would be either $X_{n+1} = X_n + L$ or $X_{n+1} = X_n - L$. The question is, what is the fan's location after $n = N$ steps? The answer will depend on the sequence of random plus and minus steps generated. Statistically, however, the average displacement of the fan from the pub door would average out to zero. The root mean square displacement D, however, will not average out to zero, but instead will be given by $D = \sqrt{N}\,L$ after N steps. The argument leading to this latter expression is as follows.

After the nth step, the displacement squared is either $X_{n+1}^2 = (X_n + L)^2 = X_n^2 + 2\,X_n\,L + L^2$, or $X_{n+1}^2 = (X_n - L)^2 = X_n^2 - 2\,X_n\,L + L^2$. We add the two possible results and assume that for equal probabilities for movement to the right or to the left, the plus and minus contributions will cancel on the average. Then, on dividing the sum by two, and using angle brackets to denote the statistical average, we obtain $\langle X_{n+1}^2 \rangle = \langle X_n^2 \rangle + L^2$. But for $n = 0$, we have $\langle X_1^2 \rangle = \langle X_0^2 \rangle + L^2 = L^2$, for $n = 1$, $\langle X_2^2 \rangle = \langle X_1^2 \rangle + L^2 = 2\,L^2$, and by induction, $\langle X_N^2 \rangle = N\,L^2$. The root mean square (rms) distance then is $D \equiv \sqrt{\langle X_N^2 \rangle} = \sqrt{N}\,L$. The argument leading to $D \propto \sqrt{N}$ is, however, a statistical one and the actual root mean square distance may differ at a given N value from the theoretical prediction for a particular random walk.

Using a random-number generator, let's now simulate the one-dimensional random walk and look at the actual behavior of the root mean square distance D as a function of N. The plots and Statistics packages are loaded because we shall be using the `display` and `ScatterPlot` commands.

```
> restart: with(plots): with(Statistics):
```

To obtain a good statistical average, we shall average over the random walks of many fans, so the displacement variable will have two subscripts, one (n) to keep track of which step has been executed for a given fan, the other (k) to keep track of which fan it is. The subscript k is referred to as the "trial" number.

For the kth fan, the displacement algorithm will be written in the form

$$X_{n+1,k} = X_{n,k} + d \tag{7.2}$$

with d taking on the value $d1 = +L$ or $d2 = -L$. This corresponds to equal-size steps to the right and to the left. A random number will be generated on each step for the kth fan. If this number lies between 0 and p, then $d1$ is selected, while if it is between p and 1, then $d2$ is chosen. For our example, we take $p = \frac{1}{2}$, so that there is an equal probability of a step to the right or to the left. The alteration of the code to handle unequal step sizes and unequal probabilities is clearly easy to implement. The step size is taken to be $L = 1$ length unit.

```
> L:=1; d1:=L; d2:=-L; p:=1/2;
```

$$L := 1 \quad d1 := 1 \quad d2 := -1 \quad p := \frac{1}{2}$$

The maximum number of steps is taken to be $N = 50$, and 1500 fans (trials) are considered. If you have sufficient speed and memory on your computer, you may want to increase the number of steps. Fewer trials, on the other hand, will in general lead to a less-accurate statistical average.

```
> N:=50: trials:=1500:
```

The call `randomize()`, which is now entered, resets the random number seed according to the system clock. On each step, a new random integer will be generated in the specified range. If this command line is not entered, the same seed will be used each time the program is run and one will not obtain a new set of random numbers. If desired, you can omit this line and check that this is the case.

```
> randomize():
```

To gauge the efficiency of a numerical simulation, the beginning time and ending time to execute the major portion, or all, of a program can be recorded. The command `time()` returns the total CPU time in seconds used since the start of the Maple session.

```
> begin_time:=time();
```

$$begin\_time := 5.047$$

Thus, in this case 5 seconds have elapsed since the authors started this particular Maple session. Now, we present the heart of the code, which involves two do loops. The first, or "outer," loop will iterate over the number k of fans.

```
> for k from 1 to trials do
```

Each fan is positioned at step $n = 0$ at the origin (the pub door, say).

```
> X[0,k]:=0;
```

The second, or "inner," do loop is now applied to the displacement of the kth fan, who undergoes a maximum number of $N = 50$ steps.

```
> for n from 0 to N do
```

To create a random fractional number between 0 and 1, the `rand()` command with no argument specified is first used to produce a random 12-digit nonnegative number, and then this number is divided by 10^{12}. If the random number is less than or equal to p, then $d = d1$ is chosen. Otherwise, $d = d2$ is selected. This is accomplished through the following conditional "if...then" statement.

```
> if rand()/10^12<=p then d:=d1 else d:=d2 end if;
```

Equation (7.2) is entered,

```
> X[n+1,k]:=X[n,k]+d;
```

and the inner do loop ended.

```
> end do:
```

In the outer loop, the sequence of $X_{n,k}^2$ from $n = 0$ to $n = N$ is formed into a Maple list and labeled as S_k, a list being produced for each fan.

```
> S[k]:=[seq((X[n,k]^2),n=0..N)];
> end do:
```

On completion of the outer loop, we might have wished to insert one or more additional command lines somewhere inside the loops without typing the do loop structure over again. To insert a command line inside a loop, place the cursor at the start (just after Maple prompt >) of the next command line following the desired point of insertion of the new line. Then click on the left keyboard arrow to move the cursor to the left of > and then press Enter. This process will insert a blank command line above the cursor.

We now add up all the lists S_k and divide by the number of fans (trials) to give us a list of average mean square displacements, each entry corresponding to a different n value up to $n = N$.

```
> Pts:=evalf(add(S[k],k=1..trials)/trials):
```

The map command is used to take the square root of each entry in the list Pts. The new list, labeled Pts2, now contains the (average) root mean square displacements, D, for different total numbers of steps.

```
> Pts2:=map(sqrt,Pts):
```

A list of the step numbers from $n = 0$ to $n = N$ is formed.

```
> n_coords:=[seq(n,n=0..N)]:
```

All that remains is to plot the data. How much CPU time has the calculation taken? By asking for the end_time,

```
> end_time:=time();
```

$$end\_time := 7.290$$

and subtracting begin_time from this number,

```
> elapsed_time:=end_time-begin_time;
```

$$elapsed\_time := 2.243$$

we see that a total of about 2 seconds of CPU time has elapsed. The CPU time can be used as a benchmark against which any attempts to improve the efficiency of a given program on the same computer can be compared. The CPU times quoted in this chapter were obtained with a 3 GHz Pentium IV personal computer. We have tried to keep the CPU times of most of the recipes short, but be warned that some of the problems require considerably longer CPU times to obtain accurate statistical results. In these cases you might wish to "play" with the recipes in order to make them run faster.

A graph of the numerical points is now created with the ScatterPlot command, but not displayed. A point style is chosen, the numerical points being represented by size-14 circles.

```
> gr1:=ScatterPlot(n_coords,Pts2,style=point,
        symbol=circle,symbolsize=14):
```

The theoretically predicted behavior of D as a function of the step number, which was derived earlier, is plotted as a thick blue line.

```
> gr2:=plot(L*sqrt(x),x=0..N,color=blue,thickness=2):
```

The two graphs are superimposed with the display command,

```
> display({gr1,gr2},tickmarks=[3,3],labels=["N","D"]);
```

the result being shown in Figure 7.3.

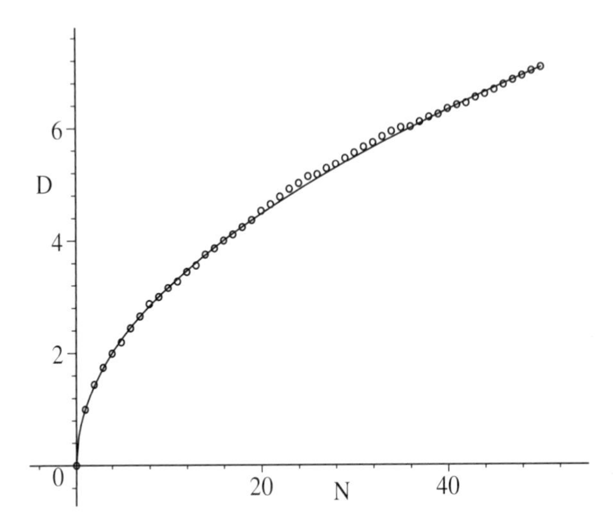

Figure 7.3: Root mean square distance D versus number of steps N. Solid line: theoretical prediction. Circles: numerical points.

The numerical simulation data are clearly in excellent agreement with the theoretically derived formula $D = L\sqrt{N}$.

Of course, for an actual inebriated soccer fan, the step sizes to the right and to the left will undoubtedly be unequal, as will the probabilities. Further, the random walk will not normally be confined to one dimension, so the model must be modified to take these various factors into account. Our goal is not to make a lifelong study of the random stagger of inebriated soccer fans, so we leave these aspects for the interested reader to explore. An example of a two-dimensional random walk with unequal probabilities in different directions is given in a different context in the next story.

PROBLEMS:

Problem 7-5: Unequal probabilities

Suppose that in the text recipe the probability p of a step to the right is three times the probability q of a step to the left, all step sizes being exactly the same. It can be shown, using statistical arguments, that more generally,

$$D = \sqrt{(p-q)^2\, N^2 + 4\, p\, q\, N}.$$

Confirm D by superimposing it on a plot of the numerical simulation data.

Problem 7-6: Different L values

Confirm that the (average) rms distance correctly scales with L as well as N.

Problem 7-7: Number of trials

In the text recipe, about how many trials are needed to obtain D to approximately 5% accuracy for $N = 9$ steps? for $N = 49$ steps?

Problem 7-8: Unequal step sizes

Suppose for the drunken soccer fans in the text recipe that $d1 = 2\,L$ and $d2 = L$, all other parameters being the same.

(a) Numerically determine the average displacement $\langle X(N) \rangle$, which will now be nonzero, and plot the result.

(b) Numerically determine the average mean square displacement $\langle X(N)^2 \rangle$ and plot the result.

(c) Calculate $\sqrt{\langle X(n)^2 \rangle - \langle X(N) \rangle^2}$ and plot the result. What does this quantity represent?

7.1.2 Blowin' in the Wind

How many times must a man look up, Before he can see the sky? Yes 'n' how many ears must one man have, Before he can hear people cry? ... The answer, my friend, is blowin' in the wind, ...
Bob Dylan, American folk singer and songwriter (1941–)

In 1962 Bob Dylan composed the popular folk song "Blowin' in the Wind," whose powerful lyrics and stirring tune became identified with the civil rights movement in the United States. In 1964 the well-known trio Peter, Paul, and Mary received "Grammy Awards" for their recording of this song.

In this section, recalling the swirling events of his childhood and inspired by the lyrics of Dylan's song, Russell (the engineer) decides to simulate the random walk of a raindrop falling from a rain cloud and being buffeted by the swirling winds of Rainbow County.

```
> restart: with(plots):
```

The bottom of the rain cloud is taken to be 1 km, or $h = 1000$ meters, above the ground. For simplicity, each random step (displacement) of the raindrop will be chosen to be of the same length, say, $d = 1$ meter. The reader can, of course, change the value of d if so desired to see what effect step size has on the raindrop's trajectory.

```
> h:=1000: d:=1:
```

In Rainbow County, the prevailing wind tends to gust from the west, but due to its swirling nature it occasionally reverses direction at different altitudes. Russell treats the random-walk problem as two-dimensional, taking the horizontal direction to be labeled as x, the vertical direction as y. The origin ($x = 0$, $y = 0$) is chosen at a point on the ground directly below the initial position ($x = 0$, $y = h$) of the raindrop in the cloud. Positive x corresponds to being east of the initial position, negative x to the west. The extension of the model calculation to include north and south gusts is easy to implement.

On a given step, Russell assumes that the probability of the particle moving vertically upward a distance d due to an updraft is 0.1, of falling vertically downward is 0.7, of being blown to the right (wind from the west) is 0.15, and of being blown to the left (wind from the east) is 0.05. The position of the raindrop at step $n+1$ is related to its position at the nth step by the relations

$$x_{n+1} = x_n + a_i, \quad y_{n+1} = y_n + b_i, \tag{7.3}$$

with i taking on the values 1, 2, 3, 4. On each of the n steps, a random number r lying between 0 and 1 is generated. If $r < p_1 = 0.1$, then $i = 1$ is chosen; otherwise, if $r < p_2 = 0.8$, then $i = 2$ is selected; otherwise, if $r < p_3 = 0.95$, then $i = 3$ is chosen; otherwise, $i = 4$ is selected. The values of the a_i, b_i, and p_i are now entered.

```
> a[1]:=0: a[2]:=0: a[3]:=d: a[4]:=-d:
> b[1]:=d: b[2]:=-d: b[3]:=0: b[4]:=0:
> p[1]:=0.1: p[2]:=0.8: p[3]:=0.95:
```

If, for example, the random number $r = 0.55$ is generated on a given step n, then the coordinates of the raindrop on the next step are $x_{n+1} = x_n + a_2 = x_n$, $y_{n+1} = y_n + b_2 = y_n - d$. In this step, the raindrop falls vertically a distance d.

Since Russell has chosen $d = 1$ and $h = 1000$, it will take at least 1000 steps for a raindrop to strike the ground, because there are small, but not negligible, probabilities of horizontal and even upward displacement. In this run, Russell allows the calculation to proceed to a maximum of $N = 2000$ steps. The input coordinates ($x_0 = 0$, $y_0 = h$) of the raindrop, as it leaves the bottom of the cloud, are entered separately and as a list.

```
> N:=2000; x[0]:=0: y[0]:=h: pnt[0]:=[x[0],y[0]];
```

$$N := 2000 \qquad pnt_0 := [0, 1000]$$

The random number seed is set, and the starting time for the body of the program recorded, but not displayed.

```
> randomize(): begin_time:=time():
```

The zeroth step is entered,

```
> n:=0:
```

and the conditional loop begins. Russell would like to know how many steps it takes on a given numerical run for the raindrop to strike the ground, this number varying from one run to the next. To find the total number of steps and also stop the program when the raindrop hits the ground, he inserts a conditional while statement. The iteration of the do loop will continue only while $n \leq N$ and $y_n \geq 0$.

```
> while (n<=N and y[n]>=0) do
```

A random number, expressed in decimal form, is generated between 0 and 1,

```
> r:=evalf(rand()/10^12);
```

and the "if then" probability statement is entered,

```
> if r<p[1] then i:=1 elif r<p[2] then i:=2
  elif r<p[3] then i:=3 else i:=4 end if;
```

as well as Equation (7.3).

```
> x[n+1]:=x[n]+a[i]; y[n+1]:=y[n]+b[i];
```

The raindrop coordinates on step $n+1$ are formed into a list,

```
> pnt[n+1]:=[x[n+1],y[n+1]];
```

and the value of n incremented by one.

```
> n:=n+1;
> end do:
```

On completion of the do loop, the total number of steps for the raindrop to strike the ground, the final vertical position (which should be $y = 0$) and its horizontal displacement are displayed.

```
> Total_steps:=n-1; Vertical_position:=y[n-1];
  Horizontal_displacement:=x[n-1];
```

$Total\_steps := 1697 \quad Vertical\_position := 0 \quad Horizontal\_displacement := 181$

For this particular run, it took 1697 steps for the raindrop to hit the ground, its final horizontal position being 181 m to the right (to the east) of its initial horizontal position when it left the cloud. That the raindrop would likely land to the east was expected on probabilistic grounds.

The end time is recorded, and the elapsed CPU time (in s) evaluated.

```
> end_time:=time():
> elapsed_time:=%-begin_time;
```

$$elapsed\_time := 0.100$$

This calculation took only a fraction of a second to execute.

Feeling rather whimsical and, maybe, slightly nostalgic for the innocence of his lost youth, Russell decides to use the `polygonplot` command to pictorially create the bottom portion of the rain cloud. Making use of `style=patch` and an appropriate shade of blue, a graph of the cloud is formed.

```
> gr1:=polygonplot([[-500,1000],[500,1000],[500,1050],
        [-500,1050]],style=patch,color=COLOR(RGB,0.2,0.3,0.6)):
```

The proportions of red (R), green (G), and blue (B) are controlled through the numerical values inserted into the color command. The reader who prefers a different-colored cloud can adjust the numbers according to his or her taste. The next command line creates a graph of the raindrop's trajectory, the random steps being represented as straight-line segments, all of equal length. The raindrop is assigned the same color as the cloud from which it originated. In actuality, this may not be a realistic thing to do, so once again feel free to choose your own color scheme.

```
> gr2:=pointplot([seq(pnt[j],j=0..Total_steps)],
        style=line,color=COLOR(RGB,0.2,0.3,0.6)): #raindrop
```

Making use of the `display` command, the falling raindrop and the rain cloud are shown in Figure 7.4. To keep the scaling in the horizontal and vertical directions the same, Russell uses the `scaling=constrained` plot option.

```
> display({gr1,gr2},axes=boxed,scaling=constrained,
  view=[-500..500,0..1050],labels=["x","y"],tickmarks=[3,4]);
```

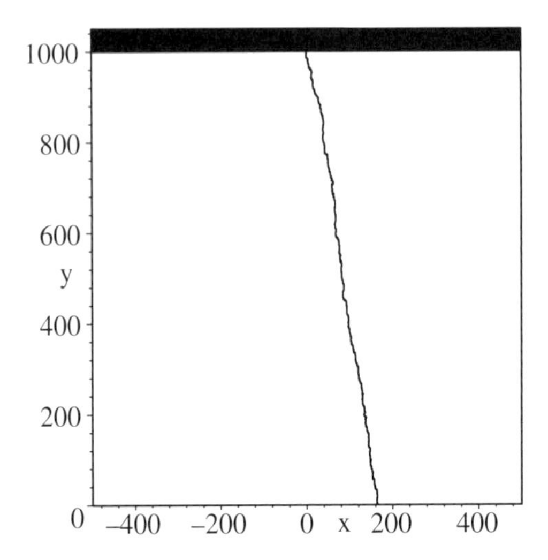

Figure 7.4: Random walk of the falling, wind-buffeted raindrop.

The curious reader might wonder what it was in Bob Dylan's song that triggered Russell to indulge in his raindropmodel calculation. When asked this question, Russell declined to be specific but was heard to reply enigmatically, "The answer my friend, is blowin' in the wind."

PROBLEMS:

Problem 7-9: Calculating averages

By averaging over a large number of numerical runs, determine the average number $\langle n \rangle$ of steps it takes the raindrop to hit the ground and the average horizontal displacement D. How do $\langle n \rangle$ and D depend on h?

Problem 7-10: Different wind velocities

In the text calculation, the displacement was taken to be the same on each step whether due to a downdraft, updraft, or a sidedraft. Explore the effect of unequal displacements due to different wind velocities in different directions.

Problem 7-11: Three-dimensional motion

Alter the recipe to allow for three-dimensional motion of the falling raindrop and create a three-dimensional plot of the raindrop's trajectory. Alter the probabilities to values that you think are reasonable.

Problem 7-12: A rain-buffeted skyscraper

In the text recipe, make the following modifications:

(a) Using the `polygonplot` command, create a graph of a solidly colored skyscraper 800 meters high and occupying the region between $x = 150$ and 500 meters. Color the skyscraper brown by taking the color combination $(\text{RGB}, 0.5, 0.5, 0.3)$. Include this graph in the `display` command.

(b) Modify the `while` statement to terminate the program when a raindrop strikes the skyscraper.

(c) Record and display the vertical height above the ground at which the raindrop hits the skyscraper.

(d) By averaging over a large number of numerical runs, determine the average height above the ground at which the raindrop hits the skyscraper.

(e) What percentage of raindrops make it all the way to the ground without hitting the skyscraper?

7.1.3 Flight of Penelope Jitter Bug

Of all the thirty-six alternatives, running away is best.
Chinese proverb

In a certain computer fantasy game, Penelope Jitter Bug is being pursued by that odious reptile, Snide Lee Lizard. To avoid capture, Penelope is allowed to move erratically on a three-dimensional rectangular lattice. The rules are that she can move randomly in the x-, y- and z-directions with a possible step of $+1$, 0, or -1 in each direction. Thus, if Penelope starts out at the origin, ($x=0$, $y=0$, $z=0$), after the first step she could end up at $(0,0,0)$, i.e., she could fake Snide Lee out and not move at all, or Penelope could end up at one of 26 other neighboring positions, e.g., $(1,0,0)$, $(0,-1,0)$, $(1,-1,1)$, etc. Maple's random-number generator will be used to randomly select 1, 0, or -1 for each of the x, y, and z steps.

```
> restart: with(plots):
```

The starting point is the origin, which is labeled `step||0`, and $n = 1000$ time steps are considered. The random number seed is set,

```
> step||0:=[0,0,0]: n:=1000: randomize():
```

and the starting time recorded.

```
> begin:=time():
```

A do loop repeats the calculation n times.

```
> for i from 0 to n-1 do
```

The `rand(-1..1)` command allows the values $-1, 0, +1$ to be randomly generated for each of the x, y, and z steps.

```
> xstep:=rand(-1..1):
> ystep:=rand(-1..1):
> zstep:=rand(-1..1):
```

The triplet, `xstep()`, `ystep()`, `zstep()` of random numbers is then added to the coordinates of `step||i` to give the new coordinates at `step||(i+1)`.

```
> step||(i+1):=step||i+[xstep(),ystep(),zstep()];
> end do:
```

```
> cpu_time:=(time()-begin)*seconds;
```

$$cpu\_time := 0.461\ seconds$$

On completion of the do loop, which took only a fraction of a second to complete, the random path traced out by Penelope Jitter Bug is plotted using the pointplot3d command, with the sequence of steps connected by lines of length 0, 1, $\sqrt{2}$, or $\sqrt{3}$. To avoid possible confusion, a boxed set of axes is chosen.

```
> pl:=pointplot3d([seq(step||i,i=0..n)],axes=boxed,style=line,
      tickmarks=[3,3,3],labels= ["x","y","z"]):
```

To find the x-, y-, and z-coordinates of the last step, step||n is selected,

```
> last:=step||n;
```

$$last := [-23, 19, 8]$$

and the op command used.

```
> x:=op(1,last); y:=op(2,last); z:=op(3,last);
```

$$x := -23 \qquad y := 19 \qquad z := 8$$

Thus, for this particular run Penelope ends up at $x = -23$, $y = 19$, and $z = 8$ after 1000 steps. The distance from the origin is $\sqrt{(-23)^2 + (19)^2 + (8)^2} = 30.9$, which is not that far from $\sqrt{n} = \sqrt{1000} = 31.6$. Of course, this was only one run, the coordinates of the last step varying from one run to the next. However, by generalizing the theoretical argument given in the one-dimensional soccer fan story, we could have predicted that the average rms distance also scales in three dimensions with the square root of n.

Before displaying Penelope's random walk, textplot3d is used to add the words "start" (colored red) and "end" (colored blue) to the graph.

```
> tt:=textplot3d([0,0,0,"start"],color=red):
> tt2:=textplot3d([x,y,z,"end"],color=blue):
```

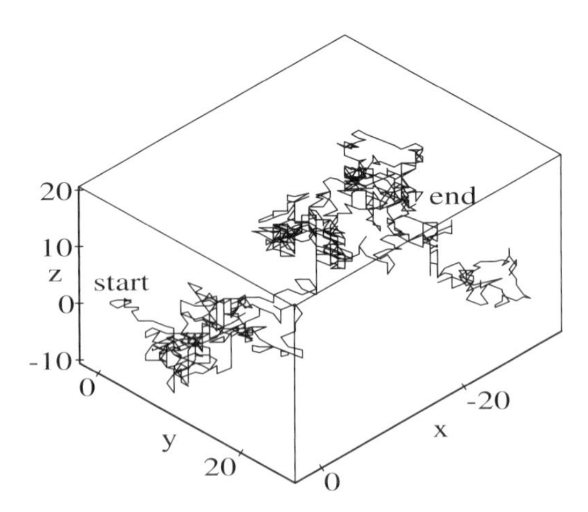

Figure 7.5: Random walk of Penelope Jitter Bug.

Finally, with the `display` command and constrained scaling,

```
> display({pl,tt,tt2},scaling=constrained);
```

the erratic path of Penelope Jitter Bug is revealed in Figure 7.5. Each time the code is run, the reader will observe a different path traced out by Penelope Jitter Bug, with a different set of final coordinates after n steps.

PROBLEMS:

Problem 7-13: Scaling of average distance with n

By carrying out a number of numerical runs for each n value, determine the average final distance $\langle d \rangle$ as a function of n and make a plot of your results. What functional form gives a good fit to your data? What can you conclude?

Problem 7-14: Penelope really starts jittering

Suppose that Penelope Jitter Bug can move randomly in the x-, y-, and z-directions with a possible step of $+2$, $+1$, 0, -1, -2 in each direction. If Penelope starts at the origin, how many neighboring positions are possible after the first step? What step lengths are possible? Explore the text file for this situation and try to compare the observed behavior with that which occurred when the possible steps were $+1$, 0, -1.

7.1.4 That Meandering Perfume Molecule

I cannot talk with civet in the room,
A fine puss-gentleman that's all perfume.
William Cowper, English poet (1731–1800)

In the previous recipe, Penelope Jitter Bug was confined to moving along a three-dimensional rectangular grid or lattice. In contrast, a diffusing perfume molecule, such as the malodorous one referred to by William Cowper, can move in any angular direction in three-dimensional space.

To describe its motion, it is necessary in our model calculation to introduce spherical polar coordinates. In this coordinate system, two angles must be specified, the angles θ and ϕ of the displacement (magnitude r) with respect to the z-and x-axes, respectively. The relation of spherical coordinates to the Cartesian coordinates x, y, z is given by

$$x = r \sin\theta \cos\phi, \quad y = r \sin\theta \sin\phi, \quad z = r \cos\theta, \tag{7.4}$$

where, by convention, θ ranges from 0 to π and ϕ from 0 to 2π radians.

The average distance that the perfume molecule travels between collisions with the surrounding air molecules is called the mean free path. At standard temperature (0 °C) and pressure (1 atmosphere), the mean free path is on the order of 10^{-7} meters.

```
> restart: with(plots):
```

For simplicity, in our simulation let's take $r = 1$, corresponding to the perfume molecule traveling one mean free path between each collision. The molecule is started at the origin and will be allowed to undergo $n = 1000$ random steps. The numerical value of π is calculated and labeled for future use as p.

```
> step||0:=[0,0,0]: r:=1: p:=evalf(Pi): n:=1000: randomize():
```

The starting time is recorded,

```
> start_time:=time():
```

and a do loop used to generate the n positions of the perfume molecule during its random walk.

```
> for i from 0 to n-1 do
```

For the angular part of the random walk, a uniform angular distribution is required. In spherical polar coordinates, the volume element is given by

$$dV = r^2 \sin\theta\, d\theta\, d\phi\, dr = \sin\theta\, d\theta\, d\phi\, dr, \tag{7.5}$$

since $r = 1$ here. The angular part of dV is $\sin\theta\, d\theta\, d\phi = -d(\cos\theta)\, d\phi \equiv -dg\, d\phi$, where $g \equiv \cos\theta$. Since any volume element one mean free path away is equally likely after a collision, a uniform distribution is desired for ϕ and g. To create a random decimal number for ϕ in the range 0 to 2π, the following command line, labeled phistep, is entered.

```
> phistep:=2.0*p*rand()/10^12:
```

The variable g will vary from -1 to $+1$ for $\theta = \pi$ to 0. The following command line therefore produces a random decimal number in the range $g = -1$ to 1.

```
> gstep:=-1+2.0*rand()/10^12:
```

Noting that $\sin\theta = \sqrt{1 - \cos^2\theta} = \sqrt{1 - g^2}$, and using Equation (7.4), we will determine the x, y, and z steps using the relations

$$x = r\, \cos\phi\, \sqrt{1 - g^2}, \quad y = r\, \sin\phi\, \sqrt{1 - g^2}, \quad z = r\, g, \tag{7.6}$$

which are now entered.

```
> xstep:=r*cos(phistep)*sqrt(1-gstep^2):
> ystep:=r*sin(phistep)*sqrt(1-gstep^2):
> zstep:=r*gstep:
```

The coordinates of the perfume molecule at step $(i + 1)$ are determined.

```
> step||(i+1):=step||i+[xstep,ystep,zstep];
> end do:
```

On completion of the do loop, the perfume molecule's x- y-, and z-coordinates are determined for the last (nth) step,

```
> last:=step||n:
> x:=op(1,last); y:=op(2,last); z:=op(3,last);
```

$$x := 9.373913264 \quad y := -17.39656991 \quad z := -25.78312563$$

and the molecule's distance $R = \sqrt{x^2 + y^2 + z^2}$ from the starting point

```
> R:=sqrt(x^2+y^2+z^2);
```

$$R := 32.48508061$$

calculated. For this particular run, the perfume molecule has moved approximately 32 mean free paths from its starting point at the origin. The end time is recorded, and the CPU time for the run found to be less than a second.

```
> end_time:=time():
> CPU_time:=end_time-start_time;
```

$$CPU\_time := 0.561$$

The sequence of 1000 steps is plotted in three dimensions and the points joined by straight-line segments.

```
> pl:=pointplot3d([seq(step||i,i=0..n)],axes=boxed,style=line,
      tickmarks=[3,3,3],labels= ["x","y","z"]):
```

The word "start" is placed at the origin, and the word "end" at (x, y, z).

```
> tt:=textplot3d([0,0,0,"start"],color=red):
> tt2:=textplot3d([x,y,z,"end"],color=blue):
```

The display command is used with constrained scaling to superimpose the plots.

```
> display({pl,tt,tt2},scaling=constrained);
```

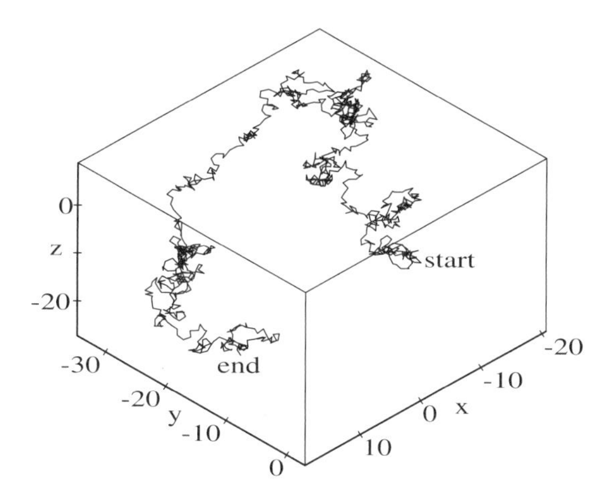

Figure 7.6: Three-dimensional random walk of the perfume molecule.

The random walk of the perfume molecule for this particular run is shown in Figure 7.6. The path traced out by the molecule and the distance r from the origin will vary from one run to the next. The viewing box can be rotated so that the path and labels may be viewed from different perspectives.

PROBLEMS:

Problem 7-15: Neutron diffusion

The diffusion of neutrons through the lead shielding wall of an atomic reactor can be simulated in a similar manner to the meandering perfume molecule.

Assume for simplicity that each neutron enters one of the two walls of a planar lead slab perpendicularly and travels one mean free path (mfp) inside the slab before suffering its first collision with a lead atom. The neutron is then scattered 1 mfp in a random direction before suffering its second collision, where it is scattered 1 mfp in a random direction once again, and so on. Suppose that each neutron can withstand only 10 such collisions. If the slab is 3 mfp thick, what proportion of entering neutrons will be able to escape through the opposite wall? Does your answer make intuitive sense? Explain. How thick should the slab be to reduce the proportion to about 1%?

7.2 Monte Carlo Integration

Another important application of the Monte Carlo random-number generator approach is to the evaluation of two-and three-dimensional integrals with complicated or irregular boundaries and to still higher-dimensional integrals that can arise in such areas as statistical mechanics. Monte Carlo techniques are used when more conventional numerical techniques are either difficult or almost impossible to apply.

However, to illustrate the Monte Carlo integration method, the recipes and problems of this section deal mainly with one-dimensional integrals so that the answer can be compared with the result obtained using standard numerical techniques. To evaluate the one-dimensional integral

$$I = \int_a^b f(x)\,dx \tag{7.7}$$

using a Monte Carlo method, we note that the area under the curve $f(x)$ between $x\!=\!a$ and $x\!=\!b$ can be written as

$$\int_a^b f(x)\,dx = (b-a)\,\langle f\rangle, \tag{7.8}$$

where $\langle f\rangle$ is the average value of $f(x)$. But if we are able to generate a uniform distribution of x_i between $x=a$ and $x=b$ with a random-number generator, then each $f(x_i)$ can be evaluated, and the average is

$$\langle f\rangle = \frac{1}{n}\sum_{i=1}^{n} f(x_i). \tag{7.9}$$

The Monte Carlo estimate of the integral will then be given by

$$I = \int_a^b f(x)\;dx = \frac{(b-a)}{n}\sum_{i=1}^{n} f(x_i), \tag{7.10}$$

the accuracy of the numerical value depending on the size of the number n.

As with the random walk examples, for large n it can be shown that there is a distribution of Monte Carlo values for the integral centered at the correct answer, the width of the distribution decreasing as $1/\sqrt{n}$. Comparing this error with that for the standard integration rules, the Monte Carlo method is really not competitive for one-dimensional integrals. Very much larger n values have to be used in the Monte Carlo method to attain similar accuracy to that attained by, for example, Simpson's rule, which has an error proportional to $1/n^4$. However, the Monte Carlo integration error turns out to be independent of the dimensionality d of the integral, whereas the error in, say, Simpson's rule when generalized to d dimensions is $1/n^{4/d}$. In this case, the Monte Carlo method would become more accurate for dimensions higher than $d=8$. You might snort that you are not likely to run into integrals with such a high dimensionality. If you are a physics major you might! According to deVries [DeV94], integrals similar to the 9-dimensional integral (with integrations over all space)

$$I = \int\int\cdots\int \frac{da_x\, da_y\, da_z\, db_x\, db_y\, db_z\, dc_x\, dc_y\, dc_z}{(\vec{a}+\vec{b})\cdot\vec{c}}, \tag{7.11}$$

appear in the study of electron plasmas. The Monte Carlo random-number method can be used to give an approximate estimate of the value of this integral while Simpson's rule, for example, is not feasible.

Before demonstrating the Monte Carlo integration technique, we should remind the reader of some of the common numerical methods for evaluating one-dimensional definite integrals. This is done in the following recipe.

7.2.1 Numerical Integration Methods

An effective human being is a whole that is greater than the sum of its parts.
Ida P. Rolf, American biochemist, (1896–1979)

Consider a definite integral of the general structure $I = \int_a^b f(x)\,dx$. To integrate I by standard numerical techniques, the integration range a to b is divided into n equal intervals $\Delta x = (b-a)/n$. Then the points in the a to b range are labeled as $x_i = x_0 + i\,\Delta x$ with $x_0 = a$ and $x_n = b$. Three elementary numerical integration formulas, or "rules", which make use of different linear combinations of the $f(x_i)$, are

$$\begin{aligned} I_n &= \sum_{i=0}^{n-1} f(x_i)\,\Delta x, \qquad \text{rectangular,} \\ I_n &= \frac{1}{2}\left[f(x_0) + 2\sum_{i=1}^{n-1} f(x_i) + f(x_n)\right]\Delta x, \qquad \text{trapezoidal,} \\ I_n &= \frac{1}{3}[f(x_0) + 4f(x_1) + 2f(x_2) + 4f(x_3) + \cdots \\ &\quad + 2f(x_{n-2}) + 4f(x_{n-1}) + f(x_n)]\,\Delta x, \qquad \text{Simpson.} \end{aligned} \tag{7.12}$$

For Simpson's rule the number of intervals n must be even. It can be shown that for the rectangular, trapezoidal, and Simpson's numerical integration formulas, the error (difference between the exact and approximate answers) has (approximately) a $1/n$, $1/n^2$, and $1/n^4$ dependence, respectively [GT96].

As a specific example we shall integrate $I = \int_0^1 e^{-x^2}\, dx$ using the trapezoidal rule and determine the dependence of the error on n using the least squares fitting method. The following recipe can also be used to study other numerical integration formulas, such as Simpson's rule.

Maple's default numerical integration algorithm is the Clenshaw–Curtis quadrature, which is discussed in standard scientific computing texts, for example [Hea02]. However, other numerical integration methods can be invoked by loading the Student library package, with the Calculus1 subpackage. To perform a least squares fit for the error, the statistical package is also loaded.

```
> restart: with(Student[Calculus1]): with(Statistics):
```

The following line will allow us to select the method M employed in the recipe. If you wish to use Simpson's rule, change the entry `trapezoid` to `simpson`.

```
> alias(M=trapezoid): #can change method
```

The integrand is entered,

```
> f:=exp(-x^2);
```

$$f := e^{(-x^2)}$$

and the exact value obtained for the integration range $x = 0$ to 1.

```
> Exact:=int(f,x=0..1);
```

$$Exact := \frac{1}{2}\,\mathrm{erf}(1)\,\sqrt{\pi}$$

The answer is just the *error function* evaluated at $x = 1$, multiplied by $\sqrt{\pi}/2$. The answer is now expressed in floating-point form, for later comparison.

```
> evalf(Exact);
```

$$0.7468241330$$

The `ApproximateInt` command can be used to apply a particular numerical integration method M. A functional operator is created to apply this command, the number n of partitions to be specified. Various output options are available. The entry `sum` allows the formal sum to be produced in the output. Other output options are available.

```
> Approx:=n->ApproximateInt(f,x=0..1,method=M,partition=n,
                  output=sum):
```

For example, the trapezoidal value of the integral is obtained for $n = 1$.

```
> Approx(1)=evalf(Approx(1));
```

$$\frac{1}{2}\left(\sum_{i=0}^{0}\left(e^{(-i^2)} + e^{(-(i+1)^2)}\right)\right) = 0.6839397205$$

The difference between the trapezoidal and exact values is quite large in this case. The error can be reduced by increasing n. We will now determine how the error varies with n for our integral. From our preliminary remarks, the error should be approximately given by a power law of the form $\text{Error} = k\, n^b$, with k a constant and $b = -2$ for the trapezoidal rule. Leaving the value of b undetermined for the moment, taking the natural logarithm of both sides of the Error yields $\ln(\text{Error}) = a + b\,\ln(n)$, with $a \equiv \ln(k)$. So, if we plot the log of the Error vs. the log of n, a straight line should result with slope b. The value of b can then be obtained by finding the best least-squares straight-line fit.

To implement this, we first calculate $\ln(n)$ for, say, $n = 1$ to 10. The sequence of numbers is put into a Maple list X.

```
> X:=[seq(evalf(ln(n)),n=1..10)]:
```

The log of the absolute value of the error divided by the exact numerical value is then calculated in Y for the same range of n.

```
> Y:=[seq(evalf(ln((abs(Exact-Approx(n)))/Exact)),n=1..10)]:
```

For plotting purposes, the ith entry from each list is joined in XY into a list of plotting points $[X_i, Y_i]$, with $i = 1$ to 10.

```
> XY:=[seq([X[i],Y[i]],i=1..10)]:
```

XY is plotted using a point style, the points being represented by size-14 circles.

```
> gr1:=plot(XY,style=point,symbol=circle,symbolsize=14):
```

The best-fitting straight line, $y = a + b\,x$, to the numerical points is obtained.

```
> y:=Fit(a+b*x,X,Y,x);
```

$$y := -2.481398442777759636 - 2.009331453529703 99\,x$$

The slope, which is the coefficient of x in y, is very close to the expected value of -2. The error does indeed seem to scale like $1/n^2$ for the trapezoidal rule. Then y is plotted as a blue line over the range from the minimum value in the X list to the maximum value.

```
> gr2:=plot(y,x=min(X[])..max(X[]),color=blue):
```

The graphs are superimposed, the resulting picture being shown in Figure 7.7.

```
> plots[display]({gr1,gr2},labels=["log(n)","log(Error)"]);
```

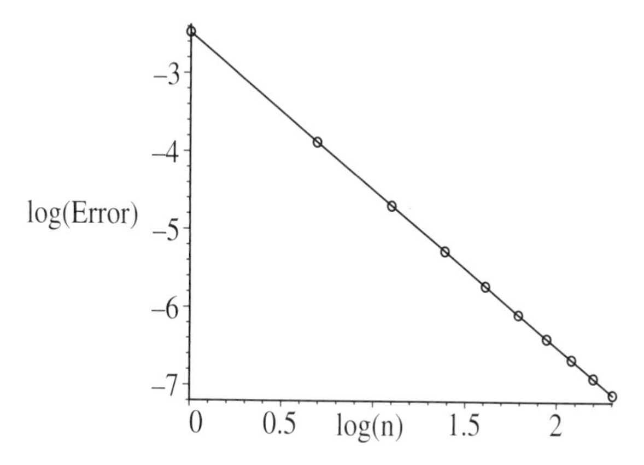

Figure 7.7: Circles: logs of the Error values. Solid: least squares formula F.

The straight line fits the numerical points extremely well. The values of b and a are now extracted (evaluated to 10 digits) from y.

```
> b:=evalf(coeff(y,x),10); a:=remove(has,y,x);
```

$$b := -2.009331454 \qquad a := -2.481398443$$

The error as a function of n is given by $e^a\, n^b$, which is now calculated,

```
> Error:=exp(a)*n^b;
```

$$Error := \frac{0.08362619731}{n^{2.009331454}}$$

completing our recipe.

PROBLEMS:

Problem 7-16: Simpson's rule

Rerun the text recipe, but now using Simpson's rule. How does the derived power law for the Error compare with the theoretically expected result, $1/n^4$? Quantify your answer to this question by calculating the standard deviation.

Problem 7-17: Trapezoidal rule

Evaluate the following integrals using the trapezoidal rule.

$$\textbf{(a)}\ I = \int_0^1 e^{-x^3}\, dx; \qquad \textbf{(b)}\ I = \int_0^{\pi/2} \sin(x^2)\, dx.$$

Determine the analytic form of each integral and then apply the floating-point evaluation to the outputs. What value of n is required using the trapezoidal rule to obtain four-figure agreement with each of the above answers? Confirm the $1/n^2$ power law for the error. (Hint: Choose the n range carefully.)

Problem 7-18: Comparison of trapezoidal and Simpson's rules

Consider the integral $I = \int_1^2 \ln x\, dx$.

- **(a)** Analytically evaluate the integral and also give the decimal value.
- **(b)** Evaluate the integral with the trapezoidal rule, using a sufficiently large n to give five-decimal agreement with the exact answer.
- **(c)** Evaluate the integral with Simpson's rule, using a sufficiently large n to give five-decimal agreement with the exact answer.
- **(d)** Relate your answers to the error discussion in the text.

Problem 7-19: Viscous drag on a boat

A toy boat of mass $m = 10$ kg, initially moving through the water with speed $v(0) = 10$ m/s, is subjected to a viscous drag $F_{\rm drag} = -v\sqrt{v}$ newtons.

- **(a)** Write out the integral expression for the time t it takes the boat to slow down to a speed $v(t)$.
- **(b)** Calculate the exact time for the boat to slow down to 5 m/s.
- **(c)** Taking $\Delta v = 0.25$, estimate the time in **(b)** using the trapezoidal rule.
- **(d)** Taking $\Delta v = 0.25$, estimate the time in **(b)** using Simpson's rule.

(e) Calculate the percentage differences between the exact and approximate answers for each case.

Problem 7-20: How long is that track?
A race car completes one lap of a track in 84 seconds. Using a radar gun, the speed of the car is measured in m/s at 6-second intervals and the results are given in Table 7.1. Plot the speed versus time, joining the data points with lines. Use Simpson's rule to estimate the length of the race track.

Table 7.1: Time and speed data for the race-car problem.

Time	0	6	12	18	24	30	36	42
Speed	37.2	40.2	44.4	46.8	44.1	39.9	36.3	32.7
Time	48	54	60	66	72	78	84	
Speed	29.7	25.5	23.4	26.7	31.2	34.8	36.9	

Problem 7-21: Disk brakes
To simulate the temperature characteristics of disk brakes, Secrist and

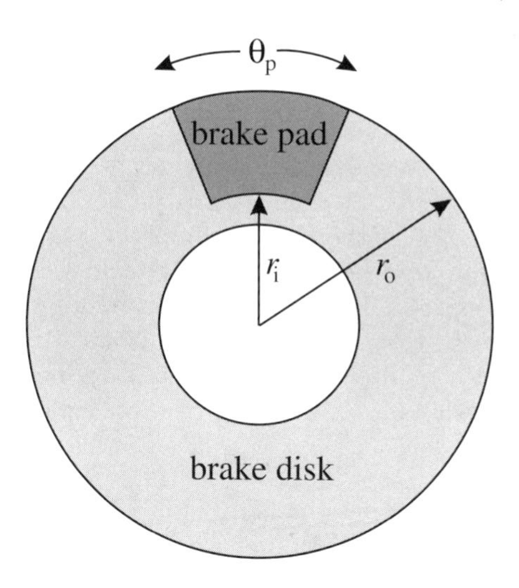

Figure 7.8: Geometry for the disk brake problem.

Hornbeck [SH76] numerically calculated the *area-averaged lining temperature* $\langle T \rangle$ of the brake pad, where $\langle T \rangle$ is given by the equation

$$\langle T \rangle = \int_{r_i}^{r_o} T(r)\, r\, \theta_p\, dr \Big/ \int_{r_i}^{r_o} r\, \theta_p\, dr,$$

where $r_i = 9.4$ cm and $r_o = 14.6$ cm are the inner and outer radius at which pad–disk contact takes place (see Figure 7.8), $\theta_p = 0.705$ radians is the angle subtended by the brake pad, and $T(r)$ is the temperature in degrees Celsius of the pad at radius r cm. Table 7.2 gives T as a function of r.

Table 7.2: Disk brake data.

r	9.4	9.9	10.4	10.9	11.5	12.0	12.5	13.0	13.5	14.1	14.6
T	338	423	474	506	557	573	619	622	651	661	671

(a) Plot $T(r)$, joining the data points with straight lines.

(b) Use Simpson's rule to calculate $\langle T \rangle$.

7.2.2 Wait and Buy Later!

The buyer needs a hundred eyes, the seller not one.
George Herbert, English poet and clergyman (1593–1633)

As the reader is undoubtedly aware, new computer models are initially priced high and then their price tends to drop substantially as even newer models with faster chips come on stream. The patient computer buyer, who doesn't try to be the first to have the latest model, can often find some good bargains by waiting until the price is right. From the computer company's viewpoint, their revenue, and therefore their profit, per computer is greater at the beginning of the sales campaign than some months later. In this hypothetical example, we look at the total revenue generated by the **DALE** computer company, which has introduced its new computer with the revolutionary Hexium chip by **OUTEL**. The price of such a computer is initially \$2000, but as more units are sold and the competition from rivals increases, the price is dropped by the manufacturer. If q is the quantity of computers sold ($q=1$ equals 1 million computers) and $p=1$ corresponds to the initial price, the p–q relationship is found to be given by

$$p = e^{-q^{2.3}}. \tag{7.13}$$

The total normalized revenue for the first one million computers sold will just be the area under the $p(q)$ curve between $q=0$ and $q=1$, i.e., equal to the integral $\int_0^1 p(q)\,dq$. This integral doesn't have an analytic solution. It can, of course, be evaluated numerically using standard methods, but it is instructive to show how the area can be estimated using the Monte Carlo approach.

The plots library package is loaded, and the endpoints $a=0$ and $b=1.0$ of the integration range are entered, the latter being given in decimal form to force Maple to numerically evaluate the integral.

```
> restart: with(plots):
> a:=0: b:=1.0:
```

The price function is now entered,

```
> p:=exp(-q^2.3);
```
and plotted as a thick black line.
```
> plot1:=plot(p,q=a..b,color=black,thickness=2):
```
The area below the curve is filled with a green hue,
```
> plot2:=plot(p,q=a..b,color=green,filled=true):
```
and the resulting shaded p versus q curve displayed in Figure 7.9.
```
> display({plot1,plot2},labels=["q","p"],tickmarks=[3,4]);
```

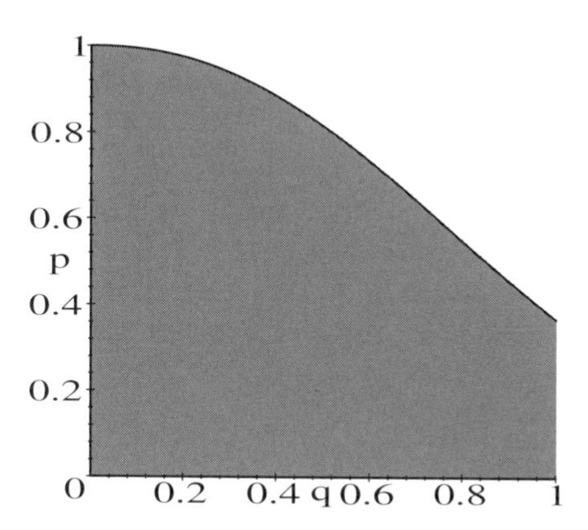

Figure 7.9: Price (p) versus quantity (q) of Hexium computers sold.

The shaded area under the normalized price curve is equal to the total normalized revenue generated by the sale of the first 1 million Hexium computers. The area is first evaluated numerically with Maple's default numerical integrator and labeled as EXV to denote the "exact" value as opposed to the Monte Carlo value (MCV), which we will be generating next.
```
> EXV:=int(p,q=a..b);
```
$$EXV := 0.7686600683$$
The "exact" value of the integral is 0.76866 ... Knowing this number will allow us to determine the error in our Monte Carlo estimate.

Now we make a Monte Carlo estimate of the normalized area between $q=a=0$ and $q=b=1$ using the expression

$$\int_a^b p(q)\,dq = \frac{(b-a)}{n}\sum_{i=1}^{n} p(q_i) \tag{7.14}$$

and generating a uniform distribution of q_i between $q=0$ and $q=1$ with a random-number generator. The random-seed call is invoked.
```
> randomize():
```
A total of $n=1600$ random numbers are generated in each experiment and 10 experiments carried out. The beginning time for the double do loop is recorded.
```
> n:=1600: Expts:=10: begin:=time():
```

In the outer do loop, which runs over the total number of experiments,

```
> for j from 1 to Expts do
```

we set the sum of all the prices, sump, equal to zero.

```
> sump:=0:
```

In the inner do loop, the sum $\sum_{i=1}^{n} p(q_i)$ is calculated for each experiment, where q_i is a random number between 0 and 1.

```
> for i from 1 to n do
> sump:=sump+exp(-(evalf(rand()/10^12))^2.3);
> end do:
```

To calculate the Monte Carlo value, MCV, of the integral in the jth experiment, sump is multiplied by $(b-a)/n$ and the outer loop ended.

```
> MCV[j]:=(b-a)*sump/n;
> end do:
```

The Statistics package is loaded so that a mean and standard deviation can be calculated for the Monte Carlo estimates obtained from the different numerical experiments.

```
> with(Statistics):
```

The Monte Carlo estimates of the integral value generated by each experiment are put into a list format and displayed.

```
> data:=[seq(MCV[j],j=1..Expts)];
```

$$data := [0.7679573881, 0.7655670588, 0.7669357200, 0.7748553394, 0.7671948456, 0.7635738425, 0.7699460706, 0.7763241906, 0.7680766425, 0.7627424581]$$

The mean Monte Carlo value $\langle MCV \rangle$ of the integral, averaged over the 10 experiments, will be given by

$$\langle MCV \rangle = \frac{1}{\text{Expts}} \sum_{j=1}^{\text{Expts}} MCV_j. \tag{7.15}$$

This average is calculated for the *data* list using the Mean command.

```
> <MCV>:=Mean(data);
```

$$\langle MCV \rangle := 0.7683173556$$

The mean Monte Carlo value of the integral obtained here is 0.7683, which compares favorably with the "exact" value of 0.7687 calculated earlier. The total CPU time in seconds for the iterative procedure is now determined,

```
> cpu_time:=(time()-begin)*seconds;
```

$$cpu\_time := 4.926\, seconds$$

and is found to be about 5 seconds. As with the Monte Carlo estimates themselves, this CPU time varies slightly from one run to the next. The percent deviation of the Monte Carlo estimate from the exact value

```
> PercentDeviation:=100*(<MCV>-EXV)/EXV;
```

$$PercentDeviation := -0.04458572965$$

is about 0.04%, on the low side. The standard deviation of the 10 Monte Carlo experiments will be given by the standard deviation

$$\sigma = \sqrt{\langle MCV^2 \rangle - \langle MCV \rangle^2}, \text{ where } \langle MCV^2 \rangle \equiv \frac{1}{\text{Expts}} \sum_{j=1}^{\text{Expts}} (MCV_j)^2.$$

Applying the `StandardDeviation` command to the data list

```
> sigma:=StandardDeviation(data);
```

$$\sigma := 0.004171430781$$

yields $\sigma \approx 0.0042$. In the following command lines, the Monte Carlo value for the integral, the Monte Carlo value plus and minus one sigma (one standard deviation), and the exact value of the integral are summarized.

```
> MC_value:=<MCV>; MC_value_plus_one_stand_dev:=<MCV>+sigma;
  MC_value_minus_one_stand_dev:=<MCV>-sigma; Exact_value:=EXV;
```

$$MC\_value := 0.7683173556$$

$$MC\_value\_plus\_one\_stand\_dev := 0.7724887864$$

$$MC\_value\_minus\_one\_stand\_dev := 0.7641459248$$

$$Exact\_value := 0.7686600683$$

According to standard statistical theory, which assumes that the data are distributed normally (see, for example, Gould and Tobochnik [GT96]), a single Monte Carlo measurement has a 68% chance of being within one standard deviation of the "true" mean and a 95% chance of being within two standard deviations of the mean. Examine the rather short `data` list to determine how many of the 10 Monte Carlo estimates for this particular run lie within one σ of the mean value and within $2\,\sigma$. On your own computer, you should increase the number of experiments and see what results you get for this recipe.

Since our Monte Carlo estimate of the integral value is quite close to the exact value, the Monte Carlo estimate of the total revenue obtained from selling the first 1 million computers is also very close to the exact estimate of the total revenue. The exact estimate of the total revenue is

```
> Total_revenue:=EXV*2000*10^6;
```

$$Total\_revenue := 0.1537320137\,10^{10}$$

i.e., about 1537 million dollars.

PROBLEMS:

Problem 7-22: DALE revenue

Assuming that the p-q relation given in the text prevails:

(a) Plot the integrand $p(q)$ over the range $q=0$ to 2 with the area under the curve shaded red.

(b) Calculate the "exact" numerical value of the integral over this range.

(c) Taking $n = 1600$ and 10 experiments, calculate the mean value and thus make a Monte Carlo estimate of the integral over this range.

(d) Calculate the standard deviation of the Monte Carlo estimates.

(e) Discuss the accuracy of your Monte Carlo mean value.

(f) What is the Monte Carlo estimated total revenue from the sale of the first two million computers?

Problem 7-23: Monte Carlo integration
Use the Monte Carlo integration approach with reasonably large n and a reasonably large number of experiments to evaluate the integral

$$I = 4\int_0^1 \sqrt{1-x^2}\,dx.$$

By what percentage does the mean value of the experiments differ from the exact value of the integral? What is the standard deviation for your experiments?

Problem 7-24: 6-dimensional integral
By generating random numbers in the interval 0 to 1 in groups of six, evaluate the following 6-dimensional integral and investigate how fast your answer converges to the exact value as the number of trials is increased.

$$\int_0^1\int_0^1\int_0^1\int_0^1\int_0^1\int_0^1 \frac{1}{1+u+v+w+x+y+z}\,du\,dv\,dw\,dx\,dy\,dz.$$

7.2.3 Wait and Buy Later! The Sequel

Nowadays people know the price of everything and the value of nothing.
Oscar Wilde, Anglo-Irish Writer (1854–1900)

In the previous subsection, the total revenue generated by the **DALE** computer company from the sale of its first million computers containing the revolutionary Hexium chip was estimated using the Monte Carlo approach. In normalized units, the pricing function was taken to be $p(q) = e^{-q^{2.3}}$, where $q = 1$ corresponded to one million computers and $p = 1$ to two thousand dollars. The total revenue for the first million computers is the area under the price curve between $q = 0$ and $q = 1$, i.e., given by the integral $I = \int_0^1 p(q)\,dq$. The Monte Carlo estimate of I was found to be in reasonable agreement with the "exact" numerical result obtained using Maple's numerical integrator, but would take considerably longer to carry out if greater accuracy were desired. Although the Monte Carlo approach is not meant as a serious competitor to standard numerical techniques for calculating 1-dimensional integrals, the question does arise as to how the Monte Carlo approach can either be speeded up while maintaining the same accuracy, or alternatively be made more accurate for approximately the same CPU time. Such considerations can become important when it is desired, for example, to numerically evaluate a multidimensional integral with a complicated boundary where the Monte Carlo approach may be the only viable method to use. One important approach to dealing with these issues is to make use of so-called *importance sampling*, which is now discussed.

One chooses a function, called the *importance sampling function*, that resembles the integrand of the integral being considered and that can be easily integrated analytically. For example, in the present case we might select the function $u(q) = e^{-q}$, which qualitatively resembles $p(q) = e^{-q^{2.3}}$. Then, the integral I is rewritten as

$$I = \int_0^1 \frac{p(q)}{u(q)}\, u(q)\, dq. \tag{7.16}$$

But $q = -\ln u$, so that I becomes

$$I = -\int_1^{e^{-1}} \frac{e^{-(-\ln u)^{2.3}}}{u}\, du = \int_{1/e}^{1} \frac{e^{-(-\ln u)^{2.3}}}{u}\, du. \tag{7.17}$$

The same Monte Carlo approach as in the last story can be used, but now with a new integrand and a new sampling range. The new integrand will display less variation with u than the old integrand did with q and as a consequence fewer sampling points, i.e., fewer values of n, can be used to obtain approximately the same accuracy as before. Smaller n leads to a faster CPU time.

To illustrate the method, the above computer sales example is solved again, the Monte Carlo procedure borrowing heavily from the last algorithm. The plots and Student[Calculus1] packages are loaded,

```
> restart: with(plots): with(Student[Calculus1]):
```

and the interval endpoints a and b,

```
> a:=0: b:=1:
```

and the normalized price function $p = e^{-q^{2.3}}$ are entered.

```
> p:=exp(-q^2.3):
```

The inert form is used to display the integral to be evaluated.

```
> Integral:=Int(p,q=a..b);
```

$$Integral := \int_0^1 e^{(-q^{2.3})}\, dq$$

Contained within the Student[Calculus1] package are "rule" commands for changing the integration variable and limits. We change from the old integration variable q to the new variable u, using the transformation $e^{-q} = u$.

```
> Integral2:=Rule[change,exp(-q)=u](Integral);
```

$$Integral2 := \int_0^1 e^{(-q^{2.3})}\, dq = \int_1^{e^{(-1)}} -\frac{e^{(-(-\ln(u))^{2.3})}}{u}\, du$$

The limits in the new integral are then "flipped."

```
> Integral3:=Rule[flip](Integral2);
```

$$Integral3 := \int_0^1 e^{(-q^{2.3})}\, dq = -\int_{e^{(-1)}}^{1} -\frac{e^{(-(-\ln(u))^{2.3})}}{u}\, du$$

As expected, the resulting integral on the rhs of *Integral3* is the same as in Equation (7.17). The double minus sign on the rhs of *Integral3* can be removed by applying the `expand` command.

```
> Integral4:=expand(rhs(Integral3));
```

$$Integral4 := \int_{e^{(-1)}}^{1} \frac{1}{e^{((-\ln(u))^{2.3})}\, u}\, du$$

The `Integrand` command, also part of the Student[Calculus1] package, allows us to extract the integrand, labeled P, of the above integral.

```
> P:=Integrand(Integral4);
```

$$P := \frac{1}{e^{((-\ln(u))^{2.3})}\, u}$$

The new limits on the integral are labeled as *anew* and *bnew*.

```
> anew:=evalf(exp(-1)); bnew:=1.0;
```

$$anew := .3678794412 \qquad bnew := 1.0$$

The curve P is plotted as a thick black line between $u = anew$ and $bnew$,

```
> plot1:=plot(P,u=anew..bnew,color=black,thickness=2):
```

and the region below the curve filled in with an aquamarine color.

```
> plot2:=plot(P,u=anew..bnew,color=aquamarine,filled=true):
```

The new integrand P is displayed in Figure 7.10.

```
> display({plot1,plot2},labels=["q","P"],tickmarks=[3,3]);
```

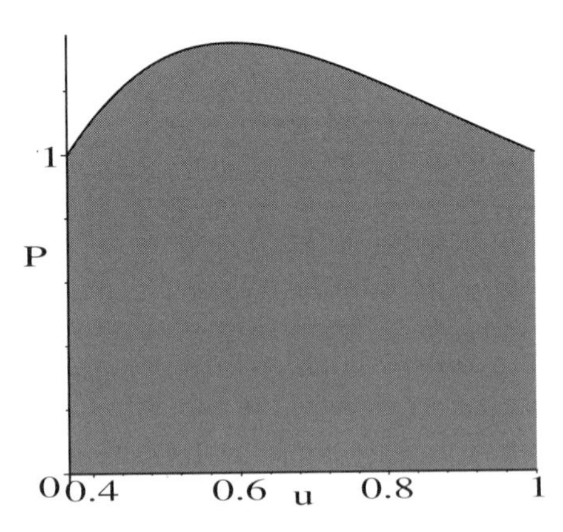

Figure 7.10: Variation of new integrand P with u.

If this plot is compared with Figure 7.9, it is seen that $P(u)$ shows less variation with u than $p(q)$ did with q. Of course, the area under the $P(u)$ curve between $u = 1/e$ and $u = 1$, and therefore the value of the integral, should be the same as the area under the original $p(q)$ curve between $q = 0$ and 1. As a check, the "exact" value is numerically obtained for the new form of the integral.

```
> EXV:=evalf(Integral4);
```

$$EXV := 0.7686600683$$

As the reader can readily check, the answer in the output is identical to the exact numerical value of the integral previously obtained, as it should be. Now the Monte Carlo estimate of the integral expressed in its new form is carried out, the random number seed being set according to the computer clock.

```
> randomize():
```

Previously, 10 numerical experiments, each with $n = 1600$, were carried out. Here we will consider 20 experiments but with only $n = 400$ in each experiment. The CPU time will again be monitored on the same PC as that used in the earlier Monte Carlo integral calculation so a fair comparison of the running time can be made. The double do loop structure that follows is similar to that previously used, except for minor modifications related to the fact that our integrand and the range are different.

```
> n:=400: Expts:=20: begin:=time():
> for j from 1 to Expts do
> sump:=0;
> for i from 1 to n do
```

A random number **u** lying between *anew* and *bnew* is created.

```
> u:=anew+(bnew-anew)*rand()/10^12;
```

The new integrand is used in the next command line.

```
> sump:=sump+P
> end do:
> MCV[j]:=sump*(bnew-anew)/n;
> end do:
```

The statistical package is loaded,

```
> with(Statistics):
```

and the list of 20 Monte Carlo estimates displayed.

```
> data:=[seq(MCV[j],j=1..Expts)];
```

$$data := [0.7670788498, 0.7682175292, 0.7623540762, 0.7673922182, 0.7691164685, 0.7684995240, 0.7680226728, 0.7685383508, 0.7692022982, 0.7714850918, 0.7708542478, 0.7675425145, 0.7670399112, 0.7722572638, 0.7682834112, 0.7700104492, 0.7670294725, 0.7688806968, 0.7715536368, 0.7725692748]$$

The mean value of the Monte Carlo estimates is calculated

```
> <MCV>:=Mean(data);
```

$$\langle MCV \rangle := 0.7687963980$$

and found to be 0.76880, quite close to the exact value, 0.76866. The elapsed CPU time is about 3 seconds, compared to roughly 5 seconds earlier.

```
> cpu_time:=(time()-begin)*seconds;
```

$$cpu\_time := 3.195\ seconds$$

The percent deviation of the mean Monte Carlo value from the exact value

```
> PercentDeviation:=100*(<MCV>-EXV)/EXV;
```

$$PercentDeviation := 0.01773601955$$

is about 0.02%. The standard deviation

```
> sigma:=StandardDeviation(data);
```

$$\sigma := .002264761800$$

is $\sigma \approx 0.0023$, about one-half of the standard deviation previously obtained. So, in this example importance sampling has reduced the CPU time and increased the accuracy of the Monte Carlo estimate.

Again, a summary of the main results is presented, giving the Monte Carlo and exact estimates of the integral and the integral values corresponding to plus and minus one σ from the Monte Carlo estimate.

```
> MC_value:=<MCV>; MC_value_plus_one_stand_dev:=<MCV>+sigma;
  MC_value_minus_one_stand_dev:=<MCV>-sigma; Exact_value:=EXV;
```

$$MC\_value := 0.7687963980$$

$$MC\_value\_plus\_one\_stand\_dev := 0.7710611598$$

$$MC\_value\_minus\_one\_stand\_dev := 0.7665316362$$

$$Exact\_value := 0.7686600683$$

In the data list, 15 of the 20 Monte Carlo estimates, or 75%, lie within the one-σ bounds, consistent with what would be expected from statistical theory.

PROBLEMS:

Problem 7-25: Monte Carlo estimate of integral value

Consider the integral

$$I = \int_0^3 x^{3/2} e^{-x}\, dx.$$

(a) Analytically evaluate I and then express the result in floating-point form.

(b) Make a Monte Carlo estimate of I without using importance sampling.

(c) Make a Monte Carlo estimate of I using the importance sampling function (integration variable transformation) $u(x) = e^{-x}$.

(d) Compare and discuss the various numerical values of I.

Problem 7-26: Another Monte Carlo estimate

Consider the integral

$$I = \int_0^{\pi} \frac{1}{x^2 + \cos^2 x}\, dx.$$

(a) Can I be evaluated analytically, i.e., in closed form?

(b) Numerically evaluate I.

(c) Make a Monte Carlo estimate of I without using importance sampling.

(d) Make Monte Carlo estimates of I using the importance sampling function $u(x) = e^{-a\,x}$ and choosing different values of the parameter a.

(e) Determine the value of a, to two digits, that minimizes the standard deviation in the Monte Carlo estimate.

7.2.4 Estimating π

God does not play dice with the universe.
Albert Einstein objected to the random element at the heart of modern quantum mechanics.

Probably the oldest documented application of the Monte Carlo technique is due to Comte de Buffon in 1773. Buffon considered the problem of randomly throwing a needle of length L onto a horizontal plane ruled with straight parallel lines a distance D, greater than L, apart. He was able to demonstrate mathematically, and verify experimentally, that the probability P of the needle intersecting a line is given by $P = 2\,L/(\pi\,D)$. In principle one could carry out the needle experiment, repeatedly throwing the needle a very large number of times in a random manner, to deduce the value of π if it were not known by other means. However, this approach is not very practical or very accurate.

Instead of following Buffon's method, we shall estimate π by considering a circle of radius R inscribed inside a square of sides of length $2\,R$. The ratio of the area of the circle to the area of the square is $\pi\,R^2/(2\,R)^2 = \pi/4$. Now imagine repeatedly throwing a dart randomly at the square and assume that you never miss. Sometimes the dart will land inside the circle, other times not, the probability of landing inside being the ratio of areas. By measuring the ratio of hits inside the circle to the total number of darts thrown and multiplying the result by 4, an estimate of π is possible. Instead of actually throwing darts one can again use a random-number generator, this time to produce random x, y coordinates inside the square.

This dart-throwing technique is useful for measuring the area or volume of a region $\mathcal{R}$ that has a complicated or irregular boundary, a problem that might be quite difficult to solve by more standard numerical techniques. Simply enclose the region $\mathcal{R}$ with a larger region $\mathcal{S}$ whose area or volume is known. Randomly throw darts at the region $\mathcal{S}$ and record the fraction of darts that land inside $\mathcal{R}$. Multiplying the area or volume of $\mathcal{S}$ by this fraction will give the area or volume of $\mathcal{R}$. The same approach can also be applied to calculating other quantities associated with $\mathcal{R}$ such as, e.g., the center of mass coordinates.

The recipe for estimating π begins by setting the random-number seed.

```
> restart: randomize():
```

The variable *number* will be used to tell us how many times the random coordinate is inside the circle. We start this number at zero at the beginning of the numerical run. The number N of random coordinates generated is taken to be fairly large, namely $N = 100{,}000$, and only one experiment will be performed.

```
> startnumber:=0; number:=startnumber; N:=100000;
```

$$startnumber := 0 \qquad number := 0 \qquad N := 100000$$

The starting time is recorded prior to the beginning of the do loop.

```
> begin:=time():
> for j from 1 to N do
```

Instead of working with a whole circle, it suffices to consider a quarter-circle of radius $R = 1.0$ inscribed inside a square with sides of length R. The ratio of the quarter-circle's area to that of the square is still $\pi/4$. The range of both the x and y coordinates will be from 0 to 1. Recall that the command `rand()` with no argument specified will generate a random positive 12-digit number. The x (and similarly for y) random-number coordinate can be kept between 0 and 1 by dividing the `rand()` operator by 10^{12} and multiplying by 1.0 to give a decimal result.

```
> x:=1.0*rand()/10^12;
> y:=1.0*rand()/10^12;
```

If $x^2 + y^2 < 1$, i.e., the square of the circle's radius is less than one, we count it as a "hit" inside the quarter-circle, and each time this occurs the value of *number* is incremented by one. If the radius is greater than or equal to one, the contribution does not get counted.

```
> if x^2+y^2<1 then
> number:=number+1;
> end if;
> end do:
```

When the do loop is completed, the ratio of *number* to N will be a Monte Carlo estimate of $\pi/4$. Thus, the Monte Carlo estimate of π, labeled *pie*, will be 4 times this ratio.

```
> pie:=4.0*number/N;
```

$$pie := 3.140600000$$

For this run, the value of *pie* is very close to the "exact" (numerical) value (EXV) of π, which is now determined.

```
> EXV:=evalf(Pi);
```

$$EXV := 3.141592654$$

The percentage error is easily calculated,

```
> percent_error:=(1-pie/EXV)*100;
```

$$percent\_error := .03159716$$

and the Monte Carlo estimate is found to differ from the exact result by a little over 0.03%. The accuracy can, of course, be increased by increasing N and considering a large number of experiments. On a PC, the CPU time involved can become quite substantial. For the present single experiment, the CPU time

```
> cpu_time:=(time()-begin)*seconds;
```

$$cpu\_time := 4.547\ seconds$$

was about 5 seconds.

The interested reader might like to apply the techniques of this section to estimate the volume of a 4-dimensional hypersphere in the following problem.

PROBLEMS:

Problem 7-27: Volume of a 4-dimensional hypersphere

By suitably modifying the Monte Carlo program for estimating π, estimate the volume of a 4-dimensional hypersphere of unit radius. Hint: The interior of the hypersphere is defined by the condition $x_1^2 + x_2^2 + x_3^2 + x_4^2 \leq 1$. Enclose the hypersphere within a 4-dimensional box with sides 2 units in length. Use the random-number generator and calculate the fraction of random numbers inside the box that lie inside the hypersphere. Multiply this fraction by the volume of the box to obtain the volume of the hypersphere. For your estimate, take 10^4 trials in each experiment and average the volume estimate over 10 experiments. Compare your estimate with the exact value.

Problem 7-28: Error variation with trial number

Modify the text program to keep track of the estimated numerical value V_N of π as a function of the number of trials N. Consider a large number of exeriments for each N. Make a log-log plot of the difference $|V_N - \pi|$ as a function of N. What is the approximate dependence of the error on N for very large N?

Problem 7-29: Take me out to the ball game

A major league baseball player has a "300" batting average. That is to say, he averages 300 hits for each 1000 times at bat. Assuming that he comes to bat 4 times a game, what are his chances of getting 0, 1, 2, 3, 4 hits? Make use of a random-number generator, assume that a season is 150 games long, and average your results over 5 seasons. Compare the answers you obtain with what would be predicted by elementary probability theory. (See Section 7.3.)

7.2.5 Chariot of Fire and Destruction

An archaeologist is the best husband any woman can have: the older she gets, the more interested he is in her.
Agatha Christie, British Mystery Writer (1891–1976)

Leaving Vectoria behind to work in the MIT physics department, Mike has had the good fortune to land a summer job with an archaeological dig in a remote area of Asia. While excavating around the shattered battlements of an ancient city, the archaeologists unearth the remnants of what appears to be a siege tower. A siege tower is a medieval engine of war for storming operations, consisting of a tower on wheels having several platforms. The lowest platform was generally occupied by a battering ram, with the highest platform containing archers with flaming arrows and armed soldiers with scaling ladders. A massive segment, originally of toroidal shape with major radius 4 meters and minor radius 2 meters, is uncovered among the siege tower remains.

Being a mathematics major, Mike was able to deduce that the segment is a piece of a torus bounded by the intersection of two planes. Specifically, he finds that its shape is defined by the three inequalities

$$z^2 + \left(\sqrt{x^2+y^2} - 3\right)^2 \leq 1, \quad x \geq 1, \quad y \geq -3. \tag{7.18}$$

To create a 3-dimensional plot of the shape, Mike loads the plottools package.

```
> restart: with(plottools):
```

He uses the `semitorus` command to create a 3-dimensional semitorus centered at [0, 0, 0], with a meridian of radius 1 meter and a distance from the center of the meridian to the center of the semitorus of 3 meters. The angular range spans π radians.

```
> segment:=semitorus([0,0,0],0..Pi,1,3):
```

By choosing an appropriate view in the following `plots[display]` command,[3] the segment is effectively cut by the two desired intersecting planes.

```
> plots[display](segment,scaling=constrained,style=patch,
  orientation=[0,-179.99],axes=normal,view=[-3..4,1..4,-1..1],
  tickmarks=[3,2,2],labels=["y","x","z"]);
```

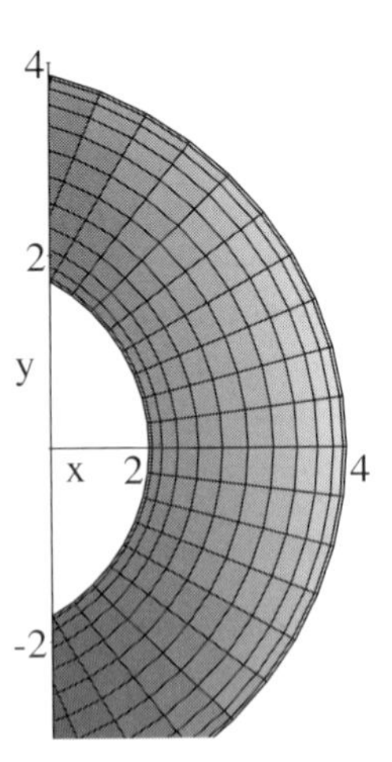

Figure 7.11: The toroidal segment.

[3]This shorthand syntax is equivalent to first loading the plots package before using the `display` command.

The toroidal segment, which appears to be made of bolted and shaped oak planks, is displayed in Figure 7.11. When the viewing box is rotated, the segment looks like a portion of a great wheel, but it seems much too big, even for a massive siege tower. Was it some other part of the siege tower or did it have some other altogether different purpose? While the great archaeological minds stew over this question, Mike passes the time thinking about some mathematical and physical aspects of the toroidal artifact. If the density ρ of oak is known, can the mass of the segment be estimated solely from its shape? Further, can the location of the center of mass also be determined without actually performing an experimental measurement?

For a 3-dimensional object, the mass m would be given in Cartesian coordinates by the volume integral

$$m = \int \rho \, dv = \int \rho \, dx \, dy \, dz,$$

while the three center of mass coordinates are defined by

$$x_{\rm cm} = \int x \, \rho \, dx \, dy \, dz / m, \quad y_{\rm cm} = \int y \, \rho \, dx \, dy \, dz / m, \quad z_{\rm cm} = \int z \, \rho \, dx \, dy \, dz / m.$$

Because of the nature of the boundaries involved here, it is not a trivial task to perform the necessary integrations. A Monte Carlo approach, however, is relatively easy to implement. One can think of the toroidal segment as being enclosed by a rectangular box stretching from $x = 1$ to 4, $y = -3$ to 4, and $z = -1$ to 1. The basic idea is to generate random coordinates (triplets of numbers) inside the box. If the coordinates lie inside the toroidal segment, then the contribution to the mass or the center of mass is counted, otherwise not. The answers are determined by the fraction of the total generated points inside the box that lie inside the toroid.

Before implementing this numerical procedure, Mike decides to use an analytic approach to put some upper and lower bounds on his Monte Carlo estimates of the toroidal mass and center of mass coordinates. First, he considers a full half-torus defined by

$$z^2 + \left(\sqrt{x^2 + y^2} - 3\right)^2 \leq 1, \quad x \geq 0, \quad y \geq -4.$$

The mass of the half-torus will be larger than the mass of the toroidal segment, while $x_{\rm cm}$ should be less for the half-torus than for the segment. By symmetry, one would also expect that $y_{\rm cm} = z_{\rm cm} = 0$ for the half-torus.

Assuming that the density ρ is constant, the z integration from

$$z = -\sqrt{1 - \left(\sqrt{x^2 + y^2} - 3\right)^2} \quad \text{to} \quad z = \sqrt{1 - \left(\sqrt{x^2 + y^2} - 3\right)^2}$$

in the mass integral is easily done, yielding a 2-dimensional integral I,

$$I = \int \int 2 \, \rho \sqrt{1 - \left(\sqrt{x^2 + y^2} - 3\right)^2} \, dx \, dy = \int \int 2 \, \rho \sqrt{1 - (r - 3)^2} \; r \, dr \, d\theta,$$

where, in the last step, the polar coordinates r and θ have been introduced. For the half-torus, the radial coordinate will range from the inner radius $r = 2$ to the outer radius $r=4$ and Mike takes the angular range to be from $\theta=-\Theta=-\pi/2$ to $\Theta=\pi/2$. He sets $f=2\sqrt{1-(r-3)^2}$ and, by consulting a handbook, finds that the density of oak is $\rho=750$ kg/m^3.

```
> f:=2*sqrt(1-(r-3)^2); rho:=750; Theta:=Pi/2;
```

$$f := 2\sqrt{-8-r^2+6\,r} \qquad \rho := 750 \qquad \Theta := \frac{\pi}{2}$$

The double integration is carried out.

```
> totalmass:=Int(Int(rho*f*r,r=2..4),theta=-Theta..Theta);
```

$$totalmass := \int_{-\frac{\pi}{2}}^{\frac{\pi}{2}} \int_2^4 1500\sqrt{-8-r^2+6\,r}\,r\,dr\,d\theta$$

To evaluate the inert form of the integral, the `value` command is applied

```
> totalmass:=value(%);
```

$$totalmass := 2250\,\pi^2$$

which can be converted into a decimal form.

```
> totalmass:=evalf(%);
```

$$totalmass := 22206.60991$$

The half-torus has a mass of 22,207 kg, or about 22 metric tons. Since the toroidal segment is clearly larger than a quarter-torus, Mike expects his numerical estimate of its mass to lie somewhere between 11 and 22 tons. Next, he determines the center of mass coordinates of the half-torus. First he decides to check whether $y_{\rm cm}=0$. Setting $y=r\sin\theta$ in polar coordinates, the y coordinate of the center of mass is evaluated

```
> y[cm]:=Int(Int(rho*f*r*sin(theta)*r,r=2..4),
            theta=-Theta..Theta)/totalmass;
```

$$y_{cm} := 0.00004503163716 \int_{-\frac{\pi}{2}}^{\frac{\pi}{2}} \int_2^4 1500\sqrt{-8-r^2+6\,r}\,r^2\sin(\theta)\,dr\,d\theta$$

```
> y[cm]:=value(y[cm]);
```

$$y_{cm} := 0.$$

and is found to be indeed equal to zero. What about $x_{\rm cm}$? Since $x=r\cos\theta$, the integral for $x_{\rm cm}$ is given by

```
> x[cm]:=Int(Int(rho*f*r*cos(theta)*r,r=2..4),
            theta=-Theta..Theta)/totalmass;
```

$$x_{cm} := 0.00004503163716 \int_{-\frac{\pi}{2}}^{\frac{\pi}{2}} \int_2^4 1500\sqrt{-8-r^2+6\,r}\,r^2\cos(\theta)\,dr\,d\theta$$

which when numerically evaluated yields

```
> x[cm]:=evalf(value(x[cm]));
```

$$x_{cm} := 1.962910964$$

$x_{\rm cm}\approx 1.96$ m. As the reader may easily verify by changing the angular limits to range from $\theta=-\pi/4$ to $\theta=\pi/4$, for a quarter-torus $x_{\rm cm}\approx 2.78$ m. So the Monte

Carlo estimate of $x_{\rm cm}$ for the toroidal segment should lie somewhere between these limits.

Having obtained some lower and upper bounds on his numerical estimates, Mike now carries out a Monte Carlo calculation of the total mass, $x_{\rm cm}$, $y_{\rm cm}$, and $z_{\rm cm}$. The random seed call is made and the density entered once again because the worksheet has been restarted.

```
> restart: randomize(): rho:=750:
```

Using the defining integrals, the mass and x, y, and z center of mass moments per unit volume are incremented point by point starting with zero values.

```
> startmass:=0: mass:=startmass:
> xmoment:=0: ymoment:=0: zmoment:=0:
```

The volume of the rectangular box is also given

```
> vol:=2.0*3.0*7.0: #torus is inside this volume
```

as well as the total number of random points N.

```
> N:=25000:
```

The starting time, before commencement of the do loop, is recorded.

```
> begin:=time():
```

Mimicking the procedure for the random walk of the perfume molecule, the command line `x := 1+3.0*rand()/10^12:` below generates random decimal numbers in the range $x = 1$ to 4. Similar command lines are inserted for the y and z directions. Next a conditional "if then" statement is included, which increments the mass and the center of mass moments if the random point falls inside the toroidal segment enclosed by the box.

```
> for j from 1 to N do
> x:=1+3.0*rand()/10^12:
> y:=-3+7.0*rand()/10^12:
> z:=-1+2.0*rand()/10^12:
> if z^2+(sqrt(x^2+y^2)-3)^2<1 then
> mass:=mass+rho;
> xmoment:=xmoment+x*rho;
> ymoment:=ymoment+y*rho;
> zmoment:=zmoment+z*rho;
> end if;
> end do:
```

Since the calculated mass of the toroid is per unit volume and represents the sum total of the contribution from each random number generated inside the toroid, the total mass is simply the volume of the box times the mass divided by the total number of points.

```
> totalmass:=vol*mass/N;
```

$$totalmass := 16605.54000$$

Thus the mass of the toroidal segment is about $16\frac{1}{2}$ thousand kilograms, or $16\frac{1}{2}$ metric tons. This value lies roughly midway between the lower and upper bounds calculated above, so it makes sense. The value of $x_{\rm cm}$ is now calculated

```
> x[cm]:=(vol*xmoment/N)/totalmass;
```

$$x_{cm} := 2.410563299$$

and found to be 2.41 m, again lying appropriately between the previously calculated bounds. Because the toroidal segment has a small portion removed between $y = -3$ and -4 (see the figure), thus destroying the symmetry in the y-direction, Mike anticipates that $y_{\rm cm}$ should lie slightly above zero.

```
> y[cm]:=(vol*ymoment/N)/totalmass;
```

$$y_{cm} := 0.1932795233$$

He finds that $y_{\rm cm} \approx 0.19$ m. Finally, as a partial check on the accuracy, Mike calculates $z_{\rm cm}$ which theoretically should be equal to zero on symmetry grounds.

```
> z[cm]:=(vol*zmoment/N)/totalmass;
```

$$z_{cm} := 0.0009712943084$$

The value of $z_{\rm cm}$ is quite close to zero, giving Mike some confidence in his estimated values. Of course, the estimated values will vary slightly from one run to the next, and an averaging procedure over many runs could be implemented.

```
> cpu_time:=(time()-begin)*seconds;
```

$$cpu\_time := 3.254\, seconds$$

The CPU time for this run was about 3 seconds.

In this example, Mike took the density of the toroidal segment to be constant. If necessary, it is easy to insert a variable density expression inside the do loop and carry out the Monte Carlo estimates. In this case, the importance sampling technique discussed earlier could prove quite useful.

PROBLEMS:

Problem 7-30: Placing the center of mass

Using `style=wireframe` and `textplot3d`, place the labeled center of mass at the proper location of the toroidal segment.

Problem 7-31: A different shape

Suppose that the region corresponding to x greater than 3 is missing from the toroidal segment in the text recipe.

(a) Create a 3-dimensional plot of the new segment.

(b) Use the Monte Carlo approach of the recipe to estimate the segment's volume and locate its center of mass.

(c) Place the labeled center of mass (represented by a colored circle) on the 3-dimensional plot of part (a).

7.3 Probability Distributions

7.3.1 Of Nuts and Bolts and Hospital Beds Too

Statistical thinking will one day be as necessary for efficient citizenship as the ability to read and write.
H. G. Wells, English novelist, historian, and sociologist (1866–1946)

Recall that Colleen, the manager of the ladies' leisure section of the Glitz department store was able to successfully model cumulative sales numbers for bikini swimsuits with the logistic curve. Because of her business graduate background, and her success in improving sales numbers in her section and keeping inventory costs down with such statistical approaches as that employed for the swimsuits, she has been promoted to head the store's statistical analysis division. It's not long before Mel, the head of hardware, approaches her with an interesting statistical question related to his department.

"Colleen," Mel begins, "in our hardware section, we have a kit available for assembling a garden shed. The shed is quite stylish but there is a problem. The construction of the shed calls for the use of 10 nut and bolt sets, but the quality of the hardware available from the supplier is such that only 80% of the nut and bolt sets are functional. I could change suppliers, but the current supplier has agreed to include a few extra nut and bolt sets in the kit. I would like to know how many nut and bolt sets should we insist that the supplier include in each kit in order that 95% of the kits have enough functional nut and bolt sets? Obviously, the number must be greater than 10 and I could guess at what it should be, but I would rather have a more precise estimate. Can you help me?"

"I think so," Colleen replies. "This is a classic example of what is known in statistical analysis as a Bernoulli[4] trial. If one picks any nut and bolt set at random, there are only two mutually exclusive outcomes. Either the nut and bolt set is OK (functional) or it is not (nonfunctional). Based on a large number of nut and bolt sets, the probability of a set being OK is determined to be $p=0.8$, while the chance of the set being defective is $q=1-p=0.2$.

Assuming that each trial is independent, that the probability is the same on each trial, and that there are only two possible outcomes on each trial, it can be shown [AL79] that the probability P_n of having n functional nut and bolt sets (or number of heads for the flipped coin) out of a total N sets is given by the *binomial probability distribution*,

$$P_n = C_n^N \, p^n \, q^{N-n}, \quad \text{where} \quad C_n^N = \frac{N!}{n!\,(N-n)!} \tag{7.19}$$

is the *binomial coefficient*. Although I could grind out an answer with a pocket calculator, I have the Maple system on my computer that will make our work much easier. First I will load the plots and Statistics packages into the worksheet and record the starting time for our code.

[4] Jacob Bernoulli was an eighteenth-century Swiss mathematician.

```
> restart: with(plots): with(Statistics):
> begin:=time():
```

As you said, Mel, since $p=0.8 < 1$, the total number N of nut and bolt sets that should be included in each kit should clearly be larger than 10. My strategy is to increase N from 10 by successive integers until the sum of the probabilities from 10 to N is greater than 0.95."

After a few minutes of trial and error, Colleen finds that $N=16$ will suffice. "OK, Mel, I will show you what happens for $N=16$. I will enter this value and the probability $p=0.80$ of obtaining a functioning set into the program.

```
> N:=16: p:=0.80:
```

The following functional operator will generate the binomial probability distribution (7.19) for a specified value of n.

```
> P:=n->ProbabilityFunction(Binomial(N,p),n):
```

Making use of P, we can add the probabilities P_n from $n=10$ to $n=N=16$.

```
> add(P(n),n=10..N);
```

$$0.9733426686$$

If 16 sets are included in each kit, there is a 97% probability that 10 or more sets will be functional. If only $N=15$ sets are included, this probability drops to $94\frac{1}{2}\%$. So, I would recommend that you insist that the supplier include 16 nut and bolt sets in each kit, or he should improve his quality control.

I should point out that the same answer can be obtained using the *cumulative distribution function* command,[5] CDF, and subtracting the cumulative probability up to $n=9$ from that for $n=N=16$.

```
> CDF(Binomial(N,p),N)-CDF(Binomial(N,p),9);
```

$$0.9733426685$$

Before you go, Mel, you might be interested in what the binomial distribution of probabilities looks like for your problem. Let's form the sequence of n values and corresponding P_n into two separate lists.

```
> number:=[seq(n,n=0..N)]: prob:=[seq(P(n),n=0..N)]:
```

Given any two lists L1 and L2, we can create a functional operator SP to apply the ScatterPlot command to the two lists.

```
> SP:=(L1,L2)->ScatterPlot(L1,L2,style=point,symbol=circle,
        symbolsize=14,labels=["n","P"],tickmarks=[4,3]):
```

Then, entering

```
> SP(number,prob);
```

produces the following picture (see Figure 7.12) on the computer screen.

Note how the probability remains essentially equal to zero up to $n = 7$, peaks at $n = 13$, and has dropped back toward zero at $n = 16$. Because the value of p is not equal to one-half here, as it would be for a flipped coin, the binomial distribution is not symmetric around its maximum. If you have any other statistical questions, don't hesitate to contact me."

[5]The "long form" of this command is CumulativeDistributionFunction.

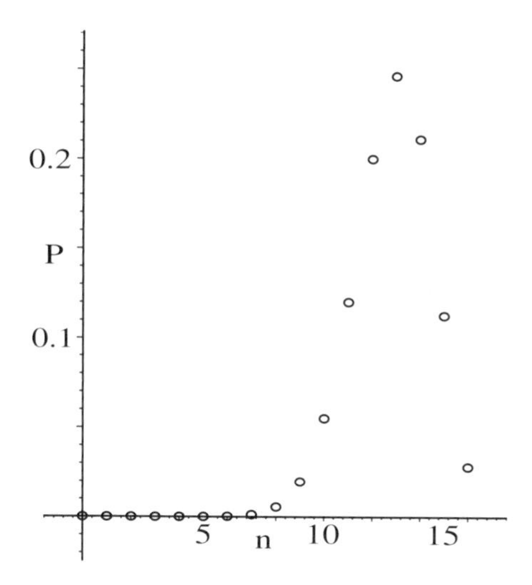

Figure 7.12: Probability P that n nut and bolt sets are functional.

"Thanks for your help, Colleen," Mel replies, "let me treat you to a coffee down in the cafeteria."

While sipping her cappuccino with Mel, who is glancing at last night's sports scores in the *Metropolis Daily News*, Colleen peruses the business section. There is an article about the severe budget constraints that are confronting local hospital administrators and the consequent reduction of numbers of available beds, even in emergency wards. On returning to her office after the coffee break and motivated by her success with the nuts and bolts problem, she considers the following possible hospital scenario.

Suppose that the probability of a patient staying more than 24 hours in an emergency ward, and thus requiring a bed, is $p = 0.32$. On a certain day a total of $N = 60$ patients are admitted to the emergency award. The hospital administrator has proposed to keep 20 beds open. What is the probability that at most 20 of these patients will stay more than 24 hours and require a bed?

Again, this is an example of a Bernoulli trial, there being only two possible outcomes. There is a 32% chance that a patient will have to stay more than 24 hours and a 68% probability that the patient will not. So, again the binomial probability distribution should apply. Rather than start a new worksheet, Colleen unassigns the values of N and p in the worksheet that she used for the nuts and bolts example and enters the new values $N=60$ and $p=0.32$.

```
> N:='N': p:='p': N:=60: p:=0.32:
```

Colleen decides to first plot the probabilities, forming new lists for n and P_n,

```
> number2:=[seq(n,n=0..N)]: prob2:=[seq(P(n),n=0..N)]:
```

and applying the scatter plot command to these lists.

```
> pl1:=SP(number2,prob2): pl1;
```

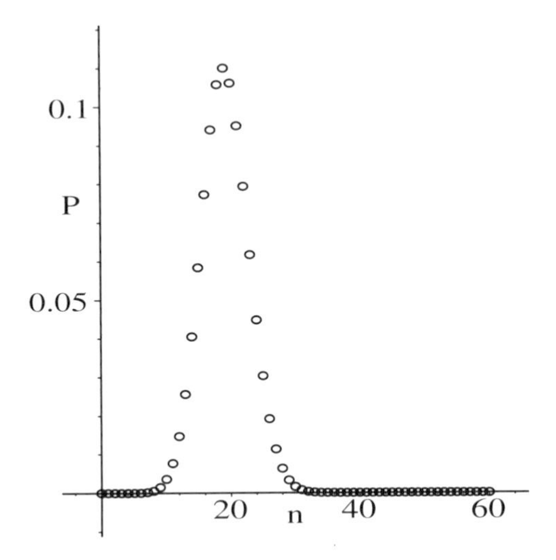

Figure 7.13: Probability that n patients will stay more than 24 hours.

The resulting picture, reproduced in Figure 7.13, shows the probability that n emergency patients will stay more than 24 hours in the hospital.

Even though p is again substantially different from 1/2, the binomial distribution in this case appears to be more symmetrical than in the previous example. This is because the new value of N is quite a bit larger than the value it had in the garden shed example. As N is made very large and if the mean value $N\,p$ is large, it can be shown [MW70] that the binomial distribution approaches the continuous symmetric *normal (Gaussian) probability distribution* $p(x)$, where

$$p(x) = \frac{1}{\sqrt{2\,\pi\,\sigma^2}} e^{-(x-\langle x\rangle)^2/(2\,\sigma^2)}, \text{ with } \langle x\rangle = N\,p,\ \sigma = \sqrt{N\,p\,(1-p)},$$

being the mean value of x and standard deviation, respectively.

Before checking how good an approximation $p(x)$ is to the binomial distribution for this example, Colleen calculates the cumulative probability up to and including $n = 20$.

```
> CDF(Binomial(N,p),20);
```

$$0.6457478779$$

The probability of at most 20 patients staying more than 24 hours is about $64\frac{1}{2}\%$. So, the probability that more than 20 beds would be required is $35\frac{1}{2}\%$.

To check how good an approximation the normal distribution is here, Colleen calculates the mean value of x for the normal curve and the standard deviation.

```
> <x>:=N*p; sigma:=sqrt(N*p*(1-p));
```

$$\langle x\rangle := 19.20 \qquad \sigma := 3.613308733$$

She finds that $\langle x\rangle = 19.2$ and $\sigma \approx 3.6$. The mean value of the normal curve distribution is very slightly higher than the $n\!=\!19$ value at which the maximum in the binomial distribution occurs. The mean number of patients to be hospitalized is slightly less than the number of beds that the administrator wants to leave open. For the normal distribution, there is a 68% probability of x lying within one standard deviation of the mean, i.e., in the range $\langle x \rangle - \sigma = 15.6$ to $\langle x \rangle + \sigma = 22.8$ for the present example.

Colleen next forms a functional operator `P2` to plot the normal distribution $p(x)$ as a function of $X \equiv x$. She uses the "short form," `PDF`, of the `ProbabilityDensityFunction` command.

```
> P2:=X->PDF(Normal(<x>,sigma),X):
```

The normal probability distribution is plotted as a blue curve, but not displayed.

```
> pl2:=plot(P2(X),X=0..N,color=blue):
```

Colleen wishes to show the locations of the mean, $\langle x \rangle$, and of $\langle x \rangle \pm \sigma$. She forms a graphing function `Gr` to plot a green, dashed (`linestyle=3`), vertical line at the horizontal coordinate `A`.

```
> Gr:=A->plot([[A,0],[A,0.12]],linestyle=3,color=green):
```

Using `Gr` with $A = \langle x \rangle$, $A = \langle x \rangle - \sigma$, and $A = \langle x \rangle + \sigma$ and superimposing these three graphs on `pl1` and `pl2` generates Figure 7.14.

```
> display(pl1,pl2,Gr(<x>),Gr(<x>-sigma),Gr(<x>+sigma),
  labels=["x","P"]);
```

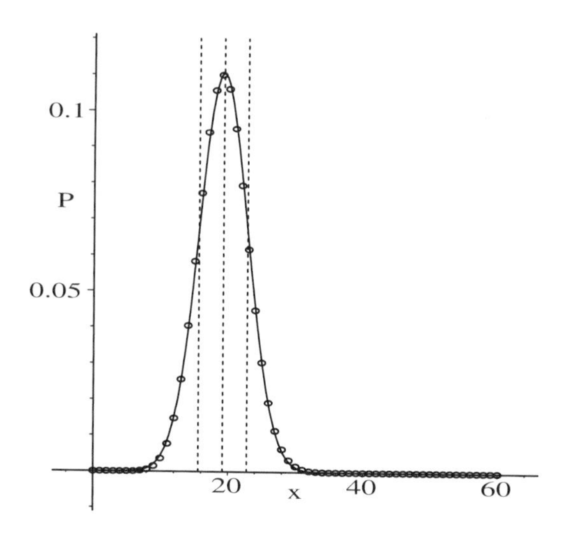

Figure 7.14: Superposition of normal probability density curve on binomial distribution data. Vertical lines are at the mean $\langle x \rangle$ and at $\langle x \rangle \pm \sigma$.

The continuous normal probability distribution does quite a good job of fitting the discrete data points obtained from the exact binomial distribution. As a further check on the accuracy of the normal curve, Colleen calculates the probability of no more than 20 patients being hospitalized for more than 24 hours using the short form of the `CumulativeDistributionFunction` command.

```
> CDF(Normal(<x>,sigma),20.5);
```

$$0.640494561237072046$$

Note that Colleen took x to be 20.5 here, rather than 20. This is because the binomial probability P_n, which is defined only at integer values, is approximately the area under the continuous normal curve between $n - 1/2$ and $n + 1/2$. The normal curve estimate of the cumulative probability, 64%, is in good agreement with the exact result of $64\frac{1}{2}$% obtained earlier for the binomial distribution.

Although it wasn't really necessary to introduce the normal distribution in this hospital example, Colleen knows from her college statistics course that the occurrence of the normal or Gaussian distribution is much more universal in statistical analysis than merely being just a large-N approximation to the binomial distribution. A wide variety of phenomena in nature approximately obey a normal distribution for large N, independent of the particular underlying probability distribution.

Finally, Colleen notes that the total CPU time for the complete code

```
> cpu_time:=(time()-begin)*seconds;
```

$$cpu\_time := 0.210\, seconds$$

took only a fraction of a second.

PROBLEMS:

Problem 7-32: Boys or girls

If the chance of having a boy in any birth averages out to about $52\frac{1}{2}$% for the population of Erehwon as a whole, what proportion of families with six children on Erehwon would be expected to have: **(a)** Three boys and three girls? **(b)** Six boys and no girls? **(c)** Four boys or more and no girls?

Problem 7-33: One smart parrot

Polly parrot is trained to touch, on command, one of two levers, A or B. The probability of touching lever A is 75%. If Polly's responses to the commands given in different trials are independent, what is the probability that out of 5 tries, Polly touches lever A 3 or 4 times? Plot the probability distribution.

Problem 7-34: Chance of being left-handed

In the large city of Metropolis, it is known that 12% of the people are left-handed. If 500 citizens are selected from Metropolis, what is the probability that there are at most 45 of them who are left-handed. You may assume a normal distribution. Plot the probability distribution, indicating the mean and the standard deviation on the graph.

Problem 7-35: Lung disease

In a large population of smokers, it is found that 20% have some sort of lung disease. A sample of 400 smokers is taken from this population and tested for lung disease. Assuming that the normal probability distribution $p(x)$ prevails:

(a) Calculate the mean number $\langle x \rangle$ of smokers in this sample who have a lung disease. Calculate the standard deviation σ.

(b) Plot $p(x)$ with $\langle x \rangle$ and $\langle x \rangle \pm \sigma$ superimposed.

(c) Calculate the probability that (i) at least 100 smokers have a lung disease; (ii) at least 70 but not more than 95 smokers have a lung disease; (iii) at most 75 smokers have a lung disease.

7.3.2 The Ice Wines of Rainbow County

When asked what wine he liked to drink, he replied, "That which belongs to another."
Diogenes Laertius, Greek philosopher (c. A.D. 200)

When not on special photographic assignments for the *National Geographic* magazine, Sheelo sells a vintage ice wine produced in her uncle's vineyard located along the Columbia River in Rainbow County. Ice wine is produced from a small fraction of the total grape crop that has been left unpicked and exposed to the first frosts of fall, to increase the sugar content of the grapes. For Sheelo, selling the quite expensive, limited quantity, ice wine is strictly a hobby, rather than a business, since she sells only five bottles on average each week to close friends and neighbors. To be safe, each Monday she ensures that her stock of this wine is eight bottles. On a recent occasion, however, she had no bottles left by the end of the week and ended up disappointing a customer.

Although she could simply add several more bottles at the beginning of the week to avoid this problem, she is curious as to what the probability of running out actually is with a stock of eight bottles. Further, how many bottles should she start the week with to reduce the probability of running out to about 1%? This is assuming that her business does not grow and the average number of bottles sold per week remains at five. To answer these questions, Sheelo consults her sister Colleen, who has developed a strong interest in applying statistical distributions to practical situations. Colleen points out that Sheelo's ice wine questions can be answered by assuming that a Poisson distribution applies.

"What is a Poisson distribution," Sheelo asks, "and under what circumstances does such a distribution occur?"

"Since you are not a mathematician, I won't go into the mathematical aspects of statistical distributions. Instead, let me explain by giving you a simple example of a Poisson distribution," Colleen replies. "Suppose that the police are trying to catch speeders by placing a photo radar unit at a certain point adjacent to the south Metropolis freeway. Because most motorists are aware of the radar unit's location, only a handful of the thousands of motorists passing the unit each week receive a ticket. They were either daydreaming, talking on their cell phones, or just tourists. Relative to the large number N of motorists, the mean number λ of tickets issued each week, averaged over a number of weeks, is small. In this case, the probability $P(n)$ of n tickets being issued in any week is given by the *Poisson distribution*

$$P(n) = \frac{\lambda^n}{n!} e^{-\lambda}. \tag{7.20}$$

Similarly, in your case the mean number of bottles, namely five, that you sell each week is small relative to the large number of people that you know who could be potential customers.

To answer your question, let's explore the problem on the computer. The Poisson distribution (7.20) is entered.

```
> restart:
> PD:=lambda^n*exp(-lambda)/n!;
```

$$PD := \frac{\lambda^n e^{(-\lambda)}}{n!}$$

It's instructive to first check some features of the Poisson distribution for $N = \infty$. The total probability should sum up to one, which we now confirm.

```
> "Total probability"=sum(PD,n=0..infinity);
```

$$\text{“Total probability”} = 1$$

We can also check that the mean value $\langle n \rangle = \sum_{n=0}^{N=\infty} n\,P(n)$ is equal to λ.

```
> <n>:=sum(n*PD,n=0..infinity);
```

$$\langle n \rangle := \lambda$$

To get a feeling for the "width" of the Poisson probability distribution, which is a measure of the spread in n values around the mean, we can calculate the root mean square deviation, $\sigma = \sqrt{\langle n^2 \rangle - \langle n \rangle^2}$, with $\langle n^2 \rangle = \sum_{n=0}^{\infty} n^2\,P(n)$. Let's first calculate $\langle n^2 \rangle$,

```
> <n^2>:=sum(n^2*PD,n=0..infinity);
```

$$\langle n^2 \rangle := \lambda\,(\lambda + 1)$$

and then use this result and that for $\langle n \rangle$ to determine σ.

```
> sigma:=sqrt(<n^2>-<n>^2);
```

$$\sigma := \sqrt{\lambda}$$

Loading the Statistics and plots library packages, which will be needed shortly, I will now set $\lambda = 5$, the mean number of bottles sold per week, and the possible number n of bottles sold per week will be allowed to range up to $N = 16$. You may object to the fact that this value of N is not terribly large, but as I will show you, taking N larger will not alter the results appreciably.

```
> with(Statistics): with(plots): lambda:=5: N:=16:
```

To see this, let's compare the values of the total probability, $\langle n \rangle$, and σ for $N = \infty$ and $N = 16$. For $N = \infty$, the mean value of n and the root mean square deviation take on the following values.

```
> <n>[infinity]:=lambda; sigma[infinity]:=evalf(sigma);
```

$$\langle n \rangle_\infty := 5 \qquad \sigma_\infty := 2.236067977$$

Although we could use *PD* to calculate the corresponding values for $N = 16$, an alternative way is to form the following functional operator P to evaluate the Poisson distribution for a given x value. By default, the probability function is computed using exact arithmetic. To compute the probability function numerically, the `numeric` option is included here.

```
> P:=x->ProbabilityFunction(Poisson(lambda),x,numeric):
```

Using P, we now calculate the total probability TP, the mean $\langle n \rangle$, and σ.

```
> TP:=add(P(n),n=0..N); <n>:=add(n*P(n),n=0..N);
  sigma:=sqrt(add(n^2*P(n),n=0..N)-<n>^2);
```

$$TP := 0.9999801308 \qquad \langle n \rangle := 4.999654959 \qquad \sigma := 2.235497461$$

All three values are extremely close to those for $N = \infty$, confirming what I said earlier. I will work with the $N = 16$ values from now on.

The probability of selling exactly n bottles of ice wine is determined for n ranging from 0 to N.

```
> Prob:=seq(P(n),n=0..N);
```

$$Prob := 0.006737946999, 0.03368973500, 0.08422433749, 0.1403738958, 0.1754673698, 0.1754673698, 0.1462228081, 0.1044448629, 0.06527803935, 0.03626557742, 0.01813278870, 0.008242176687, 0.003434240286, 0.001320861649, 0.0004717363030, 0.0001572454343, 0.00004913919822$$

The first entry in the list is the probability (0.0067 or 0.67%) of selling zero bottles, the second entry is the probability (about 3%) of selling exactly one bottle, and so on. The probability of selling, e.g., exactly eight bottles could be obtained by similarly inspecting the above list, or by entering `Prob[9]`, or by simply calculating $P(8)$ directly as is done in the next line.

```
> "Probabilility of selling exactly 8 bottles"=P(8);
```

"Probabilility of selling exactly 8 bottles" = 0.06527803935

The probability of selling exactly eight bottles is about 0.065, or $6\frac{1}{2}$%.

An operator `Gr1` is formed to plot a point A as a size-12 blue circle.

```
> Gr1:=A->pointplot(A,symbol=circle,symbolsize=12,color=blue,
          tickmarks=[3,3]):
```

A second graphing operator `Gr2` is created to plot a dashed (`linestyle=3`), green, vertical line at the horizontal coordinate C.

```
> Gr2:=C->plot([[C,0],[C,0.17]],linestyle=3,color=green):
```

`Gr1` and `Gr2` are used in the following command line. The first entry in the Maple set plots $P(n)$ for integer values from $n = 0$ to N, while the second, third, and fourth entries plot dashed vertical green lines at $\langle n \rangle$, $\langle n \rangle + \sigma$, and $\langle n \rangle - \sigma$. The four graphs are superimposed with the `display` command,

```
> display({Gr1({seq([n,P(n)],n=0..N)}),Gr2(<n>),Gr2(<n>+sigma),
  Gr2(<n>-sigma)},view=[0..14,0..0.2],labels=["n","P"]);
```

the resulting picture being shown in Figure 7.15.

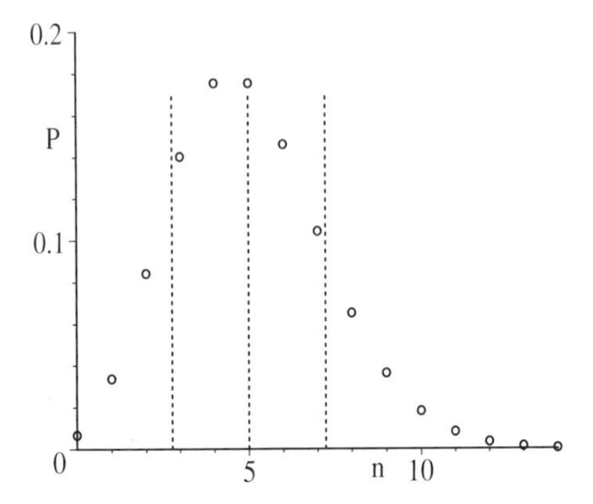

Figure 7.15: Poisson probability P of selling exactly n bottles of ice wine.

The locations of $\langle n \rangle$, $\langle n \rangle + \sigma$, and $\langle n \rangle - \sigma$ are indicated by the dashed lines. Notice how the Poisson distribution is asymmetric around the mean at $n = 5$ and has a relatively long tail for n values above the mean.

We can calculate the cumulative probability (CP) of selling n or fewer bottles of wine in a given week by forming the following functional operator `CP`. The short form of the `CumulativeDistributionFunction` command is used.

```
> CP:=x->CDF(Poisson(lambda),x,numeric):
```

The cumulative probability CP is then plotted using `CP`,

```
> display(Gr1({seq([n,CP(n)],n=0..N)}),labels=["n","CP"]);
```

the result being shown on the left of Figure 7.16.

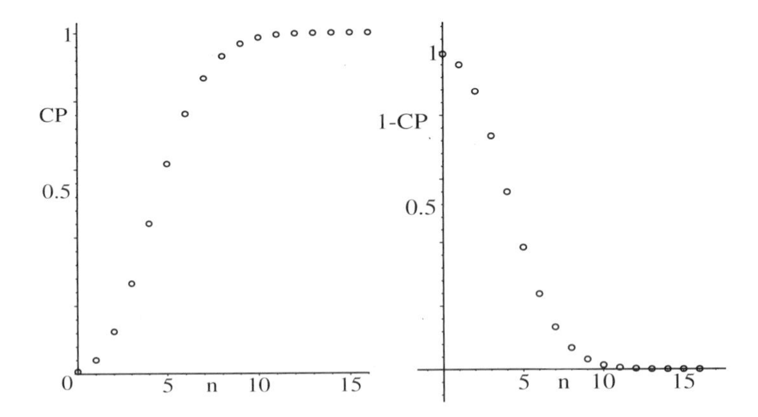

Figure 7.16: Left: CP vs. n. Right: Probability of selling more than n bottles.

In Figure 7.15, the range $\langle n \rangle \pm \sigma$ spans the integer values from $n = 3$ to 7. Using the cumulative probability operator `CP`, the probability of selling 3 to 7 bottles of wine is seen to be equal to the difference `CP(7)-CP(2)`.

```
> "Probability of selling 3 to 7 bottles"=CP(7)-CP(2);
```

"Probability of selling 3 to 7 bottles" = 0.7419763103

The probability is 0.74 or 74% that you will sell a number of bottles in this range. Now let's answer your original questions by calculating the probability of selling more than n bottles in a week. This probability is simply obtained by forming $1 - CP$. This new probability distribution is now plotted, the resulting picture being shown on the right of Figure 7.16.

```
> display(Gr1({seq([n,1-CP(n)],n=0..N)}),labels=["n","1-CP"]);
```

The probability distribution drops rapidly to zero at large n. The probability of selling more than 8 bottles is

```
> "Probability of selling more than 8 bottles"=1-CP(8);
```

"Probability of selling more than 8 bottles" = 0.0680936308

about 0.068, i.e., about 7%. So the chance is fairly small that a potential customer will be disappointed if you stock eight bottles at the beginning of each week. However, you said that it did happen to you recently, so let's increase the number. By increasing the number to ten bottles,

```
> "Probability of selling more than 10 bottles"=1-CP(10);
```

"Probability of selling more than 10 bottles" = 0.0136952678

your chances of running out drops to about 1%. So, I would recommend that you stock up with 10 or 11 bottles at the beginning of each week, unless you contemplate expanding your wine business."

"Thanks, Colleen, I will do that. With my *National Geographic* assignments, I am pretty busy at times and not looking to increase my business, particularly since my uncle has only a limited supply that he can send to me. And the supply varies from year to year, depending on the timing, duration, and severity of the frosts along the Columbia River gorge. But it would be nice not to disappoint my regular customers. Most of these are close friends who really enjoy this unique ice wine. Say, I don't believe that you have tasted this wine yet. How about joining me and splitting a bottle before you go. It's the least I can do for all the help you have given me."

PROBLEMS:

Problem 7-36: Typographical errors

Assume that typographical errors, committed by an anonymous author in writing the first draft of a textbook on symbolic computation, occur completely at random. Suppose that the book of 600 pages contains 600 such errors. Assuming that the Poisson distribution holds:

(a) Calculate the probability that a page contains no errors.

(b) Create a plot of the probability distribution spanning the range from zero to five errors on a page. Is the plot symmetric or asymmetric?

(c) Given the above range, calculate σ for the probability distribution. By what percentage does σ differ from σ_∞?

(d) Create a new graph showing not only the probability distribution, but the mean and the mean $\pm\, \sigma$.

(e) Calculate the probability that a page contains at least three errors.

(f) Create a plot of the probability that a page contains at least n errors, with $n = 0$ to 5.

Problem 7-37: Alpha decay
The emission of α particles by a radioactive source during some time interval can be described by a Poisson distribution [Rei65]. Suppose that for a particular source, the mean number of disintegrations per minute is 24. One disintegration corresponds to the emission of one α particle. For a time interval of 10 s:

(a) What is the mean number of α particles emitted?

(b) What is the probability of emitting two α particles? five α particles?

(c) Plot the probability of observing n alpha particles for n=0 to 10.

(d) Calculate the width of the probability distribution, and place the width and mean on the same graph as the probabilities.

(e) Calculate the probability that at least eight α particles are emitted.

(f) Create a plot of the cumulative probability.

(g) Create a plot of the probability of observing at least n alpha particles, with n=0 to 10.

7.4 Monte Carlo Statistical Distributions

7.4.1 Estimating e

Not everything that can be counted, counts.
Not everything that counts, can be counted.
Albert Einstein, Nobel laureate in physics (1879–1955)

The numerical value of e can be determined with a number of different Monte Carlo approaches, one of the most efficient according to Mohazzabi [Moh98] being the random number equivalent of the dart method. Consider a dartboard that has been divided into R equal-size regions. We randomly throw N darts at the board and assume that we never miss the board. (Ha! Ha!) On any given throw the probability p that a dart lands in a given region is $p = 1/R$. The probability of missing this region is $q = 1-p$. Since there are only two mutually exclusive outcomes (hit or miss) on a given throw, the probability $P(n)$ of finding n darts in a given region after N throws is given by the binomial distribution

$$P(n) = C_n^N\, p^n\, q^{N-n}, \quad \text{with} \quad C_n^N = N!/(n!\,(N-n)!). \tag{7.21}$$

Thus the probability of finding an empty region is

$$P(0) = C_0^N \, p^0 \, q^N = q^N = (1-p)^N. \tag{7.22}$$

Suppose that we can make $p = 1/N$, so that

$$P(0) = \left(1 - \frac{1}{N}\right)^N. \tag{7.23}$$

Now, the exponential function e^{-x} is given by the limit

$$e^{-x} = \lim_{N\to\infty} \left(1 - \frac{x}{N}\right)^N, \tag{7.24}$$

so that

$$e^{-1} = \frac{1}{e} = \lim_{N\to\infty} \left(1 - \frac{1}{N}\right)^N. \tag{7.25}$$

Thus, for sufficiently large N, $P(0)$ gives an estimate of $1/e$. How do we make $p = 1/N$? For the binomial distribution, the mean, or expectation, value of n is given by $\langle n \rangle = p\,N$. If we take the number of darts thrown equal to the number of regions, i.e., $N = R$, then the mean number of darts hitting a region must be $\langle n \rangle = 1 = p\,N$, so that $p = 1/N$ as desired.

So the recipe to estimate the value of e is to randomly throw N darts at a dartboard divided into N equally sized regions, where N is taken to be very large. If $N(0)$ is the number of empty cells, where no dart has struck, then

$$\frac{1}{e} \approx \frac{N(0)}{N}, \quad \text{or} \quad e \approx \frac{N}{N(0)}. \tag{7.26}$$

Since it turns out that N must be very large to get a reasonable estimate of e and it is difficult to really throw darts randomly, we turn to a Monte Carlo computer approach that simulates the dart throwing. We can number the regions by integer values from 1 to N. A random-number generator is used to generate N random numbers in this range. A number n in this random sequence of numbers corresponds to hitting region n with a dart. By counting the number $N(0)$ of regions that were not hit, the value of $e \approx N/N(0)$ can be determined. To improve the estimate, the numerical experiment will be repeated a number of times and an average value of e calculated.

We begin the recipe by loading the Statistics and plots library packages.

```
> restart: with(Statistics): with(plots): begin:=time():
```

Suppose that in each experiment, the number of darts thrown, which is equal to the number of regions, is taken to be $N{=}5000$ and 600 experiments are carried out. This is equivalent to throwing three million darts! Depending on the speed and memory of your PC, you may have to adjust these input values. Decreasing N leads to a poorer estimate of e for a given experiment. For statistical analysis purposes, histograms of the data will be created, the 600 values of e being divided into 12 equally spaced bins. For a large number of experiments, the binomial distribution can be replaced by the normal distribution, which will be used here. The call `randomize()` sets the random-number seed.

```
> N:=5000: Expts:=600: bins:=12: randomize():
```

The do loop, which now begins, runs from $j=1$ to 600 here.

```
> for j from 1 to Expts do
```

For each experiment, a random integer is generated in the range 1 to $N=5000$.

```
> r:=rand(1..N):
```

By using the sequence[6] command and summing over k from 1 to N, a list of N random integers in the desired range is produced for the jth experiment. The `Tally` command is applied to count the number of occurrences of each integer. The number of operands is then determined by applying the **nops** command. Subtracting the resulting number from N finally determines the number of missing regions (missing integers).

```
> Missing[j]:=N-nops(Tally([seq(r(),k=1..N)])):
```

The value of e is then calculated for the jth experiment.

```
> e_estimate[j]:=evalf(N/Missing[j]):
> end do:
```

Ending the do loop, the 600 estimates of e are put into a data list.

```
> data:=[seq(e_estimate[j],j=1..Expts)]:
```

The mean value of e is calculated and compared with the exact numerical value.

```
> <e>:=Mean(data); exact_e:=evalf(exp(1));
```

$$\langle e \rangle := 2.719467815 \qquad exact\_e := 2.718281828$$

The percentage deviation of the Monte Carlo mean, $\langle e \rangle$, from the exact e value is determined and found to be 0.04% for this particular run.

```
> PercentDeviation:=100*(<e>-exact_e)/exact_e;
```

$$PercentDeviation := 0.04363002349$$

Next, the standard deviation, $\sigma = \sqrt{\langle e^2 \rangle - \langle e \rangle^2}$, is calculated,

```
> sigma:=StandardDeviation(data);
```

$$\sigma := 0.03269196410$$

yielding $\sigma \approx 0.033$.

For large N and a large number of experiments, the distribution of e estimates will be approximated by the normal (Gaussian) probability distribution,

$$p(x) = e^{-(x-\langle e \rangle)^2/(2\,\sigma^2)}/\sqrt{2\,\pi\,\sigma^2}.$$

For the normal distribution there is a 68% chance of finding an e value within one σ of the mean and a 95% chance of finding an e value within two σ.

We shall plot the Monte Carlo e value estimates as histograms and superimpose the normal distribution in the same graph as well as locate the positions of $\langle e \rangle$ and $\langle e \rangle \pm \sigma$. First, we extract the minimum (a) and maximum (b) Monte

[6]The **seq** command is more efficient than using a do loop. Using **seq**, the cpu time increases linearly with the length of the sequence, but increases quadratically for the do loop.

Carlo e values from the `data` list. The width δ of a histogram bin is equal to the range $b - a$ divided by the number of bins.

```
> a:=min(data[]); b:=max(data[]); delta:=(b-a)/bins;
```

$$a := 2.624671916 \qquad b := 2.818489290 \qquad \delta := 0.01615144783$$

An operator for calculating the histogram bin boundaries is formed.

```
> x:=i->a+delta*(i-1):
```

The `TallyInto` command is used to determine the number of estimated e values in the data list that lie within each histogram bin.

```
> data2:=TallyInto(data,[seq(x(i)..x(i+1),i=1..bins)]):
```

An operator for calculating the normalized height (normalized so the area under the complete histogram curve is 1) of the ith histogram is created.

```
> y:=i->op([i,2],data2)/(Expts*delta):
```

An operator is formed to plot the ith histogram (colored cyan),

```
> p:=i->polygonplot([[x(i),0],[x(i),y(i)],[x(i+1),y(i)],
          [x(i+1),0]],color=cyan):
```

which is used to generate the complete histogram plot.

```
> h:=display(seq(p(i),i=1..bins)):
```

The normal probability distribution is plotted over the range a to b.

```
> pp:=plot(PDF(Normal(<e>,sigma),x),x=a..b,
        colour=red,thickness=3):
```

A functional operator `Gr` is formed to plot a thick red vertical line of height 12 at the horizontal coordinate x.

```
> Gr:=x->plot([[x,0],[x,12]],style=line,color=red,thickness=3):
```

Entering the following command line produces Figure 7.17,

```
> display({h,pp,Gr(<e>),Gr(<e>-sigma),Gr(<e>+sigma)},
  tickmarks=[3,3]);
```

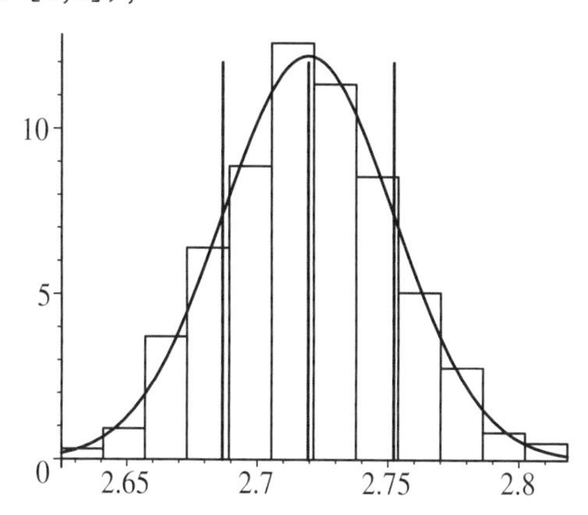

Figure 7.17: Comparison of Monte Carlo histogram with normal distribution.

containing the histogram plot, the normal distribution (solid curve), and vertical lines at $\langle e \rangle$, $\langle e \rangle - \sigma$, and $\langle e \rangle + \sigma$.

If the number of experiments is increased, the width of the histogram bins can be decreased. The width of the probability distribution, as measured by the standard deviation σ, will decrease (σ scaling as $1/\sqrt{N}$) as the number N of darts thrown in each experiment is increased.

```
> cpu_time:=(time()-begin)*seconds;
```

$$cpu\_time := 10.336\, seconds$$

The CPU time for this run was about 10 seconds.

PROBLEMS:

Problem 7-38: Scaling of σ with increasing N

Holding all other parameters the same as in the text recipe, double the number N of darts to 10,000 and then to 20,000. How does the width of the probability distribution, as measured by σ, scale as the number N is doubled? Depending on the speed of your computer, this calculation may take considerable CPU time.

7.4.2 Vapor Deposition

They [atoms] move in the void and catching each other up jostle together, and some recoil in any direction that may chance, and others become entangled with one another in various degrees according to the symmetry of their shapes and sizes and positions and order, and they remain together and thus the coming into being of composite things is effected.

Simplicius, *De Caelo, 242, 15* (490–560)

While Mike is off on his archaeological dig, Vectoria is in the process of learning about various probability distributions as part of her summer job in the MIT physics department. While thumbing through the first chapter of Reif's statistical physics text [Rei65], she encounters a reference to the Poisson probability distribution as well as several related problems. She reads that the Poisson distribution is a limiting case of the binomial probability distribution,

$$P_n = \frac{N!}{n!\,(N-n)!}\, p^n\, (1-p)^{N-n}. \tag{7.27}$$

Here P_n is the probability that an event characterized by a probability p occurs n times in N trials. The Poisson distribution results in the limit that $p \to 0$ (p is small), $N \to \infty$ (N is large compared to n), and the product $N\,p \equiv \lambda$ remains finite. In (7.27),

$$\frac{N!}{(N-n)!} = N\,(N-1)\,(N-2)\cdots(N-n+1) \to N^n,$$

$$(1-p)^{N-n} = \left(1 - \frac{\lambda}{N}\right)^{N-n} \to \left(1 - \frac{\lambda}{N}\right)^{N} \to e^{-\lambda}.$$

So in this limit, Equation (7.27) reduces to the *Poisson probability distribution*

$$P_n = \frac{\lambda^n}{n!}\, e^{-\lambda}. \tag{7.28}$$

The quantity $\lambda \equiv N\,p$ is equal to the mean number of events.[7]

Vectoria has studied the Monte Carlo examples presented earlier and decides to perform a numerical simulation whose inspiration is based on a Poisson distribution question mentioned in Reif. The problem is one involving the evaporation of metal atoms from a hot filament in vacuum. The emitted metal atoms are incident on a quartz plate located some distance away and form a thin metal film on the plate. The quartz plate is held at a sufficiently low temperature that any incident metal atom sticks at the place of contact with the plate and doesn't migrate away from this contact point. It is assumed that the metal atoms are equally likely to hit any region of the plate. The vapor deposition of thin metallic films on various substrates is of great interest to the material scientists in the lab where she is working.

In the problem, Vectoria is asked to answer a number of relevant questions. If b is the diameter of the metal atom and one considers a substrate area element of size b^2, show that the number of metal atoms piled up on this area should be distributed according to a Poisson distribution. If enough metal atoms are evaporated to form a film of mean thickness corresponding to 6 atomic layers:

(a) What fraction of the substrate is not covered by metal at all?

(b) What fraction is covered by metal layers three atoms thick?

(c) What fraction is covered by metal layers six atoms thick?

Vectoria's self-appointed task is to carry out a simple Monte Carlo simulation of the vapor deposition process, answer the above questions, and compare the experimental (numerical) results with the theoretical (statistical) predictions.

The plots and statistical packages are loaded.

```
> restart: with(plots): with(Statistics):
```

Vectoria assumes for the sake of definiteness that there are $N1 = 20$ thousand possible substrate sites on which an emitted metal atom can land. Since any site is equally probable, the probability of landing on a given site is $p = 1/N1 = 1/20000$. Enough atoms are evaporated to form a film of mean thickness corresponding to $L = 6$ atomic layers. The total number of atoms emitted, therefore, is $N = N1\ L = 120{,}000$ atoms. The mean number of layers is $\lambda = p\,N = 6$.

```
> N1:=20000: L:=6: p:=1/N1; N:=N1*L; lambda:=p*N;
```

$$p := \frac{1}{20000} \qquad N := 120000 \qquad \lambda := 6$$

The starting time is recorded, and the random-number seed is entered.

---

[7]Recall the ice wine story.

```
> begin:=time(): randomize():
```

The count number c_i (number of atoms on site i) is initialized to zero for each of the $N1 = 20$ thousand possible target sites.

```
> for i from 1 to N1 do; c[i]:=0; end do:
```

In the following do loop, random numbers between 1 and $N1 = 20$ thousand are generated for $N = 120$ thousand atoms. These atoms can land randomly on any one of the possible target sites. Each time an atom lands on a given site, the count number of atoms on that site is increased by one.

```
> rn:=rand(1..N1):
> for i from 1 to N do;
> r:=rn();
> c[r]:=c[r]+1;
> end do:
```

The count number for each of the $N1$ target sites is formed into a list and sorted so the count numbers are in ascending order. The maximum count number M in the data list is extracted.

```
> data:=sort([seq(c[i],i=1..N1)]): M:=max(data[]);
```

$$M := 19$$

The `Tally` command is used to tally or count the number of sites that have received a given number of atoms.

```
> data2:=Tally(data);
```

$$data2 := [0 = 47,\ 1 = 301,\ 2 = 908,\ 3 = 1774,\ 5 = 3161,\ 4 = 2706,\ 7 = 2801,\ 6 = 3150,\ 10 = 818,\ 11 = 426,\ 8 = 2131,\ 9 = 1379,\ 15 = 21,\ 14 = 45,\ 13 = 105,\ 12 = 212,\ 16 = 11,\ 17 = 3,\ 19 = 1]$$

The first entry "$0 = 47$" in *data2* informs Vectoria that 47 sites received zero atoms, the second entry that 301 sites received one atom, etc. The maximum number, $M = 19$ atoms, was achieved on one site. The fraction of sites receiving zero atoms, one atom, two atoms, etc., is calculated and put into a list.

```
> list1:=[seq(nops(select(has,data,i))/N1,i=0..M)];
```

$$list1 := \left[\frac{47}{20000}, \frac{301}{20000}, \frac{227}{5000}, \frac{887}{10000}, \frac{1353}{10000}, \frac{3161}{20000}, \frac{63}{400}, \frac{2801}{20000}, \frac{2131}{20000}, \frac{1379}{20000}, \frac{409}{10000}, \frac{213}{10000}, \frac{53}{5000}, \frac{21}{4000}, \frac{9}{4000}, \frac{21}{20000}, \frac{11}{20000}, \frac{3}{20000}, 0, \frac{1}{20000}\right]$$

A list of count numbers up to the maximum number M is formed.

```
> list2:=[seq(i,i=0..M)]:
```

The `ScatterPlot` command is used to plot the Monte Carlo experimental data, the data point coordinates being given in the two lists, *list1* and *list2*.

```
> plot1:=display(ScatterPlot(list2,list1,symbol=circle,
          symbolsize=12),color=red):
```

The following functional operator calculates the numerical value of the Poisson probability, with $\lambda = 6$, for a given n.

```
> P:=n->ProbabilityFunction(Poisson(lambda),n,numeric):
```

Using P, a list of the Poisson probabilities from $n=0$ to M is produced.

```
> list3:=[seq(P(n),n=0..M)];
```

$$list3 := [0.002478752177, 0.01487251306, 0.04461753919, 0.08923507837, \ldots]$$

To make a comparison with the Monte Carlo results, Vectoria creates a plot of the theoretical values predicted by the Poisson statistical distribution, choosing to represent the theoretical points graphically with blue crosses.

```
> plot2:=display(ScatterPlot(list2,list3,symbol=cross,
          symbolsize=12),color=blue):
```

The experimental (Monte Carlo) points are superimposed on the same graph as the theoretical (Poisson distribution) points,

```
> display({plot1,plot2},labels=["n","f"]);
```

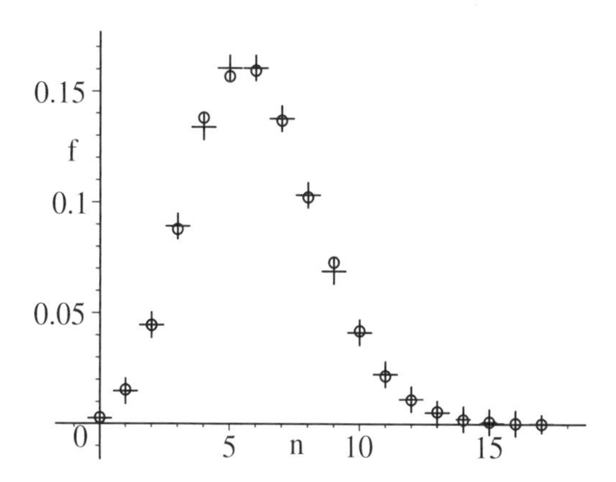

Figure 7.18: Fraction f of sites receiving n atoms during vapor deposition. Circles: Monte Carlo. Crosses: Poisson formula.

the resulting picture being displayed in Figure 7.18. The experimental results are in excellent agreement with what would be expected on the basis of statistical theory.

Vectoria is now in a position to answer the questions about layer coverage on the substrate sites. Although she could click on the computer graph to get approximate values, more precise answers follow from looking at the various lists that have been generated. From *list1* she finds that experimentally the fraction of sites receiving zero atoms, three atoms, and six atoms is 47/20000, 887/10000, and 63/400, or about 0.24%, 8.9%, and 15.8%. From *list3*, she notes that the Poisson distribution predicts 0.25%, 8.9%, and 16.1%. Vectoria is pleased with how easy it has been to simulate the experiment suggested by Reif's problem and even more pleased with how well the Monte Carlo experimental results are accounted for by the theoretical Poisson distribution.

```
> cpu_time:=(time()-begin)*seconds;
```

$$cpu\_time := 1.212\,seconds$$

She further notes that the elapsed CPU time was only about 1 second for the vapor deposition simulation. Looking at her watch to see what time it actually is, she realizes that Mike is supposed to phone shortly, which will round off what has been a good day.

PROBLEMS:

Problem 7-39: Alpha decay

A radioactive source emits α particles during a time interval T. Now imagine that T is divided into many very small time intervals ΔT. Since α particles are emitted at random times, the probability of a disintegration occurring in a particular ΔT is independent of the probability of a disintegration in another small time interval. If ΔT is sufficiently short, the probability of more than one disintegration in ΔT is negligible. Thus if p is the probability of a disintegration in ΔT, then the probability of no disintegration is $1-p$. Thus each time interval ΔT will be an independent Bernoulli trial and there will be $N = T/\Delta T$ trials during the time T.

(a) For a particular radioactive source, the mean number of disintegrations per minute is 24. Assuming that a Poisson distribution prevails, what is the probability of n counts occurring in a 10-second time interval? For n, choose a range of integers from 0 to 8.

(b) Carry out a Monte Carlo simulation of the α decay in part (a).

(c) Plot the experimental and theoretical values for the probabilities of n counts occurring in the 10-second time interval in the same graph, and discuss the accuracy of your simulation.

Chapter 8

Fractal Patterns

Art is the imposing of a pattern on experience, and our aesthetic enjoyment is recognition of the pattern.
Alfred North Whitehead, English philosopher and mathematician (1861–1945)

Patterns pervade the natural world as well as the world of the intellect. In the biological realm, we are quite aware that when we mentally visualize a zebra we probably first think of its most prominent feature, its stripes. When we look at certain butterflies, it is usually the colorful markings on the wings that grab our attention. If we study magnified ice crystals, our interest is captivated by the richness and regularity of the patterns displayed. If we go into a wallpaper store to shop for our home, we can be overwhelmed by the artistic choices available. If we listen to a piece by Beethoven we are struck by the musical tapestry that one of the world's greatest composers has woven. If we talk to a scientist we will soon find that his or her goal in life is usually to discover (impose?) some underlying pattern to the phenomena under investigation. Clearly, patterns are important in many different ways. As a consequence, the scientific study of pattern formation is a very large field, and any attempt to systematically cover the topic is far beyond the aim or scope of this text.

Therefore, we have asked our MIT mathematics faculty friend Jennifer whether she could provide us with some recipes that produce artistic masterpieces of mathematical pattern formation based on some common theme. She has graciously agreed and has elected to show us a few examples of so-called *fractal* patterns. A fractal structure is characterized by a *noninteger (fractal) dimension*, the usual concept of dimension (which is limited to integer values) being extended to describe geometric objects with jagged boundaries (e.g., clouds, coastlines, ferns), planar objects with holes in them, and so on.

How is a fractal dimension defined? The reader undoubtedly knows that a point has zero dimension, a smooth, continuous, line has one dimension, a filled-in planar object has two dimensions, and so on. When one has patterns with jagged boundaries or made up of lines and planar objects with holes in them, one can generalize the concept of dimension to describe such geometrical objects. There are several different ways [PC89] of doing this, but one simple definition of fractal dimension is the so-called *capacity dimension*, D_C.

To mathematically develop a formula for calculating D_C, first consider a continuous straight line (or more generally a smooth curve) of length L as shown in Figure 8.1. This line is covered by $N(\varepsilon)$ one-dimensional segments,

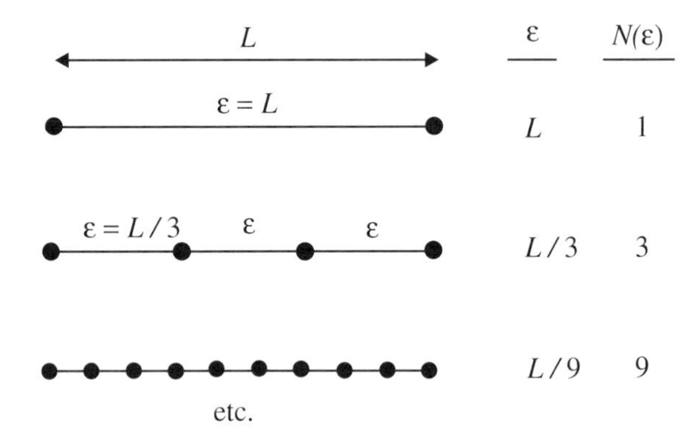

Figure 8.1: Covering a line of length L with line segments of length ε.

each of length ε, the segment boundaries being indicated by dots. On the top line, $\varepsilon = L$ and $N(\varepsilon) = 1$. At the next level, let's arbitrarily divide the line into three segments. In this case, $\varepsilon = L/3$ and $N(\varepsilon) = 3 = L/\varepsilon$. Quite generally, for any line subdivided in the same manner, $N(\varepsilon) = L/\varepsilon$.

Next, consider a two-dimensional square of side L, as shown in Figure 8.2. The square is covered with identical boxes of side ε and again $N(\varepsilon)$, the number

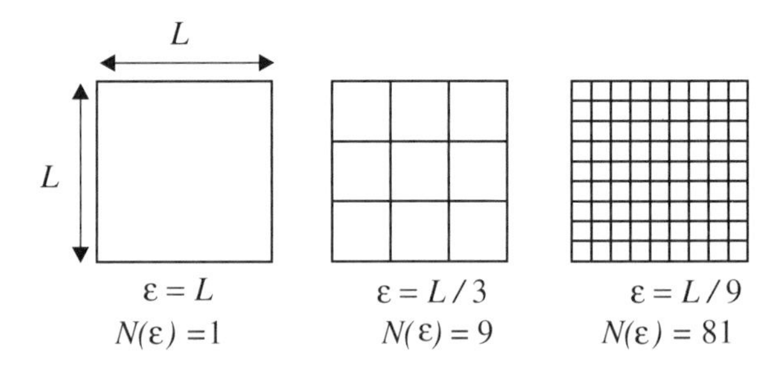

Figure 8.2: Covering a square of side L with boxes of side ε.

of boxes needed to fill the square, is determined. In this case, $N(\varepsilon) = L^2/\varepsilon^2$. In three dimensions, one clearly obtains $N(\varepsilon) = L^3/\varepsilon^3$ and, generalizing, in D-dimensions, $N(\varepsilon) = L^D/\varepsilon^D$. Taking the logarithm and solving for D yields

$$D = \frac{\ln N(\varepsilon)}{\ln L + \ln(1/\varepsilon)}. \tag{8.1}$$

As $\varepsilon \to 0$, then $\ln(1/\varepsilon) \gg \ln L$ and the capacity dimension is defined by

$$D_C = \lim_{\varepsilon \to 0} \frac{\ln N(\varepsilon)}{\ln(1/\varepsilon)}. \tag{8.2}$$

So, D_C agrees with the "normal" concept of dimension for the examples above involving continuous lines, planar objects, etc.

Let us now return to Figure 8.1 and throw away the middle third at each step as in Figure 8.3. Taking $L = 1$ for simplicity, let's count the number of

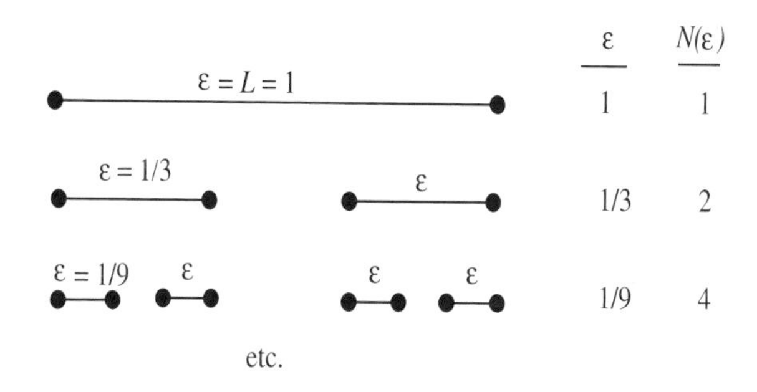

Figure 8.3: The Cantor set.

line segments $N(\varepsilon)$ needed to cover the unit interval, i.e., the empty segments are not counted. On the kth step, $\varepsilon = (1/3)^k$ and $N(\varepsilon) = 2^k$, thus yielding a capacity dimension

$$D_C = \lim_{k\to\infty} (\ln 2^k / \ln 3^k) = \ln 2/\ln 3 = 0.6309\cdots. \tag{8.3}$$

The segmented line with gaps in Figure 8.3 is referred to as a *Cantor set*. With a capacity dimension $D_C \approx 0.63$, it has a fractal dimension intermediate between a point (zero dimensions) and a continuous line (one dimension). The Cantor set has a fractal dimension, which makes intuitive sense, since it is "more" than a point but not quite a solid line.

Now let's see what fractal patterns Jennifer has created for us.

PROBLEMS:

Problem 8-1: The Koch triadic curve

Consider a line of length 1 unit. Instead of throwing away the middle third as in the Cantor set, form an equilateral triangle in the middle third. Each line segment has length $\varepsilon = 1/3$. Repeat the process with each new segment in step

L=1

ε ε
ε=1/3 ε

1 to produce step 2. Each segment now has length 1/9. Repeating this process indefinitely, determine D_C. Does your answer make intuitive sense? Explain.

Problem 8-2: The middle-half Cantor set
The Cantor set is also known as the middle-third Cantor set, since on each step the middle third of each remaining line segment is thrown away. In the middle-half Cantor set, the line is initially divided into quarters and the inner two quarters (the middle half) are thrown away. If this action is repeated indefinitely with the remaining line segments, what is the capacity dimension of the middle-half Cantor set? If you compare this dimension with that for the middle-third Cantor set, does your answer make intuitive sense? Explain.

8.1 Difference Equations

8.1.1 Wallpaper for the Mind

Either that wallpaper goes, or I do.
Oscar Wilde (1854–1900), last words as he lay dying in a drab hotel room

In the September 1986 issue of the magazine *Scientific American* the cover featured intricate computer-generated designs which were referred to as "Wallpaper for the Mind." An example of such a wallpaper pattern is that generated by the following pair of coupled nonlinear difference equations:

$$x_{n+1} = y_n - \text{signum}(x_n)\sqrt{|\,b\,x_n - c\,|}, \quad y_{n+1} = a - x_n, \tag{8.4}$$

where the *signum function* is defined by

$$\text{signum}(x) = x/|\,x\,|, \text{ for } x < 0 \text{ and } x > 0.$$

Thus signum(x) is a *step* function, equal to -1 for $x < 0$ and $+1$ for $x > 0$.

A delicate and pretty lace pattern occurs for the parameter values $a = 3.14$, $b = 0.3$, $c = 0.3$ and the initial values $x(0) = y(0) = 0.2$. The `plots` and `plottools` packages are loaded. The latter is needed because the `rotate` command will be applied to the graph.

```
> restart: with(plots): with(plottools):
```

How much CPU time is used is important in some examples of pattern formation, so the beginning time is recorded.

```
> begin:=time():
```

The initial values,

```
> x[0]:=0.2: y[0]:=0.2:
```

and the parameter values are entered.

```
> a:=3.14: b:=0.3: c:=0.3: N:=30000:
```

To obtain a wallpaper design with considerable detail, $N = 30$ thousand iterations will be carried out.

The difference equations are then iterated from $n = 0$ to N.

```
> for n from 0 to N do
> x[n+1]:=y[n]-signum(x[n])*sqrt(abs(b*x[n]-c));
> y[n+1]:=a-x[n];
> end do:
```

The sequence command produces a list of lists for the plotting points.

```
> plotpoints:=[seq([x[n+1],y[n+1]],n=0..N)]:
```

The `pointplot` command is used to create the basic wallpaper design,

```
> pl:=pointplot(plotpoints,symbol=POINT):
```

which is rotated through $-\pi/4$ radians, and displayed,

```
> pl2:=rotate(pl,-Pi/4):
> display(pl2,axes=boxed,tickmarks=[0,0],color=red,
  scaling=constrained);
```

producing the wallpaper pattern in Figure 8.4.

Figure 8.4: A wallpaper design.

The CPU time is about 7 seconds on a 3 GHz personal computer.

```
> cpu_time:=(time()-begin)*seconds;
```

$$cpu\_time := 7.162\, seconds$$

PROBLEMS:

Problem 8-3: Different parameter values

Explore the nonlinear map given in the text for other values of the parameters and see whether you can find any other suitable wallpaper patterns.

Problem 8-4: Altering the model

Keeping all parameter values the same as in the text recipe, explore the effect of altering the model. For example, you might insert a factor of 2 in front of the signum function, or take the absolute value of x_n in the y equation, or try some other form. When executing a new model for the first time it is a good idea to reduce the N value so that you do not tie your PC up in the do loop.

8.1.2 Sierpinski's Fractal Gasket

Great fleas have little fleas upon their backs to bite 'em,
And little fleas have lesser fleas, and so ad infinitum.
And the great fleas ... in turn, have greater fleas to go on;
While these again have greater still ... and so on.
Augustus De Morgan, English mathematician (1806–1871)

Difference equations are useful for generating patterns that mimic those observed in nature, e.g., the triangular array seen on the conus seashell. As an illustrative example of a triangular design, Jennifer will now consider the following numerical simulation, which reproduces a pattern commonly referred to as *Sierpinski's gasket.* It involves still another use of the random-number generator discussed in Chapter 7. Because the dynamics involve a random, or *stochastic,* component, the difference equations in this example are not deterministic.

A call is made to the plots package and the random-number seed initialized.

```
> restart: with(plots): randomize(): begin:=time():
```

Jennifer enters three points (planar coordinates $(A[i], B[i])$ with $i = 0, 1, 2$) that lie at the vertices of an equilateral triangle whose sides are of length 2.

```
> A[0]:=0: B[0]:=0: A[1]:=1: B[1]:=1.732: A[2]:=2: B[2]:=0:
```

Starting at, say, the origin $x[0]=y[0]=0$, the following two-dimensional map

$$x[n+1] = x[n] + \varepsilon\,(A[i] - x[n]), \quad y[n+1] = y[n] + \varepsilon\,(B[i] - y[n]), \quad (8.5)$$

with $0 < \varepsilon < 1$, is iterated N times. Jennifer takes $N = 4000$ and $\varepsilon = 0.5$. On each step, a random-number generator will be used to randomly select from among the values 0, 1, and 2 for the index i.

```
> x[0]:=0: y[0]:=0: epsilon:=0.5: N:=4000: r2:=rand(0..2):
```

The do loop,

```
> for n from 0 to N do
```

begins with r2() for randomly selecting the values 0, 1, and 2 for the index i.

```
> sel:=r2():
```

The two-dimensional map is inserted into the do loop,

```
> x[n+1]:=x[n]+epsilon*(A[sel]-x[n]):
> y[n+1]:=y[n]+epsilon*(B[sel]-y[n]):
```

and a point plot formed, the points being represented by diamonds.

```
> pl[n]:=pointplot([x[n+1],y[n+1]],symbol=diamond):
> end do:
```

The entire sequence of 4000 points is displayed in a single graph.

```
> display(seq(pl[j],j=0..N),tickmarks=[3,2],labels=["x","y"],
  scaling=constrained);
```

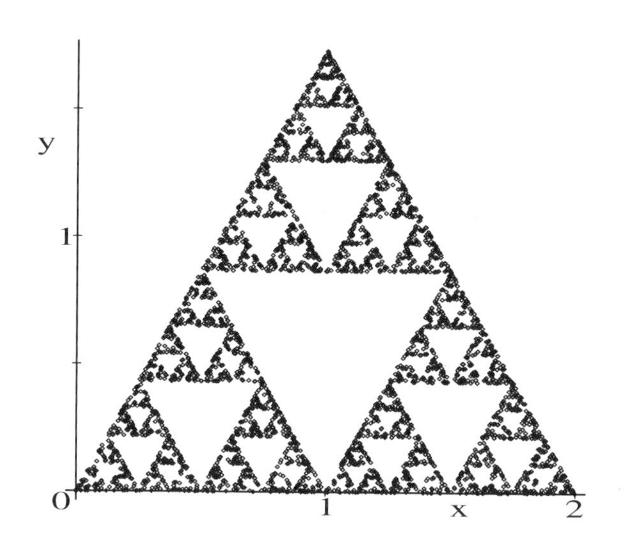

Figure 8.5: Dynamical generation of Sierpinski's gasket.

Figure 8.5 is a numerical simulation of a geometrical pattern known as Sierpinski's gasket. The CPU time to produce this pattern is about 3 seconds.

```
> cpu_time:=(time()-begin)*seconds;
```

$$cpu\_time := 2.914\ seconds$$

Sierpinski's gasket is traditionally created by carrying out the following geometrical construction. Consider an upright, black equilateral triangle. Remove an inverted equilateral triangle inscribed inside the black triangle with vertex points bisecting the sides of the black triangle. One will now have an inverted white triangle with three smaller upright black triangles adjacent to its three sides. Then, repeat this superposition process inside each of the three new black triangles, and so on.

Although this geometrical procedure can be easily carried out by hand, it soon becomes tedious as one goes to finer and finer scale. Therefore we have asked Jennifer to develop a computer algebra recipe that will do the job.

Here is her recipe. To make the computer plot more picturesque, Jennifer has colored the triangles, replacing black with blue, and white with red. She has also used the process of superposition of triangles, rather than removal.

The plots and plottools library packages are loaded. The latter is needed in order to use the Maple commands scale and translate. The value of N determines how many times the construction process is to be repeated.

```
> restart: with(plots): with(plottools): N:=5:
```

The vertex coordinates of the original upright triangle are specified,

```
> v0:=[[0,0],[1,1.732],[2,0]]:
```

and a solid blue triangle with these vertices created with polygonplot.

```
> p0:=polygonplot(v0,color=blue):
```

The vertex coordinates of the inscribed inverted triangle are given,

```
> v1:=[[1/2,0.866],[1.5,0.866],[1,0]]:
```

and this triangle is plotted with a solid red color.

```
> p1:=polygonplot(v1,color=red):
```

The first step of the Sierpinski gasket construction can be now accomplished by superimposing the inverted red triangle on top of the upright blue triangle using the following display command. In Figure 8.6, black corresponds to blue and white to red. Note that the plots p1 and p0 are placed in a list and that the order of the entries is important. If the order is reversed, the larger black (blue) triangle will completely cover the inverted white (red) triangle.

```
> display([p1,p0],tickmarks=[4,2],scaling=constrained);
```

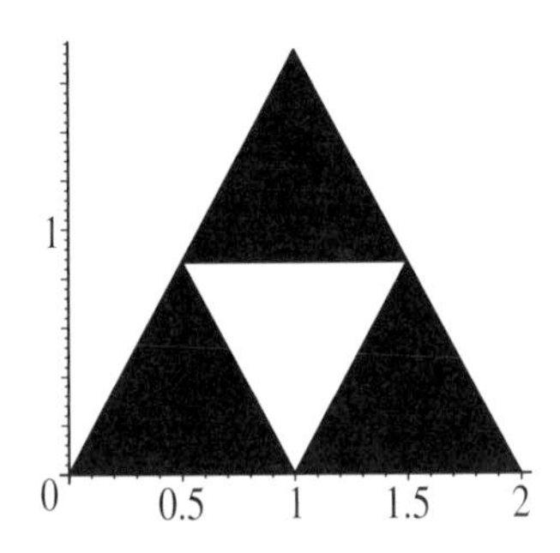

Figure 8.6: First step in constructing Sierpinski's gasket.

The above first step mimics a hand calculation. Now Jennifer automates the remainder of the gasket construction. The following scaling operator S will produce scaled-down replicas of the central inverted red triangle when k, which takes on positive integer values, is specified. The scale factor is $1/2^k$. So, e.g., $k=1$ generates a red triangle one-half the size of the central red triangle.

```
> S:=k->scale(p1,1/2^k,1/2^k):
```

The vertex operator V is used to reduce the coordinate values of the vertices of the original upright blue triangle by a factor 2^k, where k is a positive integer.

```
> V:=k->map(x->x/2^k,v0):
```

Now an appropriate number of scaled-down replicas of the central inverted red triangle must be generated on each step and translated to the right locations. An operator P is formed to do this for a specified positive integer k.

```
> P:=k->seq(translate(S(k),T||k[j][]),j=1..3^k):
```

For $k = 1$, three half-sized (since one has S(1)) replicas of the original inverted red triangle will be translated to the appropriate locations determined by T||1[j][], with $j=1$, 2, 3. For each j value, two numbers are given that specify the translations in the horizontal and vertical directions. These numbers still have to be determined. For $k=2$, 3, etc., $3^2=9$ quarter-sized, $3^3=27$ one-eighth-sized, etc., replicas will be translated to their proper positions.

Taking the initial translation of the three vertices to be zero,

```
> T||0:=[[0,0],[0,0],[0,0]]:
```

the following do loop determines the translations of the replicas.

```
> for k from 0 to N do
> T||(k+1):=[seq(seq(T||(k)[j]+V(k+1)[n],j=1..3^k),n=1..3)]:
> end do:
```

Using P, all the plots are superimposed to produce Sierpinski's gasket.

```
> display([seq(P(k),k=1..N),p1,p0],tickmarks=[3,2],
  scaling=constrained);
```

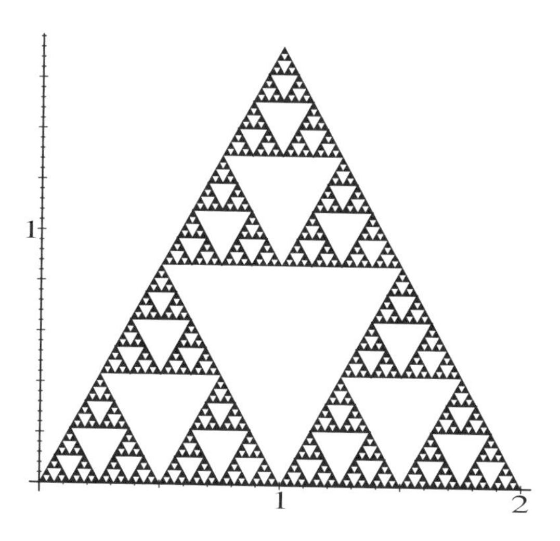

Figure 8.7: Geometrical formation of Sierpinski's gasket.

A black-and-white version of the colorful computer picture is shown in Figure 8.7. Clearly, the numerical simulation mimics the geometrical procedure. "So what!" the reader might exclaim. Mathematical biologists, in trying to understand the microscopic origin of macroscopic patterns such as the spots on a leopard, the stripes on a zebra, or the spiral array of triangles on a conus seashell, postulate dynamical rules involving either difference or differential equations that they hope will ultimately lead to deeper understanding.

Mathematicians, such as Jennifer, are also interested in patterns such as Sierpinski's gasket, because it has a fractal structure, i.e., has a noninteger dimension. In the black-and-white version, the white regions can be regarded as holes in a black background. What is the dimension of this gasket? Here $\varepsilon = (1/2)^k$ and $N(\varepsilon) = 3^k$, so that in the limit $k \to \infty$, $D_{\rm C} = \ln 3 / \ln 2 \approx 1.585$. Because of the white triangular holes, the black-and-white Sierpinski gasket has a "dimension" intermediate to those of a smooth black line and a solid black triangle. Once again, this result makes intuitive sense.

Finally, Jennifer wants to point out that the Cantor set and Sierpinski's gasket are also referred to as *self-similar* fractals, since the basic geometric pattern in each case is repeated indefinitely on a finer and finer scale as N is increased. Not all fractal patterns are self-similar.

PROBLEMS:

Problem 8-5: Another pattern

In the dynamical simulation recipe:

(a) Input the following vertex points:
$A[0]=0$, $B[0]=10$, $A[1]=20$, $B[1]=10$, $A[2]=15$, $B[2]=17.33$, $A[3]=5$, $B[3]=17.33$, $A[4]=15$, $B[4]=0$, $A[5]=5$, $B[5]=0$, $A[6]=10$, $B[6]=10$.
Take $\varepsilon = 0.8$, replace 2 with 6 in the `rand` command, and set $N = 2000$. What is the symmetry of the resultant pattern on executing the new recipe?

(b) Experiment with different values of ε and different integers in the `rand` command in part (a).

Problem 8-6: Sierpinski's carpet

A black square with sides of unit length is divided into nine smaller equal squares and the central square is colored white. Then this process is repeated for each of the eight remaining black squares, and so on. Create a recipe that geometrically produces the Sierpinski "carpet" that results after at least five such iterations. Determine the fractal dimension of Sierpinski's carpet and comment on whether the answer makes intuitive sense. Is the carpet a self-similar fractal?

8.1.3 Barnsley's Fern

All the effects of nature are only the mathematical consequences of a small number of immutable laws.
Pierre Simon Laplace, French astronomer and mathematician (1749–1827)

Another example of using a finite difference algorithm to produce a picture that resembles a real-life pattern is due to the mathematician Michael Barnsley, who pioneered the use of simple sets of equations to generate fractal ferns. By producing a random number r between 0 and 1, and iterating the two-dimensional piecewise map

$$(x_{n+1}, y_{n+1}) = \begin{cases} (0,\, 0.16\, y_n), & 0.00 < r < 0.01, \\ (0.2\, x_n - 0.26\, y_n,\, 0.23\, x_n + 0.22\, y_n + 0.2), & 0.01 < r < 0.08, \\ (-0.15\, x_n + 0.28\, y_n,\, 0.26\, x_n + 0.24\, y_n + 0.2), & 0.08 < r < 0.15, \\ (0.85\, x_n + 0.04\, y_n,\, -0.04\, x_n + 0.85\, y_n + 0.2), & 0.15 < r < 1.00, \end{cases}$$

a fern is "created" that resembles the black spleenwort (*Asplenium adiantum-nigrum*). Still other pictures of ferns can be produced that bear a close resemblance to actual species occurring in nature by changing the coefficient values.

These ferns are all characterized by having fractal boundaries. The fractal dimension of a given fern can be estimated by recalling that the capacity dimension D_C is defined in the limit as $\varepsilon \to 0$ through the relation

$$\ln N(\varepsilon) = D_C \ln(1/\varepsilon) + D_C \ln L \equiv D_C \ln \delta + b, \qquad (8.6)$$

where $\delta = 1/\varepsilon$ and b is a constant. This is the equation of a straight line if $\ln N(\varepsilon)$ is plotted as a function of $\ln \delta$, with slope D_C and intercept b.

To apply Equation (8.6), a box-counting approach is used as follows. First, the algorithm is iterated a large number of times to produce the fern. The two-dimensional picture then is covered with a reasonably fine grid of squares, each square being of length ε along a side. Then the number of squares that have one or more data points inside are counted, giving us $N(\varepsilon)$ for a given δ value. The process is then repeated with finer and finer grids. Ideally, one could proceed by successively halving the value of ε a large number of times, but this may not be practical on a PC, where there is usually a limitation to the total number of points that it is feasible to generate in a reasonable length of time.

Calculating $\ln N$ for each δ value, a least squares routine similar to that employed in Chapter 2 is then used to find the best-fitting straight line to the data points. From Equation (8.6) the slope of this line then yields D_C.

Jennifer will now produce Barnsley's fern for us and estimate its fractal dimension. Calls are made to the plots and Statistics packages. The latter is required so that the best-fitting straight line to the data points can be found.

```
> restart: with(plots): with(Statistics):
```

Since the symbol `D` will be used to represent the fractal dimension, it is necessary to unprotect the symbol from its Maple meaning as the differential operator.

```
> unprotect(D): begin:=time():
```

To make the programming of the algorithm a little neater, the given piecewise relation is written as

$$x_{n+1} = a[\,i\,]\,x_n + b[\,i\,]\,y_n + e[\,i\,], \quad y_{n+1} = c[\,i\,]\,x_n + d[\,i\,]\,y_n + f[\,i\,],$$

where the first branch of the piecewise map corresponds to $i = 1$ and is selected if the random number r is less than $p[\,1\,]=0.01$, the second branch corresponds to $i=2$, and so on. The various coefficients are now specified:

```
> a[1]:=0: a[2]:=0.2: a[3]:=-0.15: a[4]:=0.85:
> b[1]:=0: b[2]:=-0.26: b[3]:=0.28: b[4]:=0.04:
> c[1]:=0: c[2]:=0.23: c[3]:=0.26: c[4]:=-0.04:
> d[1]:=0.16: d[2]:=0.22: d[3]:=0.24: d[4]:=0.85:
> e[1]:=0: e[2]:=0: e[3]:=0: e[4]:=0:
> f[1]:=0: f[2]:=0.2: f[3]:=0.2: f[4]:=0.2:
> p[1]:=0.01: p[2]:=0.08: p[3]:=0.15:
```

Jennifer takes $x_0 = y_0 = 0$ as the starting coordinates and will carry out $N = 10$ thousand iterations.

```
> N:=10000: x[0]:=0: y[0]:=0:
```

The `randomize()` command sets the random number seed for the random-number generator.

```
> randomize():
```

Making use of the random-number generator command, the map is iterated.

```
> for n from 0 to N do
> r[n]:=rand()/10^12; #random number between 0 and 1
> if r[n]<p[1] then i:=1 elif r[n]<p[2] then i:=2
  elif r[n]<p[3] then i:=3 else i:=4 end if;
> x[n+1]:=a[i]*x[n]+b[i]*y[n]+e[i];
> y[n+1]:=c[i]*x[n]+d[i]*y[n]+f[i];
> pnt[n+1]:=[x[n+1],y[n+1]];
> end do:
```

A plot of the fractal fern is created but not shown. The options `view` and `scaling=constrained` are used to keep the picture correctly proportioned.

```
> p:=pointplot([seq(pnt[n],n=1..N)],symbol=point,
     labels=["x","y"],tickmarks=[3,3],scaling=constrained,
     axes=boxed,view=[-0.75..0.75,0..1.5],color=green):
```

A grid of identical square boxes is to be superimposed on top of the fractal fern graph `p`. The total number T of boxes along each side of the picture is taken to be, for example, $T = 6$. In this case, $6 \times 6 = 36$ boxes are created. From the `view` command, the fractal fern picture generated in the above `pointplot` command line is square with length 3/2 along each side. Thus, each of the 36 grid boxes has an edge of length $\varepsilon = 3/(2\,T) = 1/4$, so that $\delta = 1/\varepsilon = 4$.

```
> T:=6: epsilon:=3/(2*T); delta:=1/epsilon;
```

$$\varepsilon := \frac{1}{4} \qquad \delta := 4$$

The `polygonplot` command is used in a double sequence to generate the square grid. By increasing T, a finer grid can be produced. But remember that N should also be increased, thus leading to longer CPU times.

```
> gr:=seq(seq(polygonplot([[-0.75+epsilon*i,epsilon*j],[-0.75
    +epsilon*(i+1),epsilon*j],[-0.75+epsilon*(i+1),epsilon*
    (j+1)],[-0.75+epsilon*i,epsilon*(j+1)]]),i=0..T),j=0..T):
```

Using the `display` command, Figure 8.8 is produced showing Barnsley's fern with the square grid superimposed.

```
> display({p,gr}); cpu_time:=(time()-begin)*seconds;
```

$$cpu\_time := 1.231\ seconds$$

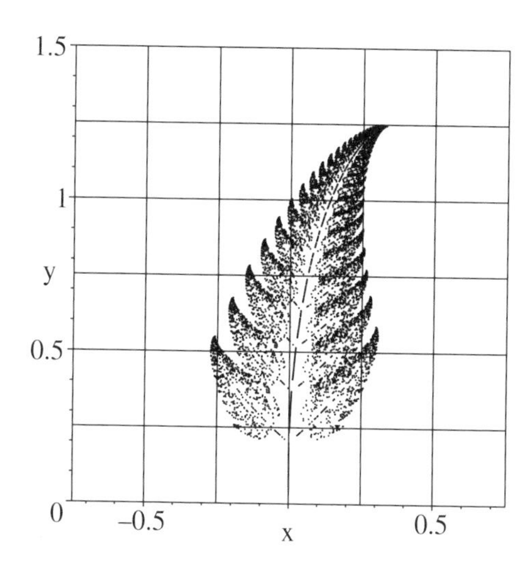

Figure 8.8: Barnsley's fractal fern with a square grid superimposed.

Careful examination of the figure reveals that 15 boxes contain one or more points. So, one data point entry will be $[\ln 4.0, \ln 15.0]$, the decimal point being added so that the output will be expressed in decimal form. Adding the zero after the decimal point does not indicate some mysterious increase in accuracy.

Altering the grid size by setting T = 9, 12, 15, so that δ = 6, 8, and 10, Jennifer has found that $N(\varepsilon)$ = 27, 39, and 58, respectively. With these values, she forms two Maple lists of the logarithms of the δ and $N(\varepsilon)$,

```
> ln_delta:=[ln(4.0),ln(6.0),ln(8.0),ln(10.0)];
```

$$ln\_delta := [1.386294361,\ 1.791759469,\ 2.079441542,\ 2.302585093]$$

```
> ln_number:=[ln(15.0),ln(27.0),ln(39.0),ln(58.0)];
```

$$ln\_number := [2.708050201,\ 3.295836866,\ 3.663561646,\ 4.060443011]$$

The least squares `Fit` command generates the best-fitting straight line to the data points.

```
> eq:=Fit(a*X+b,ln_delta,ln_number,X);
```

$$eq := 0.68969452307214O518 + 1.45092551362301392\,X$$

To show how well the straight line fits the data, the data points are plotted along with `eq`, and displayed in Figure 8.9.

```
> gr2:=ScatterPlot(ln_delta,ln_number,style=point,
      symbol=CIRCLE,symbolsize=14):
> gr3:=plot(eq,x=1..2.5):
> display({gr2,gr3},tickmarks=[2,2],
  labels=["ln_delta","ln_number"]);
```

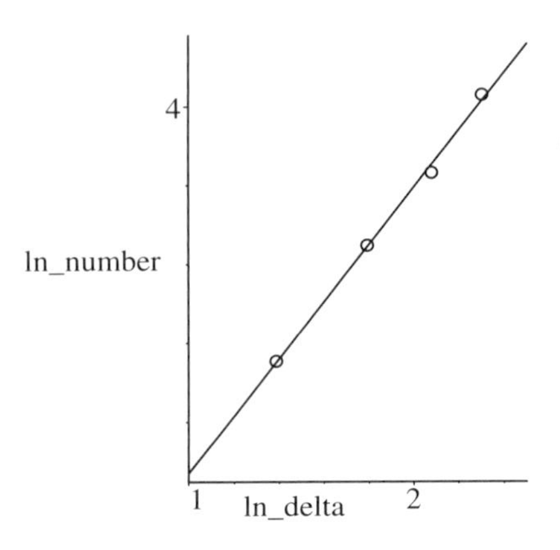

Figure 8.9: Slope of least squares straight line yields a fractal dimension.

Although ideally, the range of data points should be extended to smaller values of ε to obtain a more accurate answer, Jennifer can see that the least squares line does a good job of fitting the data points obtained. The slope of the straight line yields an estimate of the fractal dimension D_C. By taking the coefficient of X in `eq`,

```
> D[C]:=evalf(coeff(eq,X),3);
```

$$D_C := 1.45$$

Jennifer finds that $D_C \approx 1.45$ for Barnsley's fern.

PROBLEMS:

Problem 8-7: An impressionist's tree

In the world of art, impressionism refers to a painting style developed by Manet, Monet, Renoir, Degas, Pissarro, etc. The chief aim of their works was to reproduce only the immediate and overall impression made by the subject on the artist, without much attention to detail. By iterating the following map, and selecting the appropriate branch on each iteration according to the random number generated between 0 and 1, you will become a computer artist of the impressionistic school:

$$(x_{n+1},\, y_{n+1}) = \begin{cases} (0.05\, x_n,\; 0.60\, y_n), & 0.0 < r < 0.1, \\ (0.05\, x_n,\; -0.50\, y_n + 1.0), & 0.1 < r < 0.2, \\ (0.46\, x_n - 0.15\, y_n,\; 0.39\, x_n + 0.38\, y_n + 0.60), & 0.2 < r < 0.4, \\ (0.47\, x_n - 0.15\, y_n,\; 0.17\, x_n + 0.42\, y_n + 1.1), & 0.4 < r < 0.6, \\ (0.43\, x_n + 0.28\, y_n,\; -0.25\, x_n + 0.45\, y_n + 1.0), & 0.6 < r < 0.8, \\ (0.42\, x_n + 0.26\, y_n,\; -0.35\, x_n + 0.31\, y_n + 0.70), & 0.8 < r < 1.0. \end{cases}$$

Take ($x_0 = 0.5$, $y_0 = 0$) and $N = 25{,}000$.

Problem 8-8: Fishbone fern

Taking $N = 10{,}000$ and $x[0] = y[0] = 0$, estimate the fractal dimension of the *fishbone* fern generated by replacing the parameters in Barnsley's fern with the following values:

$$\begin{array}{llll} a[1] = 0, & a[2] = 0.95, & a[3] = 0.035, & a[4] = -0.04; \\ b[1] = 0, & b[2] = 0.002, & b[3] = -0.11, & b[4] = 0.11; \\ c[1] = 0, & c[2] = -0.002, & c[3] = 0.27, & c[4] = 0.27; \\ d[1] = 0.25, & d[2] = 0.93, & d[3] = 0.01, & d[4] = 0.01; \\ e[1] = 0, & e[2] = -0.002, & e[3] = -0.05, & e[4] = 0.047; \\ f[1] = -0.4, & f[2] = 0.5, & f[3] = 0.005, & f[4] = 0.06; \\ p[1] = 0.02, & p[2] = 0.86, & p[3] = 0.93. & \end{array}$$

Problem 8-9: Cyclosorus fern

Taking $x[0] = y[0] = 0$ and as large an N value as possible, estimate the fractal dimension of the *Cyclosorus* fern generated by replacing the parameters in Barnsley's fern with the following values:

$$\begin{array}{llll} a[1] = 0, & a[2] = 0.95, & a[3] = 0.035, & a[4] = -0.04; \\ b[1] = 0, & b[2] = 0.005, & b[3] = -0.2, & b[4] = 0.2; \\ c[1] = 0, & c[2] = -0.005, & c[3] = 0.16, & c[4] = 0.16; \\ d[1] = 0.25, & d[2] = 0.93, & d[3] = 0.04, & d[4] = 0.04; \\ e[1] = 0, & e[2] = -0.002, & e[3] = -0.09, & e[4] = 0.083; \\ f[1] = -0.4, & f[2] = 0.5, & f[3] = 0.02, & f[4] = 0.12; \\ p[1] = 0.02, & p[2] = 0.86, & p[3] = 0.93. & \end{array}$$

Problem 8-10: Dissecting the fern

For Barnsley's fern, determine what each branch of the piecewise algorithm contributes to the overall fractal picture.

8.1.4 Douady's Rabbit and Other Fauna and Flora

Predictions of the future are never anything but projections of present automatic processes and procedures ...
Hannah Arendt, political philosopher (1906–1975)

One of the best-known two-dimensional maps is due to the mathematician Benoit Mandelbrot. The Mandelbrot map is obtained by iterating the equation

$$z_{n+1} = z_n^2 + c, \tag{8.7}$$

where $z = x + i\,y$ is a complex variable and $c = p + i\,q$ is a complex constant. Separating (8.7) into real and imaginary parts yields the two-dimensional map

$$x_{n+1} = x_n^2 - y_n^2 + p, \quad y_{n+1} = 2\,x_n\,y_n + q. \tag{8.8}$$

If a particular set of values is chosen for p and q and N iterations are carried out, where N is large, the iterated x, y values either diverge to infinity or converge to a small finite value of x_n, y_n. For concreteness, Jennifer takes $p = -0.12$, $q = -0.74$, and $N = 25$.

```
> restart: begin:=time():
> p:=-0.12: q:=-0.74: N:=25:
```

To carry out the iteration of the Mandelbrot map, Jennifer defines a Maple procedure to which she gives the name JULIA. Gaston Julia was a French mathematician who studied the structure of the complicated boundaries generated between the regions of convergence and divergence. In his honor, the sets of points lying on such boundaries are now called *Julia sets*. The procedure begins with the command proc(x,y) and terminates with end proc. When values of x and y are specified, the procedure will carry out $N = 25$ iterations according to the prescribed algorithm.

```
> JULIA:=proc(x,y)
```

Within the body of the procedure, new local variables X, Y and a copy of X, labeled COPY_X, are introduced. Local variables have meaning only inside the procedure. X and Y are obtained by evaluating the input values x and y.

```
> local X,Y,COPY_X;
> X:=evalf(x): Y:=evalf(y):
```

If the radius squared, i.e., $X^2 + Y^2$, exceeds 4, it is assumed that the values of X, Y are going to diverge. Therefore, a while condition is introduced into the do loop that allows iterations of the two-dimensional map as long as the radius squared is less than or equal to 4. The copy of X is carried out first and used in the evaluation of Y. Can you see why this is necessary?

```
> to N while X^2+Y^2<=4 do
> COPY_X:=X: X:=X^2-Y^2+p: Y:=2*COPY_X*Y+q:
> end do:
```

On completion of the do loop, Jennifer inserts an `if..then..else` statement to assign the value 1 to regions of divergence (i.e., when $X^2 + Y^2 > 4$) and 0 to regions of convergence.

```
> if X^2+Y^2>4 then 1 else 0 end if:
> end proc:
```

If specific x and y values are now given, then the Julia function defined by the above procedure is evaluated. For example,

```
> JULIA(0,0); JULIA(1,0);
```

$$0 \quad 1$$

so that the input point $(0,0)$ converges while the point $(1,0)$ diverges.

The Julia function is now plotted for the range $x = -1.2$ to 1.2, $y = -1.2$ to 1.2 and the three-dimensional plot oriented to be viewed from above.

```
> plot3d(JULIA,-1.2..1.2,-1.2..1.2,grid=[150,150],orientation
  =[-90,0],scaling=constrained,style=patchnogrid,shading=zhue,
  lightmodel=light3);
```

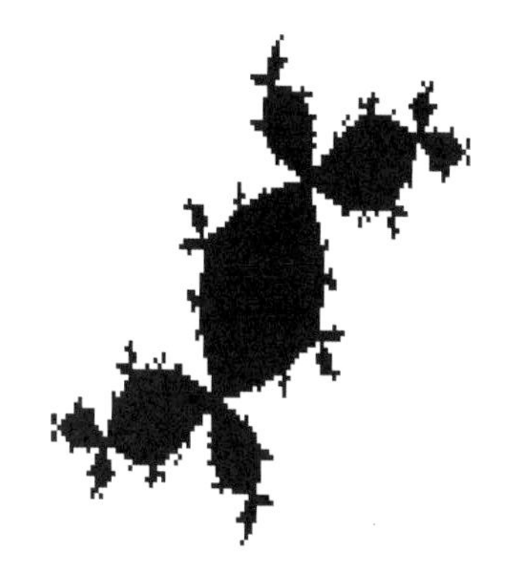

Figure 8.10: Douady's rabbit.

The boundary between the two regions (black on a white background here in the text) can be clearly seen in Figure 8.10. The points on the boundary form the Julia set for the Mandelbrot map. If you mentally rotate the Julia set slightly, and have a good imagination,[1] you may be able to see Douady's "rabbit." The complicated boundary formed by the Julia set is another example of a fractal structure. Other geometrically interesting Julia sets can be generated for appropriate choices of p and q.

```
> cpu_time:=(time()-begin)*seconds;
```

$$cpu\_time := 0.711\ seconds$$

[1]The kind of imagination needed to see animal shapes in clouds and in Rorschach tests administered by psychologists.

The CPU time for the Julia procedure was less than a second.

Instead of choosing a particular p and q and sweeping through x and y values, one can do the opposite, i.e., choose a particular x, y, e.g., the origin, and systematically sweep through different p and q values. This procedure, which Jennifer produces below, generates the *Mandelbrot set* of points.

```
> restart: N:=25: begin:=time():
> MANDELBROT:=proc(p,q)
> local z,n;
```

In the following two command lines, Jennifer starts with $x=y=0$ and $n=0$.

```
> z:=evalf(p+I*q); n:=0;
```

Again, a radius of $|z| = 2$ units is used as the boundary between diverging and converging $z = x + iy$ values.

```
> to N while abs(z)<2 do
> z:=z^2+(p+I*q); n:=n+1;
> end do:
> n; end proc:
```

A three-dimensional plot is now created with $p = -1.5$ to 1, $q = -1$ to 1, and the output n values being the third axis. Points that escape rapidly to infinity will be characterized by small n values while points that escape slowly to infinity or not at all (i.e., attracted to a fixed point at finite x, y) will have large n values. The orientation chosen in the `plot3d` structure shows the p–q plane, but the figure can be rotated to show the three-dimensional character. The `zhue` shading is used to color the different output n values.

```
> plot3d(MANDELBROT,-1.5..1,-1..1,grid=[100,100],
  orientation=[-90,0],style=patchnogrid,shading=zhue);
```

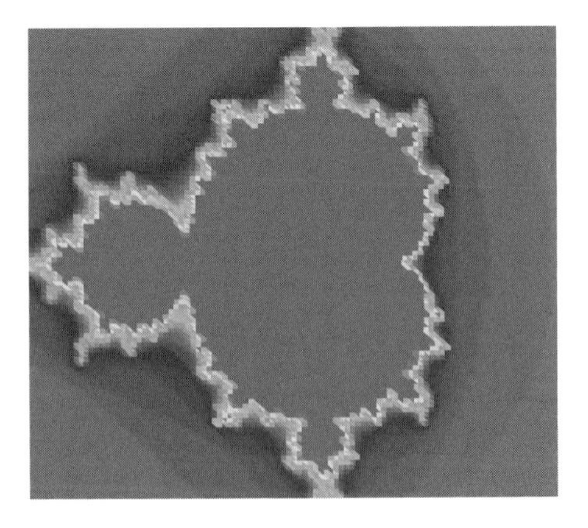

Figure 8.11: Mandelbrot set.

The resulting picture is shown in Figure 8.11. Does your imagination suggest a possible flora or fauna name for the outline of this region?

The CPU time for the Mandelbrot procedure

```
> cpu_time:=(time()-begin)*seconds;
```

$$cpu\_time := 0.200\, seconds$$

is even shorter than for the Julia procedure.

Here are a few problems involving the Mandelbrot map that Jennifer has created for you to try. She suggests that you adjust the shading and lighting to your own taste.

PROBLEMS:

Problem 8-11: The San Marco attractor

Generate the so-called *San Marco attractor*, which results from taking $p = -0.75$, $q = 0$ in the Julia set procedure.

Problem 8-12: The octopus

Generate the "*octopus*," which results from taking $p = 0.27334$, $q = 0.00742$ in the Julia procedure. Take $N = 100$ and both x and y varying from -1.1 to 1.1.

Problem 8-13: Other Julia sets

Generate the Julia sets corresponding to:

(**a**) $p = -1$, $q = 0$; (**b**) $p = 0.32$, $q = 0.043$.

Problem 8-14: Complex Julia input

Reformulate the Julia procedure with complex numbers and explore the recipe.

Problem 8-15: Different scales

Explore the Julia and Mandelbrot sets at different scales For example, plot the Mandelbrot set in the range $x = -1.5$ to -1.3, $y = -0.1$ to 0.1.

Problem 8-16: The fern

In the Mandelbrot set procedure, generate a fernlike object by taking $p = -0.745385$ to -0.745468, $q = 0.112979$ to 0.113039, $N = 200$, and $|z| < 10$.

Problem 8-17: Variations on the Mandelbrot set

In the Mandelbrot set procedure, create new figures by taking:

(**a**) $z_{n+1} = z_n^3 + c$; (**b**) $z_{n+1} = z_n^4 + c$;

(**c**) $z_{n+1} = z_n^5 + c$; (**d**) $z_{n+1} = \dfrac{z_n^2}{1 + z_n^2} + c$.

Adjust the viewing scale and orientation to include the entire figure and to give the "best" work of art.

Problem 8-18: The Beauty of Fractals

Go to your college library and obtain a copy of *The Beauty of Fractals*, by Peitgen and Richter. [PR86] See how many of the figures therein you can generate. Tables of the parameter values and viewing ranges can be found at the end of that text.

8.1.5 The Rings of Saturn

It is marvelous indeed to watch on television the rings of Saturn close; and to speculate on what we may yet find at galaxy's edge.
Gore Vidal, U.S. novelist (1925–)

As an example of pattern formation in nature, Jennifer will now consider a difference equation model that produces planar planetary rings qualitatively similar to those for Saturn. Saturn's nearly planar rings are shown in the NASA photograph reproduced in Figure 8.12. Although Saturn's rings span

Figure 8.12: Saturn's rings with the Cassini gap clearly evident.

Table 8.1: Classification of Saturn's rings.

Ring	**Distance** (10^3 km)	**Width** (10^3 km)	**Mass** (kg)
D	66.9	7.6	?
C	74.7	17.3	1.1×10^{18}
B	92.0	25.6	2.8×10^{19}
Cassini gap			
A	122.2	14.6	6.2×10^{18}
F	140.2	0.5	?
G	170.0	5.0	1×10^{7} ?
E	181.0	302.0	?

more than 250,000 km in diameter, they are very thin, some rings being only tens of meters thick. The particles making up the rings are composed primarily of water ice, but may include ice-coated rocks. The particles range in size from centimeters to several meters, with a few kilometer size rocks likely. The historical designation of the more prominent rings is given in Table 8.1. The data is extracted from the NASA web site (nssdc.gsfc.nasa.gov). The distances are from Saturn's center to the ring's inner edge. The large gap between the A and B rings is called the Cassini gap, named after Cassini, who observed the rings and discovered several of Saturn's moons in the late 1600s. The gaps are not entirely empty and there are further variations within the rings.

In addition to the rings, Saturn has 34 named moons and 13 unamed satellites. Table 8.2 lists the inner seven and several of the outer named moons. Note that although many of the moons were discovered long ago, the very

Table 8.2: Some of Saturn's moons.

Moon	**Distance** (10^3 km)	**Radius** (km)	**Density** (kg/m^3)	**Discoverer** (date)
Pan	134	10	630	Showalter (1990)
Atlas	138	19×17×14	630	Terrile (1980)
Prometheus	139	74×50×34	630	Collins (1980)
Pandora	142	55×44×31	630	Collins (1980)
Epimetheus	151	69×55×55	600	Walker (1980)
Janus	151	97×95×77	650	Dollfus (1966)
Mimas	186	209×196×191	1140	Herschel (1789)
....	.....	.....	.....	.....
Titan	1222	2575	1881	Huygens (1655)
Hyperion	1481	185×140×113	1500	Bond (1848)
Iapetus	3561	718	1020	Cassini (1671)
Phoebe	12952	115×110×105	1300	Pickering (1898)

inner moons, which are small in radius, were observed only in very recent times. Probably because of its larger mass compared to the other inner moons, Mimas[2] plays an important role in the organization of Saturn's inner rings. Mimas, in itself, is an interesting moon. Figure 8.13 shows a NASA photograph of Mimas that is dominated by the Herschel impact crater 130 km across, which is about one-third of Mimas's diameter. From the length of the shadow cast by the central peak inside the crater, one can deduce that the crater walls are about 5 km high and the central peak rises 6 km from the crater floor, parts of which are 10 km deep. From Table 8.2, Mimas is seen to have a density of 1140 kg/m^3, indicating that it is composed mainly of ice with only a small amount of rock.

There are theoretical reasons, first suggested by the French scientist Edouard Roche in 1848, for thinking that there is an inner limiting radius, inside of which moons cannot exist for any planet. He argued that within a critical distance from a planet's center, now called the *Roche limit*, any orbiting moon would break up because the tidal force on the moon due to the planet would be larger

[2]Mimas was one of the Titans slain by Hercules.

Figure 8.13: Mimas.

than the gravitational force holding the moon together. The moon would be shredded into smaller particles such as those found in the inner rings of Saturn.

Jennifer will first derive the formula for the Roche limit for a small rigid satellite moon of mass m, radius a, density ρ_s orbiting at a distance $r \gg a$ from the center of a planet of mass M, radius R, and density ρ_p. She will then use the formula to estimate the "rigid Roche limit" for the moon Pan.

First, the mathematical form of the tidal force must be determined. Consider two identical "test" masses (mass μ) located at distances r and $r - a$ from the planet's center. The test mass that is closer to the planet will feel a stronger planetary pull than the mass that is further away. The tidal force F_t is the difference between the planetary forces on the two masses. Using Newton's law of gravitation, F_t is given by

```
> restart:
> F[t]:=G*M*mu/(r-a)^2-G*M*mu/r^2;
```

$$F_t := \frac{G\,M\,\mu}{(r-a)^2} - \frac{G\,M\,\mu}{r^2}$$

where G is the gravitational constant. Since $a \ll r$, F_t can be Taylor expanded about $a = 0$ to second order and the "order of" term removed.

```
> F[t]:=taylor(F[t],a=0,2);
```

$$F_t := \frac{2\,G\,M\,\mu}{r^3}\,a + \mathrm{O}(a^2)$$

```
> F[t]:=convert(%,polynom);
```

$$F_t := \frac{2\,G\,M\,\mu\,a}{r^3}$$

The tidal force varies as $1/r^3$. Now, the gravitational force F_g exerted by the satellite on a single test mass μ at its surface is

```
> F[g]:=G*m*mu/a^2;
```

$$F_g := \frac{G\, m\, \mu}{a^2}$$

If the tidal and gravitational forces on this test mass balance, the mass will not be "ripped" off the satellite. The value of r at which this occurs is the Roche limit. Equating the tidal and gravitational forces yields the balance equation.

```
> balance_eq:=F[t]=F[g];
```

$$balance\_eq := \frac{2\, G\, M\, \mu\, a}{r^3} = \frac{G\, m\, \mu}{a^2}$$

Assuming that both the planet and satellite are spherical, their masses are related to their densities by $M = (4/3)\, \pi\, R^3\, \rho_p$ and $m = (4/3)\, \pi\, a^3\, \rho_s$.

```
> M:=4*Pi*R^3*rho[p]/3; m:=4*Pi*a^3*rho[s]/3;
```

$$M := \frac{4}{3}\, \pi\, R^3\, \rho_p \qquad m := \frac{4}{3}\, \pi\, a^3\, \rho_s$$

Knowing that the mass relations will be automatically substituted into the balance equation, Jennifer uses the isolate command to isolate r^3, the cube of the Roche limit, to the lhs of the balance equation.

```
> sol:=isolate(balance_eq,r^3);
```

$$sol := r^3 = \frac{2\, R^3\, \rho_p}{\rho_s}$$

The Roche limit for a rigid satellite then follows on taking the cube root of the rhs of *sol* and simplifying with the assumption that $R > 0$.

```
> Roche(rigid):=simplify(rhs(sol)^(1/3)) assuming R>0;
```

$$Roche(rigid) := 2^{(1/3)}\, R\, \left(\frac{\rho_p}{\rho_s}\right)^{(1/3)}$$

To estimate the rigid Roche limit for Pan, the following parameter values are entered: $\rho_p = 687$ kg/m^3, $\rho_s = 630$ kg/m^3, and $R = 60.4$ thousand km.

```
> pars:=rho[p]=687,rho[s]=630,R=60.4:
```

Then the rigid Roche limit for Pan is numerically evaluated,

```
> Roche(rigid):=evalf(eval(Roche(rigid),{pars}));
```

$$\text{Roche}(rigid) := 78.32835390$$

and found to be about 78 thousand km, well within Pan's orbital radius of 134 thousand km. However, this derivation assumed that the satellite is rigid.

For a fluid satellite, the tidal force causes the satellite to be deformed from its spherical shape, being elongated in the direction of the planet. If the satellite is tidally locked and deformed into a prolate spheroid, the "fluid Roche limit" is (see en.wikipedia.org/wiki/Roche_limit) given approximately by

```
> Roche(fluid):=2.423*R*(rho[p]/rho[s])^(1/3);
```

$$\text{Roche}(fluid) := 2.423\, R\, \left(\frac{\rho_p}{\rho_s}\right)^{(1/3)}$$

Taking the same parameter values as before, the fluid Roche limit is evaluated,

```
> Roche(fluid):=evalf(eval(Roche(fluid),{pars}));
```

$$\text{Roche}(\mathit{fluid}) := 150.6361065$$

and found to be about 151 thousand km, somewhat higher than Pan's actual orbital radius. Undoubtedly, Pan is neither perfectly rigid nor perfectly fluid.

Now Jennifer will consider a nonlinear difference equation model that qualitatively produces the rings of Saturn. This model was developed by Fröyland [Fro92] and discussed by Gould and Tobochnik [GT96]. Letting σ be the radial distance of Mimas from Saturn's center, r_n the radial distance of a ring particle from Saturn's center after the nth revolution, and θ_n the angular position of a ring particle with respect to Mimas after n revolutions, the coupled model equations are of the form

$$\theta_{n+1} = \theta_n + 2\,\pi\left(\frac{\sigma}{r_n}\right)^{3/2}, \qquad r_{n+1} = 2\,r_n - r_{n-1} - A\,\frac{\cos\theta_n}{(r_n-\sigma)^2}, \tag{8.9}$$

with A a positive parameter that remains to be estimated.

To understand the structure of Equations (8.9), Jennifer will now briefly discuss their physical origin. Her arguments are a combination of fundamental physical principles and some "hand-waving." With so many other moons present, the detailed calculation of the entire banded ring structure is quite complicated and beyond the scope of this text. In the model there are two major influences on the ring particles: the dominant effect of Saturn's gravitational force and the perturbing influence of Mimas. The effect of Saturn is included as follows. Each time Mimas completes an orbit of radius σ with period T_σ, it undergoes an angular change of $2\,\pi$ radians. If T_n is the period for any other satellite object on its nth revolution, the angle θ that the object makes with respect to Mimas on revolution $n+1$ will be given[3] by

$$\theta_{n+1} = \theta_n + 2\,\pi\,(T_\sigma/T_n). \tag{8.10}$$

But Kepler's third law for planetary orbits states that the period T of an object orbiting a planet of mass $M_{\rm p}$ in a circular[4] orbit of radius r is given by

$$T^2 = \frac{4\,\pi^2}{G\,M_{\rm p}}\,r^3. \tag{8.11}$$

Letting r_n be the distance of a ring particle from Saturn's center after n revolutions, the angular equation in (8.9) immediately follows on using the square root of (8.11) to calculate the ratio T_σ/T_n.

The effect of Mimas is to perturb the radial distance r of a ring particle, causing the distance to change from one orbit to the next. By Newton's second law, a particle's radial acceleration will be given by

$$\ddot{r} = \frac{F_r}{m}, \tag{8.12}$$

[3]To within a term $2\,\pi\,n$, whose omission doesn't affect the results.

[4]For an elliptical orbit, the radius is replaced in the third law with the semimajor axis.

where F_r is the radial component of the gravitational force between Mimas and the particle of mass m. Jennifer approximates $\ddot{r}(t)$ by

$$\ddot{r}(t) \approx \frac{r(t+\Delta t) - 2\,r(t) + r(t-\Delta t)}{(\Delta t)^2}, \tag{8.13}$$

where Δt is a short time interval at t. This result is easily confirmed by Taylor expanding the numerator of the rhs of (8.13) to order $(\Delta t)^2$.

Averaging over one complete period T_σ of Mimas, she replaces Equation (8.13) with $(r_{n+1} - 2\,r_n + r_{n-1})/(T_\sigma)^2$ and evaluates the right-hand side of (8.12) at the end of the nth revolution. With these approximations, Equation (8.12) reduces to

$$r_{n+1} - 2\,r_n + r_{n-1} = f(r_n, \theta_n), \tag{8.14}$$

with the radial force function $f(r_n, \theta_n) \equiv T_\sigma^2\, F_r(r_n, \theta_n)/m$ still to be established.

According to Gould and Tobochnik, the form of $f(r_n, \theta_n)$ is very complicated, particularly if the perturbing effects of other moons are also included. Following their lead, Jennifer assumes that f is given by

$$f = -A\,\frac{g(\theta_n)}{(r_n - \sigma)^2}, \tag{8.15}$$

with

$$A \equiv G\,M_\sigma\,(T_\sigma)^2 = 4\,\pi^2\,\sigma^3\,\frac{M_\sigma}{M_s}, \tag{8.16}$$

M_s being the mass of Saturn, and the angular dependence $g(\theta_n)$ still not specified. However, by symmetry, the function g should be an even function of θ_n. For simplicity, rather than detailed realism, Gould and Tobochnik take the angular dependence to be given by $g = \cos(\theta_n)$, thus resulting in the radial equation of (8.9).

Jennifer is not entirely happy with this angular form but does note that the cosine term can undergo a sign change when θ_n varies, the effect being to pull particles in (a "bunching" effect) toward Mimas when they are nearby and to push particles away from Mimas's orbit when they are on the opposite side of Saturn. This latter scenario would perhaps reflect the weak gravitational effect of Mimas and a stronger influence of other moons neglected in the analysis.

Noting that Saturn has a mass $M_s = 5.68 \times 10^{26}$ kg and expressing radial distances r_n in thousands of kilometers, Jennifer estimates that $A \approx 17$. However, since the approximation of the force law is very crude, Gould and Tobochnik suggest that there is considerable latitude in choosing the value of A.

Jennifer will now iterate the basic difference equations (8.9) and use them to plot the resulting planar ring structure superimposed on a colored sphere to represent Saturn. The sphere will be produced with the `sphere` command contained in the plottools library package.

```
> restart: with(plots): with(plottools): begin:=time():
```

Equations (8.9) are entered, with π being numerically evaluated so that a conditional "if then" statement contained in the recipe will work.

```
> eq1:=theta[n+1]=theta[n]+2*evalf(Pi)*(sigma/r[n])^(3/2);
```

$$eq1 := \theta_{n+1} = \theta_n + 6.283185308\left(\frac{\sigma}{r_n}\right)^{(3/2)}$$

```
> eq2:=r[n+1]=2*r[n]-r[n-1]-A*cos(theta[n])/(r[n]-sigma)^2;
```

$$eq2 := r_{n+1} = 2\,r_n - r_{n-1} - \frac{A\cos(\theta_n)}{(r_n-\sigma)^2}$$

Since *eq2* is a second-order difference equation, both r_0 and r_1 must be inputted. The initial orbital radii of the particles must, of course, lie outside the surface of Saturn. In the following simulation, $Ns+1=21$ input radii are chosen, and $N=4000$ iterations are performed for each of these initial particle orbital radii.

```
> N:=4000: Ns:=20: #N points per radial step Ns
```

The orbital radius ($\sigma=185.7$ thousand km) of Mimas is entered and the proportionality constant A in *eq2* taken to be $A=15$. How the ring structure changes with A is left as a problem.

```
> sigma:=185.7: A:=15: #radius of Mimas and force coefficient
```

The smallest input orbital radius for the particles in the following do loop is taken to be $rs=70.0$ thousand km, which lies outside the 60.4 thousand km radius of Saturn. The count parameter c, which is set to zero initially, counts the number of output graphs obtained.

```
> rs:=70.0: c:=0: #smallest radius and count initialization
```

The following outer do loop is iterated from 0 to Ns.

```
> for j from 0 to Ns do
```

The input orbital radii r_0 are incremented in steps of 5 thousand km, from 70 thousand km out to a maximum of $70+5\times 20=170$ thousand km, the latter being well inside Mimas's orbital radius. As input values for the radial and angle variables, Jennifer sets $r_1=r_0$ and $\theta_0=0$. For each j value, the iteration parameter n is reset to $n=0$.

```
> r[0]:=rs+5*j: r[1]:=r[0]; theta[0]:=0; n:=0;
```

Then θ_1 is equal to the right-hand side of eq1.

```
> theta[1]:=rhs(eq1);
```

A second inner do loop iterates n from 1 to N as long as the orbital radius lies outside Saturn and remains less than $5\times 70=350$ thousand km from Saturn's center. The upper cutoff is included to prevent numerical overflow if a particle escapes from the vicinity of Saturn and its moons.

```
> for n from 1 to N while r[n]>60.4 and r[n]<5*rs do
```

The Cartesian coordinates $x_n=r_n\cos(\theta_n)$, $y_n=r_n\sin(\theta_n)$ of a particle after the nth iteration are calculated.

```
> x[n]:=r[n]*cos(theta[n]); y[n]:=r[n]*sin(theta[n]);
```

Then θ_{n+1} and r_{n+1} on step $n+1$ are determined and the inner loop ended.

```
> theta[n+1]:=rhs(eq1); r[n+1]:=rhs(eq2);
> end do:
```

Only those particles that do not wander outside the specified range before the total number N of iterations is completed are to be counted and their coordinates graphed as 3-dimensional colored points. The three random number commands in the COLOR option generate random decimal numbers between 0 and 1, which set the fraction of red (R), green (G), and blue (B) in the color coding. This coloring scheme generates a very pretty ring pattern.

```
> if n>N then gr[c]:=pointplot3d({seq([x[i],y[i],0],i=1..N-1)},
  color=COLOR(RGB,rand()/10^12,rand()/10^12,rand()/10^12),
  symbol=point):
```

The count number is increased by 1, the conditional statement ended, and the outer do loop completed.

```
> c:=c+1 end if:
> end do:
```

A colored sphere (shaded in the z direction) of radius 60.4 thousand km is plotted to represent Saturn. The style option patchnogrid removes the default grid that would otherwise appear on the sphere's surface.

```
> gr[0]:=sphere([0,0,0],60.4,shading=z,style=patchnogrid):
```

Saturn and its vividly colored banded ring structure are then displayed with constrained scaling. A particular orientation has been chosen for the resulting three-dimensional picture, which can be rotated on the computer screen by dragging with the mouse. The "magnify" comment suggests that the reader should increase the magnification, say to 200 percent, in the tool bar.

```
> display(seq(gr[i],i=0..c-1),scaling=constrained,
  orientation=[-20,55],shading=z); #magnify
```

On excuting the above command line, a picture similar to that on the cover of this text will result. Finally, the CPU time is recorded.

```
> cpu_time:=(time()-begin)*seconds;
```

$$cpu\_time := 88.688\ seconds$$

Given the various approximations and assumptions made in obtaining the model equations, Jennifer strongly emphasizes that the model is intended to show how the ring particles could be organized into a ring pattern with gaps, rather than being an accurate predictor of Saturn's actual detailed ring pattern.

PROBLEMS: Problem 8-19: Different ring structure

Plot the banded ring structure for the rings of Saturn taking $A=150$ and 1500. Explore other values for the parameters and ranges.

Problem 8-20: Other angular dependencies

Investigate whether banded ring structures are produced for other simple angular dependencies, e.g., $\sin\theta$, $\cos^2(\theta)$, $\cos^3\theta$, $\sin^2\theta$. Discuss your results.

8.2 Cellular Automata

Complex patterns can be produced on square cellular lattices by postulating simple rules governing the evolution of some initial configuration of "live" (or "excited") and "dead" (or "quiescent") cells. Such dynamical systems are called *cellular automata* and were first investigated by John Von Neumann and Stan Ulam, and some years later by Stephen Wolfram [Wol86][Wol02]. With dead cells assigned the value 0 and live cells the value 1, the `listdensityplot` command can be used to produce a black (corresponding to zero) and white (corresponding to one) pattern generated by application of a given rule.

Jennifer will now provide two cellular automata recipes, the first for creating a geometric pattern reminiscent of that seen on Navaho rugs, the second commonly referred to as the *one out of eight rule*. The mathematically inclined reader is referred to E. Atlee Jackson's text [Jac90] for many more examples of cellular automata and a discussion of their classification and properties.

8.2.1 A Navaho Rug Design

It is here in mathematics that the artist has the fullest scope of his imagination.
Havelock Ellis, English psychologist, scientist, and author (1859–1939)

After hiking down into the Grand Canyon, coauthor Richard stopped sometime later at the Cameron Trading Post to refuel his weary body with a large, mouth-watering Navaho taco. In the gift shop outside the restaurant, an elderly Navaho woman was observed to be weaving an intricate geometric-patterned rug on a large loom. Starting with the bottom row, she progressed slowly upward row by row, creating a complex geometrical design that probably had been handed down from generation to generation in her family.

Motivated by this rug-weaving episode, we have asked Jennifer to mathematically "weave" a geometric pattern, given some initial configuration of black (dead) and white (live) cells on the first row. We have left the choice of rule for proceeding from row to row up to her, but have asked her to keep the rule simple and use the same rule on all rows.

Jennifer begins by first loading the LinearAlgebra and plots packages,

```
> restart: with(LinearAlgebra): with(plots):
```

and takes the starting row of the pattern to be $N = 100$ cells long.

```
> N:=100: begin:=time():
```

To create the initial row, first a list of N zeros is entered, the zeros being colored black in the final pattern. This list is assigned the name *initialization*.

```
> initialization:=[seq(0,i=1..N)]:
```

So that some sort of pattern may be generated, some of the black cells in *initialization* must be converted to ones, so that they will be ultimately colored white. To accomplish this, Jennifer uses the arrow operator to indicate that the value 1 is to be assigned to cell $(N/2+r)$. For example, since $N=100$, then the command rhs(c(1))

```
> c:=r->N/2+r=1; cell[lhs(c(1))]:=rhs(c(1));
```

$$c := r \to \frac{N}{2} + r = 1 \qquad cell_{51} := 1$$

tells us that cell 51 is to be colored white. To achieve the conversion of selected cells from black to white, the subsop function is used to replace specified operands in *initialization* with the new values 1. For example, in row_0 below, Jennifer enters c(-1) and c(1) in order to replace the zeros in cells 49 and 51 with ones.

```
> row[0]:=subsop(c(-1),c(1),initialization):
```

To generate the second and subsequent rows, a simple rule is introduced for proceeding from one row to the next. Using the following arrow operator,

```
> s:=i->op(i,row[j-1]);
```

$$s := i \to \mathrm{op}(i,\ row_{j-1})$$

the ith operand of the $(j-1)$ row will be given by $s(i)$. The $s(i)$ can take on only the values 0 and 1. The color of the ith element in the jth row is determined by forming the function

$$F = s(i-1) + s(i) + s(i+1),$$

i.e., the sum of the s values for the three nearest neighbors in the previous row. Altering this function F will generate a different rule and in general a different pattern.

```
> F:=s(i-1)+s(i)+s(i+1);
```

$$F := \mathrm{op}(i-1,\ row_{j-1}) + \mathrm{op}(i,\ row_{j-1}) + \mathrm{op}(i+1,\ row_{j-1})$$

If the sum F is zero, the ith element of the jth row will be zero and therefore black. If the sum is one, the ith element will be colored white. To keep the sum always equal to zero or one for every element, the modulo-2 (mod 2) condition is imposed on F in the following do loop. The number of cells in each subsequent row is kept the same as in the initial row by including a zero at both ends of each new list (row) that is generated. To avoid difficulties at the "edges" of the rug (ends of each list), the do loop is terminated at $N/2-5$ (45 here).

```
> for j from 1 to N/2-5 do
> row[j]:=[0,seq(F mod 2,i=2..N-1),0];
> end do:
```

A matrix *M1* is formed with the sequence of rows from 0 to $N/2-5$, a total of 46 rows here.

```
> M1:=Matrix([seq(row[i],i=0..N/2-5)]):
```

To make a symmetric Navaho rug, a second matrix *M2* is created with the same rows placed in reverse order.

```
> M2:=Matrix([seq(row[N/2-5-i],i=0..N/2-5)]):
```

The two matrices are joined together in the following command line to form a 92×100 matrix M.

```
> M:=<M1,M2>:
```

The geometrical pattern of the rug is then revealed using the `listdensityplot` command with constrained scaling and no coordinate axes.

```
> listdensityplot(M,scaling=constrained,axes=NONE);
```

Figure 8.14: A Navaho rug design.

The geometrical pattern shown in Figure 8.14 is somewhat reminiscent of Sierpinski's gasket and is clearly fractal in nature. Note that in the picture the "rows" are actually running vertically. To orient the rug with the rows running horizontally, the `transpose` command should be applied to the matrix M. By changing the rule (i.e., the function F) and initial configuration (the input row), other "rug patterns" can be generated in a matter of seconds,

```
> cpu_time:=(time()-begin)*seconds;
```

$$cpu\_time := 1.262\ seconds$$

considerably less time than the many days it takes to produce a Navaho rug by hand in real life.

Although the pattern that was produced in Figure 8.14 is two-dimensional in appearance, it is actually a simple example of one-dimensional cellular automaton growth, growing in one direction only, row by row. In the following recipe, Jennifer will present an example of two-dimensional cellular automaton growth.

PROBLEMS:

Problem 8-21: Changing mod

Explore the effect on the pattern in the text recipe of the values of mod.

Problem 8-22: Another geometric pattern

In the text recipe, generate another interesting pattern by making the following changes. In `row[0]`, take cells corresponding to `c(-1)`, `c(1)`, and `c(2)` to be white (the remainder black) and use the function $F = s(i-1) + s(i+1)$ mod 2. How does this pattern compare with that in the text example?

Problem 8-23: More nearest-neighbor contributions

By considering various configurations for the input row, explore the patterns generated by the function $F = \sum_{r=-2}^{2} s(i+r)$ mod 2. Take N as large as possible and be careful to not let the pattern reach the edges.

8.2.2 The One out of Eight Rule

Rules are not necessarily sacred, principles are.
Franklin D. Roosevelt, U.S. president (1882–1945)

Jennifer now considers a square lattice of "length" $L = 61$ cells by 61 cells and assumes that initially only one central cell (square) is alive, all other cells being dead. Dead cells are brought to life according to the *one out of eight rule*, which states that a cell comes alive if exactly one of its eight immediate neighbors is alive. Otherwise it remains unchanged. As in the last example, live cells are assigned the value 1 and dead cells the value 0, thus allowing the pattern of live cells after a certain number of steps to be plotted as white squares on a black background using the `listdensityplot` command.

i–1, j–1	i–1, j	i–1, j+1
i, j–1	i, j	i, j+1
i+1, j–1	i+1, j	i+1, j+1

Figure 8.15: Immediate neighbors of cell (i, j).

With the given size of lattice, Jennifer will now determine the pattern of live cells after $N = 29$ steps, i.e., 29 applications of the rule. To formulate the algorithm, consider Figure 8.15, which shows a representative cell (i, j) in row i and column j and its eight immediate neighbors. To avoid difficulties at the edges of the $L \times L$ lattice, both i and j are allowed to range only from 2 to $L - 1$. For the same reasons, L must be increased if N is made larger.

Since matrices are to be used, a call is made to the LinearAlgebra package. The plots package is also invoked, since a listdensity plot is to be made.

```
> restart: with(LinearAlgebra): with(plots): begin:=time():
```

The lattice "length" $L=61$ and number $N=29$ steps are entered.

```
> L:=61: N:=29:
```

On each step a new $L \times L$ matrix is to be created from the old one. The matrices *New* and *Old* are first initialized using the arrow command to have all matrix elements equal to zero. This may be checked by replacing the colons with semicolons in the following two command lines. The zero values will correspond to dead cells. The original (old) live cell is placed at $i=31$, $j=31$ and assigned the value 1. Any live cells created from dead cells by the one out of eight rule will also be given the value 1.

```
> New:=Matrix(L,L,(i,j)->0):
> Old:=Matrix(L,L,(i,j)->0): Old[31,31]:=1:
```

Jennifer's code involves three do loops, the "outer" loop to increase the step number from $k=1$ to N, the inner two loops to run over i and j from 2 to $L-1$ for a given k value.

```
> for k from 1 to N do
> for i from 2 to L-1 do
> for j from 2 to L-1 do
```

She sets the number n of nearest neighbors that are initially alive to 0.

```
> n:=0;
```

The following command line checks the eight nearest neighbors to cell (i, j) and adds up the number that are alive in the old matrix. For example, if $n=3$ then three nearest neighbors are alive.

```
> n:=n+Old[i+1,j-1]+Old[i+1,j]+Old[i+1,j+1]+Old[i,j-1]
     +Old[i,j+1]+Old[i-1,j-1]+Old[i-1,j]+Old[i-1,j+1];
```

For a given (i, j) value, n could in principle have one of the values $0, 1, 2, \ldots, 8$. If n is exactly equal to 1, the cell (i, j) in the new matrix is assigned the value 1 (i.e., is alive); otherwise, it is assigned its old value. In the latter case, if it was dead it remains dead, and if it was alive it is still alive.

```
> if n=1 then New[i,j]:=1; else New[i,j]:=Old[i,j]; end if;
> end do; end do;
```

On completion of the i and j do loops, the new matrix elements are used to form the old matrix for the next iteration of the outer loop.

```
> Old:=Matrix(L,L,(i,j)->New[i,j]);
```

```
> end do;
```

With the N iterations completed, Jennifer graphs the pattern in Figure 8.16.

```
> listdensityplot(Old);
```

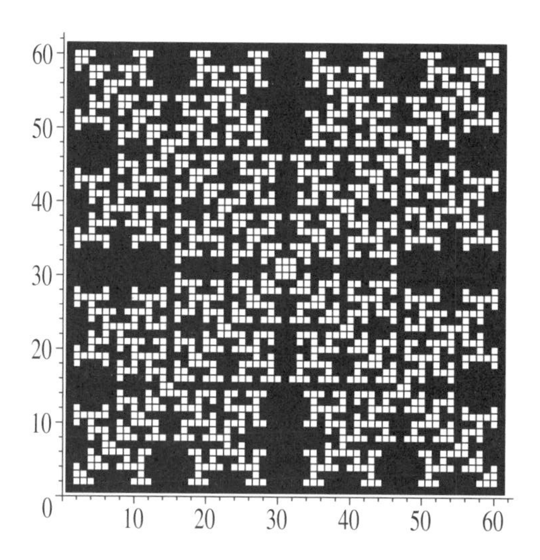

Figure 8.16: Pattern generated by the one out of eight rule after 29 steps.

```
> cpu_time:=(time()-begin)*seconds;
```

$$cpu\_time := 1.172\, seconds$$

As the cellular automata rules are changed, many interesting patterns can be generated. Try the problems that follow or try to invent your own rules.

PROBLEMS:

Problem 8-24: Four live cells

Using the 1 out of 8 rule determine the pattern that evolves after 29 steps if there are initially 4 live cells located at $(30, 25)$, $(30, 35)$, $(25, 30)$, and $(35, 30)$. Note: Adjust L so that the boundary of the lattice is not reached.

Problem 8-25: Great grandma's lace

Modify the 1 out of 8 rule to a 2 out of 8 rule and determine the lacy patterns, reminiscent of those fashionable in great grandma's era, that evolve after 29 steps if the following initial conditions prevail:

(a) cells $(30, 30)$ and $(31, 31)$ are initially alive;

(b) cells $(29, 30)$, $(30, 30)$, $(31, 30)$, and $(32, 30)$ are initially alive.

8.3 Strange Attractors

Jennifer will conclude this all-too brief excursion into the world of fractal patterns with an example of a *strange attractor* solution of a system of three coupled nonlinear ODEs. When plotted in the 3-dimensional *phase space*, the trajectory is attracted to a localized region where it traces out a chaotic fractal pattern, a behavior that when first observed was regarded as "strange."

8.3.1 Lorenz's Butterfly

Does the flap of a butterfly's wings in Brazil set off a tornado in Texas?
E. N. Lorenz, subtitle of his famous conference paper on predictability (1917–)

In 1963, Edward Lorenz published a classic paper [Lor63] on the practicability of very-long-range weather forecasting. Starting with the Navier–Stokes equations, a set of nonlinear partial differential equations (PDEs) used to describe fluid flow, Lorenz attempted to model thermally driven convection in the earth's atmosphere. In his model, the earth's atmosphere is treated as a flat fluid layer that is heated from below by the surface of the earth, which absorbs sunlight and is cooled from above due to the radiation of heat from the atmosphere into outer space. Lorenz managed to approximate the original set of PDEs by the following coupled nonlinear ODE system:

$$\dot{x} = \sigma\,(y - x), \qquad \dot{y} = r\,x - y - x\,z, \qquad \dot{z} = x\,y - b\,z. \tag{8.17}$$

Here x is proportional to the convective velocity, y to the temperature difference between ascending and descending flows, and z to the mean convective heat flow. The positive coefficients σ and r are the Prandtl and reduced Rayleigh numbers, respectively, and $b > 0$ is related to the wave number.

On attempting to solve this set of equations numerically, Lorenz discovered that very small changes in initial conditions could lead to dramatically different long-term behavior of the numerical solutions. One day he tried to continue a computer calculation for the equations, starting with the (x, y, z) values that occured partway through an earlier numerical run. Much to his surprise, he found that after a short time his numerical plots became distinctly different from those previously obtained. He traced the problem down to the fact that in the new run he had entered the input (x, y, z) values to fewer decimal places than in the original data. In effect, Lorenz had slightly changed the initial conditions at the point where the second numerical run began. Lorenz found that this sensitivity to initial conditions was a general feature of nonlinear systems displaying irregular, or *chaotic*, oscillations.

Although Lorenz's model was a drastic oversimplification of the convective behavior of the Earth's atmosphere, he realized the implications for long-range weather forecasting. In his 1963 article, he stated,

"When our results ... are applied to the atmosphere ... they indicate that prediction of the sufficiently distant future is impossible by any method, unless the present conditions are known exactly. In view of the inevitable inaccuracy

and incompleteness of weather observations, precise very long-range forecasting would seem to be nonexistent."

In the popular literature, this phenomenon is referred to as the *butterfly effect*. Although a hyperbolic overstatement, the mere beating of an unknown butterfly's wings deep in the Amazon jungle could change the initial conditions and thus the very long-range weather patterns.

Jennifer will now show us a very famous strange attractor solution of the Lorenz model equations, the attractor resembling the wings of a butterfly. First, she loads the DEtools library package, because it contains the `DEplot3d` command, which is a generalization to three dimensions of the `phaseportrait` command encountered in Chapter 5.

```
> restart: with(DEtools):
```

The initial condition is taken to be $x(0) = 2$, $y(0) = 5$, $z(0) = 5$, and the coefficients $r=28$, $b=8/3$, $\sigma=10$. The total time is $T=100$ time units.

```
> ic:=x(0)=2,y(0)=5,z(0)=5:
```

```
> r:=28: b:=8/3: sigma:=10: T:=100:
```

The system of three coupled ODEs is now entered,

```
> odes:=diff(x(t),t)=sigma*(y(t)-x(t)),diff(y(t),t)
        =r*x(t)-y(t)-x(t)*z(t),diff(z(t),t)=x(t)*y(t)-b*z(t):
```

and the numerical solution plotted with the `DEplot3d` command. Using the

```
> DEplot3d([odes],[x(t),y(t),z(t)],t=0..T,[[ic]],
  stepsize=0.02,scene=[x,y,z],axes=framed,tickmarks=[3,3,3],
  orientation=[-65,74],shading=zhue,thickness=1);
```

`scene=[x,y,z]` option generates the phase space trajectory shown on the left of Figure 8.17. The two vividly colored lobes, which resemble the wings of a

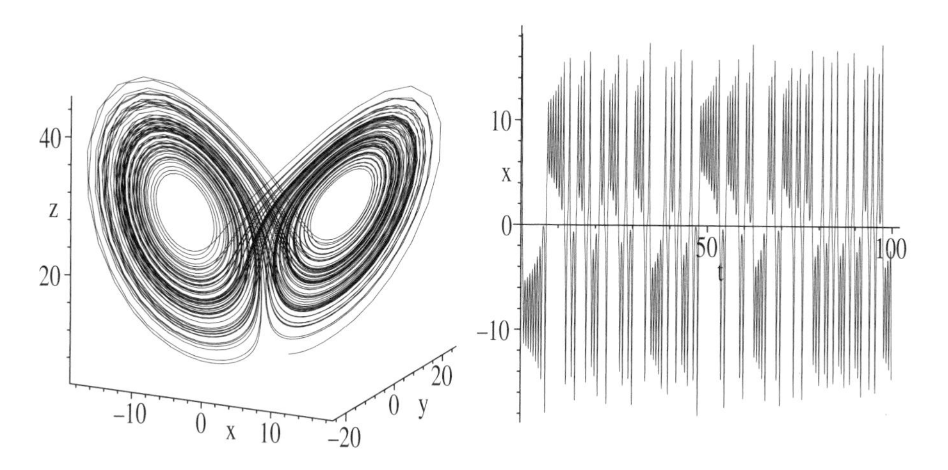

Figure 8.17: Left: Lorenz's butterfly. Right: Corresponding chaotic $x(t)$.

butterfly, are actually located in different planes, which can be confirmed by rotating the 3-dimensional viewing box. The behavior of, say, x versus t can be observed by changing the scene option to `[t,x,y]` and the orientation to `[-90,0]`. The resulting picture is shown on the right of Figure 8.17. The time sequence does not repeat, although over certain time intervals it sometimes looks like it might. The sensitivity of the chaotic solution to very slight changes in initial conditions is left for the reader to explore as a problem.

If a plane, oriented perpendicular to the plane of a given wing, is drawn through the wing, a pattern reminiscent of the previously discussed Cantor set results. Lorenz determined that the three-dimensional butterfly attractor has a fractal dimension $D_C = 2.06 \pm 0.01$, which is closer to two than to three.

PROBLEMS:

Problem 8-26: Sensitivity to initial conditions

Take $x(0) = 2.0001$ in the text recipe with all other parameters the same and superimpose the new solution on that for $x(0) = 2.0$. Does your plot support the idea that the asymptotic solution is sensitive to initial conditions?

Problem 8-27: Varying r

Explore how the solution of the Lorenz system changes as r is varied.

Problem 8-28: Rössler's strange attractor

Taking $a=0.2$, $b=0.2$, $c=5.7$, show that Rössler's ODE system [Ro76]

$$\dot{x}(t) = -y - z, \quad \dot{y}(t) = x + a\,y, \quad \dot{z}(t) = b + z\,(x - c),$$

has a strange attractor. Take $x(0)=1$, $y(0)=0$, $z(0)=0$ and a time of 200.

Epilogue

Now this is not the end. It is not even the beginning of the end. But it is, perhaps, the end of the beginning.

Winston Churchill, wartime speech (1942)

Now we must reluctantly end this introductory computer algebra guide to the mathematical models of science. We trust that you have enjoyed the diverse selection of intellectually delicious recipes that have been presented, as well as the "stories" that accompanied them. But to echo Winston Churchill, this is not really the end of our survey of the mathematical models of science, just the end of the beginning. The recipes and stories continue in **Computer Algebra Recipes: An Advanced Guide to the Mathematical Models of Science** with Jennifer, Mike, Vectoria, and their friends exploring a wide variety of interesting linear and nonlinear ODE and PDE models. On their behalf, we invite you to join us there.

Richard and George, Your CAS chefs

Bibliography

[AL79] J. C. Arya and R. W. Lardner. *Mathematics for the Biological Sciences.* Prentice-Hall, Englewood Cliffs, NJ, 1979.

[AM82] R. Anderson and R. May. The logic of vaccination. *New Scientist,* November, 1982.

[ASW87] D. R. Anderson, D. J. Sweeney, and T. A. Williams. *Statistics for Business and Economics.* West Publishing, St. Paul, MN, 1987.

[BBS79] E. Batschelet, L. Brand, and A. Steiner. On the kinetics of lead in the human body. *J. Math. Biol.,* 8:15, 1979.

[Bec69] M. G. Becker. *Introduction to Terrain Vehicle Systems.* University of Michigan Press, Ann Arbor, MI, 1969.

[Ber41] H. Bernadelli. Population waves. *J. Burma Res. Soc.,* 31:1, 1941.

[BF89] R. L. Burden and J. D. Faires. *Numerical Analysis,* 4th ed. PWS-KENT, Boston, MA, 1989.

[BHT90] M. Begon, J. L. Harper, and C. R. Townsend. *Ecology, Individuals, Populations and Communities.* Blackwell Scientific, Cambridge, MA, 1990.

[Boa83] M. L. Boas. *Mathmatical Methods in the Physical Sciences.* John Wiley, New York, 1983.

[Bro28a] R. Brown. *Phil. Mag.,* 4:161, 1828.

[Bro28b] R. Brown. *Ann. Phys. Chem.,* 14:294, 1828.

[CLR90] T. H. Cormen, C. E. Leiserson, and R. L. Rivest. *Introduction to Algorithms.* MIT Press, Boston, 1990.

[DeV94] P. L. DeVries. *A First Course in Computational Physics.* Wiley, New York, 1994.

[DG95] J. Downes and J. E. Goodmand. *Dictionary of Financial and Investment Terms.* Barrons Educational Series, Hauppauge, NY, 1995.

[Dur54] W. Durant. *Our Oriental Heritage.* Simon and Schuster, New York, 1954.

[Ein56] A. Einstein. *Investigations on the Theory of the Brownian Movement.* Dover, New York, 1956.

[EK88] L. Edelstein-Keshet. *Mathematical Models in Biology.* Birkhäuser, Boston, MA, 1988.

[EM00] R. H. Enns and G. C. McGuire. *Nonlinear Physics with Maple for Scientists and Engineers,* 2nd ed. Birkhäuser, Boston, MA, 2000.

[FC99] G. R. Fowles and G. L Cassiday. *Analytic Mechanics.* Saunders College, Orlando, FL, 1999.

[Fei78] M. J. Feigenbaum. Quantitative universality for a class of nonlinear transformations. *J. Statist. Phys.*, 19:25, 1978.

[FLS64] R. P. Feynman, R. B. Leighton, and M. Sands. *The Feynman Lectures on Physics.* Addison-Wesley, Reading, MA, 1964.

[Fro92] J. Fröyland. *Introduction to Chaos and Coherence.* Institute of Physics, 1992.

[Gou88] M. Gouy. *J. Phys. (Paris)*, 7:561, 1888.

[GT96] H. Gould and J. Tobochnik. *An Introduction to Computer Simulation Methods.* Addison-Wesley, Reading, MA, 1996.

[Hea02] M. T. Heath. *Scientific Computing: An Introductory Survey,* 2nd ed. McGraw-Hill, New York, 2002.

[Jac90] E. A. Jackson. *Perspectives of Nonlinear Dynamics,* Vol. 2. Cambridge University Press, Cambridge, 1990.

[KE96] L. D. Kiel and E. Elliot. *Chaos Theory in the Social Sciences.* University of Michigan Press, Ann Arbor, MI, 1996.

[KW86] M. H. Kalos and P. A. Whitlock. *Monte Carlo Methods.* Wiley, New York, 1986.

[LC90] P. Lorrain and D. R. Corson. *Electromagnetism.* W. H. Freeman, New York, 1990.

[LFHC95] D. R. LaTorre, I. B. Fetta, C. R. Harris, and L. L. Capenter. *Calculus Concepts.* D. C. Heath, Lexington, MA, 1995.

[Lin82] H. Lin. Fundamentals of zoological scaling. *Amer. J. Phys.*, 50:72, 1982.

[Lor63] E. N. Lorenz. Deterministic nonperiodic flow. *J. Atmospheric Sci.*, 20:130, 1963.

[LP97] R. H. Landau and M. J. Páez. *Computational Physics.* Wiley, New York, 1997.

[Map05] Maplesoft. *Maple 10 User Manual.* Waterloo Maple, Waterloo, Canada, 2005.

[May76] R. M. May. Simple mathematical models with very complicated dynamics. *Nature*, 261:459, 1976.

[McM73] T. McMahon. *Science*, 179:1209, 1973.

[MGH+05] M. B. Monagan, K. O. Geddes, K. M. Heal, G. Labahn, S. M. Vorkoetter, J. McCarron, and P. DeMarco. *Maple 10 Introductory (Advanced) Programming Guide.* Waterloo Maple, Waterloo, Canada, 2005.

[Moh98] P. Mohazzabi. Monte Carlo estimations of e. *Amer. J. Phys.*, 66:138, 1998.

[Mor48] P. M. Morse. *Vibration and Sound.* McGraw-Hill, New York, 1948.

[MS74] J. Maynard-Smith. *Models in Ecology.* Cambridge University Press, Cambridge, 1974.

[MSS73] J. Maynard-Smith and M. Slatkin. The stability of predator–prey systems. *Ecology*, 54:384, 1973.

[MT95] J. B. Marion and S. T. Thornton. *Classical Dynamics of Particles and Systems,* 4th ed. Saunders College, Orlando, FL, 1995.

[Mur89] J. D. Murray. *Mathematical Biology.* Springer-Verlag, New York, 1989.

[MW70] J. Mathews and R. L. Walker. *Mathematical Methods of Physics,* 2nd ed. W. A. Benjamin, New York, 1970.

[Oha85] H. C. Ohanian. *Physics.* W. W. Norton, New York, 1985.

[PC89] T. S. Parker and L. O. Chua. *Practical Numerical Algorithms for Chaotic Systems.* Springer-Verlag, New York, 1989.

[PFTV89] W. H. Press, B. P. Flannery, S. A. Teukolsky, and W. T. Vetterling. *Numerical Recipes.* Cambridge University Press, Cambridge, 1989.

[PLA92] M. Peastrel, R. Lynch, and A. Armenti. Terminal velocity of a shuttlecock in vertical fall. In Angelo Armenti Jr., editor, *The Physics of Sports*. American Institute of Physics, New York, 1992.

[PR86] H. O. Peitgen and P. H. Richter. *The Beauty of Fractals*. Springer-Verlag, New York, 1986.

[Pur85] E. M. Purcell. *Electricity and Magnetism*. McGraw-Hill, New York, 1985.

[Ro76] O. E. Rössler. An equation for continous chaos. *Phys. Lett. A*, 57:397, 1976.

[Rei65] F. Reif. *Fundamentals of Statistical and Thermal Physics*. McGraw-Hill, New York, 1965.

[RRKW86] J. L. Riggs, W. F. Rentz, A. L. Kahl, and T. M. West. *Engineering Economics*. McGraw-Hill Ryerson, Toronto, 1986.

[Sap84] A. M. Saperstein. Chaos—a model for the outbreak of war. *Nature*, 309:303, 1984.

[SAU94] *Statistical Abstract of the U.S.* Bernan Press, 1994.

[Sea58] F. W. Sears. *Mechanics, Heat and Sound*. Addison-Wesley, Reading, MA, 1958.

[SH76] D. A. Secrist and R. W. Hornbeck. An analysis of heat transfer and fade in disk brakes. *Trans. ASME, J. Eng. Industry*, 18:385, 1976.

[Ste87] J. Stewart. *Calculus*. Brooks/Cole, Pacific Grove, CA, 1987.

[Str85] J. Strnad. Physics of long-distance running. *Amer. J. Phys.*, 53:371, 1985.

[Str88] S. H. Strogatz. Love affairs and differential equations. *Math. Mag.*, 61:35, 1988.

[Str94] S. H. Strogatz. *Nonlinear Dynamics and Chaos*. Addison-Wesley, Reading, MA, 1994.

[Tip91] P. A. Tipler. *Physics for Scientists and Engineers*. Worth, New York, 1991.

[Wie73] S. Wieder. *The Foundations of Quantum Theory*. Academic Press, New York, 1973.

[Wol86] S. Wolfram. *Theory and Applications of Cellular Automata*. World Scientific, Singapore, 1986.

[Wol02] S. Wolfram. *A New Kind of Science*. Wolfram Media, Champaign, Illinois, 2002.

Index

AC circuit, 131
air resistance, 23
alpha decay, 372
angle of incidence/refraction, 152
angstrom, 29
Aristotle, 152
arms expenditures, 310
associative law, 176
average unit cost, 146
Avogadro's number, 322, 323

Bacon, Roger, 152
badminton, 22
Barnsley's fern, 391
beats, 59
Bernoulli trial, 361, 372
Bernoulli, Jacob, 361
Bernoulli, John/Daniel, 286
Bessel functions, modified, 263
binomial coefficient, 361, 372
binomial distribution, 361, 372, 376
binomial function, 49
board foot, 38
boiling point, 98
Boltzmann constant, 322
bombs versus schools, 122
bone strength, 43
brachiosaur heart rate, 81
brain weight, 98
break-even point, 144
bridge design costs, 5
Brown, Robert, 321
Brownian motion, 321
butterfly effect, 415

calorie, 161
Cantor set, 383
Cantor set, middle-half, 384
capacitance, 131
capacity dimension, 383
carrying capacity, 295
Cassini gap, 400
cellular automata, 408
center of gravity, 137
center of mass, 163, 171, 357
centripetal force, 266
chaos, 301
charge density, 190
Chebyshev polynomials, 281
Cheops (Khufu), 156
chi-square, 69, 106
chimpanzee brain volume, 36, 82
circulation, 195
closed path, 195
cloverleaf orbits, 260
cobweb diagram, 301
competition, interspecific, 292
competition, intraspecific, 292
compound interest, 30
computer algebra system (CAS), 1
Comte de Buffon, 353
conservative field, 189, 194
consumer prices, 16
consumption matrix, 49
corn yield, 67, 75
Coulomb potential, 15
Coulomb's law, 140, 187
CPU time, 326, 327
critical angle, 185
critical damping, 226
cross product, 175
cubit, 163

cumulative bikini sales, 88
cumulative distribution function, 366
cumulative Poisson distribution, 370
cumulative probability, 362

data, 15
data storage matrices, 48
DC circuit, 130, 135
Descartes, René, 152
devotion, 310
difference equation
  first-order/second-order, 272
  inhomogeneous, 272
  linear, 271
  nonlinear, 272, 297
  second-order, 287
  third-order, 282
differential cross section, 102
diffusion, 322
Dijkstra's algorithm, 34
dinosaur, brachiosaur, 77
disposable income, 71
divergence theorem, 201
do loop, 93
dot product, 173
Douady's rabbit, 397
Dow Jones index, 91
drug concentration, 126

earthquake, 127
eccentricity, 258
ecology, 292
Edelstein-Keshet, Leah, 283
eigenfrequencies, 61, 287
eigenvalue, 288
Einstein equation, 322, 324
Einstein, Albert, 321
electric dipole potential, 55
electric field, 55, 189
electric force, 188
electric quadrupole potential, 55
electrocardiograph, 251
electron volts, 29
electrostatic potential, 48, 189
ellipse, 258
Enon, 54, 76
envelope of safety, 148, 149
equipartition of energy, 322
equipotential surface, 191
Erehwon, 30, 122, 151
  aardwolves, 227
  Institute of Technology, 216
erosion, 59
error function, 166

Feigenbaum constant, 305
Ferdinand, Archduke Francis, 310
Fermat's principle, 35, 155
Fermi, Enrico, 286
Fermi–Pasta–Ulam, 286
fern, Black Spleenwort, 391
fern, cyclosorus, 395
fern, fishbone, 395
Fibonacci (Leonardo of Pisa), 279
Fick's law, 235
fixed point, 313
Flamsteed, John, 255
floating-point number, 72, 78
flux, 199, 201
focal/spiral point, 224
Fourier series, 57, 251
fractal dimension, 383, 390, 394
  box counting, 391
free body diagram, 137
free space, 190

Gauss's law, 190, 201
Gaussian distribution, 374
geometric similarity, 39, 41
global variables, 114
GNP, 71, 310
Gouy, M., 321
gradient operator symbol, 51
gradient/divergence/curl, 188
graphs, functions of, 13
gravitational acceleration, 98
gravitational constant, 99
great flu epidemic, 21, 90
ground-state energy, 128
Gulliver's Travels, 41

Halley, Edmund, 255
hard spring equation, 220
heat loss, 42
helix, 181
Herodotus, 156
histogram, 47
Hooke's law, 287
human surface area, 44
hydrogen atom motion, 29
hyperbolic function, 24
hypersphere, volume of, 355

ideal gas constant, 322
ideal gas law, 14
impedance, 131
importance sampling, 349
inclined plane, 141
inductance, 131
inflation, 15, 71
inflection point, 30
initial condition sensitivity, 416
integral, mass, 357
integration
    Clenshaw–Curtis method, 340
    error, 340
    Monte Carlo, 348
    Monte Carlo error, 340
    rectangular rule, 340
    Simpson's rule, 340
    trapezoidal rule, 340
inverse cube law, 259
inverse square law, 255
irrotational field, 193

Julia set, 396
Julia, Gaston, 396

Kepler's third law, 15, 82, 404
Khrushchev's purge, 86
Kirchhoff's rules, 130, 222
Koch curve, 384

ladder problem, 136
Laplacian operator, 195
lattice, rectangular, 333
lead poisoning, 236
least squares, 65, 70, 106
Legendre function, 267
Legendre polynomials, 282
Lennard–Jones potential, 15
lifetime, 82
Lin, Herbert, 81
linear correlation coefficient, 66, 115
local variables, 114, 396
log-log plot, 36, 77
logistic curve, 88, 90, 157
logistic map, 299
Lorenz equations, 414
lung disease, 366

Machu Pichu, 46
Mandelbrot set, 398
Mandelbrot, Benoit, 396
map
    first/second iterate, 302
    fixed point, 297
    fixed point stability, 303
    logistic, 299
    Mandelbrot, 396
    nonlinear, 297
    octopus, 399
    one-dimensional, 297
    San Marco attractor, 399
    standard, 308
    third iterate, 302
    trajectory, 297
Maple
    animation, 56, 57
    arrow (functional) operator, 62
    assignment operator, 6
    case sensitive, 27
    clearing internal memory, 6, 16
    comment, 16, 17
    composition operator, 302
    concatenation, 163
    context bar, 26
    copying examples, 9
    default accuracy, 8
    differentiation, 8
    ditto operator, 27
    double integral, 163

Help, Full Text Search, 9
Help, Topic Search, 9
Help, using, 9
integration, 29, 160
left quotes, 176
library packages, 16
list, 16, 23, 68
mouse dragging, 125
plot options, 19
procedure, 114
protected symbols, 23
removing warnings, 19
set, 20
string, 7, 25
trailing tilde removal, 25
triple integral, 163
viewing box rotation, 51
Maple Command
^, 6
*, 6
+, 6
-, 6
/, 6
:=, 6
:, 6, 17
;, 6, 17
?, 10
#, 17
%, 27, 158
− >, 37, 43, 72
<< , , >>, 103, 206
< | >, 104
., 175, 206
&x, 176
<>, 184
BandMatrix, 288
Binomial, 362
CDF, 362, 366, 370
CharacteristicMatrix, 209
CharacteristicPolynomial, 210
Column, 203, 211
Correlation, 68
CrossProduct, 175
CumulativeDistributionFunction, 362
Curl, 189, 194, 196
DEplot3d, 415
DEplot, 239
DeleteColumn, 204
DeleteRow, 204
Del, 189
Determinant, 203
Digits, 25, 77, 84, 308
Dimension, 203
Divergence, 190, 200
DotProduct, 175, 180
Eigenvalues, 210, 289
Eigenvectors, 210
ExponentialFit, 85
Fit, 73, 78
Gradient, 51, 194
Integrand, 350
Int, 27, 165
Laplacian, 195
LineInt, 191
LinearSolve, 207, 210
Line, 191
MapToBasis, 196
MatrixInverse, 204, 207
Matrix, 46, 49, 103, 203
Mean, 346, 351, 374
Multiply, 204, 208
NonlinearFit, 89
Normal, 365, 375
PDF, 365, 375
PLOT3D(POLYGONS), 200
PLOT, 63
Pi, 27, 56
Poisson, 369, 379
ProbabilityDensityFunction, 365
ProbabilityFunction, 362, 369
REplot, 274, 277, 293
RandomMatrix, 104
Re, 133
RootOf, 138
Row, 105, 203
Rule[change], 349
Rule[flip], 350
ScatterPlot, 17, 23, 68, 88

SetCoordinates, 189, 195
StandardDeviation, 352, 374
TIMES,ITALIC, 25
TallyInto, 375
Tally, 320, 378
Trace, 203
Transpose, 47, 49, 203
VectorField, 189, 195, 199
VectorNorm, 180
Vector, 174
about, 239
abs, 176
addvertex, 32
add, 57, 104, 251
alias, 32
align=LEFT, 200
align=RIGHT, 164
allpairs, 33
allvalues, 138
alpha, 60
animate, 57
arccos, 175
arcsin, 155
arctan, 53
arc, 196
array, 224
arrows=MEDIUM, 217
arrows=THICK, 52, 191
arrow, 164, 177, 196
assign, 131, 184, 219
assume, 25, 133
assuming, 176, 180, 201
axes=NONE, 63, 200
axes=boxed, 47, 108
axes=framed, 44, 168
axes=normal, 124, 181
background, 63
beta, 251
binomial, 49
cartesian[x,y,z], 196
coeff, 85
collect, 246, 252, 300
color=COLOR(RGB), 331
color, 19, 23
combine(ln), 122
combine(symbolic), 122
combine(trig), 180
connect, 32
contourplot3d, 168
contourplot, 52
contours, 47, 190
convert(array), 47
convert(degrees), 139
convert(polynom), 25
convert(radians,degrees), 175
cosh, 25
cos, 50, 60, 184
cylindrical[r,theta,z], 195
dchange, 222
delete, 34
delta, 56, 153
dfieldplot, 217
diff, 8, 26
dirgrid, 217, 223
display, 20, 25
draw, 33
dsolve(laplace), 261
dsolve(numeric), 269
dsolve, 219, 226
edges, 33
end proc, 115
epsilon, 190
evalc, 133, 240, 253
evalf, 27
eval, 7, 158
eweight, 33
expand, 195
exp, 50, 60, 79
factor, 154, 280
fibonacci, 281
fieldplot3d, 191, 199
fieldplot, 52, 196
filled=true, 47, 163, 168
firint, 243
for...do...end do, 93, 104
frac, 308
frames, 57
fsolve, 8, 27
gamma, 221
generator=rand, 104

global, 114
grid, 52, 191, 196
heights=histogram, 47
if...then...end if, 105
implicitplot3d, 191
infinity, 219
infolevel[dsolve], 243
interface, 19, 46
intfactor, 242
int, 27
isolate, 154, 243, 403
kappa, 288
labelfont, 25
labels, 7, 17, 25, 44
lambda, 56
lhs, 154
lightmodel=light2, 168
lightmodel=light3, 49
lightmodel=light4, 50
limit, 219
linecolor, 218, 223
listcontplot3d, 47
listcontplot, 47
listdensityplot, 47
ln, 78, 84, 122
local, 114
loglogplot, 37, 43
logplot, 40
map, 230, 289, 327, 389
matrixplot, 47, 49, 204
mu, 137
new, 32
nops, 91, 107
numpoints, 57, 181
omega, 132, 226
op, 133
orientation, 47, 49
output=listprocedure, 269
paraminfo=false, 60, 63
phaseportrait, 218, 223
phi, 153
piecewise, 157, 185
plot3d, 50, 108, 190
plot, 7, 25, 38
pointplot3d, 109, 309, 407
pointplot, 62, 186
polarplot, 259
polar, 133
polygonplot, 331, 375, 388
proc, 114, 396
radnormal, 133, 280, 289
radsimp, 241
rand(), 320
randomize(), 103, 320, 326
rectangle, 63
remove, 211, 267, 290
restart, 6, 23
rhs, 154
rotate, 385
round, 8, 86, 290
rsolve, 273, 277
rtablesize=infinity, 46
rtablesize, 203
scale, 389
scaling=constrained, 50
scene, 223
select, 240
semitorus, 356
seq, 19, 49, 84
shading=xyz, 49, 60
shading=zgreyscale, 50, 191
shading=zhue, 47, 191
shading=z, 168
shortpathtree, 34
signum, 385
simplify(symbolic), 175
simplify, 6, 138
sin, 56, 138, 153, 184
sol[], 147
sort, 289, 378
spacecurve, 124, 181, 246, 297
sphere, 407
spherical[r,theta,phi], 194
stepsize, 218, 223
style=hidden, 192
style=patchcontour, 190
style=patchnogrid, 108
style=patch, 47, 49, 331
style=point, 37, 43
subsop, 409

subs, 27
symbol=box, 17, 37
symbol=circle, 23, 62
symbol=cross, 19
symbolsize, 17, 19, 37
tau, 222
taylor, 25
textplot3d, 168
textplot, 63, 144, 164, 186
theta, 138, 153
thickness, 19
tickmarks, 7, 17, 25
time(), 326
title, 25, 101, 144
translate, 389
trunc, 303
type, 105
unapply, 74
unassign, 139, 150, 233
unprotect, 23, 221, 230
value, 165
view, 17
warnlevel=0, 19
weights, 32
while, 330, 406
with(DEtools), 216, 221, 242
with(LREtools), 274, 276
with(LinearAlgebra), 46, 49
with(PDEtools), 221
with(Statistics), 16, 23, 68
with(Student[Calculus1]), 340
with(VectorCalculus), 49, 174
with(combinat,fibonacci), 281
with(networks), 32
with(plots), 19, 20, 23
with(plottools), 63, 163, 174
zip, 37, 72, 230

matrix
add/subtract/multiply, 202
adjoint, 202
characteristic, 209
characteristic polynomial, 209
cofactors, 202
complex conjugate, 202
consumption, 206
determinant of, 202
diagonalization, 211
eigenvalue/eigenvector, 209
Hermitian, 202
Hermitian conjugate, 202
identity, 207
inverse, 202
joining, 104
nonsingular, 202
principal diagonal, 202
row/column, 202
shorthand syntax, 206
solving linear equations, 207
square, 202
transpose, 202
tridiagonal, 288
unit, 202

maxima/minima, 51
mean free path, 335
Metropolis, 16
microton accelerator, 308
Mimas, 401
minimizing travel time, 31
MIT, 22
model
aardwolves, 227
Anderson–May, 316
arms race, 310
bactericide, 86
bandicoots, 278
Barnsley's fern, 391
birds munch aphids, 207
blood alcohol level, 86
blood CO_2, 285
bombs versus schools, 122
bouncing ball, 306
brainteaser, 187
bumpy road, 253
capacitor charging, 242
CO_2 vibration, 291
competition, 292
DALE computer, 344, 348
dating game, 103
dimensional scaling, 40
diver, 249

Douady's rabbit, 396
Dow Jones, 91
drug exchange, 227
drunkard's walk, 324
economics, 142
electric circuit, 130
envelope of safety, 148
epileptic seizures, 305
erosion, 59
exponential, 84
falling burger, 249
falling raindrop, 329
Fibonacci, 279
Fick's law, 235
fir tree yield, 38
fish population, 276
flu epidemic, 282
fly, 179
gnat, 272
gnus, 278
goat population, 282
gravitational acceleration, 98
Halley's comet, 255
heart rate, 77
ice wine, 367
inclined ladder, 136
infectious disease, 234
Intel processor chip, 87
Khrushchev's purge, 86
lace, 413
lead poisoning, 236
logistic, 88, 299
Lorenz's butterfly, 414
matrix, 202
Maynard-Smith/Slatkin, 294
Monte Carlo π, 353
Navaho rug, 408
Newton's law of cooling, 229
normal mode, 285
nuts and bolts, 361
one of eight rule, 411
orbital precession, 260
Penelope Jitter Bug, 333
perfume diffusion, 335
population growth, 235
power law, 36, 77
predator–prey, 237
price and supply, 237
puffin explosion, 275
pursuit, 238
pyramid, 156
Rössler, 416
rainbow, 152
real estate, 107
red blood cell, 283
RLC circuit, 221
Romeo and Juliet, 228
rotating wheel, 260
safety envelope, 245
Saturn's rings, 400
Senate renewal, 84
ship hull, 162
Sierpinski gasket, 386
ski hill, 49
standard map, 309
straight-line, 73
swamp fever, 275
swimsuit, 102
tripolar arms race, 315
tug of war, 127
vapor deposition, 376
vibrating heart, 251
wallpaper pattern, 384
water skier, 238
weedeater, 266
well depth, 248
whale harvesting, 278
wheel, 355
yeast growth, 237
model equations, 15, 65, 106
modulus, 133
moment of a force, 138
moment of inertia, 171
Monte Carlo dart method, 353, 372
Moore's law, 87
multiple regression, 106

Newton's cooling law, 40, 229
Newton's gravitation law, 99
Newton's method, 126

Newton's resistance law, 24
Newton's second law, 24, 99
Newton, Isaac, 152, 286
Noah's ark, 162
normal distribution, 364
normal modes, 61, 266, 286

ODE
  autonomous/nonautonomous, 214
  damped SHO, 223
  overdamped SHO, 225
  steady-state solution, 252
  transient solution, 252
  underdamped SHO, 224
Ohm's law, 130
open surface, 195
Operophtera brumata, 297
optical path length, 155
order of, 25
outbreak of war, 310

parabolic cylinder, 200
parabolic trajectory, 148
paraboloid, 168
parallelepiped volume, 176, 179
Paris gun, 248
partial derivative, 51
perihelion, 258
period four, 301
period one, 301
period two, 301
permittivity, 140, 190
Perrin, Jean, 322
phase, 307
phase angle, 133
phase plane, 214
phase-plane portrait, 223
phase-plane trajectory, 214
Physics of Sports, 22
piecewise function, 157
piecewise linear, 19
planets, 258
Poisson distribution, 367, 376
polar form, 133
polar plot, 182
polio epidemic, 90
potential function, 191
power law, 24, 40
purchasing power, 16, 17, 69
Pyramid, Great, 156

radial field, 193
radioactive chain, 236
rainbow, 152
raindrop, 153
random walk, 319, 321, 324
  perfume molecule, 335
  three-dimensional, 332
random-number generator, 319, 333
range, 249
Rayleigh's criterion, 266
real estate appraisal, 107
rectangular table, 46
red blood cells, 283
refractive index, 152
regression analysis, 69, 106
relative velocity, 186
resistance, 131
revenue curve, 147
Richter scale, 127
right-hand rule, 173
RLC circuit, 221
Roche limit, 401
Roche, Edouard, 401
rocket flight, 162
root mean square deviation, 368
root mean square distance, 324
rotation of Earth, 99
rotational equilibrium, 138
rowing times, 39
rule of 72, 129

Saperstein, Alvin, 310
Saturn's moons, 401
scaling, 39, 40, 77, 81
self-similar, 390
semilog plot, 84
semimajor axis, 258
semitorus, 356
sequoias, 45

sewage treatment plant, 143
siege tower, 356
Sierpinski's carpet, 390
Sierpinski's gasket, 386
signum function, 240
simple harmonic motion, 52
simple interest, 30
simple pendulum equation, 220
skeletal bone weight, 43
Snell's law, 35, 152, 185
soccer, 324
solid tetrahedron, 172
Spanish flu epidemic, 21
spherical polar coordinates, 194, 335
spleen/bone marrow, 283
stability, 168
standard deviation, 70, 94, 107, 374
standard temperature/pressure, 335
static equilibrium, 137
static friction, coefficient of, 137
Statistical Abstract of the U.S., 16
steady state, 297
Stegobium panaceum, 297
stellerator, 308
stochastic, 321, 386
stock market, 91, 92
Stokes's resistance law, 27, 245
Stokes's theorem, 195
strange attractor, 414
surface integral, 195
suspension bridge, 140
Swift, Jonathan, 41

tangent field, 215, 217
Taylor expansion, 27
tenure, 216
terminal velocity, 24
tetrahedron, 201
Theodoric of Freiburg, 152
tidal force, 401
time of flight, 249
toroid, 356, 357
torque, 178
transcendental equation, 128, 185
trial number, 325
triathlon, 184
Tribolium confusum/castaneum, 297
tripolar arms race, 315
turning points, 29

Ulam, Stan, 408
unit vector, 51, 178
uplift, 61
utility function, 127, 128

Van der Pol equation, 220
Van der Waals equation, 55
vapor deposition, 377
vector
    acceleration, 180
    angle between two, 178, 179
    coplanar, 178
    cross (vector) product, 173
    displacement, 184
    dot product, 173
    field, 173
    identity, 179
    position, 180
    sum, 173, 174
    velocity, 180
velocity field, 195
visible spectrum, 154
visual hallucination patterns, 48
volcanic eruption, 148
Von Neumann, John, 408
vortex point, 224

wave
    dispersion, 58
    longitudinal/transverse, 56
    standing, 56
    train, 59
    traveling, 57
Wolfram, Stephen, 408
world record, 83, 97

zero circulation, 191
zeroth law of thermodynamics, 286